Tyrosine Phosphoprotein Phosphatases

SECOND EDITION

Series Editor

Published Titles

Tyrosine phosphoprotein phosphatases (second edition). *Barry Goldstein*

Lysosomal cysteine proteinases (second edition). *Heidrun Kirschke, Alan J. Barrett and Neil Rawlings*

Helix–loop–helix transcription factors (third edition). *Trevor Littlewood and Gerard Evan*

Forthcoming Titles

Actins (fourth edition). *Peter Sheterline, Jon Clayton and John Sparrow*

Actin monomer-binding proteins. *Uno Lindberg and Clarence Schutt*

Intermediate filament proteins (third edition). *Roy Quinlan, Chris Hutchison and Birgit Lane*

Gelsolin family. *Horst Hinssen*

Kinesins (third edition). *Viki Allan and Rob Cross*

Myosins (second edition). *Jim Sellers and Holly Goodson*

Cadherins. *Sasha Bershadsky and Benny Geiger*

Integrins. *Martin Humphries*

IgCAMs (third edition). *Fritz Rathjen and Thomas Brümmendorf*

Matrix metalloproteinases. *Fred Woessner and Hideaki Nagase*

Network- and filament-forming collagens. *Michel Van der Rest and Bernard Dublet*

Nuclear receptors (second edition). *Hinrich Gronemeyer and Vincent Laudet*

EF-hand calcium-binding proteins (third edition). *Hiroshi Kawasaki and Bob Kretsinger*

Heterotrimeric G proteins (third edition). *Steve Pennington*

Serine phosphoprotein phosphatases. *Patricia Cohen and David Barford*

Signalling domains. *Bruce Mayer and Matthias Wilmann*

Proline isomerases. *Andrzej Galat*

Subtilisins. *Roland Siezen*

Thermolysins. *Rob Beynon and Ann Beaumont*

Online Edition

See `http://www.oup.co.uk/protein_profile`

Art Fein

Tyrosine Phosphoprotein Phosphatases

SECOND EDITION

Barry J. Goldstein

Dorrance H. Hamilton Research Laboratories,
Department of Medicine, Jefferson Medical College,
Thomas Jefferson University, Philadelphia, PA 19107, USA
Telephone: +1 215 503 1272
Fax: +1 215 923 7932
Email: BarryGoldstein@mail.tju.edu

OXFORD NEW YORK TOKYO
OXFORD UNIVERSITY PRESS
1998

Oxford University Press, Great Clarendon Street, Oxford OX2 6DP

Oxford New York
Athens Auckland Bangkok Bogota Bombay
Buenos Aires Calcutta Cape Town Dar es Salaam
Delhi Florence Hong Kong Istanbul Karachi
Kuala Lumpur Madras Madrid Melbourne
Mexico City Nairobi Paris Singapore
Taipei Tokyo Toronto Warsaw

and associated companies in
Berlin Ibadan

Oxford is a trade mark of Oxford University Press

Published in the United States
by Oxford University Press, Inc., New York

A catalogue record for this book is available from the British Library

Library of Congress Cataloging in Publication Data
(Data available)

ISBN 0 19 850247 8

Typeset & printed by the Alden Group, Oxford

Series preface

The Protein Profile *series has been developed from a recognition that individuals find it increasingly difficult to access readily the enormous amount of information accumulated by the international research community; information which is crucial for the efficiency and quality of their activities.*

The *Protein Profile* series aims to provide a practical, comprehensive and accessible information source on all major families of proteins. Each volume of *Protein Profile* is focused on a single family or sub-family of proteins, and contains tables and figures presenting a comprehensive accumulation of structural, kinetic and biochemical information available on that particular protein group, coupled to an extensive bibliography. Every volume will be refined and updated to provide users with a practical up-to-date single source of information by the publication of new editions approximately every two years.

Content

The text provides a brief overview of the biological context of the function of the protein group followed by an overview of available information on:

- function
- kinetic and biochemical properties
- sequences, sequence relationships and sequence features
- domain structure
- mutations
- 3-D structure
- binding sites of protein
- ligand binding sites and interactions with drugs
- derivatization sites
- proteolytic cleavage sites

Each volume follows the same format but with different emphases depending on the protein family. The text is extensively supported by tables and figures listing key information gathered from the literature with comprehensive reference to primary sources.

Available protein sequences, or where there are very large numbers, representatives from each subgroup, are aligned in an Appendix at the back of each issue.

Available references pertinent to the properties, structure and function of the proteins are listed in a full bibliography. References are numbered for access in the text, but are also arranged alphabetically under headings to allow browsing. Reviews are listed at the beginning and key papers are identified.

An **online version** of the Protein Profile series will be available during 1998. For the latest information about the series, see our web page at `http://www.oup.co.uk/protein_profile`

Preface

Investigators studying various aspects of the molecular biology and physiology of PTPases have had the very gratifying experience of witnessing the rapid growth of major developments in this field over the past 15 years. Dozens of PTPase superfamily members have been cloned and tremendous insight into their biochemical properties and their role in cellular physiology has been achieved. Structural studies have provided a detailed three-dimensional view that confirms many biochemical notions about the molecular mechanism of protein-phosphotyrosine hydrolysis. With all of this fertile groundwork, the next several years will continue to provide exciting new information on the cellular functions of the many PTPases now recognized. One may also anticipate that these discoveries will also lead to new approaches to modulate the activity of specific PTPases in the development of novel therapeutics to ameliorate a number of human diseases including immunodeficiency disorders, cancer and diabetes mellitus.

Additional comments

As with other protein families described in other issues of *Protein Profile*, our intention in the series on PTPases is to provide as comprehensive and accurate a database as possible as a resource for investigators in this field. When undertaking this type of endeavor, it is often both surprising and disappointing to realize the limitations of the available literature and sequence databases from which references and other important pieces of data can be overlooked. If readers of this series are aware of any oversight in the reference citations, perhaps as a result of inaccurate keyword coding, or inaccuracies in the tabulated data, please do not hesitate to contact the author at the address/FAX/E-mail number given at the beginning of this book. Also, suggestions for improvement or additional sections for future editions would be welcomed.

Acknowledgements

I would like to express my gratitude to Dr Peter Sheterline for his encouragement and advice and for selecting me to participate in this project. I am also grateful to Dr Ron Kahn for sharing his keen foresight and introducing me back in 1989 to what he perceived would be a very 'hot' research area. As always, special thanks go to my wonderful family for their patience during the completion of this monograph. Work in my laboratory is supported by NIH grant DK-R0143396.

Contents

List of figures

List of tables

Glossary

AB, antibody
AcP, acid phosphatase
BVH1, baculovirus VH1 homologue
BVP, baculovirus PTPase
CA, carbonic anhydrase
CC, cultured cells
Cdi-1, cyclin-dependent kinase interactor-1
ch, chromosome
CHO, chinese hamster ovary
Con A, concanavalin A
CPTP, cortex-enriched PTPase
CPTP, cytoplasmic low-molecular-weight PTPase
CSPG, chondroitin sulfate proteoglycan
D1, membrane-proximal PTPase domain
D2, distal PTPase domain
DEP, high cell density-enhanced PTPase
DMSO, dimethyl sulfoxide
E, embryonic day
EGF, epidermal growth factor
EM, electron microscopy
EPO, erythropoietin
ER, endoplasmic reticulum
FAP, Fas-associated phosphatase
FGF, fibroblast growth factor
FISH, fluorescence *in situ* hybridization
Fn-III, fibronectin type III
GLEPP, glomerular epithelial cell PTPase
GVBD, germinal vesicle breakdown
HAAP, human adipocyte acid phosphatase
HCP, haematopoietic cell PTPase
HePTP, haematopoietic PTPase
HIV, human immunodeficiency virus
ICA, islet cell antigen
Ig, immunoglobulin
IGF, insulin-like growth factor
IL, interleukin
IPP, ileal Peyer's patches
IR, insulin receptor
IRS-1, insulin receptor substrate-1
ISH, *in situ* hybridization
KAP, CDK-associated phosphatase
LAR, leukocyte common antigen-related
LC-PTP, leukocyte PTPase
LCA, leukocyte common antigen
LIP.1, LAR interacting protein-1
LPAP, lymphocyte phosphatase-associated protein
LPS, lipopolysaccharide
LRP, LCA-related phosphatase
mAb, monoclonal antibody
MAM, adhesive domain with homology to meprin, A5 and μ
MEG, megakaryoblastic PTPases
MEL, mouse erythroleukaemia
MHC, major histocompatibility complex
MKP, MAP kinase phosphatase
MSG, multicopy suppressor of *gpa1*
N-CAM, neural cell adhesion molecule
NGF, nerve growth factor
OST-PTP, osteotesticular PTPase
P, postnatal day
PAC, phosphatase of activated cells
PCR, polymerase chain reaction
PDGF, platelet-derived growth factor
PEP, PEST-domain phosphatase
PEST, Pro, Glu, Ser, Thr-rich domain
PG, proteoglycan
PHA, phytohaemagglutinin
PKC, protein kinase C
PMA, phorbol 12-myristate 13-acetate
PMA, phorbol myrisate acetate
***p*NPP,** *para*-nitrophenyl phosphate
pS, phosphoserine
pT, phosphothreonine
PTP-BAS, basophil PTPase
PTP-P1, also called LAR-PTP2 and RPTP-σ
PTP-PS, alternatively spliced form of PTP-P1
PTP-S, PTPase
PTP-SL, STEP-like PTPase
PTPase, protein-tyrosine phosphatase
pY, phosphotyrosine
RCM, reduced carboxamidomethylated and maleyated
RKPTP, rat kidney PTPase
RPTP, receptor-like (transmembrane) PTPase
RT, reverse transcriptase
RVH1, raccoonpox virus VH1
SAP, stomach cancer-associated PTP
SH-PTP, SH2 domain-containing PTPase
SHP, SH2 domain-containing PTPase
SH2, SRC-homology domain 2
SH3, SRC-homology domain 3
SHP, SH2 domain phosphatase
SPR, surface plasmon resonance
STEP, striatum-enriched PTPase
stp, small tyrosine phosphatase
TCPTP, T cell PTPase
TCR, T cell receptor
TM, transmembrane
TPA, tumour promoting agent
Tyr(P), phosphotyrosine
VH1, vaccinia open reading frame H1
VHR, VH1-related PTPase
WA, whole animal
YAC, yeast artificial chromosome
Yop, *Yersinia* outer membrane proteins
YVH1, yeast VH1

Introduction

Covalent modification of proteins by phosphorylation and dephosphorylation has long been appreciated to be an essential physiological mechanism for the control of cellular function. The extensive molecular and biochemical characterization of tyrosine kinases that has taken place through the 1980s has now been followed by studies on a rapidly emerging superfamily of complementary enzymes, the protein-tyrosine phosphatases (PTPases; EC 3.1.4.38), which reverse the effect of the kinases and serve to dephosphorylate protein-tyrosine residues of cellular substrate proteins [16, 26]. Work on reversible protein phosphorylation was highlighted by the awarding of the 1992 Nobel Prize in Physiology or Medicine to Drs Edmond Fischer and Edwin Krebs for their fundamental contributions in this area. Although the phosphorylation of protein-tyrosine residues represents only a small fraction of the overall modification of cellular proteins by phosphorylation, which occurs primarily on serine and threonine residues, this reversible protein alteration has been demonstrated to be involved in many critical aspects of cellular physiology. Reversible tyrosine phosphorylation mediates the regulation of cell growth, cell division and differentiation as well as responses to tyrosine kinases that are expressed in cells as plasma membrane receptors or as non-receptor tyrosine kinases. PTPases have been considered to balance the steady-state phosphorylation of a variety of cellular tyrosine kinase substrates and influence cellular signal transduction by attenuating the activity of activated (autophosphorylated) tyrosine receptor kinases or by participating in an on–off switching mechanism mediated by reversible tyrosine phosphorylation of substrate proteins involved in downstream signalling.

From the first recognition of protein phosphotyrosyl dephosphorylating activity in the early 1980s [736, 742, 772], the past 15 years has witnessed a profusion of new information on this important class of phosphatases. This has prompted the inclusion of 'protein-tyrosine phosphatases' as a medical subject heading in the Medline database starting in 1990, and hundreds of research papers have been published on PTPases in each of the past few years. Studies from a number of laboratories have recently provided a well-rounded body of knowledge stemming from early work on PTPase activity in cell extracts, moving into the cloning and characterization of a variety of PTPases and culminating in the recent crystallization and achievement of the detailed three-dimensional structure of several PTPase catalytic domains in 1996. Now that many individual PTPases have been cloned and expressed in various recombinant systems, ongoing work will provide exciting new insight into their role in the regulation of a variety of cellular processes as well as novel approaches to developing new pharmaceutical agents that can modulate the activity of PTPases and impact on human diseases that involve cancer, immunological defence mechanisms and diabetes mellitus.

Biochemical studies in the 1980s provided evidence for various PTPase enzymes in cell and tissue extracts. In 1988, the demonstration of the first primary structure for a PTPase (PTP1B) by Charbonneau *et al.* [484, 485] initiated the rapid accomplishment of numerous molecular cloning studies that greatly expanded the biochemical literature on PTPases. This new edition provides a comprehensive set of data and references on PTPases which have been identified either through molecular cloning studies or protein sequencing, in order to provide useful information on the structure, expression, catalytic activity and role in cellular physiology for individual members of the PTPase superfamily. A detailed summary of early biochemical characterization of PTPases is provided in the excellent review of Lau *et al.* [33]. While the tabulated information in this book was selected to include specifically identified PTPase homologues, additional reviews and individual citations covering a comprehensive listing of all published papers on this category of enzymes accumulated by careful searching of Medline and additional databases from 1979 through 1996 are presented in the bibliography listing at the end of this monograph. The text will provide a brief summary of salient points to accompany the comprehensive data provided in the tables and figures.

Protein-tyrosine phosphatases

Sequences and structures

The first descriptions of PTPases in cell and tissue extracts provided evidence that they were distinct from other protein phosphatases because of their substrate specificity, neutral pH optimum, resistance to EDTA, and unique inhibition by low concentrations of zinc or vanadate [33]. Biochemical studies have since provided evidence that PTPases are found ubiquitously in cytosolic and particulate fractions of cells not only in mammals, but in diverse organisms including invertebrates and lower organisms such as yeast, viruses and bacteria. In early work, the major soluble PTPases in human placenta and bovine brain were purified to homogeneity [403, 801]. The purified placental enzyme, called PTP1B by its elution from DEAE–cellulose, has an M_r of 37.5 kDa and was dependent on sulfhydryl compounds for activity [352]. As is characteristic of many other PTPases, PTP1B is inhibited by low concentrations of vanadate, molybdate and Zn^{2+}, but not by classical inhibitors of protein serine/threonine phosphatases, including okadaic acid and fluoride.

Molecular cloning studies of PTPases and recent work on *in vitro* and *in vivo* characterization of the recombinant enzymes have rapidly expanded our understanding of PTPases as an extensive superfamily of proteins that exert both positive and negative influences on a number of pathways of cellular signal transduction and metabolism [5, 8, 9, 32, 56, 57, 70]. Although PTP1B had no homology with other known protein phosphatases, Charbonneau *et al.* [484] discovered in 1989 that PTP1B had a striking 33–37% sequence identity to the two tandemly repeated domains within the intracellular portion of the leucocyte cell surface glycoprotein CD45. At that time, CD45 was known to consist of a family of high M_r (180–240 kDa) transmembrane glycoproteins that were expressed exclusively on haematopoietic cells and underwent a characteristic pattern of isoform expression through the T cell developmental lineage [59, 66]. The close homology to the PTP1B enzyme led Tonks *et al.* [350] to demonstrate directly that CD45 had intrinsic PTPase activity. These two enzymes could be differentiated biochemically in a number of ways, including the observation that CD45 was much less active against reduced carboxymethylated RCM-lysozyme, and by differences in their inhibition by the divalent cations Zn^{2+} and Mn^{2+}.

PTPase classification

PTPases for which the primary structure is known may be divided into two broad categories. These include *non-receptor type*, which have a single PTPase domain and additional functional protein segments (Fig. 1), and *receptor type*, which have a general structure like a membrane receptor with an extracellular domain, a single transmembrane segment and one or two tandemly conserved PTPase catalytic domains of $\sim$250 residues (Fig. 2). The recognition of these categories of PTPases has led to the proposal of at least two classification systems for PTPases, which have attempted to incorporate various structural features of members of the PTPase superfamily [16, 540]. These essential features include a predicted transmembrane or non-transmembrane localization (based on the identification of a signal peptide and a transmembrane domain), tandem or single PTPase homology domain, characteristics of the extracellular segment (including relative size and presence of identifiable domains with potential functions, such as Ig-like and fibronectin (Fn) type III repeats [756, 855, 891, 901] or other unique homology domains), and additional functional domains of the non-transmembrane PTPases. For the purposes of the data

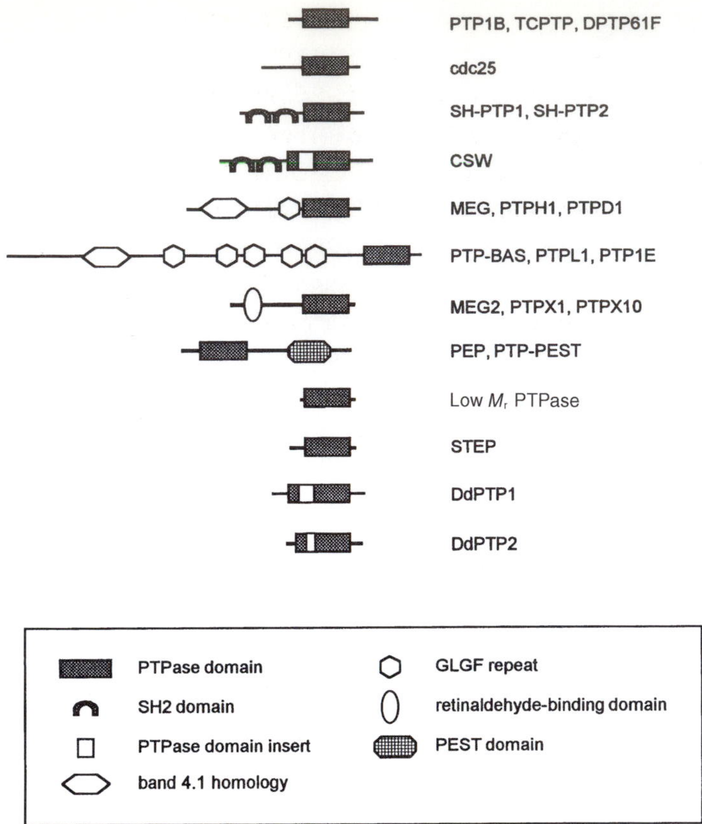

Figure 1

Schematic diagram of non-receptor-type intracellular PTPases. Representative enzymes of the PTPase subfamilies are drawn to approximate scale with symbols depicting various functional motifs as defined in the box below the PTPase drawings.

compilation presented here, a classification scheme that builds on available data from the cloned enzymes has been developed (Table 1). Detailed information including source of the cloned PTPases and their accession numbers is provided in Table 2.

Within this diverse collection of cloned PTPases, sequence homology analysis has demonstrated a series of highly conserved amino acids within the ~250 residue PTPase catalytic domain among the PTPases ranging from *Drosophila* to mammals [540], as well as a particularly high degree of identity within the discrete region that incorporates the PTPase 'signature' motif

(I/V)HCXAGXGR(S/T)G.

This motif has been confirmed to be involved in the catalytic mechanism of PTPase homologues by numerous studies using site-directed mutagenesis, enzyme derivatization and isolation of catalytic intermediates (see references in Tables). Furthermore, the recent determination of tertiary structures for three PTPase catalytic domains has provided direct evidence for the formation of a phosphate-binding 'P' loop in which the nucleophilic thiolate anion of the Cys residue coordinately binds to the phosphate of the substrate protein along with the Arg residue at position Cys + 6. The 'P' loop tertiary structure has been found in the *Yersinia* PTPase and in PTP1B, confirming the data from biochemical studies on the enzyme catalytic center [632, 635, 637, 1133].

The inclusion of a group of small, cytosolic phosphatases termed the 'low M_r PTPases' in the superfamily of PTPases presented here has recently been validated by structural studies. These enzymes were

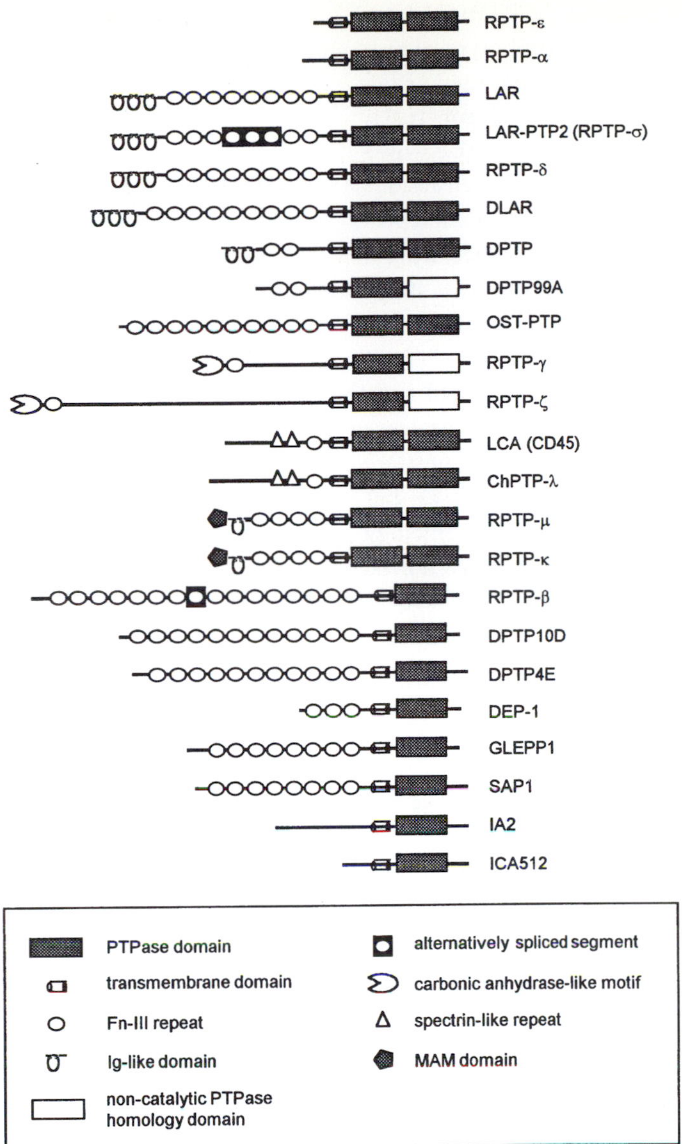

Figure 2
Schematic diagram of receptor-type transmembrane PTPases with tandem or single PTPase homology domains. Representative enzymes of the PTPase subfamilies are drawn to approximate scale with symbols depicting various functional motifs as defined in the box below the PTPase drawings.

initially identified as widely expressed cytosolic acid phosphatases. Later, they were found to have PTPase activity over a broad pH range (5.5–7.5), and as found with other PTPase homologues, they were sensitive to sulfhydryl reagents and low concentrations of vanadate. Although these enzymes lack overall sequence homology to the prototypical ~250 residue PTPase domain, they have been shown to contain a minimal 'P' loop homology domain that includes a '**CXXXXXR**' motif, and structural studies have demonstrated that the phosphate ion is cradled between the catalytic Cys and the Arg side-chains that stabilize the phosphate by hydrogen bonding. Outside these two residues, sequence differences are responsible for variations between the PTPase homologues in the stabilization of the phosphotyrosyl protein substrate binding to the enzyme and the kinetics of the reaction with various potential substrates. Thus, the PTPase superfamily is characterized by a conserved minimal homology within the seven-residue 'P' loop, and a common catalytic mechanism that proceeds through a cysteinyl-phosphate intermediate, which is unique among protein phosphatases [73].

An additional subcategory of PTPases includes those wih homology to the *Vaccinia* virus VH1 PTPase, including PAC-1, 3CH134, CL100 and the cell cycle regulator cdc25. These enzymes have high sequence homology within the extended PTPase signature motif described earlier to many other PTPases, but like the low M_r PTPases, they lack a general homology to the other enzymes within the $\sim$250 residue catalytic domain. Interestingly, these enzymes also have in common the ability to dephosphorylate not only protein-phosphotyrosine residues, but also substrates phosphorylated on serine and threonine, hence their designation as dual-specificity PTPases [909].

Mapping of PTPase genes

An imbalance in the tyrosine phosphorylation of key regulatory proteins can induce cellular proliferation and oncogenesis. By antagonizing the effects of tyrosine kinases, PTPases may thus act as potential antioncogenes and protect cells from malignant transformation. With this in mind, PTPases have been mapped in both human and rodent genomes to examine whether their respective genes might localize to known sites of potential tumour suppressor genes (Table 3). Initially, this type of analysis was done with RPTP-γ which was mapped to a region frequently deleted in lung and renal cell carcinoma

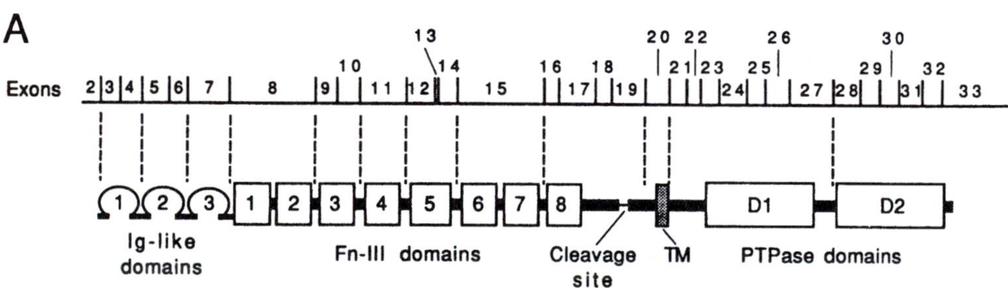

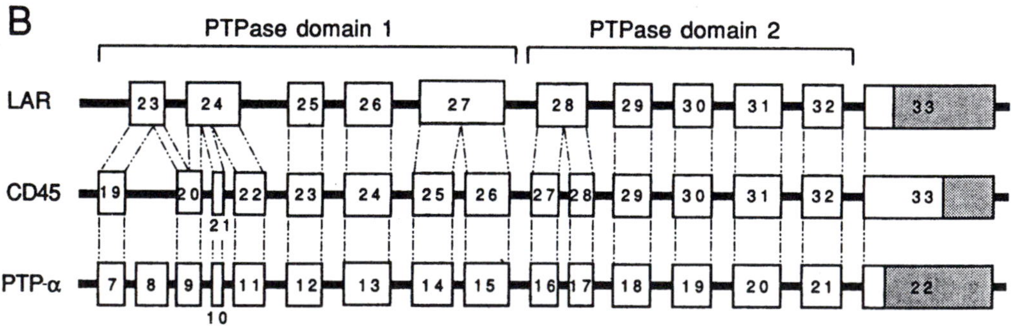

Figure 3

Exon organization of the human LAR gene. (a) The general structures of the 33 exons of the human LAR gene are correlated with structural domains of the protein, including the Ig-like and Fn-III domains, the proprotein cleavage site and the intracellular PTPase homology domains, D1 and D2. TM shows the position of the transmembrane region. (b) Comparison of the exon/intron organization of the tandem PTPase homology domains of LAR with those of CD45 and RPTP-α. Reprinted with permission from O'Grady *et al.* [573].

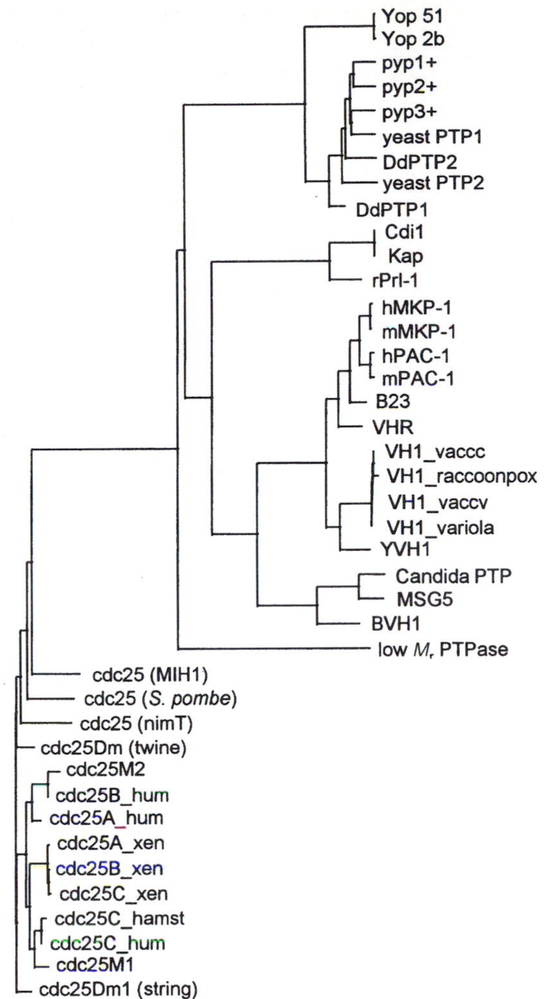

Figure 4

Phylogenetic tree of the PTPases in lower organisms and also including those in the Yop, VH1, low M_r PTPase and cdc25 homology subfamily designation. Abbreviations are given for specific enzymes described in Table 2. The tree was generated from sequences of ≈275 amino acids surrounding the catalytic Cys residue, or with the entire enzyme sequence for smaller enzymes, using version 8.0 of the GCG computer programs. Sequences were aligned with PILEUP and phylogenetic data tabulated using DISTANCES with the Kimura protein distance correction. The neighbour-joining method was used to generate the phylogenetic tree from the distance matrix using GROWTREE. Several phylogenetic clusters are evident, including the cdc25 subgroup in various species, the PTPases found in yeast and *Dictyostelium discoideum*, the low M_r PTPases and the large subgroup of VH1-like enzymes.

[542]. Recent studies have also implicated deletions in the RPTP-γ gene in the ability of mouse L-cells to proliferate into sarcomas [618]. Other PTPases have since been mapped to regions implicated in haemato-proliferative disorders and other malignancies. Mapping of SH-PTP1 (see Table 2) led to its identification as the genetic defect in the *moth-eaten* mouse, whose phenotype occurs because of tissue accumulation of abnormally proliferating macrophages and granulocytes, as well as severe underlying immunodeficiency and haematopoietic cellular dysregulation [69].

The structure of a few PTPase genes has been determined, including CD45, CL100, DPTP10D, DPTP4E, DPTP61F, DPTP99A, ERP(3CH134), PTP1B, LAR and RPTP-α (Table 3), and some promoter regions for PTPase genes are being cloned and analysed. Evaluating the gene structure for PTPases will provide additional insight into their evolutionary relationships. For example, the exon structure of the cytoplasmic domain of LAR is very similar to that of both CD45 and RPTP-α, including an interesting disruption of the highly conserved catalytic centre sequences by an intron/exon junction (Fig. 3) [573, 620].

Alternative splicing of PTPase mRNAs

As more PTPase cDNAs have been cloned and sequenced, it has been appreciated that alternative splicing commonly affects various functional domains of the PTPases both outside as well as within the catalytic domain and provides an additional level of heterogeneity in this enzyme family (see Tables 2 and 3). The alternative splicing of extracellular domains of CD45 has been appreciated for many years and recently the mechanisms underlying the cell-specific machinery involved in generating diversity in the extracellular domain of CD45 have been characterized [191, 204, 585]. As is common with other proteins having cell adhesion molecule motifs, the extracellular domains of several additional PTPases have been shown to be affected by alternative mRNA splicing.

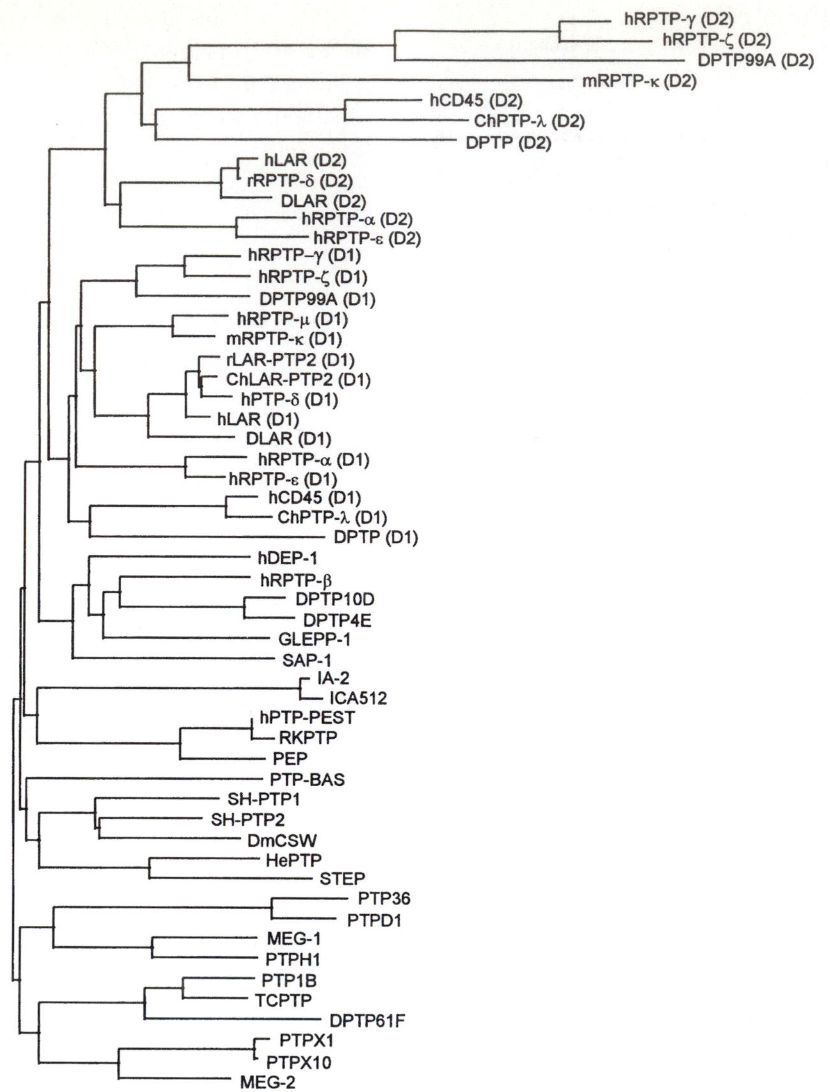

Figure 5

Phylogenetic tree of receptor and non-receptor PTPases in higher organisms including mammals and *Drosophila*. Abbreviations are given for specific enzymes described in Table 2. The tree was generated exactly as described in the legend to Fig. 4. Clustering is apparent among the second domain (D2) of the tandem domain PTPases separate from the proximal PTPase homology domain sequences (D1). The single-domain, transmembrane PTPases also exhibit a similar phylogenetic origin as do various subgroups of the single-domain, intracellular PTPases.

Sequences encoding the PTPase catalytic region have also been shown to undergo alternative splicing. The cytoplasmic domain of RPTP-α was the first to be shown to undergo alternative splicing [554]. These findings have been followed by the discovery of interesting splicing variants involving the catalytic regions of several PTPases, including PTP-PS, SH-PTP1 and Syp (see Table 2). The resulting alterations in the protein structure are likely to play an important role in determining catalytic activity, substrate specificity, or post-translational modifications of the affected enzymes *in vivo*.

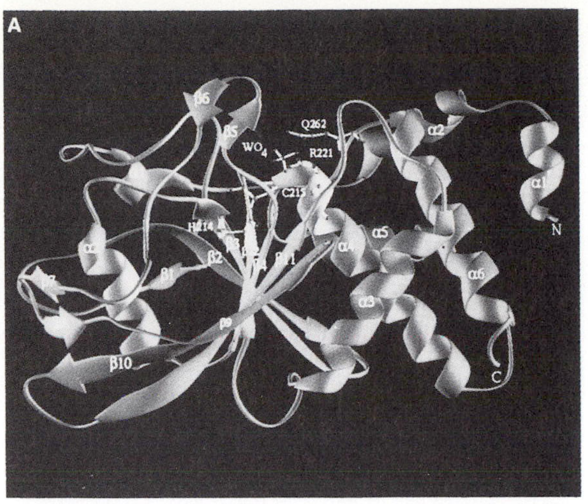

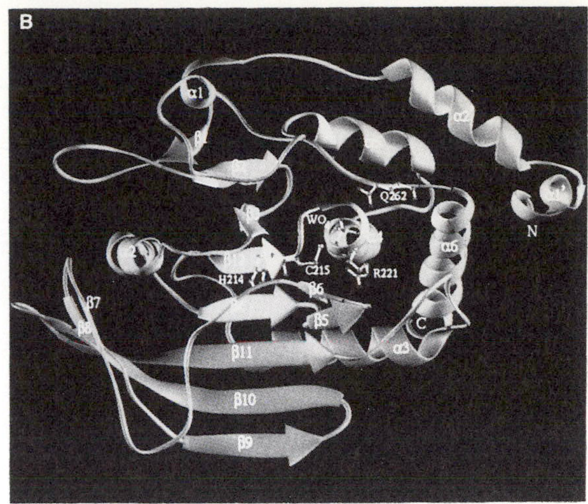

Figure 6

Three-dimensional structure of PTP1B. (a) Ribbon diagram depicting the secondary structure α-helices and β-sheet elements, the catalytic domain (containing a tungstate ion, WO_4) and certain key residues, including His-214 (H214), Cys-215 (C215), Arg-221 (R221) and Gln-262 (G262). (b) View perpendicular to A. Reproduced with permission from Barford *et al.* [632].

(a)

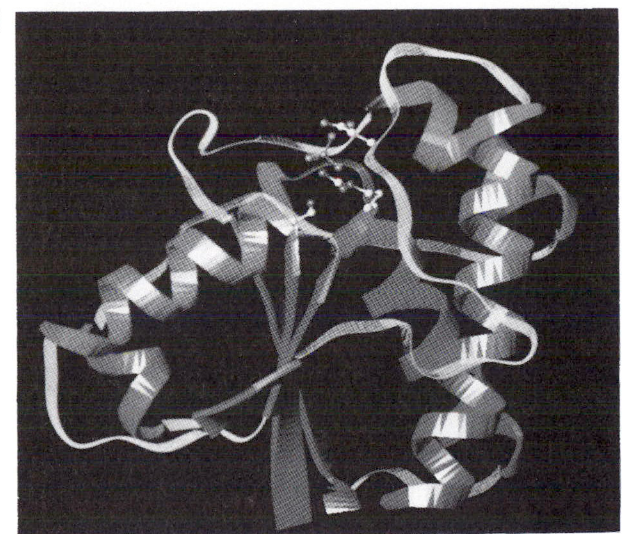

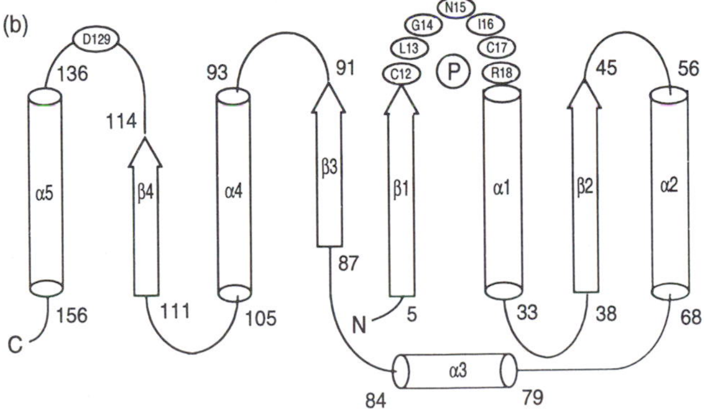

Figure 7

Three-dimensional structure of the low M_r PTPase. (a) Ribbon diagram demonstrating the central twisted β-sheet and the surrounding helices. The loop encircling the active site is shown in the background, surrounding the active-site residues Cys-12, Arg-18 and Asp-129. (b) Schematic diagram of the secondary structure of the low M_r PTPase. The β-sheets are depicted as arrows and the α-helices are drawn as cylinders. The phosphate-binding loop, encompassing the CXXXXXR motif, is labelled 'P'. Reproduced with permission from Su *et al.* [638].

Evolutionary relationships within the PTPase superfamily

PTPases have been identified in various organisms spanning the phylogenetic tree from bacteria and viruses to humans. Several lower species, including the invertebrate *Styela plicata,* the cyanobacterium *Nostoc commune* and the slime mold *Dictyostelium discoideum* express PTPases. These enzymes arise in antiquity and it is likely that they have evolved into unique physiological roles over time to this day when they serve to regulate diverse cellular functions affecting various aspects of cell function and development in higher organisms. The diverse nature of the structure and expression of PTPases in various organisms will require studies of each homologue in an appropriate cellular or whole animal system to characterize fully the impact that these enzymes have on cell signalling. Phylogenetic trees that include representative sequences from each of the PTPase homologues are presented in Figs 4 and 5.

Structural analysis of PTPase proteins

High-level expression of recombinant PTPase catalytic domains has facilitated their analysis by various physicochemical techniques. Isolated extracellular domains of CD45 expressed in mammalian cells have been imaged by electron microscopy [636]. Studies in which purified or recombinantly expressed PTPases have been crystallized are summarized in Table 4, along with data from two-dimensional nuclear magnetic resonance (NMR) and X-ray crystallography, which to date have provided high-resolution structural analysis for PTP1B, *Yersinia* PTPase and the low M_r PTPase. Despite overall sequence divergence among these three PTPase homologues, they have been shown to have a conserved phosphate-binding 'P' loop structure, as described earlier. Cocrystallization of the PTPase inhibitor tungstate as a catalytic site ligand has elucidated the hydrogen bonding within the 'P' loop that is likely also to engage the phosphotyrosyl protein substrate [637]. These studies have provided important confirmatory data for these structural relationships that have been predicted from work on evaluating kinetic parameters of PTPase substrate dephosphorylation and the effects of various inhibitors. Studies using site-directed mutagenesis of active site residues and the trapping of catalytic intermediates have also served to identify the precise mechanism followed by various PTPase homologues (see later; Tables 9 and 10). Representative PTPase structural features from crystallography data are presented in Figs 6 and 7 for PTP1B and the low M_r PTPase, respectively.

Table 1 PTPase structural classification and family designation

Class	Characteristics	Subclass designation (prototype PTPase)	Additional homology/functional domains	Representative PTPases (synonyms)*	SA†
Class I	Non-transmembrane, single PTPase domain	Class IA (PTP1B)	Localization directed by C-terminal segment	PTP1B, TCPTP (PTP-S, MPTP), DPTP61F	A
		Class IB (VH1)	Small, dual-specificity PTPases	B23, BVH1 (BVP), MKP1 (3CH134; ERP, CL-100), MSG5,PAC-1, RVH1, HVH1, VHR, YVH1	B
		Class 1C (cdc25)	Small, dual specificity PTPases	cdc25A, cdc25B, cdc25C, cdc25Dm (*string*), cdc25Dm (*twine*), cdc25Hu2, cdc25M1, cdc25M2, $cdc25^+$	C
		Class 1D (SH-PTP1)	Two N-terminal SH2 domains	*corkscrew*, SH-PTP1 (HCP, PTP1C, SHP), SH-PTP2 (PTP-SHβ, PTPL1, PTP1D/Di, PTP2C/Ci, SH-PTP3, Syp)	D
		Class 1E	Cytoskeletal protein motif (ezrin, band 4.1, talin)		E
		Class 1E1 (MEG1)	Smaller enzymes without additional homology domains	MEG1, PTPH1, PTP36, PTPD1	
		Class 1E2 (PTP-BAS)	Larger enzymes with GLGF motif as found in junction-associated guanylate kinases and multiple PEST motifs	PTP-BAS, PTP1E, PTPL1	
		Class 1F (MEG2)	Retinaldehyde-binding protein motif	MEG2, PTPX1, PTPX10	F
		Class 1G (PEST)	'PEST' domain at C terminus	PEP, PTP-PEST (P19-PTP, PTPG1, PTPty43)	G
		Class LMWP (low M_r PTP)	Very small, unique homology	low M_r PTPasem (HAAPα/β, CPTP1/2, AcP1/2)	H
		Class Yeast	*S. pombe* and *S. cerevisiae* enzymes	$pyp1^+$, $pyp2^+$, $pyp3^+$, yeast PTP1, yeast PTP2	I
		Class Dd (DdPTP)	*Dictyostelium* PTPases	DdPTP1, DdPTP2	I
		Class *Yersinia* (Yop)	*Yersinia* PTPases	Yop51 (Yop H), Yop2b, (Yop H2)	J
		Unclassified		Cdi1, HePTP, IphP, KAP, PRL-1, RKPTP, STEP	J
Class II	Transmembrane, receptor-type, single PTPase domain	Class IIA (RPTP-β)	Only Fn-III domains in extracellular segment	RPTP-β; DEP1, DPTP10D, DPTP4E, GLEPP1, SAP1	K
		Class IIB (ICA512)	Neuroendocrine cell PTPases	ICA512, IA-2	L
Class III	Transmembrane, receptor-type, tandem PTPase domains	Class IIIA (RPTP-α)	Short extracellular domain	RPTP-α, RPTP-ϵ	M
		Class IIIB (LAR)	Large extracellular domain; Ig-like and Fn-III repeats	LAR, LAR-PTP2/2B (RPTP-σ, CRYPα1/2, PTP-P1/PS, PTPNE3), RPTP-δ, DPTP and DLAR	N
		Class IIIC (RPTP-γ)	In D2, catalytic Cys is replaced by Asp; also may have non-catalytic carbonic anhydrase-like motif at N terminus (PTP-γ, PTP-ζ)	PTP-γ, PTP-ζ (ζ/β)‡, DPTP99A, PTP18	O
		Class IIID (CD45)	Single Fn-III homology domain, spectrin repeats in extracellular domain	CD45, ChPTPλ	P
		Class IIIE (RPTP-μ)	N-terminal adhesive MAM (meprin-A5-μ) motif with an Ig-like domain and Fn-III repeats	RPTP-μ, RPTP-κ	Q
		Class IIIF (OST-PTP)	Fn-III repeats only in extracellular domain	OST-PTP	

* See Table two for detailed sequence information.
† Sequence alignments are found in the Appendix at the back of the book.
‡ Since two groups named different PTPase enzymes as 'RPTP-β', the literature has maintained the designation of RPTP-ζ (or RPTP-ζ/β) for the homologue initially named RPTP-β by Kaplan *et al.* [536].

Table 2 PTPase sequence data and accession numbers

PTPase	SA*	cDNA/protein source	PIR no	SwissProt name SwissProt no	AA/Mass (Da)	Comments	EMBL no	Refs
B23	B5	B5/589 human mammary epithelial cell cDNA			397/44 k	Small, single domain PTPase induced by serum and heat shock, most similar to MKP-1	U15932	525
BVH1 (BVP)	B15	*Autographa californica* genomic DNA	A45431	PTP_NPVAC P24656	167/19 223	Unique homology		335, 516
***Candida* PTP**	I8	*Candida albicans*	S43743	597/65637			L01038	
CD45		Human	A29449		1143/130 897	One Fn-III module and spectrin-like repeats identified in comparison to ChPTPλ	Y00062	500
CD45	P3	Human λ GT11 cDNA Library		CD45_HUMAN P08575	1304/147 253	Large extracellular domain, single TM segment, tandem intracellular PTPase homology domains	Y00638	606
CD45	P2	Rat T-cell cDNA library	A60241	CD45_RAT P04157	1255/141 208		Y00065 M25820 M25821 M25822 M25823	476, 611
CD45		Human lymphocyte cDNA library				Three cDNA variants detected in N-terminal region		587
CD45 (Ly-5)	P1	Mouse B and T leukaemia cells	A28334 A29381 A61180 A60933	CD45_MOUSE P06800	1291/144 603 1152/200 k		M22455 M14342	592, 601, 615
cdc14		*Saccharomyces cerevisiae* cDNA library	A42784	CC14_YEAST Q00684	423/47 231	May be involved in chromosome assembly	M61194	617
cdc25		*Xenopus laevis*		MPI0_XENLA P30308	550/62 308	Virtually identical to MPI1_XENLA dual catalytic specificity PTPase	M96857	92
cdc25	C6 C7 C8	*Xenopus laevis*	A42679 B42679 C42679	MPI1_XENLA P30309 MPI2_XENLA P30310 MPI3_XENLA P30311	550/62 182 (A) 599/67 645 (B) 572/64 413 (C)	Three closely related isoforms	M94262 M94263 M94264	292
cdc25		Porcine cDNA library						1098
cdc25 (MIH1)	C14	*Saccharomyces cerevisiae* cDNA library	A32386	MPIP_YEAST P23748	474/54 376		J04846	590
cdc25 (*nimT*)	C12	*Aspergillus nidulans* cDNA library	S24395	MPIP_EMENI P30303	556/61 586		X64601	572
cdc25⁺	C13	*Schizosaccharomyces pombe* cDNA library	A25301	MPIP_SCHPO P06652	580/65 125		M13158	591
cdc25A	C5	Human cDNA library	A41648	MPI1_HUMAN P30304	523/58 796		M81933	503
cdc25A		Rat kidney (NRK-49) expression cDNA library			525/59 k		D6236	161

cdc25B	C3	Human cDNA library	B41648	MPI2_HUMAN P30305	566/63 442		M81934 S78187	503
cdc25B		Rat kidney (NRK-49) expression cDNA library			574		D6237	161
cdc25C	C11	Human	A38874	MPI3_HUMAN P30307	473/53 311		M34065	195
cdc25C		HeLa cell cDNA library			473/53 k	Virtually identical to cdc25C (MPI3_HUMAN)	M34065	195
cdc25C	C10	Hamster BHK 21/13 cell cDNA library			421		D10878	599
cdc25Dm *(twine)*	C1	*Drosophila melanogaster* cDNA library	S26692	TWIN_DROME Q03019	395/44 901	Identical to isolate described in ref. 530	M69018	489
cdc25Dm *(twine)*		*Drosophila melanogaster* cDNA library			431			473
cdc25Dm1 *(string)*	C2	*Drosophila melanogaster* cDNA library	A32290 S12008	MPIP_DROME P20483	479/54 124		M24909 X57495	498, 530
cdc25Hu2		HeLa cell cDNA library			566/63.5 k	Virtually identical to cdc25B (MPI2_HUMAN)	S78187	568
cdc25M1	C9	Mouse pre-B cell cDNA library			465/51.8 k		L16926	569
cdc25M2	C4	Mouse P19 teratocarcinoma cell cDNA library	A42236	MPI2_MOUSE P30306	576/65 490	Mouse homologue of human cdc25Hu2	S93521	535
cdi1	J6	HeLa cell cDNA expression library in yeast interaction trap (two-hybrid system)			212/21 k	Isolated by interactions with cyclin-dependent kinases cdc2, cdk2, cdk3 but not cdk4; only weak similarity to known PTPases; 'PEST' motif at N terminus	U02681	515
ChPTP-λ	P4	Chicken brain cDNA library and pre-B cells	A54080		1237/139 319	TM receptor-type PTPase with Ser/Thr/Pro-rich domain, one Fn-III module and spectrin-like repeats (similar to CD45); Five alternatively spliced forms varying near N terminus, combination of deleted segments designated I–IV	L13285	500
CL100	B1	Human homologue of 3CH134	A53052 S29090	PTN7_HUMAN P28562	367/39 297	VH1 family	X68277	537
CL100 (XCL100)		*Xenopus laevis* cDNA library				Homologue of the human CL100 dual specificity MAP kinase phosphatase		1108
CSW	D6	0–4 h embryonic *Drosophila melanogaster* cDNA library	A43254	CSW_DROME P29349	841/92 430	Non-receptor type PTPase with two N-terminal region SH2 domains and a single PTPase homology domain; PTPase domain is unusual in that it is disrupted by a hydrophilic insert of 150 AA rich in Ser and Cys	M94730	580
cdc25		Embryonic rat brain by RT-PCR				A novel PTPase cDNA with close homology to *Drosophila* DPTP10D; expression is specific to brain; expressed at high levels in embryonal and neonatal stages but scarcely in adults	D30749	970

Table 2 Continued

PTPase	SA*	cDNA/protein source	PIR no	SwissProt name SwissProt no	AA/Mass (Da)	Comments	EMBL no	Refs
DdPTP1	I5	*Dictyostelium discoideum* cDNA library	A44267	PTP1_DICDI P34137	521/59 427	Intracellular single domain PTPase; has N-terminal myristoylation consensus (MGXXXS); contains a unique 99 AA insert in the catalytic domain; this domain has a stretch of 23 residues containing 19 Asp	L07125	277
DdPTP2 (DdPTPa)	I6	*Dictyostelium discoideum* amoeba cDNA library		PTPX_DICDI P34138	377/43 488	Intracellular single domain PTPase; under low stringency conditions, Southern blot shows weak hybridization to additional bands, suggesting multigene family of PTPases in *D. discoideum*	L15420	588
DdPTP2		*Dictyostelium discoideum* cDNA library			377/43 k	Intracellular single domain PTPase; has 58 AA insertion located at position 24 of the catalytic domain, and which contains 10 Asp and 12 Asn –similar to CD45 D2, which also has an Asp-rich insert of 20 AA at position 33 of the PTPase domain		523
DEP-1	K3	HeLa cDNA library			1337	Extracellular segment consisting of three Fn-III repeats, a single TM segment and a single cytoplasmic PTPase domain	U10886	576
DLAR	N6	*Drosophila melanogaster* embryo cDNA library	A36182	LAR_DROME P16621	2029/229 026	Large TM proteins with tandem cytoplasmic PTPase domains; DLAR extracellular domain has three Ig-like domains and nine Fn-III repeats	M27700	606
DPTP (DPTP69D)	P5	*Drosophila melanogaster* embryo cDNA library	B36182	PTP_DROME P16620	1462/167 410	Large TM proteins with tandem cytoplasmic PTPase domains; extracellular domain consists of two Ig-like domains and two Fn-III repeats	M27699	606
DPTP10D	K1	*Drosophila melanogaster* embryo cDNA library	C41214 D41214	PTP1_DROME P35992	1630/184 861 1557/177 147	Receptor-type PTPase with 12 Fn-III repeats, a single TM domain and a single cytoplasmic PTPase domain; produces alternative protein forms which differ in their C-terminal tails; two isoforms ± exon 10; DPTP-10DA is 1631 AA, DPTP-10DB is 1558 AA	M80465	625
DPTP10D		9–12 h *Drosophila melanogaster* embryo cDNA library	A41215		1558/177 293	Receptor-type PTPase consisting of 12 Fn-III repeats, a single TM domain and a single PTPase domain, as found in RPTP-β	M80538	614
DPTP4E	K2	Isolated from 4–8 and 8–12 h pNB40 embryonic *Drosophila melanogaster* libraries			DPTP4E-A 1767/200 k DPTP4E-B 1615/183 k	Receptor-type PTPase for which two cDNA forms isolated: DPTP4E-A includes all 10 exons, and predicts a 1767 AA protein, and DPTP4E-B which lacks exon nine (508 bp) at C terminus and predicts a 1615 AA protein; extracellular domain after residue 52, a hydrophobic domain of 24 amino acids which may correspond to an internal signal peptide and confer an intracellular localization signal for the N terminus of the molecule, the extracellular domain also contains 11 Fn-III repeats, an additional tm domain of 20 residues at 80 AA N-terminal to the PTPase domain and a single cytoplasmic PTPase domain; N-terminal region to AA 52 rich in His and Gln due to repeats of CAX triplets, known as	L20894	575

						OPA or M repeats; C-terminal region downstream of PTPase domain has an acidic region with Asp and Glu (AA 1669–1683) in exon nine that are absent in the DPTP4E-B due to splicing		
DPTP61F	A7	*Drosophila melanogaster* genomic library			'm' isoform 548/62 k 'n' isoform 536/61 k	Intracellular PTPase that undergoes alternative splicing to encode two non-receptor-like proteins, denoted DPTP61Fm or DPTP61Fn that differ at C terminus; C-terminal sequences govern the targeting of each PTPase either to a cytoplasmic membrane or to the nucleus; mRNA sizes 2.8 kb for 61Fm and 2.0 kb for 61Fn	L11249 L11250 L11251 L11252 L11253	558
DPTP99A	O6	*Drosophila melanogaster* embryo cDNA library	A41214 B41214	PTP9_DROME P35832	1231/137 421 1061/119 646	Receptor-type PTPase containing two Fn-III repeats in extracellular domain and tandem PTPase homology domains; produces alternative protein forms which differ in their C-terminal tails; exon 12 absent from the 6.4 kb mRNA which predicts smaller protein of 120 k	M80464	625
DPTP99A		9–12 h *Drosophila melanogaster* embryo cDNA library	B41215		1231/137 384	The first PTPase domain has high homology to RPTP-γ, and the second PTPase domain has an aspartic acid residue substituted for the catalytic cysteine, as found in RPTP-γ	M80539	614
DPTP99A		*Drosophila melanogaster* eye-antennal disc and 9–12 h embryo cDNA libraries	A41622		1301/145 336	A glutamine-rich OPA repeat near the C terminus (AA 1076–1091); proximal PTPase domains (D1) remarkably similar to human RPTP-γ (64% identity)	M81795	520
GLEPP1	K6	Rabbit glomerular cell cDNA expression library, screened with monoclonal Ab to GLEPP1	A53661		1187/135 046	Large TM PTPase with eight Fn-III repeats and single PTPase domain (homologous with PTP-β and DPTP10D); putative 36 bp alternatively spliced region just 5′ to the PTPase catalytic domain	U09490	613
GLEPP1		Human renal cortical cDNA library			Precursor is predicted to contain 1188 AA; mature protein 1159 AA	Contains a large extracellular domain, a single transmembrane domain, and a single intracellular PTPase domain		1130
HA2						PTPase HA2 cDNA showed it to be a homologue of PTPase IB		1012
HePTP	J4	Stimulated human peripheral T-cell cDNA library	A46541	PTND_HUMAN P35236	339/38 421	Single PTPase domain, small intracellular enzyme	M64322	630
HePTP (LC-PTP)		Human PEER T-cell cDNA library	JH0692 A46541 D44929 A46752		360/40 573	Variant of HePTP; possibly due to mutations in PEER cells	D11327 S78090	470
hVH-2						A novel dual specificity PTPase; has significant identity to the VH1-related family; N-terminal region also displays sequence identity to the cell cycle regulator cdc25		1100

*SA = Sequence alignment, found at the back of the book

Table 2 Continued

PTPase	SA*	cDNA/protein source	PIR no	SwissProt name SwissProt no	AA/Mass (Da)	Comments	EMBL no	Refs
hVH-3		Human placenta			384	Highest similarity to VH1 subfamily; also identified an AYLM sequence motif in VH1 family members immediately C-terminal to the CXXXXXR domain	U16996	950
hVH-5						Novel homolog of VH1 phosphatase, shares sequence similarity with PTPases that regulate MAP kinase; a unique proline-rich region distinguishes hVH-5 from other closely related PTPases		1110
IA-2	L2	Human insulinoma subtraction library (ISL-153)			979/105 k	Relatively large extracellular domain, single TM domain and single PTPase homology region	L18983	543
IA-2		Mouse brain cDNA library			979		U11812	548
ICA512	L1	Human pancreatic islet cDNA expression library	S18121		548/60 078	Small extracellular domain PTPase with a single TM segment and a single PTPase homology domain; initial experiments unable to demonstrate PTPase activity in recombinant ICA512	X62899	586
IphP	J9	Cyanobacterium *Nostoc commune* UTEX 584	A46635		294/31 005	Contains 8 AA consensus sequence for PTPases (HCXAGXXR); 1st PTPase to be found encoded by a prokaryotic genome	L11392	584
KAP	J7	HeLa cell cDNA library by yeast two-hybrid system screening			203		L27711	518
LAR	N5	Human tonsil lymphocyte cDNA library	S03841	PTPF_HUMAN P10586	1897/211 843	Original description of this large TM enzyme with three Ig-like repeats followed by eight Fn-III repeats in the extracellular domain, a single TM segment, and tandem cytoplasmic PTPase domains	Y00815	605
LAR		Rat	A41032		Fragment	Cytoplasmic domain	M60103	326
LAR		Rat liver cDNA library			1898		L11586	631
LAR		Mouse				Analysis of both phosphatase domains of mLAR and its homologues MPTP-δ and mRPTP-σ revealed a higher evolutionary conservation of the second, C-terminal domain in comparison to the first domain		1119
LAR		Rat tissues and neuronal cells				Identified five alternatively spliced RNA forms including deletions of 4th, 6th and 7th Fn-III repeats, novel cassette in 5th Fn-III domain, two novel cassette exons in juxtamembrane domain, retained intron in extracellular domain with stop codon predicting secreted isoform and transcript with up to 21 CAG repeats in 3′ untranslated region; splicing was preferential in neuromuscular tissue		980

LAR/RPTP-δ/RPTP-σ subfamily		Human fetal lung and brain RNA and brain cDNA library				Multiple isoforms generated by alternative splicing of up to four mini-exon segments encoding peptides of 4–16 AA in both intracellular and extracellular regions in tissue-specific fashion; also detected short forms of RPTP-δ and RPTP-σ having only four of eight Fn-III domains		1117
LAR-PTP2	N1	Rat liver cDNA library			1863	Overall structure identical to LAR, but product of a different gene; full-length form of LAR-PTP2B/ RPTP-σ	L11587	631
LAR-PTP2 (CRYP-α2)		Chicken embryo brain cDNA library				Alternative form of CRYP-α1 with three Ig-like repeats and eight Fn-III repeats; additional forms isolated include CRYP-α3 (truncation that eliminates most of the distal PTPase domain and CRYP-α4 (truncated protein after Ig domain 2); all encoded by single gene by pattern of Southern blots; chicken homologue of LAR-PTP2	L32781; L32782; L32783	604
LAR-PTP2 (mRPTP-σ, MPTPT9)	N2	Mouse			mRPTP-σ T 1904/	Murine homologue of LAR-PTP2 isoforms; three forms identified: mRPTP-σ T is the dominant form in the thymus and has three Ig-like and eight FN-III-like domains in the extracellular portion; compared to LAR-PTP2, RPTP-σ T has an insert of 44 AA in the extracellular domain, this enzyme is also ~400 AA longer than the LAR-PTP2B isoforms (NE-3/σ/P1/CPTP1); mRPTP-σ B is dominant in the brain and is an isoform of LAR-PTP2B and its counterparts; B has a deletion of ~400 AA compared to mRPTP-σ T; mRPTP-σ S lacks 1159 bases and has a premature stop codon in the extracellular domain that predicts a soluble protein with no TM or cytoplasmic domains	PTP-σ T D28530 PTP-σ S D28531	179
LAR-PTP2 (PTPNU-3)		Mouse embryonic kidney cDNA library			–/211 k	Murine homologue of LAR-PTP2 (RPTP-σ)		1128
LAR-PTP2B		Rat brain cDNA library			1501	Alternatively spliced form of LAR-PTP2, with deleted extracellular domain by 362 AA, effectively removes the segment encoding the 4th–6th Fn-III domains	L12329	631
LAR-PTP2B (PTP-NE3)		Rat cerebral cortex cDNA library			1501/168 k		L19933	216
LAR-PTP2B (PTP-P1)		Rat PC12 Pheochromocytoma cell library			1494		L19180	579
LAR-PTP2B (RPTP-σ)		Rat brain stem cDNA library	A49104		1501/168 336			621
LAR-PTP2B (CRYP-α1)	N3	Chicken embryo brain cDNA library			1559/	TM enzyme with three Ig-like repeats and four Fn-III repeats; chicken homologue of LAR-PTP2B	L32780	604
Low M_r PTPase	H1	Bovine heart cDNA library	A31423	PPAC_BOVIN P11064	157/17 923	First representative of a group of mammalian, low M_r PTPases	M83656	365

*SA = Sequence alignment, found at the back of the book

Table 2 Continued

PTPase	SA*	cDNA/protein source	PIR no	SwissProt name SwissProt no	AA/Mass (Da)	Comments	EMBL no	Refs
Low M_r PTPase (CPTP1/ CPTP2)	H2 H3	Human placental cDNA library	A38148	PA1F_HUMAN P24666 PA1S_HUMAN P24667	157/17 931	Two distinct cytoplasmic PTPases with unconventional catalytic domain and no apparent additional functional regions; CPTP2 is identical to CPTP1 except for a 108 bp segment in the middle of the ORF that is only 52% identical to CPTP1; PCR suggests that the difference does not arise from alternative splicing; apparently CPTP-A and -B represent the fast and slow forms of the red cell acid phosphatases that are expressed in all human tissues	M83653 M83654	495, 496, 619
Low M_r PTPase		Purified protein from bovine liver			157/17 953	No sequence homologies were found with the two known acylphosphatase isoenzymes or the metalloproteins porcine uteroferrin and purple acid phosphatase from bovine spleen (both of which have acid phosphatase activity)		482
Low M_r PTPase (HAAP-α/ HAAP-α)		Human adipocyte cDNA library			HAAP-α 146 HAAP-β 157	18 k cytoplasmic protein from 3T3-L1 cells which dephosphorylates pNPP and the phosphorylated adipocyte lipid binding protein (ALBP); initially identified as an acid phosphatase (EC 3.1.3.2) inactivated by PAO ($k_{inact} = 10\,\mu M$); composed of two forms HAAP-α and -β which are distinguished by 34 AA isoform-specific domains		600
Low M_r PTPase		Purified protein from porcine liver			157	The enzyme is acetylated at the NH_2 terminus; this isoform shows a lower homology degree with respect to rat AcP1 and human B-fast isoforms		483
Low M_r PTPase (AcP1/AcP2)		Rat liver purified protein			AcP1 157/18.0 k; AcP2 157/17.8 k	AcP1 and AcP2 are two distinct enzymes, both consist of 157 AA, are acetylated at the N terminus, and have His at the C terminus; homologous except in the region of AA 40–73, where they differ in 50% of residues (possibly by alternative splicing)		550
LTP1		*S. cerevisiae* cDNA library			161	Related to mammalian low M_r PTPase with 39% average identity with amino acid sequence		1116
MEG1	E3	Human megakaryoblastic MEG-01 cell and HUVEC cell cDNA libraries	A41105	PTN4_HUMAN P29074	926/105 910	Non-receptor type with single PTPase domain; homology with human erythrocyte protein 4.1 and ezrin, suggesting involvement in regulation of cytoskeletal events	M68941	508
MEG2	F3	Human megakaryoblastic MEG-01 cell and HUVEC cell cDNA libraries	A42690		593/68 020	Non-receptor type PTPase; region of homology with retinaldehyde-binding protein and yeast SEC14p which is involved in protein transfer through Golgi complex	M83738	507
MKP-1 (ERP)	B2	Murine NIH 3T3 cell cDNA library	S24411	PTN7_MOUSE P28563	367/39 369	VH1 family; dual-specificity phosphatase	S64851 X61940	571
MKP-1 (3CH134)		Mouse 3T3 cells				VH1 family; dual-specificity phosphatase	MM3CH-134M	486
MKP-2		PC12 cell cDNA library				Shows significant homology with MKP-1 (58.8% at the amino acid level		111

MSG5	B16	*Saccharomyces cerevisiae*	S40029 S05831	MSG5_YEAST P38590	489/54 219	Identified by suppression of arrest of *gpa1* deletion mutants in cell cycle G_1 phase; mRNA 1.8 kb	D17548 X02561	497
Multiple		*Caenorhabditis elegans*		PTP1_CAEEL to PTP4_CAEEL	Fragments			555
Multiple **Multiple**		Human colon T lymphocytes				Partial sequences obtained by PCR, CL-6 and CL-2B were unique PTPases identified by PCR in human T cells by RT-PCR and degenerate primers		474 612
Multiple		Embryonic mouse and rat brain by RT-PCR				Expression of PTPase genes in the embryonic brain of mouse and rat with PT-PCT; several receptor and cytoplasmic types of tyrosine phosphatase genes were detected; partial sequence of a novel PTPase was reported	D30748	971
Multiple novel		Protochordate *Styela plicata*		PT01_STYPL → PT27_STYPL	Fragments	RT-PCR from *S. plicata* mRNA identified 27 distinct sequences, 25 are novel	M37986 to M38012	555
Multiple novel		Human liver				Novel PTPases identified by PCR in human liver with RT-PCR and degenerate primers		570
OST-PTP		Rat osteosarcoma UMR 106 cell line cDNA library			1711/	Novel receptor-type PTPase, consists of 10 Fn-III repeats and a cytoplasmic region with two PTPase domains	L36884	557
PAC-1	B4	Mouse		PAC1_MOUSE Q05922	318/34 546	VH1 family; dual-specificity phosphatase	L11330	
PAC-1	B3	Human T-cell cDNA library		PAC1_HUMAN Q05923	314/34 399		L11329	589
PC-PTP1		Rat PC12 cell cDNA library				Novel cDNA cloned from PC12h cells and designated as PCPTP1 (gene encoding PC12 PTPase); the longest ORF encodes a 656 AA protein with a single catalytic domain; highest similarity to STEP and LC-PTP/HePTP, with 54 and 51% identity, respectively		1124
PC12-PTP1		cDNA library from NGF-treated PC12 cells			412/47 k	Non-receptor type PTPase with ~40% sequence identity to STEP	U14914	1121, 1122
PEP	G3	Mouse B-cell cDNA library	B44390 S27876	ZPEP_MOUSE P29352	802/89 713	Pro, Glu, Ser, Thr-rich ('PEST') domain at C terminus	M90388	553
Pez		Human breast tissue and breast carcinoma cells				A novel non-receptor PTPase with N-terminal sequence homology to the ezrin–band 4.1–merlin–radixin family		1125
PRL-1	J8	Insulin stimulation of quiescent H35 rat hepatoma cells			173/19.8 k	Unique PTPase without homology to other enzymes outside the catalytic domain	L27843	493
PTP-BAS	E5	Human basophil-enriched cDNA library			Type 1 2485 Type 2 2466 Type 3 2294	Each isoform consists of non-receptor type PTPase with single PTPase domain at C terminus and two distinct structural domain sequences of 300 AA homologous to membrane-binding domains of cytoskeletal proteins (band 4.1, ezrin, radixin, moesin) and homology to PTPases PTPH1 and MEG1, and 90 AA internal repetitive sequences (GLGF) found in guanylate kinase proteins; three apparent isoforms found due to in-frame deletions in the coding region	D21209 (1) D21210 (2) D21211 (3)	549

Table 2 Continued

PTPase	SA*	cDNA/protein source	PIR no	SwissProt name SwissProt no	AA/Mass (Da)	Comments	EMBL no	Refs
PTP-BAS (PTP1E)		Human breast carcinoma ZR-75-1 cDNA library	A54971		2490/277 568	Large, single domain PTPase without a membrane spanning or signal sequence; has homology to cytoskeleton and membrane-junction associated proteins, including band 4.1 family; also contains five imperfect repeats with homology to the GLGF motif of the junction associated guanylate kinases and a region with multiple PEST-like sequences; catalytic domain at C terminus; five variant coding regions were identified	U12128 U12129 U12130 U12131 U12132	475
PTP-BAS (PTPL1)		Human glioma cell line U-343 Mga 31L	S46955		2466/274 974	Large, non-receptor type PTPase with single PTPase domain at C terminus and two distinct structural domain sequences of 300 AA homologous to membrane-binding domains of cytoskeletal proteins (band 4.1, ezrin, radixin, moesin) and homology to PTPases PTPH1 and MEG1, five GLGF repeats that may allow it to form dimers and localize it to the submembranous cytoskeleton; as well as a leucine zipper motif	X80289	596
PTP-PEST		HeLa cell cDNA library			780	Pro, Glu, Ser, Thr-rich ('PEST') domain at C terminus	M93425	622, 623
PTP-PEST		Mouse 18.5 day embryonic kidney			775			1097
PTP-PEST (P19-PTP)	G1	Murine P19 EC cell cDNA library	JH0609 G61180	PTNC_MOUSE P35831	774/86 767	Thought to be a frame shift alteration of PTP-PEST (PTPG1) see ref. 610	X63440	138
PTP-PEST (PTPG1)	G2	Human colon cDNA library	JC1368 A47506	PTNC_HUMAN Q05209	780/88 092	Pro, Glu, Ser, Thr-rich ('PEST') domain at C terminus	D13380	610
PTP-PEST (PTPty43)		PCR of myeloid cell line			Fragment			626
PTP-PS		Rat PC12 phaeochromocytoma cell library			1260	Alternative splicing form of LAR-PTP2B (PTP-P1), with stop codon that truncates the distal PTPase domain	L19181	579
PTP18		Rat brain cDNA library	A40169		Fragment	TM, tandem domain PTPase with Asp replacing Cys at catalytic centre of second domain		513
PTP1B		Purified protein from human placenta			321	C-terminal truncation, 114 AA shorter than predicted by cDNA		484
PTP1B		Human placental cDNA library			435		M33689	481
PTP1B	A1	Human placental cDNA library	A35992 A33897 A37275	PTN1_HUMAN P18031	435/49 966		M31724	487
PTP1B	A3	Rat hypothalamus cDNA library	A34845	PTN1_RAT P20417	432/49 674		M33962	514
PTP1B	A2	Mouse testis cDNA library	JN0317 S40288 A61180	PTN1_MOUSE P35821	432/49 641		M97590 Z23057	562

PTP2E		Rat decidual cDNA library			1175	Non-receptor type PTPase with single catalytic domain at C terminus; N-terminal region has homology to band 4.1 proteins; overall 80% identical to human PTPD1 (ref. 564); variant with only catalytic domain also identified, termed PTP2E1		1107
PTP36	E1	Murine thymocyte cDNA			1189	C-terminal PTPase domain and an N-terminal domain with homology to cytoskeletal proteins as well as a putative SH3-binding motif RPPPPyP	D31842	597
PTPD1	E2	Human skeletal muscle cDNA library			1174/133 k	Single PTPase domain non-receptor type enzyme with N-terminal sequences homologous to ezrin, band 4.1, merlin, radixin and moesin family	X79510	564
PTPD2		Human skeletal muscle cDNA library				Single PTPase domain non-receptor type enzyme with N-terminal sequences homologous to ezrin, band 4.1, merlin, radixin and moesin; closely related to PTPD1 – 76% identical in the catalytic domain		564
PTPH1	E4	HeLa cell library	A41109	PTN3_HUMAN P26045	913/104 029	Non-receptor type with single PTPase domain; homology with human erythrocyte band 4.1, ezrin and talin, regions that might direct the association of MEG1 family PTPases with plasma membrane or their involvement in regulation of cytoskeleton	M64572	624
PTP-RL10		Regenerating mouse liver RT-PCR followed by mouse liver cDNA			1176	No transmembrane region, NH_2 terminus with homology to cytoskeletal proteins; likely mouse homologue of PTPD1; two mRNAs, larger expressed in hear and lung > testis, smaller RNA expressed in testis > lung	D37801	1104
PTP-SL		Mouse brain cDNA library			549/60 k	Similar to rat brain STEP; single PTPase domain, transmembrane PTPase; short extracellular domain with unique structure	Z30313 Z23058	1103
PTO-U2		Human kidney library			1216/140k	Has a single transmembrane domain and a single intracellular catalytic domain; extracellular domain contains 14 putative *N*-glycosylation sites and eight Fn type III-like repeats; structurally similar to HPTP-β and DPTP10D		1120
PTPX1	F1	*Xenopus laevis* ovary cDNA library	A53978		694/79 401	Non-TM PTPase with N-terminal segment region with homology to lipid-binding proteins, retinaldehyde-binding protein and SEC14p, a yeast phospholipid transferase; also has a 97 AA insert segment with homology to PSSA, a protein involved in phosphatidyl serine biosynthesis	L33098	490
PTPX10	F2	*Xenopus laevis* ovary cDNA library	B53978		597/68 842	Non-TM PTPase with N-terminal segment region with homology to lipid-binding proteins, retinaldehyde-binding protein and SEC14p, a yeast phospholipid transferase	L33099	490
pyp1^{+}	I3	*Schizosaccharomyces pombe* cDNA library	A40449	PYP1_SCHPO P27574	550/61 587	3.4 kb mRNA	M63257	577
pyp2^{+}	I1	*Schizosaccharomyces pombe* cDNA library	S28391 A45030	PYP2_SCHPO P32586	711/79 356	3.1 kb mRNA	X59599	578
pyp2^{+}		*Schizosaccharomyces pombe* genomic library			711/85 k	2.6 kb mRNA		561

*SA = Sequence alignment, found at the back of the book

Table 2 Continued

PTPase	SA*	cDNA/protein source	PIR no	SwissProt name SwissProt no	AA/Mass (Da)	Comments	EMBL no	Refs
pyp3+	I2	*Schizosaccharomyces pombe* genomic library	S28392	PYP3_SCHPO P32587	303/34 583	Identified as a suppressor of cdc25 mutation	X69994	560
RIP		MEL cell cDNA library			2450/269.8 k	Non-receptor type with single PTPase domain near C terminus; homology domain to band 4.1 and ezrin as well as a GLGF repeat; 79% homology with human PTP-BAS; multiple RNA transcript forms identified		933
RKPTP	J5	Rat kidney cDNA libraries			382/44 k	Predicted protein is small and suggested to be cytosolic with a single PTPase domain	D38072	566
RPTP-α	M1	Human kidney cDNA library	S13085	PTPA_HUMAN P18433	793/89 717		X54890	574
RPTP-α (LRP)	M2	Mouse pre-B cell cDNA library	A35501 B35501	PTPA_MOUSE P18052	830/93 883	Alternatively spliced form has 108 bp insertion at position 248 in 1st PTPase domain	M33671	554
RPTP-α (LRP)		Human HEPG2 cell library	S12905 S17371		793/89 628 793/89 644		X53364	533
RPTP-α (LRP)		Rat kidney cDNA library	JC1285		796/90 242			565
RPTP-α (RPTP-αi)		Human brainstem cDNA library	A36004		802	Alternatively spliced form has 9 AA insert located three residues upstream of the TM segment	M34668	536
RPTP-α		Mouse brain cDNA library	A36065		802/90 718		M34668	595
RPTP-α		Human placenta cDNA library	S12049		793/89 702		X54130	540
RPTP-BR7		Mouse brain cDNA library				Cytoplasmic domain similar to HePTP/LC-PTP and STEP, but is a transmembrane enzyme with a single PTPase domain; extracellular domain without known homologies	D31898	1115
RPTP-β	K5	Human placenta cDNA library	S12050	PTPB_HUMAN P23467	1977/224 267	Single PTPase domain in cytoplasmic segment extracellular region has 16 Fn-III repeats; one isolate has deletion of Fn-III domain 8, likely to represent alternative splicing	X54131	540
RPTP-β		PCR of human lung cDNA				Variant cDNA identical to RPTP-β, but truncated at the 3′ end of the coding region, in the most conserved part of the catalytic domain (called Δ3′-HPTP-β or C22); expression of the normal form of RPTP-β was > the truncated RPTP-β form in human lung; STOP codon just after 14 AA past the catalytic Cys residue in the catalytic domain, truncates the C-terminal 86 AA		492
RPTP-β		Mouse lung cDNA library					X58289	504
RPTP-γ		Human placenta cDNA library			Fragment	Highly homologous to RPTP-ζ	X54132	540

RPTP-γ	O2	Mouse brain cDNA library	B48148	PTPG_MOUSE PQ05909	1442/161 241	N-terminal CAH-like domain but has only 1/3 His residues required to ligate Zn^{+2} for CAH catalytic activity; CAH domain is preceded by potential proteolytic cleavage site; followed by one Fn-III domain and a Cys-free region with a domain rich in Ser/Thr (32%) and a domain rich in charged and polar residues (90%); single TM domain and two tandem PTPase domains; second PTPase domain has Asp in position of conserved catalytic Cys residue found in other PTPases; also 15 AA insert in second catalytic domain, unique to this class of PTPases	L09562	478
RPTP-γ	O1	Human brainstem cDNA library	A48148	PTPG_HUMAN P23470	1445/162 112	N-terminal CAH-domain family; 95% identical to murine homolog	L09247	478
RPTP-δ		Human placenta, fetal liver and fetal lung cDNA libraries	S12052	PTPD_HUMAN P23468	Fragment	Structure consists of three Ig-like domains and eight Fn-III repeats, single TM domain and two tandem PTPase domains; exceptionally high homology to LAR (87% and 96% identity in D1 and D2 PTPase domains, respectively)	X54133	49, 540
RPTP-δ	N4	Mouse brain cDNA library			Type A 1291 Type B 1691 Type C 1916	Evidence for at least three types of mRNA, differing in composition of extracellular domain: Type A (one Ig-like and four Fn-III domains); Type B (one Ig-like and eight Fn-III domains): Type C (three Ig-like and eight Fn-III domains), each with single TM domain and tandem cytoplasmic PTPase domains; Type C also has different 5′ untranslated region and leader peptide sequence	D13903 DC13904 D13905	563
RPTP-ϵ	M3	Human placenta cDNA library	S12053	PTPE_HUMAN P23469	700/80 641		X54134	540
RPTP-κ	Q3	Mouse brain cDNA library	A48066	PTPK_MOUSE P35822	1457/164 184	Single Ig-like domain, four Fn-III domains; domain near N terminus with A5 protein homology and putative neuronal recognition activity (MAM domain); extracellular consensus for furin cleavage; large juxtamembrane domain	L10106	529
RPTP-μ	Q2	Mouse lung cDNA library	S17670	PTPM_MOUSE P28828	1452/163 593	Contains signal sequence, 13 potential glycosylation sites, single TM domain, 688 AA intracellular domain, a particularly large juxtamembrane domain and tandemly repeated PTPase domains N terminus, now identified to contain the MAM homology (see below; ref. 480) followed by one IgG domain and four Fn-III repeats	X58287	504
RPTP-μ	Q1	Human mammary carcinoma cDNA library	S17669	PTPM_HUMAN P28827	1452/163 632	98.7% AA identity to the mouse homolog	X58288	504
RPTP-μ						Identification of an adhesive domain on RPTP-μ; called MAM for Meprin-A5-μ; 170 AA and contains four conserved Cys residues; meprins are cell-surface glycoproteins that contain a Zn-metalloproteinase domain involved in cleavage of a variety of polypeptides; A5 is a developmentally regulated cell-surface molecule predominantly expressed in visual centre during innervation		480
RPTP-η		Human placenta cDNA library				Class IIA, similar to RPTP-β, has 10 Fn-III repeats in extracellular domain, a single transmembrane segment and a single PTPase homology domain		1105

*SA = Sequence alignment, found at the back of the book

Table 2 Continued

PTPase	SA*	cDNA/protein source	PIR no	SwissProt name SwissProt no	AA/Mass (Da)	Comments	EMBL no	Refs
RPTP-ζ		Human placenta cDNA library			Fragment	Highly homologous to RPTP-γ and core PTPase domain residues of rat PTP18	X54135	540
RPTP-ζ/β	O5	Human fetal brain cDNA libraries	A46151	PTPZ_HUMAN P23471	2314/254 528	N-terminal region with CAH homology, 1048 residue cysteine-free region; CAH region likely to be unable to bind zinc (required His-94 and His-119 have been changed to Thr and Gln, respectively), but might bind a small soluble ligand; second tandem PTPase domain (D2) unlikely to have catalytic activity due to natural replacement of the obligatory catalytic cysteine residue in D2 with aspartate; suggests that role of D2 may be regulatory	M93426	539
RPTP-ζ/β		Human infant brainstem cDNA library	A46700		2307/253 618	Composed of large extracellular domain with 266 residue N-terminal region having striking homology to carbonic anhydrase but lacking coordinate residues for binding of Zn^{2+} (hence no CAH catalytic activity); single TM domain and tandem cytoplasmic PTPase domains; also cloned dvRPTP- ζ/β, which lacks 859 AA from extracellular domain but has intact TM and cytoplasmic domains (AA 754 joins with 1614 by in-frame deletion of 2577 bp)		546
RPTP-ζ/β		Bovine brain expression library screened with CSPG antiserum				Found 185 k core protein with 82% overall homology to human RPTP-ζ/β		106
RPTP-ζ/β (Phosphacan)	O3 O4	Rat brain cDNA library			2316/ 1616/173 k	Used antibody to phosphacan, a soluble, rat brain chondroitin sulfate proteoglycan, its protein core was purified, sequenced and found to be 76% identical to extracellular domain of human RPTP-ζ/β; occurs in two forms, a shorter extracellular variant without TM and cytoplasmic domains, and a form highly identical to human RPTP-ζ/β	U09357 U04998	477, 556
RVH1	B10	Raccoon-pox virus and Baculovirus VH1 homologues				VH1-related PTPases from raccoon-pox virus (RVH1) highly conserved with VH1	L13165	516
SAP-1	K4	Gastric and colon carcinoma cell cDNA libraries	A49724		1118/123 038	Receptor-type PTPase with a single PTPase domain similar to RPTP-β and DPTP10D, but only 49.6% identity in the PTPase domain; extracellular domain contains signal sequence, eight Fn-III repeats, and mutiple potential glycosylation sites	D15049	552
SH-PTP1		Human	S20825 S17234 A38189 S20837	PTN6_HUMAN P29350	597/67 719	Intracellular PTPase with two SH2 domains near N terminus	X62055	

SH-PTP1 (PTP1C)		Human breast carcinoma ZR-75-1 cell cDNA library			609/68 k	Predicted to be 13 AA longer than hSH-PTP1 and differs in the last 18 AA, also identity begins only at the start of the 1st SH2 domain, suggesting that PTP-1C may be a variant form in the ZR-75-1 cell line (see ref. 582).		602
SH-PTP1 (HCP)	D5	Murine monocyte cDNA library	A42031 F61180	PTN6_MOUSE P29351	595/67 561	Three nuclear localization motifs identified in C terminus	M68902	627
SH-PTP1 (HCP)	D4	Human T-lymphoid cell line	S20825	PTN6_HUMAN P29350	595/67 561		M74903	627
SH-PTP1		Human erythroleukaemia (HEL) cell and lung cDNA libraries		PTN6_HUMAN P29350	595/68 k		M77273	582
SH-PTP1 (PTP1C)		Human breast carcinoma ZR-75-1 cell cDNA library				cDNA isolate had 32 additional nucleotides in the 5′ non-coding region compared to the PTP1C clone of Shen *et al.* [602], suggesting that different 5′ exons might be used		212
SH-PTP1 (SHP)		Mouse B-cell cDNA library	A44390		595/67 559		M90389	553
SH-PTP2	D1	Human	JN0805 A46210 A47386 A47244 S27398 A44929	PTNB_HUMAN PQ06124	593/68 010	Intracellular PTPase with two SH2 domains near N terminus	L07527 X70766	
SH-PTP2		Human fetal brain cDNA library			593/68 k		L03535	502
SH-PTP2		Human skeletal muscle, T-cell and monocyte cDNA libraries			593/68 k			479
SH-PTP2	D2	Rat adipocyte cDNA library				Sequence variant of cloned SH-PTP2/Syp family intracellular PTPases	U09307	140
SH-PTP2 (PTP-SHβ; PTP-L1)		Rat liver purified protein				Partial sequence data from proteolytic fragments		522
SH-PTP2 (PTP1D)		SK-BR-3 human mammary carcinoma cell cDNA library			593/68 k			616
SH-PTP2 (PTP1D)		Rat brain cDNA library			593/68 k			301
SH-PTP2 (PTP1Di)	D3	Rat brain cDNA library			597/68 382	mRNA splice variant of SH-PTP2/PTP1D with additional 4 AA between positions 408 and 409	U05963	301
SH-PTP2 (PTP2C)		Human umbilical cord cDNA library		PTNB_HUMAN PQ06124	593/68 k		L08807	472
SH-PTP2 (PTP2Ci)		Human umbilical cord cDNA library			597	mRNA splice variant of SH-PTP2/PTP2C with additional 4 AA between positions 408 and 409		472
SH-PTP2 (SH-PTP3)		Human T-cell PEER cDNA library		PTNB_HUMAN PQ06124	593/68 k		D13540	471
SH-PTP2 (Syp)		Mouse	A46209	PTNB_MOUSE P35235	585/66 816	Splice variant of SH-PTP2 and PTP1D	L08663	501

*SA = Sequence alignment, found at the back of the book

Table 2 Continued

PTPase	SA*	cDNA/protein source	PIR no	SwissProt name SwissProt no	AA/Mass (Da)	Comments	EMBL no	Refs
STEP	J3	Rat striatal cDNA library	A41147	PTN5_RAT P35234	369/42 449	Single PTPase domain, intracellular enzyme, with potential attachment to the membrane suggested by myristoylation amino acid-consensus sequence at N terminus	M65159 S49400	547
STEP		Mouse brain				Identified two alternatively spliced transcripts, STEP(2) has a 2.8 kDa mRNA in mouse brain and predicts a 20 kDa protein that lacks the conserved tyrosine phosphatase domain; STEP(61) has a predicted molecular mass of 61 kDa and contains a single PTPase domain; the original 46 kDa STEP(46), along with STEP(2) and STEP(61), are members of a brain-enriched family of PTPases that may have distinct functions within the central nervous system		1123
stp1$^+$		*S. pombe* cDNA expression library			–/17.5 k	Identified as cone that rescued *cdc25–22* mutation; 45% identical to mammalian low M_r PTPases		1112
stp1$^+$		*S. pombe* cDNA expression library			–/17.5 k	Gene encodes a 17.4 kDa protein that is 42% identical to members of the low M_r PTPases previously known to exist only in mammalian species		1132
TCPTP	A4	Human T-cell cDNA library	A33899 C60345	PTN2_HUMAN P17706	415/48 473	Non-receptor type PTPase with hydrophobic C-terminal domain	M25393	488
TCPTP (PTP-S)	A6	Rat spleen cDNA library	S14294	PTN2_RAT P35233	363/42 235	PTP-S (PTP-spleen); rat homologue of human TCPTP; C-terminal region with homology to basic domains of Fos and Jun; compared to TCPTP, C terminus has three deletions of 19, five and 28 AA – may be due to alternative splicing		608
TCPTP (MPTP)	A5	Mouse embryo cDNA library	A38191	PTN2_MOUSE Q06180	382/44 572	Initial isolate 89% identical to human TCPTP, diverging at C terminus; also isolated a human cDNA identical to the MPTP sequence, as a TCPTP variant	M81477 M81478 S52655	567
TCPTP			A60345 B60345		387/45 244	Two alternative forms varying at C terminus		132
VH1	B10	Encoded by *Vaccinia H1L* ORF	A42514	VH01_VACCC P20495	171/19 672	Prototypical VH1 family; dual-specificity phosphatase	M35027	511
VH1	B11	Encoded by *Vaccinia H1L* ORF'	A24481	VH01_VACCV P07239	171/19 698		M13209	
Variola VH1	B12	Variola virus (Bangladesh major)	I36845 S33098	VH01_VARV P33064	171/19 746	VH1 family; dual-specificity phosphatase	X69198 X67119	516
VHR	B9	M426 human lung fibroblast cDNA library	A47196		185/20 478	Identified by expression cloning strategy for PTPases that dephosphorylate the keratinocyte growth factor receptor; has limited similarity to VH1 and 3CH134	L05147	524
Yeast PTP1	I4	*Saccharomyces cerevisiae* lambda genomic library	A39863	PTP1_YEAST P25044	335/38 868	Small, single domain PTPase with homology to mammalian PTPases; identified a nucleotide binding fold in conserved PTPase domain GXGXXG; expressed as mRNA of 1.1 kb	M64062	509
Yeast PTP2	I7	*Saccharomyces cerevisiae* lambda genomic library	A42667	PTP2_YEAST P29461	750/85 840	mRNA size 2.7 kb	M85287	512

Yeast PTP2		*Saccharomyces cerevisiae* library	S31554 S14170 JC1484	PTP2_YEAST P29461	750/85 785		M38723	528
Yeast PTP2		*Saccharomyces cerevisiae*, expression library	A41980	PTP2_YEAST P29461	750/85 868	Identified as extragenic suppressor of a synthetic lethal for the N-end rule pathway (ubiquitin-protein degradation system); also found to be the heat-inducible PTPase, PTP2; mRNA 2.7 kb	M82872	316
Yeast PTP2		*Saccharomyces cerevisiae*	S41854		281/31 549	Homolog of mammalian protein phosphatase 2C	L14593	299
Yop51 (YopH)	J1	Encoded in plasmid carried by *Yersinia enterocolitica*		YOPH_YEREN P15273	468/50 939	99% amino acid identity to Yop2b	M30457	559
Yop2b (YopH2)	J2	Encoded in plasmid carried by *Yersinia pseudotuberculosis*	S01054	YOPH_YERPS P08538	468/50 869	Gene name YOPH	Y00551	513
YVH1	B14	*Saccharomyces cerevisiae* genomic library	S31304	PVH1_YEAST Q02256	364/41 185	Small PTPase, catalytic domain is at N terminus of the protein; related to *Schizosaccharomyces pombe* cdc25 and 3CH134; mRNA 1.2 kb VH1 family, but limited to Tyr(P) phosphatase activity	L04673 M69294	510

*SA = Sequence alignment, found at the back of the book

Table 3 Gene mapping and structure for eukaryotic PTPases

PTPase	Species	Gene	Chromosomal mapping	Gene structure	Comments	Refs
BVP	*Autographa californica*				2 major mRNAs ~3.1 and 3.9 kb; fine mapping of the transcription start sites is described	538
CD45	Human			Promoter analysis	3 transcription initiation sites, one appears to direct tissue-specific expression of CD45	494
CD45	Human		Human ch 1, long arm			587
CD45	Human			Encoded by 33 exons, including the 5′ and 3′ untranslated regions	Differential usage of exons 4–6 generates at least five distinct protein isoforms	517
CD45	Human				Differential usage of three exons generates at least five and potentially eight different CD45 mRNAs, expressed as 5.0–5.6 kb transcripts in various lymphocyte cell lines	204
CD45	Mouse		Mouse ch 1	~120 kb, comprises 34 exons including untranslated regions	Differential usage of exons 5–7 generates protein isoforms	593
CD45	Rat		Rat ch 13			506
CDC25A	Human	*CDC25A*	Human ch 3, p21			491
CDC25B	Human	*CDC25B*	Human ch 20, p13			491
cdc25C	*S. pombe*	*cdc25C*		Promoter contains cell-cycle-regulated repressor element (CDE)	Mutation of CDE impairs cell cycle regulation of cdc25C transcription and causes high expression in G_0/G_1, indicating that it functions as a *cis*-acting repressor element	952
cdc25Dm (string)	*Drosophila melanogaster*	*string (stg)*	Tip of ch arm 3R at position 99A			498
cdc25 (string)	*Drosophila*	*cdc25 (string)*		Upstream elements required for patterning	Complete pattern of string transcription throughout development requires extensive *cis*-acting regulatory sequences (>15.3 kb); smaller segments can drive temporal expression in defined spatial domains	905
cdc25Dm (twine)	*Drosophila melanogaster*	*twine*	Maps to ch region 35F			489
CL100	Human		Human ch 5, q35	Gene and promoter characterized; four exons intervened by three short introns of 400–500 bp	800 bp promoter fragment confers inducibility to serum and PMA; found new homology domain between CL100 and cdc25 family enzymes – called CH2 domain; GenBank Accession U01669	541
CL100	Human		Human ch 5, q35			551

CL100	Human		Human ch 5, q34		Highly conserved sequences found in genomic DNA from mouse, chicken, *Xenopus* and *Drosophila* by Southern blotting	499
CSW	*Drosophila melanogaster*	*corkscrew (csw)*	X-linked locus that maps to ch bands 2D3–4	Covers ~20 kb of genomic DNA		580, 581
DdPTP1	*Dictyostelium discoideum*			ORF interrupted by two introns	Spans a 4.8 kb genomic fragment	277
DdPTP2	*Disctyostelium discoideum*			ORF contains four introns; three are located in the catalytic region		523
DPTP (DPTP69D)	*Drosophila melanogaster*		Cytologic position 69D			74
DPTP10D	*Drosophila melanogaster*		Cytologic position 10D	11 exons		625
DPTP4E	*Drosophila melanogaster*		Cytologic position 4E1–2 on X chromosome	10 exons	Spans 16 kb of genomic DNA	575
DPTP61F	*Drosophila melanogaster*		Localized to polytene chromosome position 61F	8 introns	Spans > 11 kb; sequences for intron/exon boundaries provided	558
DPTP99A	*Drosophila melanogaster*		Cytologic position 99A	13 exons	Mapped in this reference	625
DPTP99A	*Drosophila melanogaster*		Polytene chromosome bands 99A7-8			520
ERP	Mouse		Mouse ch 17, region A2-C	4 exons and three introns	Distributed in a gene fragment of 2900 bp, including 650 bp of 5′ promoter sequence from cap site; also noted TATA box, polyadenylation signal and other consensus sequences in the gene structure	571
GLEPP1	Human		ch 12, p12–p13		FISH technique	1130
HePTP	Human	*PTPN7*	Human ch 1, q32.1			629
HePTP (LC-PTP)	Human		Human ch 1, q32.1			470
hVH-3	Human		Human ch 10, q25	–		551
hVH-4	Human		Human ch 10, q11	–		551

Table 3 Continued

PTPase	Species	Gene	Chromosomal mapping	Gene structure	Comments	Refs
LAR	Human			33 exons	Spanning > 85 kbp; three Ig-like domains are encoded by exons 3–7 and the eight Fn-III domains are encoded by exons 8–17; RT-PCR demonstrated alternative splicing of a mini exon (exon 13) in Fn domain five and the corresponding domain of other related genes including rat *LAR*, rat *RPTP-σ* and human *RPTP-δ*; transcript; exon 13 (–) was shown to be the major mRNA species in a variety of cell lines tested; additional alternative splicing events affecting expression of Fn domains 4–7 in various combinations were also detected; exon structure shown to be very similar to that of *CD45* and *RPTP-α* genes, including the disruption of the highly conserved catalytic centre of each of the encoded PTPases by exon/intron junctions	573
LAR	Human	*LAR*	Human ch 1, p32			532
LAR	Human	*LAR*	Human ch 1, p32-33			110
LAR	Human	*LAR*		8 exons of a partial *LAR* gene isolated in ~3.5 kb of DNA	LAR exons III, IV, VII and VIII correspond precisely to exons 23, 24, 29 and 30, respectively of the *LCA (CD45)* gene	605
LAR	Human	*PTPRF*	ch 1, p32		In a small cell lung cancer line, the genes *PTPRF* and *MYCL1* encoding protein tyrosine phosphatase LAR and L-myc were both amplified	1101
LAR	Mouse	*PTPRF*	Region C6–D1 on mouse ch 4	Cytoplasmic region of mouse LAR protein is encoded by 11 exons that span only 4.5 kb of genomic DNA		1119
LC-PTP (HePTP)				11 exons, including non-coding regions	Splicing sites which encode the catalytic domain arise at similar positons as CD45 gene structure; suggesting gene duplication of an ancestral gene	468
Low M_r PTPase (AcP1/ CPTP1 and AcP2/ CPTP2)	Human		Human ch 2; where the human red cell acid phosphatase ACP1 is found	Exon structure	Alternative splicing	545, 619

LTP1	*S. cerevisiae*	*LTP1*	Right arm of chromosome XVI, near *TKL1*			1116
PAC-1				Promoter analysis	Three transcription start sites; gene comprised of three exons; exon three contains the conserved PTPase domain; N terminus may have evolved separately from the catalytic domain and may serve a regulatory function	505
PAC-1	Human		Human ch 2, q11			541, 551
PAC-1	Human	*DUSP2*	Pericentromeric region of human chromosome 2 (2p11.2–q11)		A polymorphism has been identified by SSCP in the 3′ UTR of the gene. No consistent translocations or deletions at this locus have been reported in haematopoietic neoplasias or other tumours	1131
Pez	Human		ch 1, q32.2–41		FISH technique	1125
PTP-BAS (PTP1E)	Human		Human ch 11			475
PTP-PEST (PTPG1)	Human	*PTPN12*	Human ch 7, q11.23			609
PTP-PEST	Mouse		Chromosome 5A3 to B, in agreement with mapping of the human PEST to chromosome 7q11.23, a region of synteny with the centromeric portion of mouse chromosome 5	Spans over 90 kb and is composed of 18 exons, 10 of which constitute the catalytic phosphatase domain	Detailed analysis reveals a strong conservation of the genomic organization within the PTPase gene family	1096
PTP1B	Human	*PTPN1*	human ch 20, q13.1→q13.2	5 exons	Spans ~13 kb of genomic DNA; GenBank accession no. M33688 (exon A), M33687 (exon B), M33686 (exon C), M33685 (exon D), M33684 (exon E)	481

Table 3 Continued

PTPase	Species	Gene	Chromosomal mapping	Gene structure	Comments	Refs
PTP1C (SH-PTP1)	Human	*PTPN6*	ch 12	Gene consists of 17 exons spanning 17 kb of DNA	Three non-haematopoietic PTPN6 transcripts were identified in a variety of cell lines and were transcribed from a common promoter; alternate splicing and exon skipping result in transcripts encoding the two SH2 domains and minor transcripts from which part of or the entire N-terminal SH2 domain coding sequences are missing; the haematopoietic form of PTPN6 transcript is initiated at a downstream promoter separated by 7 kb from the first which is active exclusively in cells of the haematopoietic lineage	1093
PTPA	Human	*PTPA*	Localized to 9q34 region, positioned centromeric of *c-abl* in a region embracing several genes implicated in oncogenesis	Single-copy gene with 10 exons and 9 introns; total length of about 60 kb	PTPA is the specific phosphotyrosyl phosphatase activator of the dimeric form of protein phosphatase 2A; promoter lacks a TATA sequence but contains upstream of the transcription start four Sp 1 sites; putative binding sites for NF-κB, Myb, Ets-1, Myc, and ATE: the PTPA promoter region displayed promoter activity that seems to be cell-line dependent	1127
PTPH1	Human	*PTPN3*	Human ch 9, q31			527
PTP U2	Human		ch 12, p13.2–p.13.3			1120
pyp1$^+$	*Schizosaccharomyces pombe*	*pyp1$^+$*	Both *pyp1$^+$* and *pyp2$^+$* localized to the same 530 kb fragment at tip of the right arm of ch I			578
pyp2$^+$	*Schizosaccharomyces pombe*	*pyp2$^+$*	Both *pyp1$^+$* and *pyp2$^+$* localized to the same 530 kb fragment at tip of the right arm of ch I			578
RIP	Mouse		ch 5, between *D5Mit90* and *D5Mit25*			933
RPTP-α	Human		Human ch 20, pter–q12	–		536

RPTP-α	Human	*PTPA*	Human ch 20, p13	–		531, 533
RPTP-α	Mouse		Mouse ch 2, linked to *Il-1a* and *Bmp-2a* loci	–		595
RPTP-α (LRP)	Mouse			22 exons	Spans > 75 kb, strikingly similar to size and organization of CD45 gene; putative promoter has features of 'housekeeping genes'; 5′ genomic sequences (GenBank no. L13607); 3′ genomic sequences (GenBank no. L13608)	620
RPTP-β	Human	*PTPRB*	Human ch 12, q15 → q21	–		519
RPTP-β 2	Mouse	*PTPRJ*	Middle region of ch 2			1129
RPTP-δ	Human				RT-PCR demonstrated alternative splicing of a mini exon (exon 13) in Fn domain 5; also occurs in the corresponding domain of human and rat *LAR*, and rat *RPTP-σ*	573
RPTP-δ	Human		Human ch 9, p24	–		521
RPTP-δ	Mouse		Mouse ch 4, tightly linked to the *brown* (*b*) locus	–		563
RPTP-ϵ	Human	*PTPRE*	ch10, q26		FISH technique	1126
RPTP-ϵ	Mouse	*PTPRE*	Localized to imprinted regions in the distal part of ch 7			1129
RPTP-γ	Human	*PTPG*	Human ch 3, p21	–	Region frequently deleted in renal cell and lung carcinomas	542
RPTP-γ	Mouse	*PTPG*	Mouse ch 14, centromeric region		Region syntenic with chromosomal locus of human RPTP-γ	478, 618
RPTP-γ	Human	*PTPRG*	3p 14 region		YAC contig containing 3p14 region showed *PTPRG* located 1 Mb proximal to the hereditary renal cell carcinoma 3;8 translocation break point	1094
RPTP-κ	Mouse	*PTPK*	Mouse ch 10, ~21 cM from centromere	–		529
RPTP-μ	Human		Human ch 18, pter–q11	–		504
RPTP-μ	Human	*PTPRM*	Human ch 18, p11.2		Fine mapping	607
RPTP-μ	Mouse	*PTPRM*	ch 17			1113

Table 3 Continued

PTPase	Species	Gene	Chromosomal mapping	Gene structure	Comments	Refs
RPTP-σ (LAR-PTP2B)	Mouse		Mouse ch 17, distal region			621
RPTP-σ	Rat				RT-PCR demonstrated alternative splicing of a mini exon (exon 13) in Fn domain 5; also occurs in the corresponding domain of rat and human *LAR* and human *RPTP-δ* transcripts	573
RPTP-η	Human		Short arm of ch 11, 11p11.2		Predominant mRNA of ~7.5 kb in placenta	1105
RPTP-ζ/β	Human		Human ch 7, q31–33			546
RPTP-ζ/β	Human	*PTPRZ*	ch 7, q31.3		Somatic cell hybrid mapping and FISH	1091
SAP-1	Human		Human ch 19, q13.4	–		552
SH-PTP1	(HCP)		Human ch 12, p12 → p13			627
SH-PTP1	Human			Simple Southern blot suggests relatively small, single copy gene		582
SH-PTP1	Human	*PTPN6*	Human ch 12, p13	–		583
SH-PTP1 (HCP)	Mouse		Mouse ch 6, tightly linked to the *Tnfr-2* and *Ly-4* genes	–		628
SH-PTP1	Human	*PTPN6*	ch 12, p12–13	The *PTPN6* gene possesses two alternative first exons	An intragenic deletion is described; a 1.7 kb deletion occurring in the intron between the two alternatively used first exons is the result of an illegitimate recombination between two Alu-type repeats; the deletion increases the transcriptional activity of the distal promoter	1114
SH-PTP2 (SH-PTP3)	Human	*PTPN11*	Human ch 12, q24.1	–		526
SH-PTP2 (Syp)	Mouse		Mouse ch 5, band F	–		419
STEP	Human		chromosome 11p15.2–p15.1		Conserved synteny to the mouse locus	1109
STEP	Mouse		chromosome 7B3–B5		Conserved synteny to the human locus	1109

STEP	Mouse				Exon–intron organization responsible for the novel STEP(20) and STEP(61) alternatively spliced sequences was determined in the mouse STEP genomic DNA	1123
TCPTP	Human				Complex Southern blot consistent with multiple genes in the family or a very large (>70 kbp) gene with many introns	488
TCPTP	Human	*PTPN2*	Human ch 18, p11.2 → p11.3		Pseudogenes also identified, TCPS1 (gene symbol PTPN2P1) and TCPS13 (gene symbol PTPN2P2) map to human ch 1q22–q24 and 13q12–q13	534
TCPTP	Human	*PTPT*	Human ch 18	–	Unlinked sequence found on ch 1, identified as pseudogene by ref. [534]	594
TCPTP	Mouse		Mouse ch 18	–		594
VH-3 (novel)	Human		Human ch 10, q25			541
VH-4 (novel)	Human		Human ch 10, q11			541
VHR	Human		ch 17q21		VHR coding sequence was also screened in individuals with familial breast cancer and in sporadic breast tumour and breast cancer cell lines; no mutations were detected	1106
Yeast PTP1	*Saccharomyces cerevisiae*		Physically linked to the 5′ end of a heat shock gene SSB1			509
Yeast PTP2	*Saccharomyces cerevisiae*		Closely linked to the RET1 and STE4 genes and on the right arm of ch 15			512
YVH1	*Saccharomyces cerevisiae*		Located immediately adjacent to DAL1 on ch IX			510

Table 4 Crystallization and three-dimensional structural studies on PTPases

PTPase	Protein source	Parameter	Data	Refs
CD45	Expressed extracellular portions of four different isoforms of rat CD45 in CHO cells	EM imaging	Isolated isoforms containing either all three alternative exons ABC, B alone, C alone, or no alternative exons; EM showed that the extra segments contributed to an extended structure as has been predicted from the sequence	636
PTP1B	37 k catalytic domain of PTP1B expressed from *E. coli* (residues 1–321)	X-ray crystal structure of native enzyme at 2.8 Å as well as the complex with tungstate	First PTPase crystal structure; enzyme consists of a single domain with the catalytic site at the base of a shallow cleft; phosphatase recognition site is created from a loop located at the N terminus of an α-helix; loop comprised of 11 residue sequence that is diagnostic of PTPase activity and contains the essential Cys and Arg residues for activity	632
PTP1B	Expressed in *E. coli* as truncated forms Δ321 and Δ299	Crystals obtained; compared properties to expressed protein from 293 cells	Kinetic characterization for *p*NPP (see Table 7)	634
PTP1B	37 k catalytic domain of PTP1B expressed from *E. coli*	Crystallization and demonstration of X-ray diffraction to 2.4 Å		633
PTP1B	Cys-215→Ser catalytically inactive mutant	X-ray crystal structure in complex with phosphotyrosyl peptide modelled from EGF receptor autophosphorylation site	Peptide binding is accompanied by conformational change of a surface loop that creates a phosphotyrosine recognition pocket, phosphotyrosine side chain is buried within the protein and anchors the substrate to the binding site; specificity is conferred by interaction between acidic residues and the surface of the peptide and basic residues of the enzyme surface	1133
Low M_r PTPase (bovine heart)	Recombinant bovine heart enzyme, *E. coli*	Crystallization	Two forms isolated, one diffracts to 3 Å, the other to 2.2 Å	640
Low M_r PTPase (bovine liver)	Purified from bovine liver	Crystallization	Single form, diffracts to better than 2 Å	638
Low M_r PTPase	Recombinant protein, *E. coli*	Solution structure by NMR	Has a central four-strand parallel β-pleated sheet surrounded by four α helices and a short β10-helix; the phosphate binding site, identified by use of competitive inhibitors, is in a loop region connecting the C-terminal end of the 1st β-strand with the 1st α-helix; additional residues from other α-helices and in some of the loop regions also contribute to the binding site	635
Low M_r PTPase	Bovine heart	Backbone ^{1}H, ^{13}C, and ^{15}N assignments and secondary structure	Assignments were obtained from a combination of double- and triple-resonance multidimensional NMR experiments; BHPTPase was found to consist of a four-stranded parallel β-sheet, four α-helices and one stretch of β10-helix; the secondary structure is characteristic of the β–α–β structural motif	642

Low M_r PTPase	Bovine heart	X-ray crystal structure at 2.2 Å	Contains a four-strand central parallel β-sheet with α-helices packed on both sides in a manner characteristic of a Rossmann fold; bound phosphate is localized to a loop in the 1st β–α–β motif at the C terminus of the β-sheet; structure defined for critical residues identified by mutagenesis and biochemical analyses, including Cys-12, Arg-18 and His-72; a hydrophobic crevice suitable for phosphopeptide binding is also identified	641
Low M_r PTPase	Bovine liver	X-ray crystal structure at 2.1 Å	α/β protein containing a phosphate-binding loop that includes the active site residues Cys-12 and Arg-18; similar in structure to the PTP1B enzyme crystal structure; structure suggests a role for Cys-12 as a nucleophile in an in-line associative mechanism, as well as a potential role for Asp-129 in the reaction cycle	639
SH-PTP2	Recombinant amino terminal SH2 domain, *E. coli*	2 Å–3 Å resolution crystal structures determined with various peptides and in the uncomplexed form	Sequence specificity can extend across the five residues following the phosphotyrosine; the SH2 domain's surface topography can be altered with resulting changes in specificity, while conserving the structure of the central core of the domain	1134
***Yersinia* PTPase**		X-ray crystal structure at 2.5 Å of native enzyme and of complex with tungstate at 2.6 Å	Cys-403, nucleophilic residue that forms a phosphocysteine intermediate during catalysis is positioned in the centre of a distinctive phosphate-binding loop and at the centre of hydrogen-binding arrays that may activate Cys-403 as a reactive thiolate; binding of tungstate triggers a conformational change that traps the oxyanion and swings Asp-356, also known to be important to catalysis, into the active site	637
***Yersinia* PTPase**	Inactive Cys 403→Ser mutant of the *Yersinia* PTPase	Crystal structure of complex with sulfate	Details of the conformational change in response to oxyanion binding; small perturbations occur in active site residues, especially Arg-409, and trigger the loop to close; the intrinsic loop flexibility of different PTPases may be related to their catalytic rate and play a role in the wide range of activities observed within this enzyme family	1135
Yop51	Recombinant enzyme from *E. coli* of Yop51*Δ162 (Yop51 with Cys-235 → Arg mutation and deletion of 162 residues at N terminus)	Crystallization	Similar kinetic and fluorescence characteristics to full-length molecule	371

Patterns and regulation of PTPase expression

Distribution of PTPase mRNA and proteins

While most of the cloned mammalian PTPases are expressed in a variety of tissues, there are many examples of a wide variation in the levels of relative expression, and a number of PTPases whose expression is limited to certain tissues (Table 5). For example, CD45 is expressed solely and SH-PTP1 predominantly in haematopoietic cells, STEP and RPTP-σ (LAR-PTP2B) are expressed predominantly in brain, GLEPP1 in kidney, and OST-PTP has a peculiar distribution restricted to bone and testis. The PTPases in brain typically have a characteristic neuroanatomic distribution identified by *in situ* hybridization with cRNA probes or by immunohistochemistry with specific antibodies. These patterns of expression have led to speculation about the roles of certain PTPases in neuronal migration and axonal elongation in the establishment of neural networks in the brain and spinal cord.

The subcellular localization of PTPases has also generated interest as an important parameter that will impact on the interaction of PTPases with associated proteins and available substrates [35]. Especially in considering the cellular compartmentalization of the non-receptor type PTPases, the additional functional domains present on these proteins provide some insight into their localization, and the availability of the cDNA sequences allows these hypotheses to be tested directly. In addition, the transmembrane PTPases have been shown to distribute between the plasma membrane and internal membrane fractions, and in the case of CD45, a dynamic shift in subcellular localization has been observed upon activation of T cells [175].

While some of the intracellular, or non-receptor type PTPases are predominantly cytosolic (low M_r PTPase and SH-PTP2), others are found associated with internal membrane fractions either by way of complex formation with other membrane proteins, or protein domains that target the enzymes to specific intracellular sites. The PTP1B and TCPTP enzymes have been shown to be associated, possibly as high molecular-mass complexes, with the endoplasmic reticulum of the cell; however, a significant portion of the full-length PTP1B enzyme may also be found in the cytosol of some cell types, including fibroblasts, skeletal muscle and liver [20, 118, 119, 158]. In addition, the cloning of the cDNA for PTP1B indicated that the full-length protein has a cleavable C-terminal segment downstream from the PTPase domain, since the enzyme purified from placenta and sequenced was $\sim$12 kDa shorter than the protein predicted from the cDNA [481, 487, 514]. This C-terminal region has been shown to direct the association of the native protein with the endoplasmic reticulum either through a hydrophobic interaction or by attachment to a non-catalytic subunit [222, 148, 737]. Cleavage of the C-terminal segment of PTP1B and activation of its PTPase activity occurs in platelets in response to agonist activation and releases a relatively stable 42 kDa form of the enzyme [87]. During the preparation of tissue and cell homogenates, proteolytic cleavage of the susceptible C terminus and release of soluble truncated PTP1B protein ($\sim$37 kDa) may also occur [737]. Thus, the subcellular localization of enzymes such as TCPTP and PTP1B may be regulated by physiological mechanisms in different cell types which can also modify the protein structure and influence enzyme activation or specificity of the PTPase domain for certain cellular substrates.

Subcellular localization of certain PTPases may also be regulated by alternative splicing of enzyme isoforms. Two isoforms of the TCPTP have been identified. One isoform has a hydrophobic C-terminal segment that may direct the enzyme to a particulate localization as described earlier; the other has a shorter charged segment that may enable it to be expressed in the cytoplasm [35]. DPTP61F has been shown to be expressed as two isoforms with variant C-terminal segments. One isoform has a nuclear localization motif and the other a potential membrane-spanning segment; the predicted compartmentalization of these two DPTP61F isoforms has been confirmed by expression of recombinant constructs in COS cells [558]. SH-PTP2 has recently been shown to undergo alternative splicing in multiple domains, which may be significant in the tissue-specific expression and signalling functions of this subfamily of PTPases. Published sequences of SH-PTP2 have demonstrated variations in a five-amino acid segment (positions 409–413) at a central location within the catalytic domain as well as significant sequence variation between SH-PTP2 and Syp at the C terminus (see sequence alignment D at the back of the book) [140]. The insertion of four amino acids in the catalytic domain of SH-PTP2 has also been shown to reduce the V_{max} of the recombinant enzyme by 8- to 20-fold, depending on the substrate [301].

Modulation of PTPase expression

As the number of members of the PTPase superfamily has continued to grow, so have the examples of changes in PTPase mRNA expression and protein abundance in various cells and tissues. A compilation of available data on specific PTPase homologues from this active area for current research is given in Table 6. These studies will generate insight into the potential physiological roles of various PTPases from their response to a wide range of changes involving the cell cycle, cell differentiation, or states of hormone resistance or oncogenesis.

Table 5 Distribution of PTPases

Cell/tissue/organism	mRNA	Protein	Distribution/subcellular localization	Refs
Dictyostelium discoideum		DdPTP1-lacZ fusion	Spatially localized to prestalk and anterior-like cell types	277
		DdPTP2-lacZ fusion	Expressed preferentially in prestalk and anterior-like cell types during multicellular stages of development	523
Drosophila melanogaster		DLAR	Homogeneously expressed on most or all axons within the commissures and connectives	614
		DPTP10D	Primarily localized to the anterior commissure and its junctions with the longitudinal tracts	614
		DPTP69D	Immunohistochemistry demonstration of DPTP69D in CNS axons in embryo, with restricted expression at various larval stages	1099
		DPTP99A	Diffuse immunostaining in early embryos; homogeneous expression on most or all axons within the commissures and connectives	614
		DPTP99A	Protein staining in axons of CNS but not in nerve roots or axons of peripheral nervous system	520
	cdc25Dm	(string)	Expressed in early oogenesis, predominantly in follicle cells	489
			Transcripts found in proliferating embryonic cells, in third instar larval brains and in imaginal discs; also found in nurse cells during oogenesis	473
	cdc25Dm	(twine)	Expressed exclusively in male and female gonads; not expressed in zygotes until early embryo stage where it is uniformly distributed; translocated into oocytes from nurse cells during oogenesis	489
			Abundantly expressed in nurse cells during oogenesis; in the testis, found in growing stage of premeiotic cysts	473
Adipocytes (Rat)	Multiple		By cDNA library screening, identified expression of RPTP-α, PTP1B, SH-PTP2 and LAR at an abundance of 16, 7, 6 and 3 per 10^6, respectively	140
Adrenal gland (Rat) PC12 cells Brain	PC-PTP1		Only one 3.9 kb transcript in PC12h cells; 3.9 kb mRNA also found in brain and adrenal gland, but not in other non-neuronal tissues in adult rats; two other transcripts of 3.3 and 1.7 kb were also detected in brain	1124
B lymphocytes	LAR PTP1B PTP-α BPTP-1 (unique)		In the pre-B cell NALM-6 cell line detected expression of LAR, PTP1B, PTP-α and a unique mRNA for BPTP-1 (7.2 kb)	469
Bone (Rat)	Multiple		PTPases amplified from bone included the TM RPTPs α, β, γ, δ, ϵ, ζ, LAR and OST-PTP, as well as the intracellular enzymes PTP1B, PTPH1 and PTP1C	557
	OST-PTP (5.8 kb)		Expressed in bone and testis as well as in primary cultured rat osteoblasts	557

Brain/CNS				
Chicken/chick embryos	LAR-PTP2/2B (CRYP-α1/α2) (6.6 kb, 7.7 kb)		At E16, brain and spinal cord expressed highest levels of 6.6 kb mRNA (CRYP-α1), while other tissues expressed more of the 7.7 kb mRNA encoding CRYP-α2, including retina, gut, lung, skeletal muscle, heart > liver	604
Chicken		CRYP-α	CRYP-α is strongly expressed in the embryonic nervous system; immunolocalization showed a predominant localization in axons of the central and peripheral nervous systems, suggesting that the major, early role of the enzyme is in axonal development; in sensory neurites in culture CRYP-α localizes in migrating growth cones	967
Human	PTP1B		Immunocytochemical localization of PTP1B in hippocampus to granular and pyramidal neurones and in microglia; similar distribution in regions of hippocampus to other protein (Ser/Thr) phosphatases, MAP kinase and cdc2 kinase	182
		RPTP-ζ/β (*in situ* immunolocalization)	In neonatal cerebellum, found on pial surface only, but after chondroitinase treatment strong staining detected surrounding cells in internal and external granular layers	106
	RPTP-δ		RT-PCR revealed differences in expression between brain and kidney isoforms of RPTP-δ due to alternative mRNA splicing in the extracellular domain	959
	RPTP-ζ/β (8.6 kb)		Probes from either the extracellular domain or the PTPase domains hybridize to single band on Northern blot	106
Mouse	cdc25M2		In developing brain, concentrated at ventricular surface of telencephalon	535
	IA-2 (3.8 kb)		Predominant expression in brain, weakly seen in colon, intestine, stomach and pancreas, suggesting neuroendocrine pattern	548
	LAR-PTP2 (PTPNU-3)		Two major RNA transcripts of ~6 and ~7 kb, both expressed predominantly in brain and neuronal derived cell lines; detectable levels of larger RNA in non-neuronal tissues, alternative RNA splicing of a 132 bp segment present in the larger RNA results in protein with nine Fn-III repeats	1128
	PTP-SL (3.2 and 4.2 kb)		Expression limited to brain	1103
	PTPH1		Selectively expressed in the adult thalamus; mRNA expression is restricted to the dorsal thalamus during development and can serve as a specific marker for the dorsal thalamic nuclei; may play a role in thalamic relay nuclei	963
	RPTP-BR7		Expression limited to brain, especially cerebellum	1115
	RPTP-δ *in situ*		Expressed in hippocampus, thalamic reticular nucleus and piriform cortex	563
	RPTP-ζ/β *in situ*		In adult brain, localized to the Purkinje cell layer of the cerebellum, the dentate gyrus, and the subependymal layer of the anterior horn of the lateral ventricle; in embryonic brain, was expressed in ventricular and subventricular zones of the brain and in the spinal cord, suggesting a role in nervous system development	546

Table 5 Continued

Cell/tissue/organism	mRNA	Protein	Distribution/subcellular localization	Refs
Brain/CNS				
Mouse (Cont.)	RPTP-ζ/β (8.8, 7.5 and 6.4 kb)		First PTPase found to be strictly expressed in the nervous system; expressed in a human neuroblastoma cell line, but not three different glioblastoma cell lines; Hybridization of mouse tissues with mouse probe for RPTP- ζ/β showed expression exclusively in brain and not in a series of other tissues examined; hybridization analysis showed that the 6.4 kb mRNA represented only the deleted variant of RPTP-ζ/β, which could result from alternative splicing	546
Rat		STEP (immunolocalization)	Immunoreactivity most intense in areas of the CNS receiving dopaminergic input (basal ganglia and related structures) and localized to cell bodies, dendrites, and axonal processes; 46 kDa was major band in whole brain and not seen outside the CNS; in striatum saw all three protein bands, lowest band was only in striatum and larger two bands found in stratum and cerebral cortex	169
	CL100		ISH in rat brain revealed high expression in cingulate gyrus, ventral and medial divisions of anterior thalamus and medial geniculate nucleus; also in parietal and temporal cortex localized to non-mitotic cells; distribution of ERK, a potential physiological substrate, was different from CL100 in brain – ERK1 expression was ubiquitous in brain	541
	LAR-PTP2B (PTP-NE3) (6.75 kb; 5.6 kb)		Larger mRNA detected at low levels in heart, lung, testes; smaller form exclusive found in brain and olfactory neuroepithelium	216
	LAR-PTP2B (PTP-NE3) *in situ*		Highly expressed in hippocampal pyramidal cells and dentate gyrus of adult brain with lower levels in the cortex, thalamus, and granule cell layer of the cerebellum; expression is restricted to neurones, where in the olfactory neuroepithelium, a high level is found in immature neurones and a lower level in mature sensory neurones	216
	LAR-PTP2B (RPTP-σ) *in situ*		In rat embryos was highly expressed in brain, spinal cord and dorsal root ganglia, low expression in lung but no apparent expression in other tissues; in adult rat brain, expressed in olfactory bulb, cerebellum and hippocampus (especially pyramidal cell layer and granular layer of the dentate gyrus)	621
	MKP-2		Expressed in many areas of the brain including hippocampus, pyriform cortex and suprachiasmatic nucleus	1111
	RPTP-δ, LAR, LAR-PTP2, RPTP-α and PTP1B		Identified expression in fetal rat brain (E18)	965
	RPTP-γ *in situ*		In newborns, highest expression in hippocampal formation, septal and midline thalamic nuclei and in cortex; in adult rat brain, highly expressed in hippocampal formation but not in the thalamic nuclei or in the cortex	478

	RPTP ζ/β *in situ*		In embryos, expressed in ventricular and subventricular zones of the brain and spinal cord; in adult brain, expressed in Purkinje layer of the cerebellum, the dentate gyrus and the subependymal layer of the anterior horn of the lateral ventricle	478
	RPTP-ζ/β *in situ*		Highly expressed in radial glia and other glial cells during development; localized to radial processes of glial cells, suggesting that it can function in neuronal migration and axonal elongation	130
	RPTP-ζ/β *in situ*		Very high levels expressed in nerve fibre layer of olfactory bulb and accessory olfactory tract, a region in which neurones continue to grow throughout life	130
		SH-PTP2	SH-PTP2 is highly expressed in the rat brain; by immunohistochemistry, specific immunoreactivity was widely distributed, most abundant in neuropil and absent from white matter; intense labelling is observed on synapses; after fractionation of cells; SH-PTP2 was mainly observed in the particulate fraction, particularly in myelin and synaptosomes	968
	STEP (3 and 4.4 kb)		Both mRNAs are seletively expressed in brain, rare in adrenal, kidney and heart; 3 kb mRNA highly enriched in striatum relative to other brain areas; 4.4 kb mRNA most abundant in cerebral cortex and rare in striatum	547
		STEP	Immunohistochemical staining with antibodies against STEP was used to demonstrate the internal organization of grafts of embryonic striatal tissues implanted in lesioned neostriatum of adult rats; STEP was found in discrete patches within the grafts, providing further indication that the patch zones are indeed comprised of striatal like cell populations	939
		STEP	Multiple protein forms abundant in caudate-putamen; smaller forms (33, 37, 46 kDa) present in presynaptic axons of substantia nigra originating from caudate-putamen	1095
		LAR-PTP2 (7.5 kb; 6.0 kb)	In adult, 6.0 kb mRNA expressed highly in brain; a 7.5 kb mRNA expressed less strongly in brain and lung, also found rarely in kidney and liver; hybridization with probes to deleted extracellular domain in LAR-PTP2B and flanking sequences demonstrated that LAR-PTP2 is expressed in both brain and lung while LAR-PTP2B is expressed only in brain	193
		LAR-PTP2B (RPTP-σ) (6.9 kb; 5.7 kb)	The 5.7 kb mRNA is abundant and expressed only in brain; less abundant is the 6.9 kb mRNA found in brain as well as lung and heart $\gg$ kidney, intestine	621
	PTP1B		Intense ISH signal in pyramidal cell layer of hippocampus	514
	RPTP-κ		Abundant in cerebral cortex and hippocampus by ISH	529
Rat/mouse	RPTP-ζ/β Phosphacan (9.5, 8.4 and 6.4 kb)		Using selective probes to the extracellular and PTPase domains of RPTP-ζ/β, the 9.5 mRNA was found to correspond to the full-length molecule, the 6.4 kb mRNA to the deletion variant (dvRPTP-β) in which half of the extracellular domain has been deleted, and the 8.4 kb mRNA codes for the extracellular domain-only variant of the PTPase (Phosphacan)	77, 556

Table 5 Continued

Cell/tissue/organism	mRNA	Protein	Distribution/subcellular localization	Refs
Cultured cells				
293 cells		RPTP-α	By immunofluorescence confocal microscopy RPTP-α was shown to exist in dense intracellular granules and dispersed within the plasma membrane; TPA treatment caused redistribution of some intracellular RPTP-α to the cell surface, but without requiring phosphorylation of RPTP-α at Ser-180/Ser-204	924
BHK cells		TCPTP	Recombinant, full-length T-cell enzyme associates with particulate fraction (>600 kDa) of the cell; truncated form (37 kDa) is fully active and distributes between particulate and soluble fractions	251
(Hamster)		cdc25C	Epitope-tagged cdc25C is localized to the cytoplasm and enters the nucleus at the G_2/M phase transition	155
		cdc25C	Immunolocalization demonstrates cdc25C in the cytoplasm, prominently in the periphery of the nucleus in cells arrested in early S phase, shifts into nucleus in mitotic cells	599
CHO cells		T cell PTPase	45 kDa form, p45, and 48 kDa form, p48; p48 associates with the ER but p45 localizes in the nucleus; residues necessary for ER retention include the terminal 19 hydrophobic residues which comprise a potential membrane-spanning segment and residues 346–358 which encompass a cluster of basic amino acids that may represent another type of ER retention motif; the sequence RKRKR, which immediately precedes the splice junction, functions as a nuclear localization signal for p45	951
COS cells	MPTP	(TCPTP)	Expression of chimeric β-galactosidase with C-terminal 120 AA of MPTP conferred a localization exclusively to nuclei by immunohistochemical staining; nuclear targeting signal localized between AA 345 and 369	207
COS-1 cells	DPTP61F		C-terminal segment differs between two alternatively spliced forms and determines subcellular localization in transfected COS-1 cell DPTP61Fm form is localized to a reticular network as well as mitochondria-like organelles, DPTP61Fn form is localized to nuclei; sequences outside the alternatively spliced segments, common to both isoforms shown to play a role in localization	558
(Human and *Xenopus*)		CL100	Both CL100 proteins were localized predominantly in the nucleus in transfected COS-1 cells	1108
	PTP-PEST	PTP-PEST	mRNA is expressed throughout murine development; PTP-PEST is expressed freely in the cytosol as a 112 kDa protein; despite the presence of PEST sequence motifs, pulse-chase labelling experiments demonstrate that MPTP-PEST has a half-life of more than 4 h	1097
Endothelial cells (Human)	CD45	CD45	PCR detected constitutive CD45RO expression with an increase after IL-1 treatment; differences also noted between lymphocyte and endothelial cell CD45RO: M_r was 230 kDa in endothelial cells and 190 kDa in lymphocytes	147

Fibroblasts (Rat)		TCPTP (PTP-S)	Immunofluorescence in interphase cells shows predominant localization to nuclear compartment with some fluorescence in cytoplasm; in mitosis, was uniformly distributed throughout the cell; released from isolated nuclei by DNAse I suggesting that at least a fraction of the protein is associated with chromatin	187
		hVH-2	Immunofluorescence studies with an epitope-tagged hVH2 showed that the enzyme was localized in cell nucleus	1100
Glial cell line (C6)		TCPTP	Immunofluorescence analysis demonstrated TCPTP in perinuclear area and weakly throughout the cytoplasm	144
		PTP1B	Immunofluorescence pattern consistent with localization in the ER and staining extending throughout the cytoplasm	144
Glia (Müller cells from chick retina) (Chicken)	RPTP-α, RPTP-γ, RPTP-δ, RPTP-ζ and PTP1B		Also identified two novel PTPases, MG-PTP1 and MG-PTP2, which are related to RPTP-δ and T cell PTPase, respectively	965
Glioblastoma cells (Human)		RPTP-ζ/β	By RT-PCR, highly expressed in a glioblastoma cell line (U373MG); expression rare in other human cell lines tested	539
HeLa cells	cdc25 (p55)		Localized primarily to nucleus in interphase, speckled or punctate pattern, not in nucleoli; diffuse staining observed in mitotic cells; expression varies <2-fold throughout cell cycle	303
(Human)		PEP	Nuclear localization demonstrated by transfection of HeLa cells; directed by 18 AA segment at extreme C terminus, proven by transfer of segment to β-galactosidase	146
	cdc25Hu2		Expression is 10–100 times higher than cdc25Hu1 in HeLa and other cell lines; overall expression varies greatly among different cell lines, associated with degree of transformation	568
		CL100	Immunohistochemistry of epitope-tagged transfected enzyme demonstrated nuclear localization	950
		hVH-3	Immunohistochemistry of epitope-tagged transfected enzyme demonstrated nuclear localization	950
	PTPH1 (4.3 kb)		Low abundance in HeLa cells; other cell types not tested	624
(and other cultured cells)		PTP1B	PTP1B localized predominantly in the ER with the PTPase domain oriented toward the cytoplasm; expression of recombinant domains indicates that targeting is determined by the C-terminal 35 AA	148
Haematopoietic cell lines and various other cell lines (Human)	SH-PTP1 (HCP) (2.6 kb)		Detected in a variety of haematopoietic cell lines but also low levels seen in MCF7 breast cancer cells and A549 lung cancer cells; undetectable from fibroblast, colon epithelial or lung epithelial cell lines	627
Haematopoietic cells (Human)	HePTP (LC-PTP) (2.9 and 4 kb)		Preferentially expressed in a variety of haematopoietic cells, especially thymocytes; not expressed in several other types of cultured human cells including fibroblasts, neuroblastoma or epithelial cell lines; 2.9 kb mRNA much more abundant than 4.0 kb mRNA and follow same expression pattern	470

Table 5 Continued

Cell/tissue/organism	mRNA	Protein	Distribution/subcellular localization	Refs
Cultured cells				
HL-60 cells		SH-PTP1 (PTP1C)	With PMA-induced differentiation of HL-60 cells into macrophages, PTP1C is phosphorylated on Ser and expression is induced; also enzyme translocates from its exclusively cytosolic localization in untreated cells to 30-40% found in the plasma membrane fraction	226
		SH-PTP1 (PTP1C)	Expressed at high levels in HL-60 human promyelocytic cells, accounts for 0.15% of protein in a post-nuclear extract and ~70% of the *p*NPPase activity at pH 5.0	226
Human breast carcinoma ZR-75-1 cells	SH-PTP1 (PTP1C) (2.4 kb)			602
Human lung fibroblast M426 cells	VHR (4.1 kb mRNA)			524
Megakaryoblastic cell line (MEG-01) (Human)	MEG1 (3.7 kb)		Other cell types not tested	508
PC12 cells (Rat)	LAR-PTP2B (PTP-P1/PTP-PS)		In PC12 cells, PTP-P1 expressed in three transcripts 8, 6, and 4 kb; PTP-PS detected as single 4.8 kb transcript	579
Rat L6 myocytes		PTP1B	37% of cell homogenate PTPase activity precipitated with monoclonal antibody FG6-1G	163
Sf9 cells		TCPTP	Recombinant TCPTP in Sf9 cells found restricted to particulate fraction and released with salt (0.6 M KCl) and detergent; 11 kDa truncation off C terminus produced soluble 33 kDa protein, indicating that the C-terminal segment plays a role in regulation of the cellular localization of the protein	369
T cells (Human)		CD45	In resting cells, present in Golgi (30%) as well as on surface; upon activation with anti-T cell antigen receptor or anti-Thy-1 antibody, the Golgi is rapidly and specifically cleared of CD45, which shifts to another internal cellular compartment	175
		PAC-1	In quiescent T cells stimulated by PHA and PMA, observed shift of immunostaining from cytoplasm to nucleus by 1 h with peak at 4 h; in COS cells transiently transfected with PAC-1, showed uniform pattern of punctate staining in nucleus, none in nucleolus, also diffuse staining in cytoplasm	589
Various cell lines (Human)	LAR (8.0 kb)		mRNA expression in human kidney, prostate and T-cell lines; two close bands in Hut78 T-cells apparently differ in the 5′ end sequence	605
	MEG-2 (4.0 kb)		Widely expressed in all cell lines tested including haematopoietic, epithelial, fibroblastic and choriocarcinoma cells	507
	PTP-PEST (3.8 kb)		Widely expressed in a variety of cultured cell lines	622
	PTP-PEST (PTPG1) (4.6 kb)		Widely expressed in a variety of cell lines	610

	RPTP-γ (9.6 and 6.2 kb)		Expressed in all renal cell and lung carcinoma cell lines tested; absent from three haematopoietic cell lines	542
Various cultured cells		PRL-1	Overexpressed as a 21 kDa protein; staining in transfected NIH 3T3 cells and H35 cells in nucleus and cytoplasm, with greater staining in nucleus in H35 cells; cell fractionation showed PRL-1 primarily in insoluble fraction of nuclei	493
Epidermis (Human)	PTP1B	PTP1B	Expressed in human epidermis, cultured keratinocytes and keratinocyte cell lines; localized primarily to the basal cell layers in normal thick epidermis suggesting that it plays a role in the *in vivo* regulation of epidermal function	268
Haematopoietic cells (Monkey)	TCPTP (2.3 kb)		Abundant in thymus and spleen, lower expression in brain and human placenta	488
(Mouse)	PEP *in situ*		Diffusely expressed in lymphoid tissues in many associated cell types	146
	PEP (2.7 kb)		Abundant in spleen, thymus, lymph node and bone marrow; negligible in other tissues	553
	SH-PTP1 (SHP) (2.1 kb)		Abundant in haematopoietic cells; detectable in lung and kidney	553
	HePTP		Abundant in thymus $\gg$ spleen; not expressed in brain, liver, kidney or ovaries	630
	PAC-1		Abundant in thymus and spleen, undetectable in brain, lung, heart, kidney, liver, muscle and stomach; also expressed in various cultured lymphoid cell lines and low levels expressed in serum-stimulated human embryonic lung fibroblasts and HeLa cells	589
	cdc25M1 (3 kb)		Expressed primarily in thymus > spleen; also in pre-B lymphocytes and CD4+ helper and CD8+ cytotoxic lymphocytes	569
	SH-PTP1 (HCP) (2.6 kb)		Abundant in thymus, bone marrow, peripheral blood cells; low levels seen in other tissue perhaps as a result of contaminating blood cells in kidney, liver, lung, spleen; in cell lines, absent from fibroblasts and various epithelial and neuroblastoma cells, but readily detected in haematopoietic cell lines	627
	CD45 (4.7 and 5.0 kb)		mRNA in T cells (4.7 kb) and B cells (5.0 kb); expressed in thymocytes, lymph node cells, spleen and T- and B-cell leukaemia cells; not expressed in liver, kidney, brain, fibroblasts and sarcoma cells	601
Kidney (Rabbit)	GLEPP1 (6.5 kb)		Quantitated by RNAse protection in glomeruli > renal cortex > brain, but not in a variety of other tissues tested; protein is distributed in the renal glomerular podocyte foot processes	613
(Human)		GLEPP1	By immunohistochemistry, distribution restricted to the glomerulus in human kidney	1130
(Rat)	RPTP-α (LRP)	RPTP-α (LRP)	Large signals were found in inner medulla by Northern analysis, using RT-PCR on individual microdissected tubule segments along the nephron; the intrarenal localization of LRP mRNA was characterized; most abundant in inner medullary collecting duct	946

Table 5 Continued

Cell/tissue/organism	mRNA	Protein	Distribution/subcellular localization	Refs
Lung (Human)	RPTP-γ (9.6 and 6.2 kb)		Highly expressed as 9.6 and 6.2 kb mRNA transcripts in fetal and normal lung; found two cell lines with aberrant mRNA transcript size out of > 31 analysed; failed to identify any mutations by RNAase protection assays or SSCP analysis (restricted to cytoplasmic domain)	209
(Mouse)	ERP (MKP-1)		*In situ* hybridization demonstrates high expression in mouse pups and adult mice in lung cells lining alveoli, most likely pneumocytes; expression may be induced shortly after birth by oxidative stress	571
	RPTP-μ (5.7 kb)		Single mRNA transcript, most abundant in lung, much lower amounts in brain and heart	504
Lymphocytes (Guinea pig)		CD45	Molecular weight variation from 175 kDa on thymocytes to 230 kDa on B lymphocytes	598
(Human)		CD45 (isoforms)	180 kDa isoform present on subset of T cells but absent on B cells; 220 kDa isoform predominantly expressed on B cells and a subpopulation of T cells; the 220/205/190 kDa isoform epitope is not expressed on skin Langerhans cells or sinusoidal lymph node macrophages	184
			Alternative splicing of exons 4–6 is dependent on the length of the exon six sequence and not on intron sequences associated with exon 6	208
	CD45		Exons 4–6 are included in mRNA from B cells, but not in T cell CD45 mRNA; at least three distinct *cis*-acting elements within the exon four sequence are required for tissue-specific alternative mRNA splicing and isoform expression	205
		CD45 (isoforms)	CD45 accounts for at least 40% of the PTPase activity isolated from peripheral blood lymphocytes, in both the plasma membrane and microsomal fractions	264
(Mouse)		CD45 (isoforms)	Exons six and seven are expressed in B cell lines but not in T cell lines examined	196
(Sheep)		CD45 (isoforms)	B cells in ileal Peyer's patches (IPP) expressed a 220 kDa form at low density in contrast to peripheral B cells, which had high expression; CD45 in IPP was 200 and 190 kDa, but entirely 220 kDa in peripheral B cells, and in thymus, 210 and 190 kDa	171
Skeletal muscle (Rat)	Multiple		Biochemical purification and identification of major PTPases in particulate fraction (LAR, PTP1B and SH-PTP2) and cytosol (SH-PTP2 and PTP1B); RPTP-α was not detected by blotting purified PTPase fractions; by immunodepletion studies, LAR, SH-PTP2 and PTPase1B were shown to account for over 70% of the total activity towards RCM-lysozyme	119
			Identification of LAR and LRP (PTP-α) expression by amplification of conserved cDNA sequences	225
Testis (Mouse)	Multiple		Identified five PTPase cDNAs by PCR, including mouse cytoplasmic PTP, PTP1B, PTP-δ, PTP-ϵ, and STEP	162
Thymus (Human)		CD45	Localized *in situ* the isoforms of CD45 expressed in human thymus and correlated finding with the tissue localization of the thymic generative lineage	151

Various tissues				
Chicken	ChPTPλ (5.6 kb)		Abundant in spleen and intestine; less abundant in lung and brain; absent from liver and muscle	500
Human		LAR *in situ*	Monoclonal entibody to extracellular domain shows LAR *in situ* in a variety of human epithelial cells, smooth muscle cells and cardiac myocytes	110
	B23 (2.5 kb)		Widely expressed, highest in pancreas and brain	525
	CL100 (2.6 kb)		Abundant in human lung, highly expressed in placenta, liver and pancreas, lower expression in heart and skeletal muscle, rare in kidney	541
	CL100 (2.4 kb)		Expressed in placenta, liver, lung, heart and pancreas > skeletal muscle	950
	hVH-3 (2.4 kb)		Expressed in placenta and liver > heart, brain and kidney; also expressed in series of hepatoma cell lines including BRL3, HTC, H4 and hepG2	950
	hVH-5		Expressed predominantly in the adult brain, heart and skeletal muscle; *in situ* hybridization histochemistry of mouse embryo revealed high levels of expression and a wide distribution in the central and peripheral nervous system; specific areas of abundant hVH-5 expression included the olfactory bulb, retina, layers of the cerebral cortex, and cranial and spinal ganglia	1110
	LAR		Highly expressed in placenta, brain and pancreas, less abundant in kidney, heart, lung and liver; generally 9.0 kb RNA except brain (8.4 kb)	1117
	LAR-PTP2 (RPTP-σ)		Abundant in brain, thymus, lung, heart; weakly detected in spleen; almost undetectable in liver by RNAase protection; Northern analysis showed mRNAs of 7.0 and 5.5 kb, 7.0 kb predominant in thymus, smaller mRNA more abundant in brain	179
	Pez		By Northern blot analysis, expressed in a variety of human tissues including kidney, skeletal muscle, lung and placenta	1125
	PTP-BAS (PTP1E)		Abundant in kidney and lung, less in brain, pancreas and placenta, lowest level in heart, not detected in liver and skeletal muscle	475
	PTP-BAS (~8 kb)		Abundant in adult kidney > lung $\gg$ placenta, brain, pancreas, heart, and skeletal muscle > liver; more highly expressed in fetal compared to adult brain	549
	PTP-BAS (PTPL1) (9.5 kb)		Abundant in kidney, placenta, ovaries, testes; moderate in lung, pancreas, prostate and brain; low expression in heart, skeletal muscle, spleen, liver, small intestine and colon; absent in leucocytes	596
	PTP-PEST (PTPty43)		Widely expressed in all tissues examined	626
	PTPD1 (~7.0 kb)		Abundant in placenta, lung and normal and diabetic skeletal muscle, lower expression in kidney and colon, rare in liver and stomach, absent from spleen, relatively high expression in rhabdomyosarcoma cells	564
	PTPD2 (~11.0 kb)		Similar expression pattern to closely related PTPD1	564
	PTP-U2		By Northern blot analysis, two different transcripts for PTP-U2; in kidney and brain a 5.4 kb mRNA, and in lung and placenta as 3.5 kb; the 3.5 kb transcript was also detected in human leukaemia cell lines	1120
	PTP1B (3.5 kb)		Widely expressed: kidney > liver > placenta > heart > lung; various cell lines	487
	RPTP-α (4.3 and 6.3 kb)		Larger mRNA more prevalent in fetal tissues; abundant in fetal brain and liver, also in placenta, kidney, HUVEC and A431 cells	536

Table 5 Continued

Cell/tissue/organism	mRNA	Protein	Distribution/subcellular localization	Refs
Various tissues				
Human (Cont.)	RPTP-α (LRP) (3.0–3.5 kb)		Two transcripts in range 3.0–3.5 kb; ubiquitously expressed	533
	RPTP-α RPTP-αi		Both isoforms detected by PCR from T-cell and vascular smooth muscle libraries as well as A431 cells	84
	RPTP-δ		Multiple RNAs (7.9–11.2 kb) detected predominantly in brain and lower amounts in heart, placenta and kidney	1117
	RPTP-σ		7.7 kb RNA in all tissues except placenta and liver, brain has high levels of 6.6 kb RNA that lacks Fn-III domains F4–F7	1117
	SAP-1 (4.2 kb)		Abundant in brain and liver with a low level in heart and stomach; notably absent from lung, pancreas, colon and placenta. Also found in several, but not all, gastric cancer cell lines, and highly expressed in several colon cancer and pancreatic cancer cell lines. Absent from haematopoietic cell lines	552
	SH-PTP2 (6.0 kb)		Abundant in heart, brain, placenta, lung, liver, skeletal muscle and kidney; less abundant in pancreas	502
	SH-PTP2 (6.5 kb)		Abundant in fetal brain and heart, less abundant in adult brain, kidney, lung and liver; widely expressed in various cell lines including THP-1 monocytes, U937, HL-60, K562, Jurkat (T cell), HEPG2 cells	479
	SH-PTP2 (PTP2C) (7.0 kb)		Abundant in heart, brain and skeletal muscle. but present in all tissues examined and in a variety of cell lines	472
	SH-PTP2 (SH-PTP3) (6.0 kb)		Abundant in lung, brain, liver and pancreas; widely expressed among various cell lines tested	471
	SH-PTP1 (2.4 kb)		Highly expressed in cell lines of all haematopoietic lineages but also lower expression in a few epithelial cell lines including HepG2 hepatoma, Calu-1 lung carcinoma and HeLa cervical carcinoma cells; in human tissues, was highly expressed in megakaryocytes and platelets, as well as lung, with lower expression in kidney, liver, heart and brain; also found in primary tracheobronchial epithelial cells	582
Mouse	cdc25A		In adult mice, cdc25A transcripts are expressed at high levels in the testis > ovary, particularly in germ cells > kidney, liver, heart and muscle – a pattern distinct from that of cdc25B	976
	cdc25C cdc25B		Highest levels of cdc25C transcripts were detected in the testis (1.9 kb mRNA > 2.1 kb), localized in germ cells, specifically in late pachytene–diplotene spermatocytes and round spermatids; cdc25B mRNA was most readily detected in the somatic cells. In the ovary, cdc25C expression was apparent in cumulus granulosa cells, whereas cdc25B was detected in both growing oocytes and somatic cells, including the granulosa cells; the *cdc25* genes may have different functions in the germinal and somatic compartments of the ovary and the testis	979
	cdc25M2		Detected in most tissues, spleen > lung > brain, heart, intestine > adrenal, kidney, liver, muscle	535

	PTP1B (4.4 and 2.0 kb)	Abundant in testis (where additional mRNAs of 2.4 and 2.2 kb are expressed) and kidney, spleen, muscle, liver, heart and brain	562
	PTP36	Expressed in thymus, kidney and spleen, weakly detected in bone marrow and brain, rare in liver; also expressed at high level in thymic stroma and fibroblasts	597
	RIP	Predominant in kidney; expressed to lesser extent in lung, heart, brain and testis	933
	RPTP-α (3.0 kb)	Abundant in brain, liver and kidney; widely expressed	595
	RPTP-α (LRP) (3.2 kb)	Abundant in kidney, lung, thymus, brain; widely expressed	554
	RPTP-γ (8.5 and 5.5)	Expressed in brain, lung, kidney, heart, skeletal muscle, liver, spleen and testes; additional transcript of 3.0 kb detected in testes	478
	RPTP-κ (5. three and 7.0 kb)	Widely expressed except in testis and spleen; 5.3 kb mRNA particularly abundant in kidney and liver; also expressed in lung, heart and brain	529
	SH-PTP2 (PTP1D) (6.8 kb)	Abundant in brain, heart, kidney; less in liver, skeletal muscle, testes, lung	616
	SH-PTP2 (Syp) (7.0 kb)	Abundant in brain and heart, also in kidney, liver, lung, spleen	501
	TCPTP (MPTP) (1.9 kb)	Abundant in ovaries, testes, thymus and kidney, also found in various other tissues; in testes, an additional 1.3 kb mRNA was observed, evidently truncated at the 3′ end by hybridization analysis	567
	TCPTP (1.9 kb)	Abundant in spleen, low expression in kidney, rare in placenta, skeletal muscle and liver, not observed in fat or brain	21
	RPTP-δ ($\sim$7.0 kb)	Expressed in brain, kidney and heart cells also expressed in pre-B cell lines and some B cell lines, but not in lymphoid tissues, B cell hybridomas, T cell or macrophage cell lines	563
Rat	LAR (8.0 kb)	Abundant in liver (which also expressed two minor bands at 4.6 and 5.0 kb); 8.0 kb mRNA also expressed in kidney > heart > lung > spleen > testis > skeletal muscle; brain weakly expressed an LAR transcript at 4.6 kb	21, 631
	LAR-PTP2 (7.3 kb; 6.0 kb)	7.3 mRNA abundant in heart > testis > liver $\gg$ skeletal muscle > spleen > brain > kidney; in brain only, a 6.0 kb mRNA was highly expressed; RT-PCR demonstrated expression of both splicing forms of LAR-PTP2 (2 and 2B) in brain	631
	LAR-PTP2B (PTP-P1/ PTP-PS)	8 kb PTP-P1 mRNA detected in a variety of cell lines and tissues examined; the 6 kb mRNA is expressed mainly in neuronal cell lines and in the rat medullary thyroid carcinoma cell line W2, and within the brain, most abundant in the cortex; PTP-PS was detected in some neuronal and endocrine cells (PC12, GH4, W2 cells as well as liver and cerebral cortex	579
	MKP-2	MKP-2 differed from MKP-1 in its tissue distribution; MPK-2 expression highest in heart, lung > brain, testis $\gg$ spleen > skeletal muscle, kidney	1111
	PC12-PTP1	Three mRNAs detected in brain of 1.5, 2.6 and 3 kb; 3 kb specific for brain and is also found in PC12 cells; 2.6 kb also expressed in kidney; 1.5 kb enriched in heart, kidney and skeletal muscle > brain and lung > liver	1121

Table 5 Continued

Cell/tissue/organism	mRNA	Protein	Distribution/subcellular localization	Refs
Various tissues				
Rat (Cont.)	PTP2E		Two mRNAs of 3.1 and 5.6 kb; both abundant in adrenal > heart and skeletal muscle; only 3.1 in ovary and testis; only 5.6 in brain, uterus and lung; 3.1 kb transcript may represent smaller cDNA isolate (see Table 2); liver expressed 5.6 kb and an abundant, unique 2.4 kb mRNA	1107
	PTP1B (4.3 and 2.0 kb)		Expression shown in spleen, brain and kidney	514
	PTP1B (4.3 kb)		Expressed widely in various rat tissues with additional 1.6 kb mRNA transcript in rat liver	790
	RKPTP (2.7 kb)		Abundant expression in lung and kidney, less in spleen and brain and least in heart and liver among rat tissues tested	566
	RPTP-α (LRP) (3.0 kb)		Abundant in brain, also in heart, kidney, lung; widely expressed.	565
	SH-PTP2 (PTP1D/ PTP1Di) (6 kb)		Similar expression pattern: highest in brain, moderate in muscle and heart, less in lung and kidney; PTP1D expression generally > PTP1Di mRNA	301
	LAR (8.0 kb)		Using probes to cytoplasmic domain and 3′ untranslated region of LAR, found 8.0 kb and additional variable length mRNAs in cortex, brainstem, cerebellum, spinal cord, peripheral tissues (including lung, heart, muscle, liver, and kidney) and cultured neural, glial and phaeochromocytoma cells; ISH shows expression in brain and dorsal root ganglion neurones	170
	PRL-1 (2.7 kb)		Highly expressed in brain and skeletal muscle, detectable in spleen, kidney, thymus and liver, absent from intestine, heart, lung and stomach; highly expressed in HeLa, HepG2, and CV1 cells, less expressed in Burkitt lymphoma, and lowest in lymphoblast and fibroblast cell lines	493
	TCPTP (PTP-S) (1.7 kb)		Abundant in macrophages > lymphocytes, also expressed in thymus, spleen and brain; expression enhanced by cycloheximide	608

Table 6 Changes in PTPase mRNA expression, protein abundance or enzyme activity

Condition or disease state*	Treatment/ agonist/ condition	Tissue/cell analysed (WA = whole animal; CC = cultured cell)	PTPase	mRNA/protein	Comments	Refs
		Transient fusion of cells with different CD45 isoform expression pattern; CC	CD45		mRNA analysis using RT-PCR suggested that cells expressing the smallest of the CD45 isoforms contain negative *trans*-acting factors that allow the alternate exons to be skipped; the full-length isoform is the default pattern of splicing in agreement, incubation of thymocytes or T cell subsets with cycloheximide increased mRNA corresponding to the larger isoforms	192
	Expression of CD45 gene from YAC vector	Mouse B cells; CC	CD45		Mouse B cell line transfected with the YAC clone expressed the low M_r isoform of CD45 on the cell surface	145
	IL-1 stimulation	Human endothelial cells; CC	CD45	↑ Protein	After several days of stimulation, only the CD45RO protein isoform was detected; PCR detected constitutive CD45RO expression with an increase after IL-1 treatment	147
	IL-2	Mouse IL-3-dependent pro-B BAF-B03 and myeloid cell line 32D transfectants; CC	LC-PTP	↑ mRNA	LC-PTP mRNA induction by IL-2 required both the acidic and serine-rich regions of the IL-2 receptor β chain; overexpression of activated-lck or -Raf kinases also resulted in augmented LC-PTP mRNA expression	926
	Insulin	PC12; CC	hVH-5	↑ mRNA	Induced in a manner characteristic of an immediate–early gene, suggesting a possible role in signal transduction	1110
	Ischaemia for 20 min	Rat brain; WA, 1–48 h	CL100	↑ mRNA	↑ of CL100 mRNA in hippocampus and cortex at 1 h of reperfusion; expression of STEP, RPTP-δ and SH-PTP2 did not change	972
	Ischaemia for 30 min	Rat forebrain; WA, 6–24 h	MKP-1 (3CH 134)	↑ mRNA	↑ in neuronal and glial cells, prolonged ↑ in pyramidal cell layer of hippocampus and granule cells of dentate gyrus	978
	Ischaemia for 30 min	Rat forebrain; WA, 6–24 h	PAC-1	↑ mRNA	Induction of PAC-1 mRNA was prominent in hippocampal cells; suggests role in response to injury	977
	Retinoic acid	Swiss 3T3 fibroblasts; CC, 8 h	cdc25M2	↑ mRNA	In serum starved quiescent cells, ↑ five to 10-fold by retinoic acid; serum itself does not induce rapid induction	535

*Includes development, differentiation, etc.

Table 6 Continued

Condition or disease state*	Treatment/ agonist/ condition	Tissue/cell analysed (WA = whole animal; CC = cultured cell)	PTPase	mRNA/protein	Comments	Refs
	Antioestrogen treatment	Human MCF7 breast cancer	Membrane		Inhibition of growth factor stimulated proliferation by tamoxifen involves activation of membrane PTPase activity	149
	Actinomycin D/ cycloheximide	PC12 cells; CC	MKP-1		Under conditions that eliminated detection of MKP-1 mRNA and protein expression, the rapid deactivation of MAP kinase was unaffected, suggesting that MAP kinase deactivation is not dependent on induction of MKP-1	366
	Mating pheromone treatment	*Saccharomyces cerevisiae*; CC, 30 min	MSG5	↑ mRNA, 5-fold (1.8 kb mRNA)	Does not accumulate as a simple consequence of synchronization in G_1 phase	497
	cAMP analogues	Rat osteosarcoma UMR 106 cells; CC	OST-PTP	↑ mRNA 5–8-fold		557
	NGF	PC12	hVH-5	↑ mRNA	Induced in a manner characteristic of an immediate–early gene, suggesting a possible role in signal transduction	1110
	NGF	PC12 cells; CC, days	LAR		Induced changes in alternative splicing of multiple RNA transcripts	980
	NGF	PC12; CC, hours to days	PC12-PTP1 (3.0 kb)	↑ mRNA	↑ 2-fold by 2 h with maximal ↑ of 9-fold after 8 h that ↓ by 24 h to basal	1121
	NGF or glucocorticoids	PC12 cells; CC	PC-PTP1	↓ mRNA	NGF and glucocorticoids decreased the level of PCPTP1 mRNA in PC12h cells	1124
	Oxidative/heat stress	*Xenopus* kidney cell line; CC	CL100	↑ mRNA, ↑ protein	Expression is inducible by oxidative/heat stress; XCL100 is constitutively expressed in growing *Xenopus* oocytes	1108
	PDGF treatment	3T3 fibroblasts; CC	RPTP-α		RPTP-α (LRP) mRNA expression was modulated by PDGF treatment	932
	Phorbol ester	Cells overexpressing RPTP-α; CC, 5–15 min	RPTPα		PTPase activity ↑ due to a 2 to 3-fold ↑ in substrate affinity; enzyme is serine phosphorylated (see Table 11)	992
	Pioglitazone	Rat1 fibroblasts overexpressing the insulin receptor; CC, hours to days	PTP1B	Abrogated 2-fold rise of PTP1B protein that occurs in high glucose conditions	Piogliazone also blocked the increase in cytosolic PTPase activity that occurs in high glucose medium	953

	PTH	Rat osteosarcoma UMR 106 cells; CC	OST-PTP	↑ mRNA, 5–8-fold		557
	High glucose (27 mM)	Rat1 fibroblasts; CC, 4 days	PTP1B	↑ Protein	Full-length PTP1B protein was increased in cytosolic fraction, but not particulate fraction	158
	Serum	*Xenopus* kidney cell line; CC	CL100	↑ mRNA, ↑ protein	Expression is inducible by serum stimulation; XCL1000 is constitutively expressed in growing *Xenopus* oocytes	1108
	Thrombin activation	Platelets; CC, 1–2 min	SH-PTP1	Translocation of SH-PTP1 to cytoskeleton in activated platelets by immuno-blotting and immuno-precipitation; followed same time course as pp60$^{c\text{-}src}$ translocation	PTPase enzyme activity found to be associated with cytoskeleton of thrombin-activated platelets	168
	Thrombin stimulation; integrin α(IIb)β(3) mediated aggregation	Normal platelets or thrombasthenic platelets lacking integrin α(IIb)β(3); CC	PTP1B, PTP1C, SH-PTP2	PTPase distribution	PTP1B and its cleaved 42 kDa form were associated with the cytoskeleton in an aggregation-dependent manner; association of PTP1C with the cytoskeleton was regulated differentially both by thrombin stimulation and by α(IIb)β(3)-mediated aggregation; SH-PTP2 was distributed in fractions other than the cytoskeleton and showed no relocation on thrombin stimulation; cytoskeleton-associated PTP1C became tyrosine-phosphorylated after α(IIb)β(3)-mediated aggregation	936
	Various	HepG2 and H4 hepatoma cells; CC, 0.5–1 h	hVH-3	↑ mRNA	mRNA ↑ 5-fold by cell incubation with EGF or PMA in HepG2 cells; ↑ 15-fold by insulin and IGF-1 in H4 cells	950
Activation		T cells; CC	CD45	Altered mRNA splicing pattern	RT-PCR demonstrated novel alternative splicing pattern involving exon 7; characterized in detail the sequence of alternative splicing of exons 4–7 during T cell maturation and activation	133
		Long-term human CD4+ CD45RA+ T cell line	CD45		Regulation of CD45 isoform expression was cyclic; demonstrated that the alternative splicing pattern is highly regulated after activation; conversion from CD45RA+ to CD45RA− is potentially reversible	192
		T cells; CC, 24 h	CD45		Expression of the 5.4 kb mRNA is lost; 4.8 kb mRNA persists and increases until day 3; associated with loss of CD45R protein isoform	136

*Includes development, differentiation, etc.

Table 6 Continued

Condition or disease state*	Treatment/ agonist/ condition	Tissue/cell analysed (WA = whole animal; CC = cultured cell)	PTPase	mRNA/protein	Comments	Refs
Activation	Antigen	T cells; CC	CD45		Cell activation is accompanied by loss of the 190 kDa (B exon) dependent form with concomitant increase in the 180 kDa form	124
	Culture supernatant of a Gram (−) pathogen	Neutrophils; CC	CD45	↑ Protein	Increased CD45 protein and PTPase activity in plasma membranes	135
	Mitogen stimulation	T cells; CC	CD45		High M_r isoforms of CD45, predominantly CD45RA, ↑ during 1st two days of activation and then downregulated with ↑ appearance of the low M_r isoform, CD45RO; by 24 h, level of high M_r CD45 mRNA declines precipitously to below detectable levels; half-life of high and low M_r forms of CD45 mRNA was between 2.25 and 3.5 h in stimulated and resting cells; CD45 mRNA was superinduced by treatment of cells with cycloheximide	137
	Stimulation with IgM antibody or combination of PMA and ionomycin	Murine B cells; CC	CD45	↓ Proteins	Reduced expression of CD45R A-, B- and C-exon-dependent epitopes mediated by changes in alternative splicing of CD45 mRNA	180
Alzheimer's disease		Brain; WA	Low M_r PTPase	↓ Protein	Chromatographically isolated low M_r PTPase activity was significantly ↓ in Alzheimer's brains	202
Angiotensin II		VSMC; CC, 30 min	3CH134	↑ mRNA abundance (2.8-fold) ↑ mRNA protein synthesis of 3CH134	Maximum response at 100 nM angiotensin II; increased, but less marked to PDGF, α-thrombin, H_2O_2, PMA and ionomycin	143
Calcium ionophore treatment	Ionomycin	Thymocytes or murine T cell lines; CC	CD45		Decreased CD45 activity and decreased Ser phosphorylation of CD45; demonstration of regulation of CD45 activity *in vivo* possibly via changes in Ser phosphorylation	315

Carcinoma		Human colorectal carcinomas; WA	RPTP-α	↑ mRNA	2–10-fold ↑ in mRNA was detected in 10 of 14 tumours compared to mucosa suggest that RPTP-α overexpression could contribute to the tumorigenic process in colon carcinoma	969
	Hepatocellular	Human hepatocellular carcinoma tissue (HCC) and cell lines; WA and CC	PTPH1	↔ mRNA	Using RT-PCR, PTPH1 mRNA was detected in all HCC cell lines and tissue and adjacent non-cancerous tissues with no remarkable difference in the level of expression of PTPH1 mRNA between HCC and adjacent non-cancerous tissues; sequence analysis showed a common mutation from T to C at the third letter of codon 919 which did not lead to amino acid substitution	944
	Ovarian	Human ovarian carcinomas and cell lines; WA and CC	SH-PTP1	↑ mRNA, ↑ protein	PTPN6, (SH-PTP1) mRNA transcripts and proteins were overexpressed 2 to 4-fold in 7/8 ovarian epithelial carcinoma cell lines; also there was also a 2 to 3-fold increase in expression of PTPN6 in 10/11 invasive ovarian epithelial cancer tissues	955
Cell activation		CD4+ lymphocytes; CC	CD45	Isoform expression	The highest rate of proliferation was found among cells with a phenotype intermediate between CD4+ CD45RA+ CD45RO− and CD4 + CD45RA− CD45RO+, after 72 h of culture, the highest proliferation had shifted towards the CD4+ CD45RA− CD45RO+ population, indicating that the phenotype transition is accompanied by cell proliferation	945
Cell activation and differentiation		Cytotoxic T lymphocytes; CC	CD45	Isoform expression	Antigen-specific TCR gamma delta+ lymphocytes favour expression of the exon (45678) and exon (5678) CD45 splicing products whereas the exon (8) variant is lost; TCR alpha beta+ lymphocytes are devoid of the full-length exon (45678) splicing product	942
Cell cycle		*Xenopus* oocyte extracts			cdc25 is active at a low level throughout interphase, near onset of mitosis it becomes phosphorylated in its N-terminal region and has ↑ in activity	292

*Includes development, differentiation, etc.

Table 6 Continued

Condition or disease state*	Treatment/ agonist/ condition	Tissue/cell analysed (WA = whole animal; CC = cultured cell)	PTPase	mRNA/protein	Comments	Refs
Cell cycle		T cells	CD45		CD45RA isoform is significantly higher on T cell clones during the S, G_2 and M stages compared to G_0 and G_1; also, T cells in process of mitosis following PHA treatment coexpress both isoforms CD45RA and CD45RO	167
		Fibroblasts	cdc25		Expressed consitutively throughout the cell cycle without major changes in protein level; found localized to nucleus during interphase and early prophase; redistributed throughout the cytoplasm at time of nuclear envelope breakdown where protein remains during metaphase and anaphase	152
	Synchronized cells	HeLa cells; CC	cdc25	↑ mRNA	mRNA levels very low in G_1 and increase 4-fold in G_2 as cells progress towards M phase	195
		Schizosaccharo-myces pombe	$cdc25^+$		$cdc25^+$ is required for M-phase activation; abundance of $cdc25^+$ oscillates in the cell cycle, peaking at G_2/M transition	142
	Synchronization by serum starvation	Rat fibroblast NRK cells; CC	cdc25A		mRNA expressed in G_1 3–6 h after stimulation and peaked at 9 h, just before onset of S phase; protein begins to appear at 3 h, peaks at 6 h and decreased in early S phase; suggests that cdc25A is not a mitotic inducer, but has a function in early cell cycle	161
		Rat fibroblast NRK cells; CC	cdc25B		mRNA induced at 12 h and peaked in G_2 phase	161
		HeLa cells; CC	cdc25Hu2		Expressed throughout cell cycle with a moderate ↑ in G_2	568
		Swiss 3T3 fibroblasts; CC, 15–30 h	cdc25M2	↑ mRNA	mRNA low from G_0 to G_1, then ↑ at S phase and reaches a plateau	535
		HeLa; CC, 3–12 h	Cdi1	↑ mRNA	Expression ↑ at G_1 to S transition and forms stable complexes with cdk2	515
		NIH 3T3	TCPTP (MPTP)	Dynamic changes in mRNA levels observed	From baseline in G_0, ↑ 13-fold post-induction (at late G_1 phase), then decreased rapidly during S phase (12–18 h), followed by ↑ again in G_2; confirmed in hydroxyurea and nocodazole-arrested cell studies	207

(mitosis)		Pre-B lymphoma cells and T-lymphocyte cell lines; CC	CD45	↑ mRNA	CD45 PTPase activity ↑ two to 10-fold in late $G_2 + M$ phase cells with only minor ↑ in protein level; peak PTPase activity occurred in late mitosis or in cytokinesis	174
Cell density		Fibroblasts; CC	DEP-1	↑ mRNA	Expression ↑ dramatically with ↑ cell density with commensurate ↑ in PTPase activity in immuno-complexes; as control, PTP1B expression did not change in cell density; increase in expression is initiated before saturation cell density; DEP-1 may contribute to the mechanism of contact inhibition of cell growth	576
		HUVE cells; CC	HPTP-β	↑ mRNA	HUVE cells are contact inhibited; when harvested at high density, they displayed a 10-fold ↑ in membrane-associated PTPase activity which appeared to be specific for cell–cell-contact-directed growth arrest; PTP-γ and HPTP-β were exclusively expressed in HUVE cells; only HPTP-β displayed a pattern of expression related to the increase in PTPase activity, a 12-fold ↑ in HPTP-β mRNA expression was detected which parallelled the time course of PTPase activity	940
		3T3 fibroblasts; CC, days	LAR		Changes in alternative splicing of multiple RNA transcripts seen in cells reaching confluence	980
Cell division		HeLa cells; CC	cdc25C	↑ Protein	cdc25C immunoprecipitated from HeLa cell extracts had a 12-fold ↑ in PTPase activity in mitotic (nocodazole-arrested) than from asynchronous cells; this is due to enzyme activation, since the same amount of cdc25C protein was detected in both conditions; mitotic cdc25C had $M_r = 59$ kDa while a 56–59 kDa doublet was detected in asynchronous cells; treatment with PP-2A caused a reappearance of the 56 kDa polypeptide	213
Cell growth	EGF or isoproteronol	Rat parotid gland acinar cells; WA	SH-PTP2	↑ Protein (Activity and Tyr(P) level)	Cell lysates from EGF- or isoproterenol-stimulated parotid glands showed ↑ immunoprecipitable PTP1D activity; the immuno-precipitated protein of $M_r = 65$ kDa also had elevated levels of tyrosine phos-phorylation following isolation from cells treated to undergo proliferation	186

*Includes development, differentiation, etc.

Table 6 Continued

Condition or disease state*	Treatment/ agonist/ condition	Tissue/cell analysed (WA = whole animal; CC = cultured cell)	PTPase	mRNA/protein	Comments	Refs
Cell proliferation/ contact inhibition		Normal and v-erbB-transformed NIH 3T3 fibroblasts; CC	Low M_r PTPase (CPTP)	No change in mRNA; protein change associated with cell proliferation	Level of low M_r PTPase in normal and v-erbB-transformed NIH 3T3 fibroblasts by ELISA was inversely related to cell proliferation; normally growing cells had less enzyme than the contact-inhibited cells; v-erbB transformants had less than control cells	189
Cell proliferation	Serum	CC	PTP-S	↑ mRNA, ↑ protein	mRNA levels increased reaching a maximum during late G_1 phase and then declined; no change in mRNA levels was observed in growing cells during cell cycle; PTP-S protein levels also increased with mitogenic stimulation	956
Cell–cell contact		Mink lung epithelial cells or transfected 3T3 cells; CC	RPTP-μ	↑ Protein	RPTP-μ has a relatively short half-life (3–4 h) and undergoes post-translational cleavage into two non-covalently associated subunits; shedding of the ectodomain subunit is observed in exponentially growing cells; surface expression of RPTP-μ is restricted to regions of tight cell–cell contact; RPTP-μ surface expression increases significantly with increasing cell density and is independent of its catalytic activity and occurs post-transcriptionally; cell–cell contact appears to cause RPTP-μ to be trapped at the surface resulting in accumulation at intercellular contact regions	941
Cell trans-formation		T cell proliferative disorders	CD45		In CD4(+) T cell lymphoproliferative disorders, the CD45R(−) phenotype is prevalent	157
	Viral infection	Fibroblasts; CC	cdc25Hu2	↑ mRNA	Expression ↑ five to 10-fold in cell transformed by SV40 virus or human papilloma virus	568
Colorectal carcinoma cells		DLD-1 colon carcinoma cells; CC	PTP-PEST (PTPG1)	Altered mRNA transcript sequences by RT-PCR analysis; protein shows changed Western blot expression pattern	Three aberrant mRNAs found: one missense point mutation and 77 bp and 173 bp deletions; deletions caused frameshift and premature termination causing truncated protein; C-terminal truncated proteins may be expressed in DLD-1cells	609

Contact-inhibited cells		Swiss 3T3; CC	LAR	↑ mRNA, 2-fold (8 kb); no protein change	50% *vs.* 100% confluent cells	170
Cytokine stimulation	IL-1, IL-6	Rat mesangial cells; CC, 12 h	RPTP-α (LRP)	Marked ↓ mRNA of mRNA expression; no protein change		565
Denervation		Rat skeletal muscle; WA, weeks	LAR		Alternative splicing in the extracellular domain of multiple RNA transcripts	980
Development		Leech embryos; WA	cdc25		Lineage-dependent mechanisms of regulation of the expression of cdc25 in leech embryos; cdc25 RNA levels fluctuate during the cell cycles of the micromeres, levels peak during early G_2, due to a burst of zygotic transcription, and then decline as the cell cycles progress; cells of different lineages employ different strategies of cell cycle control	930
		Mouse embryogenesis; WA	cdc25A and cdc25B		In preimplantation embryos, cdc25B transcripts are expressed in the one-cell embryo, declined at the two-cell stage and are re-expressed at the four-cell stage; cdc25A is not expressed in the preimplantation embryo until the late blastocyst stage of development; both cdc25A and cdc25B transcripts are expressed at high levels in the inner cell mass and the trophectoderm, which proliferate rapidly prior to implantation	976
		Xenopus oocytes; CC	CL100	↑ mRNA	During early embryogenesis, levels of XCL100 mRNA are greatly increased at the mid-blastula transition, suggesting that this enzyme may be involved in the regulation of MAP kinase activity during early development	1108
		Drosophila melanogaster	DLAR		Single mRNA of 8.2 kb detectable throughout embryogenesis and present at highest level at 9–15 h (period of high axon outgrowth)	614
		Drosophila melanogaster	DPTP10D		Three major mRNAs, of 6.3, 7.1 and 8.2 kb; the 6.3 kb mRNA is expressed throughout embryogenesis and present at highest levels at 9–12 h embryos; the two larger transcripts are detectable at 9–18 h	614

*Includes development, differentiation, etc.

Table 6 Continued

Condition or disease state*	Treatment/ agonist/ condition	Tissue/cell analysed (WA = whole animal; CC = cultured cell)	PTPase	mRNA/protein	Comments	Refs
Development		*Drosophila melanogaster*	DPTP99A		Three prominent mRNAs of 4.8, 6.2 and 7.3 kb; two largest are present at highest levels at 9–15 h, similar to DLAR and DPTP10D; however, the 4.8 kb mRNA is primarily observed in embryogenesis and expressed at highest level at 0–3 h; the 4.8 kb mRNA may encode a DPTP99A lacking most or all of the extracellular domain sequences present in the 6.3 kb mRNA, thus might encode a cytoplasmic PTPase or an enzyme with an altered extracellular domain	614
		Lymphocytes; CC	CD45		Alternative isoforms selectively expressed on haematopoietic cells at various stages of development; characterized the various isoforms of CD45 in CD34+ cell subpopulations from bone marrow and long-term culture	134
		Drosophila oocytes; WA	DPTP99A, DPTP10D, *csw*, DPTP4E, DLAR, DPTP61F, DPTP69D		Diverse expression patterns were observed in the ovary; DPTP99A and *csw* transcripts are expressed in follicle cells; *csw* and DPTP10D are detected in the germline during oogenesis and localized to the oocyte during egg chamber development; DLAR and DPTP4E transcripts are found in the germline during the same developmental stages as DPTP10D transcripts, but the transcripts are not localized to the oocyte; DPTP61F transcription is detected only after stage 6 of oogenesis; DPTP69D transcripts are sequestered in the nucleus from stage 7 to stage 10, and then released to the cytoplasm	938
		T cells	CD45		Various isoforms of CD45 are expressed differentially on T cells at different stages of activation and development; suggested isoform switching pattern CD45RA(−)RO(−) or RA(+)RO(−) to RA(−)RO(+) to RA(+)RO(−) and finally from CD45RO to CD45RA as final step in thymus	150

Drosophila melanogaster; WA	cdc25Dm (string)		mRNA accumulates in latter half of interphase before mitosis phase 14 in cells destined to divide; zygotic expression occurs in a dynamic series of spatial patterns suggesting zygotically driven cell divisions	498
Mouse; WA	cdc25M2		cdc25M2 mRNA shows variation in expression in different tissues in the mouse embryo and is expressed in a developmental and cell-cycle-dependent fashion	535
Mouse; WA	cdc25M2	↑ mRNA	Peaks at day 12.5–13.5 of embryonic development	535
Drosophila melanogaster	CSW		Three major developmentally regulated mRNAs; most abundant is 4.7 kb which is expressed throughout embryonic larval, pupal, and adult stages; two larger transcripts at 6.0 and 7.2 kb are first observed during late embryogenesis; ISH shows that csw is widely expressed uniformly throughout all tissues during embryogenesis	580
Dictyostelium discoideum; CC	DdPTP1	↑ mRNA; no protein change	2.0 and 2.4 kb mRNAs, expressed at very low level in vegetative cells, are induced at four h, and maximally expressed at the tight aggregate stage and through the remainder of development; at 10–12 h; two new transcripts of 1.8 and 2.2 kb appear that were also then expressed for the remainder of development; all transcripts appeared to arise from single PTP1 gene	277
Dictyostelium discoideum; CC	DdPTP2	↑ mRNA; no protein change	Single mRNA of 2.5 kb was expressed at a moderate level in vegetative cells; induced several fold to maximal levels at 8 h and persisted throughout the remainder of development	523
Dictyostelium discoideum; CC, 1.5–19 h	DdPTP2 (DdPTPa);	↑ mRNA; no protein change	Four species of mRNA; at least three are developmentally regulated; 1.7 kb mRNA low in dormant spores, increasing from germination at 1.5 h until multicellular development at 19 h; additional mRNAs 2.3, 3.4, 5 kb, of which the 2.3 and 3.4 kb mRNAs also appeared to be developmentally regulated	588

*Includes development, differentiation, etc.

Table 6 Continued

Condition or disease state*	Treatment/ agonist/ condition	Tissue/cell analysed (WA = whole animal; CC = cultured cell)	PTPase	mRNA/protein	Comments	Refs
Development		*Drosophila melanogaster*	DPTP10D		Three developmentally regulated polyadenylated transcripts of 5.6, 6.5 and 7.8 kb; two polyadenylation sites are utilized; Exon 10 is present in the 7.8 kb mRNA but not the other two mRNAs and produced the A and B isoforms noted in Table 2; Protein expression found in the longitudinal connectives and transiently in the horizontal commissures	625
		Drosophila melanogaster	DPTP4E		Produces three major developmentally regulated mRNA transcripts of 6.5, 7.0 and 7.5 kb; widely expressed during embryogenesis, including the developing nervous system; largest transcript (7.5 kb) accumulates in the adult head and is not present elsewhere or during any other developmental stage; 7 kb mRNA accumulates to a higher level than the 7.5 or 6.5 kb mRNAs; it is present in 0–6 and 6–12 h embryos and in the adult body and is reduced in 12–24 h embryos, the second half of pupal development and in adult heads; 6.5 mRNA is present in 6–12 and 12–24 h embryos, 1st instar larvae, the second half of pupal development and adult head and body; by ISH with RNA probes, high level of ubiquitous expression was observed	575
		Drosophila melanogaster	DPTP99A		Two mRNAs of six and 6.5 kb found in imaginal discs, only the 6.0 found in 0–20 h embryos; also two transcripts of five and 3.7 kb are found in embryos and not prominent in RNA from imaginal discs, suggesting tissue-specific processing; antibodies detect protein staining in axons of CNS but not in nerve roots or axons of peripheral nervous system	520

Drosophila melanogaster	DPTP99A		Gene produces at least seven polyadenylated and developmentally regulated mRNAs, 3.9, 4.4, 5.4, 6, 6.4, 6.9 and 7.9 kb, detectable using a probe for the PTPase domains; 6 kb mRNA is specifically expressed in the adult head; the 6.9 and 5.4 kb mRNAs contain exon 12; exon 12 absent from the 6.4 kb mRNA (predicted smaller protein of 120 kDa); mRNA expression coincides with onset of axonogenesis and is expressed in several pioneer neurons for the intersegmental nerve, including aCC and RP2; and its protein is transiently expressed in the intersegmental and segmental nerves; protein expression found on axons of all major CNS nerve tracts during axon outgrowth as well as some neuronal cell bodies	625
Rat brain; WA, P0–P21	LAR	(ISH); no protein change	Developmentally regulated in a region-specific manner (relatively constant in brainstem, ↑ 6-fold in cortex and ↓ 2-fold in cerebellum during postnatal development	170
Rat brain; WA	LAR, LRP, RPTP-γ, CPTP1, PTP-1, P19-PTP, SHP		Only CPTP1 is preferentially expressed in neural tissues; the seven PTPases have overlapping, yet unique, distributions during development: LAR mRNA is highly expressed in cortical cells at E17 and in all layers of the cortex at postnatal day 4; RPTP-γ mRNA is expressed by postmitotic neurones in the embryo and predominantly by neurones in the superficial layers of the postnatal cortex	962
Fetal rat lung and selected adult rat tissues; WA	LAR, RPTP-δ		LAR mRNA and protein are expressed exclusively in the lung epithelium; in early embryonic lung (day E13 to E18) LAR is expressed by all of the epithelial cells of the forming bronchial tree; as the lung matures LAR gene expression is confined to the bronchiolar Clara cell and the alveolar type II cell; LAR gene products were also found abundantly expressed in epithelial progenitor cells of adult oesophagus, skin, and small intestine, in contract RPTP-δ is specifically expressed in the mesenchyme of the developing lung; the level of RPTP-δ mRNA decreases as the lung matures	947

*Includes development, differentiation, etc.

Table 6 Continued

Condition or disease state*	Treatment/ agonist/ condition	Tissue/cell analysed (WA = whole animal; CC = cultured cell)	PTPase	mRNA/protein	Comments	Refs
Development		Rat brain (cortex); WA	LAR-PTP2 (CPTP1 CPTP3)	7.8 and 6.5 kb mRNAs, variable; no protein change	7.8 kb mRNA is predominant embryonic transcript with highest levels at E16, falls by P4 and is relatively constant thereafter; 6.5 kb mRNA not detected in embryo, appears at birth and rises in early postnatal days to adult levels; 6.5 kb mRNA appears to be specific to neural tissue; in embryo, mRNA widely distributed, after birth find highest levels in neocortex, hippocampus, and cerebellum; most abundant in areas of cellular proliferation; expressed by progenitor cells and in developing neurons	197
		Rat lung epithelial cells; CC/ lung tissue, Gestational day 20	LAR-PTP2	↑ mRNA, ↑ protein	Expressed strongly in primary cultures of lung epithelial cells and not fibroblasts; peaks at day 20 with gradual decline just before birth; also apparent in whole lungs; protein level also ↑ to peak on day 21 by immunoblotting from epithelial cells; relatively high expression continues in epithelial cells of newborns through at least day 11, but expression of mRNA as well as protein is absent in adult lungs; not expressed in mature adult type II alveolar cells, but was found in transformed type II-like A549 cells	193
		Chicken brain; WA	LAR-PTP2 LAR-PTP2B (CRYP-α1 CRYP-α2) *in situ*	mRNA variable; no protein change	Expression high in motor neurons and in optic tectum and hypothalamus with evidence of tissue-specific and developmentally regulated exon use suggesting stage-specific neural roles	604
		Chicken brain; WA	LAR-PTP2 LAR-PTP2B (CRYP-α1 CRYP- α2)	mRNA variable; no protein change	7.7 kb mRNA (CRYP-α2) was highest between E4 and E8, then decreased rapidly; the 6.6 kb mRNA (CRYP-α1) was low before E6, peaked at E10–12, and then stabilized at an intermediate level at E14	604

Lymphocytes	LAR-PTP2B (mRPTP-σ)			Highest expression observed in CD4− CD8− cells, lower in CD4+ CD8+ cells and marginal expression seen in single positives; no Δ with Con A activation of lymphocytes	179
Lung; WA	MKP-1 (ERP)	Altered mRNA expression		Abundant in adult lung, but low in late mouse embryos; *in situ* hybridization demonstrated increase in ERP expression in lung rapidly after birth to the adult level at 1 day of age	571
	Mouse tissues and brain regions; WA	MKP-1 (ERP)		By *in situ* hybridization analysis, ERP expression is observed only in maternally derived decidual cells surrounding the developing embryo; at day 10.5, erp is weakly expressed in the embryo in the neural tube, hind gut, and other embryonic structures, but ↑ expression in 12.5-day embryos in most organs, with the highest expression restricted to the developing neural system; high levels are maintained in several parts of adult brain, such as cortical layers, thalamus, hypothala-mus, and hippocampus as well as in the cerebellar cortex, in the Purkinje cell layer, and in the granular cell layer; the expression of erp corresponds to regions undergoing terminal cell differentiation and/or regions where cell proliferation has declined	131
	Rat brain (cortex); WA	PTP-PEST (CPTP2 P19-PTP)	3.5 kb mRNA, variable; no protein change	Highest expression in developing cortex at E16, declines at birth, remains at adult levels after P4; level in adult cortex similar to that in liver, kidney and spinal cord; widely distributed in embryo with highest *in situ* signals in CNS, dorsal root ganglia and liver; after birth, neocortical expression is ↓ with expression persisting in the cerebellar cortex	197
	Rat lung epithelial cells; CC, Gestational day 20	PTP1B	↑ mRNA	Expressed equally in primary cultures of lung epithelial cells as well as fibroblasts; increase in expression was observed between day 17 and 20, but unlike LAR-PTP2, expression did not fall on days 21–22 and there was no developmental regulation in fibroblast cells	193
	Murine thymocytes; CC	PTP36	↑ mRNA	Expression higher in CD4+ CD8+ thymic cell fraction, suggesting developmental regulation	597

*Includes development, differentiation, etc.

Table 6 Continued

Condition or disease state*	Treatment/ agonist/ condition	Tissue/cell analysed (WA = whole animal; CC = cultured cell)	PTPase	mRNA/protein	Comments	Refs
Development		Mouse brain and immune system; WA, days–weeks	RPTP-δ		Four isoforms (A–D) that differ in alternative splicing of Ig and Fn-III repeats in extracellular domain are differentially expressed; in brain, only type D expressed until P7, but after P14, all isoforms are detected; in haematopoietic cell lines, only types B and C are expressed	954
		CNS; WA	RPTP-κ	Altered mRNA expression pattern; no protein change	mRNA ↑ in actively developing areas in embryogenesis and persists in adult cortex and hippocampus	529
		Rat brain; WA	RPTP-σ		By Northern blot analysis and *in situ* hybridization, the 6.9 kb mRNA transcript was more abundant during embryonic period, and the 5.7 kb mRNA was more abundant during postnatal development and in the adult; RPTP-σ was widely expressed throughout the central and peripheral nervous system during embryonic development with high levels seen in the ventricular zone, subventricular zone, cortex, dorsal root ganglia, cranial nerve ganglia, olfactory epithelium, and retina; expression generally decreased in most brain regions during postnatal development with high levels continuing in the hippocampus; the majority of RPTP-σ mRNA is expressed in neurones; cultured glial cells expressed only the 6.9 kb	975
		Mouse embryos; WA	TCPTP (MPTP)	↑ mRNA	Expression was low in embryonic stem cell line D3 and dramatically increased in 8.5 to 11.5 day embryos	567

Diabetes mellitus	Streptozotocin-induced	Skeletal muscle and liver	LAR PTP1B SH-PTP2	↓ LAR, ↑ PTP1B, SH-PTP2	Abundance of LAR was ↓ to 65% of control, parallel to changes in the measured PTPase activity in the particulate fraction; in contrast, SH-PTP2 and PTP1B were both significantly increased in diabetes; SH-PTP2 also showed an increased ratio of particulate/cytosol distribution in diabetic tissues (1.8–1.9) that was reversed after insulin treatment (0.79–0.95); Northern analysis suggested that the PTPases were regulated at a pretranslational level	118
		Diabetic KKA(y) mouse adipose tissue and liver; WA	SH-PTP2 (Syp)	↑ Protein	Syp protein was dramatically elevated in KKA(y) fat, with less striking changes in liver	931
Differentiation	Retinoic acid	Mouse embryonal carcinoma F9 cells; CC, up to 96 h		↓ SHP ↑ LRP, PTP-β2, RIP, PTP-ϵ, and PTP-μ	Expression of several PTPases was unchanged, including MPTP, CDC25C, Cdc25M2, PTP-κ, PTP-S, SH-PTP2 and STEP	210, 211
		Normal and malignant myeloid cells; CC	CD45		CD45RO (the low M_r isoform) was very dimly expressed on immature cells but became increasingly brighter beginning at approximately the myelocyte stage	129
		T cells	CD45		Clonogenic T cells are CD45 p180(−), while the majority of differentiated T cells are p180 (+) and have a commitment to intrathymic death	183
		B-cells	CD45		During B cell differentiation into plasma cells, expression of high M_r isoform shifts to low M_r isoform (180 kDa)	160
	DMSO	HL-60 leukaemia cells; CC, 1–2 d	CD45	↑ mRNA (4 to 11-fold), ↑ protein	Monocytic or granulocytic differentiation of HL60 cells is accompanied by a rapid ↑ in CD45 expression (both 180 and 200 kDa isoforms) and PTPase activity; due to ↑ gene transcription	206

*Includes development, differentiation, etc.

Table 6 Continued

Condition or disease state*	Treatment/ agonist/ condition	Tissue/cell analysed (WA = whole animal; CC = cultured cell)	PTPase	mRNA/protein	Comments	Refs
Differentiation	PHA	T cells; CC	CD45		After stimulation with PHA, T cells lose expression of CD45R (Lp220; 220 kDa isoform) without loss of CD45 isoforms detected by an antibody to determinants common to multiple CD45 isoforms; loss of CD45R (220 kDa) occurs in post-thymic T-lymphocyte differentiation; this may be associated with enhanced ability to respond to antigen stimulation	199
	PMA, 1,25-dihydroxy vitamin D3, DMSO or cyclic AMP analogues	HL-60 promyelocytic leukaemia cells, chronic myelogenous leukaemia cells, a monocytic leukaemia (THP-1), and a monoblastoid leukaemia (U937); CC	CD45		Differentiation was associated with an ↑ in total cellular PTPase activity, CD45-specific PTPase activity, CD45 cell surface expression, and synthesis of both exon B-dependent M_r 205 kDa and exon ABC-M_r 185 kDa CD45 proteins; PMA transiently reduced CD45 PTPase activity in the chronic myelogenous leukaemia cell RWLeu4 without altering the CD45 amount	126
		3T3-L1 preadipocytes	HA2		With adipocyte differentiation, expression of PTPase HA2 increases abruptly and then decreases concomitant with the transcriptional activation of adipocyte genes	1012
		Mouse erythro-leukaemia MEL cells	Multiple		Pattern of most PTPase transcripts are unchanged (including LRP, SHP, cdc25) or ↑ early then were ↓ (CD45, PTP-ϵa); transcripts for two enzymes, PTP-$\beta 2$ and RIP were induced during early differentiation and remained elevated; mutant cell lines defective in differentiation did not show induction of the PTPase transcripts	165

PMA	U937 human leukaemia cells; SS	Multiple	↑ mRNA	Cellular PTPase activity (especially in cytosol) in U937 cells ↑ up to 2-fold during the course of monocytic differentiation; using RT-PCR, cloned 13 PTPase-related gene fragments from the differentiated cells including two novel isozymes (PTP-U1 and PTP-U2); nine isozymes (both cytosolic and transmembranous) were expressed more in the differentiated cells than in the undifferentiated cells; PTP-U1 and PTP-U2 were greatly ↑ by PMA	198
	Rat osteoblasts; CC	OST-PTP	↑ mRNA	Expression ↑ in differentiating cultures and ↓ in late-stage mineralizing cells; an alternatively spliced variant of 4.8–5.0 kb, which may lack PTPase catalytic domains, is found in proliferating osteoblasts, but not other stages	557
	P19 embryonal carcinoma cells; CC	PTP-PEST (P19-PTP)	↑ mRNA expression	mRNA expression constitutively enhanced after aggregation of embryoid bodies from P19 cells, little change seen after differentiation inducing agents RA or DMSO	138
Sodium butyrate	Colon cancer Colo 320 cells; CC, 5 days	PTPH1 (4.3 kb)	↑ mRNA	PTPH1 mRNA was not detected in Colo 320 cells that overexpress c-*myc* mRNA, among the colorectal cancer cell lines examined; when Colo 320 cells were incubated with 5 mM sodium butyrate for five days, PTPH1 mRNA became detectable, concomitant with the marked decrease in the expression level of c-*myc* mRNA	
PMA, vitamin D_3, DMSO, retinoic acid	Various; CC	PTP-U2	↑ mRNA		1120
	Embryonal carcinoma and neuroblastoma cells; CC	RPTP-α	↑ mRNA expression	mRNA expression transiently enhanced during neuronal cell differentiation	139

*Includes development, differentiation, etc.

Table 6 Continued

Condition or disease state*	Treatment/ agonist/ condition	Tissue/cell analysed (WA = whole animal; CC = cultured cell)	PTPase	mRNA/protein	Comments	Refs
Differentiation	TPA (12-*O*-tetradecanoyl-phorbol-13-acetate) or 1,25-$(OH)_2D_3$	HL-60; CC, 48 h (mRNA); 72 h (protein)	SH-PTP1 (PTP1C)	↑ 3-fold (2.4 kb transcript); also PTP1C gene transcription rate (nuclear run-off) ↑ 3.5-fold; ↑ PTP1C protein immuno-precipitated from metabolically labelled cells	TPA induced HL-60 cell differentiation to monocytic/macrophage phenotype, ↑ PTP1C PTPase activity in immune-complex assays by 2.3-fold; similar treatment with 1,25-$(OH)_2D_3$ ↑ PTP1C activity 1.5-fold; in contrast, treatment with DMSO induced granulocytic differentiation without changing PTP1C PTPase activity; PTP1C mRNA half-life did not change with TPA treatment (3.3–3.6 h)	212
	PMA	HL-60; CC, 23–35 h	SH-PTP1 (PTP1C)	No mRNA change; ↑ enzyme activity and expression of PTP1C (2- to 3-fold); associated with ser-p of PTP1C and translocation from the cytosol in untreated cells to 30–40% found in the plasma membrane during differentiation	Induction and translocation of PTP1C correlates with differentiation; PMA-induced differentiation is blocked by orthovanadate, suggesting critical involvement of PTPases in this process	226
Genetic stress	Ultraviolet light; methylmethane sulfonate	HeLa cells; CC, 60–120 min	MKP-1	↑ mRNA, >10-fold	Increased expression of MPK-1 mRNA is associated with a decline in JNK activity and also decreases activation of AP-1-dependent reporter gene expression	1014
Growth factor stimulation	EGF, PDGF, bFGF, TPA, serum	Human fibroblasts; CC, 2–4 h	PTP1B	Induction of a 4.7kb mRNA transcript; protein not studied	New transcript appears in addition to typical 3.5 kb mRNA; results from retention of last intron and encodes protein with altered C-terminal seven amino acids of the human protein making the sequence identical to the known rat protein; the new mRNA is found on polyribosomes; variably expressed in human tissues	201

Growth factor treatment	NGF	PC12; CC, 3 days	LAR	↑ mRNA, 1.8-fold (8 kb); no protein change	No change in transcript size	170
Growth factor stimulation	NGF, EGF	PC12 cells; CC	MKP-2	↑ mRNA	NGF treatment ↑ MKP-2 mRNA levels to a maximum of 5-fold by 2 h; also increased 4-fold by EGF at 1 h	1111
	Insulin, TPA, PDGF, α-thrombin	CCL39, a Chinese hamster lung fibroblast cell line; CC	SH-PTP-1 (PTP1C)		PTP1C activity was significantly stimulated by insulin and TPA but was not influenced by serum, PDGF, or α-thrombin	961
	Insulin, TPA, PDGF, α-thrombin	CCL39, a Chinese hamster lung fibroblast cell line; CC	SH-PTP2 (PTP1D)		PTP1D activity was slightly stimulated by insulin and TPA but was significantly inhibited by serum, PDGF, and α-thrombin	961
Growth stimulation and arrest		3T3 fibroblasts; CC	RPTP-σ	↑ ↓ mRNA	Expression of RPTP-σ mRNA was low in actively cycling cells; in cells whose growth was arrested by contact inhibition at high cell density or by serum starvation at low cell density RPTP-σ mRNA level increased	932
Haemato-poiesis		Bone marrow cells			CD45 (T-200) had differential expression on various lineages during normal haematopoietic development	200
Heat shock		Human fibroblasts	B23	↑ mRNA	Similar to CL100	525
		Saccharomyces cerevisiae; CC, 4–10 h	Yeast PTP2	↑ mRNA	2.7 kb mRNA inducible by heating at 39°C for 30 min	316
HIV-1 infection		Lymphocytes; CC	CD45	Altered glycosylation	CD45 was partially sialylated on N- and *O*-linked carbohydrates on infected cells; may affect immune response signalling	95
IGF-I		Rat L6 myocytes; CC, 12 h	PTP1B	↑ mRNA, 12.5-fold; no protein change		163
		Rat L6 myocytes; CC, 24 h	PTP1B	↑ mRNA 6.5-fold; no protein change	Associated with increased PTPase enzyme activity to 281% of control at 32 h	163
IL-2 stimulation		L3 lymphocytes; CC	cdc25M1	↑ mRNA; no protein change	Undetectable in resting cells; ↑ 48 h after IL-2 addition (S to G_2 transition); other PTPases tested, including CD45, PEP, SHP and RPTP-α, were not regulated during the cell cycle	569

*Includes development, differentiation, etc.

Table 6 Continued

Condition or disease state*	Treatment/ agonist/ condition	Tissue/cell analysed (WA = whole animal; CC = cultured cell)	PTPase	mRNA/protein	Comments	Refs
IL-2 stimulation (Cont.)		IL-2 dependent human T cell line ILT-Mat; CC, 9 h	HePTP (LC-PTP)	↑ mRNA 6-fold; no protein change	Peak occurs simultaneously with $G_1 \rightarrow S$ phase transition; effect does not require protein synthesis; run on assays show effect is due to transcriptional activation; little change in PTP1B expression with IL-2 stimulation; suggests HePTP is an early response gene	117
IL-5 stimulation		B cells; CC	CD45		Produced a hyaluronate-adherent subpopulation of cells expressing predominantly lower M_r forms compared to the high molecular mass forms (220 kDa) expressed in unstimulated or LPS-stimulated B cells; PCR analysis demonstrated changes in alternative mRNA splicing	154
Insulin		Human A204 rhabdomyo-sarcoma cells; CC, 36 h	PTP-PEST	↑, mRNA 4-fold; no protein change	Effect persisted for five days	622
		Rat Fao hepatoma cells; CC, 3 h	PTP1B	↑ mRNA, 1.6-fold (4.3 kb transcript) and 3.1-fold (1.6 kb transcript); no protein change	Expression of mRNA for RPTP-α or LAR unchanged	153
		Rat L6 myocytes; CC, 12 h	PTP1B	↑ mRNA, 5.5-fold; no protein change		163
		Rat L6 myocytes; CC, 24 h	PTP1B	↑ Protein, 4.3-fold	Associated with increased PTPase enzyme activity to 192% of control at 32 h	163
Ionomycin treatment		Murine thymocytes; murine T cell lines; CC, <20 min	CD45		Calcium ionophore treatment ↓ the PTPase activity of CD45, associated with ↓ p-ser of CD45	315
Leucocyte activation	Stimulation with human C5a	Human peripheral leucocytes; CC	CD45	↑ Protein	In response to the anaphylatoxin C5a, increases in the membrane expression of CD45/CD45RO occurred on granulocytes and monocytes	218
Loss of tumour suppression		Various; CC	RPTP-γ	↓ mRNA; no protein change	Loss of one RPTP-γ allele occurred in 3/5 renal cell carcinoma cell lines and in 5/10 lung carcinoma tumour samples tested; can occur without loss of flanking loci	542

Lymphocyte stimulation	PHA, Con A, LPS, anti-CD3	Normal mouse lymphocytes; CC, 24–72 h	HePTP	↑ mRNA, 10- to 15-fold; no protein change		630
Malignancy						
		Rat macrophage tumour cells, AK-5; CC	TCPTP (PTP-S)	↓ mRNA, 4-fold	Genomic organization of PTP-S is unaltered in AK-5 cells; PTP1B is also downregulated	203
		Breast cancer cells; WA	cdc25	↑ mRNA	Overexpression of cdc25B was detected in 32% of human primary breast cancers tested; the cdc25 phosphatases may contribute to the development of human cancer	1000
Mammary tumours		Human tissue samples	PTP1B	↑ mRNA in 8/9 tumours tested; ↑ protein in 72.4% of 29 tumours	Also associated with overexpression of $p185^{c\text{-}erbB\text{-}2}$	220
Mitogen and growth factor stimulation	Bombesin, EGF, TPA, cAMP, FGF	NIH 3T3 cells; CC, 1 h	MKP-1 (ERP)	↑ mRNA abundance; ↑ protein abundance (closely follows mRNA stimulation)	Effect greatest with bombesin > cAMP > FGF; EGF and TPA effect more transient and weaker (30 min); $t_{1/2} = 2$ h	571
Mitogen treatment	Serum treatment of quiescent cells	NIH 3T3 fibroblasts; CC, 1 h	PRL-1	↑ mRNA		493
Myelodys-plastic syndrome		Neoplastic myelomonocytic cells; CC	HePTP	No mRNA change; ↑ protein	One patient exhibited triplication of HePTP gene associated with increased HePTP protein on Western blot analysis of neoplastic cells; blasts from some patients with acute leukaemia also demonstrated ↑ HePTP expression; normal myeloid cells show minimal HePTP expression while all haematopoietic cell lines tested show high expression	629
Neoplastic transformation		Rat ascites hepatoma cells; CC	PTP-S, PTP-1, RPTP-γ, RPTP-δ, LRP	↑ ↓ mRNA	Levels were determined in three lines of rat ascites hepatoma cells; mRNA expression of non-receptor PTPs (PTP-S and PTP-1) showed various neoplastic alterations, whereas mRNAs of receptor-like PTPs (RPTP-γ, RPTP-δ, and LRP) were lost or drastically decreased	949
Nitrogen starvation		*Schizosaccharo-myces pombe*; CC	$cdc25^{+}$	↓ mRNA	Protein expression drops to negligible levels	142

*Includes development, differentiation, etc.

Table 6 Continued

Condition or disease state*	Treatment/ agonist/ condition	Tissue/cell analysed (WA = whole animal; CC = cultured cell)	PTPase	mRNA/protein	Comments	Refs
Nitrogen starvation (Cont.)		*Saccharomyces cerevisiae*; CC	YVH1	↑ mRNA	mRNA is dramatically induced by nitrogen starvation, but not with carbon starvation	510
Non-insulin dependent diabetes		Human skeletal muscle	PTP1B	↓ Protein, 38%	Control group consisted of lean non-diabetic subjects; concluded that the PTP1B levels reflect *in vivo* insulin sensitivity and was reduced in the insulin resistant state	813
Obesity		Human adipose tissue; WA	LAR, PTP1B	↑ Protein	Obese subjects had a 1.74-fold increase in tissue PTPase activity; LAR was increased by 2.0-fold and immunodepletion of LAR from tissue homogenates normalized the PTPase activity towards the insulin receptor; the data suggest that PTPase activity is a pathogenetic factor in the insulin resistance of adipose tissue in human obesity	927
Obesity/ insulin resistance		Skeletal muscle from Zucker strain lean, obese and diabetic rats; WA	LAR PTP1B SH-PTP2	↑ Protein in obese and diabetic animals	In the solubilized particulate fraction of obese and diabetic animals, respectively, the abundance of LAR, PTP1B and SH-PTP2 ↑ by 42–69%; in the diabetic muscle, a shift of H-PTP2 mass to a plasma membrane component	120
Oncogenic cell transformation		CEF; CC	ChPTPλ	Altered mRNA abundance	Cells transformed by *ras* or *erb*A/B had ↑ expression of mRNA, if transformed by oncogenic *src* or *myc*, expression was ↓; little change on transformation by oncogenic *mos*, SV40 middle T antigen, *crk*, or *myc* plus *mil*	500
Oncogenic transformation	Transfection of activated *neu* oncogene	Non-tumourigenic breast epithelial cell line (184B5); CC	PTP1B CD45	↑ mRNA, ↑ protein	Increase in expression of specific PTPases including LAR and PTP1B observed in *neu*-transformed cell lines and their associated tumours; elevation noted in both protein and mRNA; TCPTP was only slightly elevated; level of PTPase expression correlated with neu expression level and tumourigenicity	224

Ovarian carcinomas		Human tissue samples	PTP1B	mRNA change was variable; ↑ protein in 79.6% of 54 tumours	Also associated with over-expression of $p185^{c\text{-}erbB\text{-}2}$, $p170^{EGFR}$ and $p165^{mCSFR}$	219
PKC activation						
	PMA	Human T cells; CC, 24–48 h	CD45	↑ Protein	After 24 h, CD45RA expression ↑, whereas CD45RO expression did not ↑ until 48 h; rapid ↑ in CD45RA is due to ↑ *de novo* protein synthesis of the 205 kDa, but not the 220 kDa isoform	223
	PMA	HeLa cells; CC	LAR	↑ Shedding of extracellular domain in protein	PMA induced LAR shedding in a dose-dependent manner (100 ng/ml) ↑ 9.5-fold shedding compared to untreated cells; shedding of the extracellular region could occur even if proprotein processing was blocked	105
	PMA	Human glomerular mesangial cells; CC	PTP1B	↓ PTPase activity	Immunoprecipitated PTP1B had ↓ enzymatic activity with no change in the protein level	94
	TPA	Rat Fao hepatoma cells; CC, 3 h	PTP1B	↑ mRNA, 4.5-fold (4.3 kb transcript) and 5.7-fold (1.6 kb transcript); no protein change	Half-life of PTP1B mRNA unchanged suggesting transcriptional effect; expression of mRNA for RPTP-α or LAR unchanged	153
Regeneration	Partial hepatectomy	Rat liver; CC	PTP-S	↑ mRNA	PTP-S mRNA levels increased 16-fold after 6 h (G_1 phase) and then declined; suggest a role of this phosphatase during cell proliferation	956
	Partial hepatectomy	Rat liver; WA	PTP-S, PTPH1, PTP-1, GLEPP1, LRP, PTP1D, PTPG1, RPTP-γ, RPTP-δ, LAR	↑ ↓ mRNA	mRNA levels determined during regeneration of rat liver, and expression patterns could be classified into four groups: for PTP-S and PTPH1, the mRNA levels increased rapidly, reached a maximum 7 h after partial hepatectomy, remained at a plateau for 1–2 days and then decreased gradually; for PTP-1, GLEPP1 and LRP, mRNA levels showed two peaks on days 1 and 5, and then decreased gradually; for PTP1D and PTPG1 mRNA levels increased rapidly, reached a maximum at 7 h, remained high for several days, and then did not decrease but rather increased after day 7; for RPTP-γ, RPTP-δ and LAR, mRNA levels remained constant for the first 5 days and increased over the control levels after day 7	949

[*] Includes development, differentiation, etc.

Table 6 Continued

Condition or disease state*	Treatment/ agonist/ condition	Tissue/cell analysed (WA = whole animal; CC = cultured cell)	PTPase	mRNA/protein	Comments	Refs
Regenerating liver		Rat liver; WA, 30 min	PRL-1	↑ mRNA		493
		Mouse liver; WA, hours to days	MPTP, PTPP-19, RPTP-α, LAR, PTP-RL9	mRNA all ↑ mRNA within 6 h then ↓ to normal by 24 h, then ↑ at 48–72 h		1104
		Mouse liver; WA, hours to days	PTP1B, RPTP-κ	mRNA peaked within 6 h then gradually ↓ to normal by 168 h		1104
		Mouse liver; WA, hours to days	RPTP-μ, PTP-RL10	mRNA peaked at 48–72 h then gradually ↓ to normal by 168 h		1104
Reversion of T cell proliferation	PHA	T cells; CC	CD45		Membrane PTPase activity was high in resting cells, but decreased during mitogen-induced blast transformation; when cells reverted, CD45 surface expression ↑ along with an ↑ in PTPase activity; vanadate treatment prevented cells from reverting to resting state	159
Serum stimulation		Human fibroblasts; CC, 1 h	B23	↑ mRNA, >5-fold		525
Spermato-genesis		Mouse testis	Multiple		Northern blot hybridization revealed that the transcripts of PTP-1B, mouse cytoplasmic PTP and PTP-δ were quantitatively and/or structurally regulated during germ cell development, suggesting a role in spermatogenesis.	162
		Adult rat testis; WA	OST-PTP		*In situ* hybridization revealed stage-specific expression	557
Stress	Oxidative stress and heat shock	Human skin fibroblasts; CC, 2 h	CL100	↑ mRNA abundance (up to 40-fold)	Response to oxidant agents including H_2O_2, menadione, ultraviolet A radiation, and also seen with heat shock	537
	Heat shock	*Drosophila* Schneider 2 cells; CC	MKP-1	↑ mRNA	Stress activates a MAP kinase-specific phosphatase in *Drosophila* and inhibits MAP kinase activity	934

T cell activation	Stimulation of quiescent T cells with serum, PHA and PMA	Human peripheral T cells; CC, 2–4 h	PAC-1	↑ mRNA abundance; ↑ protein synthesis in metabolically labelled cells		589
T cell activation	Concanavalin A	Rat splenic T lymphocytes; CC, 72 h	TCPTP (PTP-S)	↑ mRNA, 3-fold; ↑ protein	Half-life of mRNA increased from 25 min in resting cells to five h after mitogenic stimulation; treatment with cycloheximide also ↑ mRNA abundance (6-fold) and ↑ half-life in resting cells; no change in PTP1B expression observed in stimulated cells	188
T cell neoplasms		T lymphocytes; CC	CD45		The pattern of CD45 RA/RO antigen expression in most of T-lineage neoplasms could be determined by the respective stage of differentiation	948
Viral transformation	Epstein–Barr virus infection	Human B lymphocytes; CC	CD45	Altered ↑ isoform expression	Found heterogeneous expression of CD45RO and CD45RA isoforms; expression was unstable with eventual loss by some EBV-transformed lines, and loss followed by reappearance in others; CD45RA and CD45RO varied independently whereas CD45 remained stable and high, suggesting a fluctuation in other CD45 isoforms	173

* Includes development, differentiation, etc.

Biochemical studies of PTPase catalytic activity

Substrate preferences of PTPases

In attempts to gain insight into the nature of the PTPase catalytic reaction, as well as possible physiological substrates for the various PTPase homologues, a number of studies have utilized [^{32}P]-labelled phosphorylated peptide and protein substrates for direct comparison of the activities of recombinant PTPase enzymes (Table 7). As expected, differences in kinetic parameters towards these substrates have been demonstrated, although this analysis has been hampered by the relatively low stoichiometry of the labelled substrate and the difficulty in evaluating true initial rates of dephosphorylation and relative substrate affinities [245, 540]. These types of studies have also been helpful in mapping the residues involved in the catalytic reaction, and to distinguish between differences in the potential enzyme activity of the two domains of the tandem-domain transmembrane PTPases.

More recently, insight into methods that allow the incorporation of phosphotyrosine directly into peptide substrates for PTPases has enabled investigators to develop assays that use higher concentrations of homogeneously phosphorylated substrates for detailed kinetic analysis (Table 8). This work has shown that PTPases exhibit regioselectivity for dephosphorylation of specific phosphotyrosine sites on model peptide substrates. These results are similar to data obtained with radiolabelled phosphoprotein substrates which have also provided evidence for selective dephosphorylation of individual sites of a multiply phosphorylated substrate with functional consequences [273]. Other studies have used the dephosphorylation of non-radioactive substrates to evaluate the role of specific residues surrounding the tyr(P) site of the substrate which are involved in binding and substrate affinity [73].

Catalytic mechanism of PTPases

As discussed earlier, sequence homology studies and data from enzyme activation and inhibition by sulfhydryl reagents provided evidence for the presence of a catalytic cysteinyl residue in the enzyme active site. This residue has been identified also by numerous studies employing site-directed mutagenesis as well as chemical labeling and trapping of the enzymatic intermediates (Tables 9 and 10). Structural data have further confirmed the essential role of the active site cysteine and provided insight into reaction mechanisms that share common features between PTPases from diverse origins, including PTP1B, the *Yersinia* PTPase, and the low M_r PTPase. The proposed molecular interactions in PTPase catalysis have been summarized in the crystallography work (Table 4) as well as recent reviews by Dixon and his colleagues [56, 73]. Briefly, the catalytic cysteine appears to be stabilized as a thiolate anion with a relatively low pK_a by hydrogen bonding with neighbouring side-chains, especially that of the Arg residue that participates in the signature sequence motif. The reaction mechanism common to the PTPase superfamily involves the formation of a cysteinyl-phosphate intermediate, which is then hydrolysed. With the *Yersinia* enzyme, structural analysis has identified the involvement of additional specific residues in stabilising the enzyme intermediate including Glu-290; in the PTP1B sequence, this interaction corresponds to the invariant residue Glu-215. Zhang *et al.* [380] showed by site-directed mutagenesis with detailed kinetic analysis that both Asp-356 and Glu-290 in the *Yersinia* PTPase act as general acid and base catalysts responsible for PTPase-catalysed phosphate ester hydrolysis, and careful sequence alignments demonstrated that these residues are highly conserved among a wide range of PTPase homologues (Fig. 8). Thus, substrate binding involves a series of residues

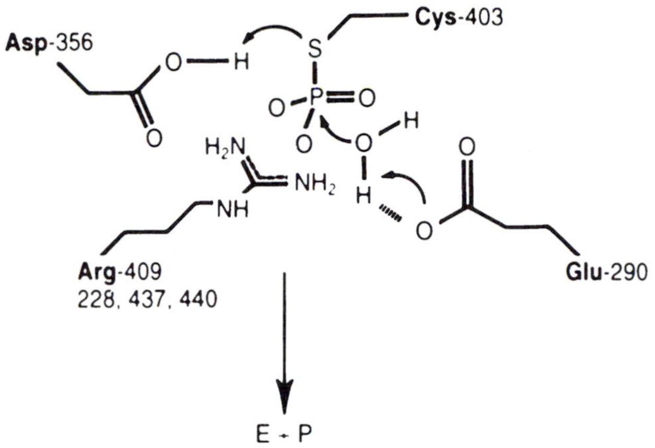

Figure 8
Model proposed for phosphate monoester hydrolysis catalysed by the *Yersinia* PTPase by Zhang *et al.* [375]. The postulated reaction mechanism involves the stabilisation of the phosphate in the binding pocket by one or more invariant Arg residues and the formation of a covalent bond between phosphorus and the nucleophilic thiol group (Cys-403) as the phosphate is transferred from the substrate to the enzyme. The phosphoryl enzyme reaction intermediate is then hydrolysed by H_2O. Glu-290 and Asp-356 also serve as general base and general acid in the catalytic reaction, enhancing protonation of the thiol phosphate to facilitate the loss of the leaving group. Reproduced with permission.

that enter the active site from a distance but are highly conserved among some members of the PTPase superfamily. Divergent sequences between these residues are undoubtedly involved in substrate recognition and determining the relative catalytic efficiency of the various enzymes *in situ*.

Of interest is the potential role that the second PTPase homology domain present in many of the receptor-type enzymes may play in substrate recognition and/or regulation of the PTPase catalytic activity. Biochemical studies have provided evidence that the second PTPase domain may completely lack activity or, in some recombinant constructs, the second domain has been demonstrated to retain a low level of activity. In the case of RPTP- and the related enzyme DPTP99A, the second PTPase homology domain has an Asp residue substituted for the catalytic cysteine, making it unlikely that this domain retains catalytic activity. It is tempting to speculate that the presence of tandem domains, only one of which contains the dominant catalytic activity, may serve to adapt the PTPase to a substrate with multiple phosphotyrosine side-chains, thereby enhancing substrate recognition and increasing the accuracy of the dephosphorylation event. By binding with high affinity to a tyrosine-phosphorylated cellular protein, a non-catalytic PTPase homology motif may also protect certain phosphotyrosine residues from dephosphorylation, as has been demonstrated with the association between SH2 domain-containing proteins and tyrosine-phosphorylated substrates [866].

Table 7 Summary of PTPase biochemical data

B23	Recombinant proteins, *E. coli*	Molecular mass; substrate dephosphorylation	*In vitro* translation product of ~44 kDa; recombinant protein expressed in *E. coli* was active PTPase, but had lower activity than VHR *in vitro;* had highest activity towards ERK-1, suggesting this might be an *in vivo* target	525
BVH1	Recombinant GST-fusion protein, *E. coli*	Substrate dephosphorylation	Efficiently hydrolysed *p*NPP and was inhibited by vanadate; however, it has 1/40 *p*NPP hydrolysing ability and 1/55 the level of Tyr(P) casein PTPase activity compared to VH1; BVH1 has especially low activity towards p-ser casein $<$ 1/2500 the activity of VH1; thus has lower, but active, p-ser hydrolysing ability	516
BVP	Cleaved, recombinant GST-fusion protein, *E. coli*		Dual-specificity PTPase like VH1 and will dephosphorylate *p*NPP with pH optimum of 8.2, with specific activity of 1.9 μmol/min/mg protein, comparable to VH1; catalytic preference demonstrated for Tyr(P) RCM-lysozyme $\gg$ Tyr(P) MBP, Tyr(P) raytide $\gg$ Tyr(P) src peptide, which was not dephosphorylated to any extent; with p-ser substrates, catalytic specificity was also demonstrated with significant activity towards RCM lysozyme $\gg$ casein, and no activity towards phosphorylase *a* or kemptide; not inhibited by okadaic acid, NaF, tetramisole or tartrate, but was blocked by vanadate and to a lesser extent, $ZnCl_2$. EDTA, spermidine and polyanionic compounds were without effect	335
	Recombinant GST-fusion protein, *E. coli*	Substrate dephosphorylation	Has PTPase activity similar to that of the Vaccinia virus H1 gene product; dephosphorylates both Tyr(P) proteins and p-ser/thr proteins; inhibited by vanadate but not by okadaic acid	538
CD45	T lymphocytes	Activation of $p56^{lck}$	*In vitro,* CD45 ↑ by $>$2-fold the kinase activity of $p56^{lck}$, effect blocked by vanadate; in cells lacking CD45 expression, $p56^{lck}$ was not activated	308
		Dephosphorylation and activation of $p56^{lck}$	$p56^{lck}$ was activated upon addition of CD45 *in vitro*, and activation was absent in T cell mutants lacking CD45; activation of lck by CD45 is time dependent, sensitive to vanadate and is accompanied by dephosphorylation of a regulatory tyrosine residue (Tyr-505) near the C terminus of lck	307
		Modulation of the binding of $p56^{lck}$ to an 11 AA Tyr(P) peptide derived from its C terminus	CD45 modulates the binding of $p56^{lck}$ to an 11 AA Tyr(P) peptide derived from the C terminus of lck; CD45 did not affect the binding of fyn, PLC-γ-1 GAP or Vav to the same Tyr(P) peptide; lck bound the peptide only after it was dephosphorylated on Tyr-505; thus, CD45 modulates the conformation of lck in a manner consistent with intramolecular regulation of the lck kinase	338
(Ly-5)	Isolated Ly-5	Substrate dephosphorylation	Using tumour cell lines expressing specific isoforms of murine Ly-5 (M_r = 180, 200 and 240 kDa), all forms of Ly-5 contain PTPase activity	339

CD45 (Ly-5)	Purified CD45	Substrate dephosphorylation	50 kDa fragment of CD45 generated by Lys-C or trypsin treatment contained < half of D1 and all of D2 and had an 8-fold increase in PTPase activity towards RCM-lysozyme; removal of 109 AA encompassing the catalytic domain of D1 produced a recombinant protein in transfected cells that was a functional PTPase; suggests that D2 may have latent PTPase activity that requires external regulation to be expressed in the native CD45 molecule	348
	Recombinant 41 kDa cytoplasmic domain of RPTP-β, *E. coli*	Substrate dephosphorylation	For RPTP-β, $K_m = 1$–$4\,\mu$M for 9–11 residue Tyr(P) peptides derived from lck, src and PLC-γ (40–200 times lower than the other PTPases tested (LAR and CD45); k_{cat} values of 30–205 s^{-1} indicating a very high catalytic efficiency; alterations at the +2 and −2 positions of model peptide substrates markedly altered K_m values and provide data for factors involved in substrate recognition	243
	Recombinant cytoplasmic domain, yeast	Substrate dephosphorylation	Dephosphorylates *p*NPP with $V_{max} = 87.5$ U/mg and $K_m = 5.5$ mM; with a Tyr(P) peptide, $V_{max} = 185$ U/mg and $K_m = 0.167$ mM	317
	Murine lymphoma cell lines	Substrate dephosphorylation	Each of the known isoforms of CD45 has equivalent basal level of PTPase activity *in vitro* with Tyr(P) angiotensin; activity associated with cytoplasmic domain; loss of CD45 increases the phosphorylation of $p56^{lck}$ on Tyr-505	314
	Purified from human spleen	Substrate dephosphorylation	Substrate preference for MBP *in vitro*; with Tyr(P) RCML, activity was ↑ up to 12-fold with spermine; Zn^{2+} is stimulatory; substrates also included IR, EGFR and $p56^{lck}$ *in vitro*	353
	Purified from human spleen	Substrate dephosphorylation	Demonstration that CD45 is a PTPase with 100-fold lower specific activity towards RCML compared to PTP1B	350
	Various isoforms from various cell lines selectively immunoprecipitated with monoclonal antibodies; recombinant cytoplasmic domains, *E. coli*	Substrate dephosphorylation	The 220 kDa CD45 isoform from human B-cell Raji cell line was an active PTPase with Tyr(P) raytide, p-ser phosphatase activity was not detected with p-ser casein; the smallest LCA isoform (180 kDa) from a mouse B-cell line (300-19) transfected with a cDNA construct encoding the 180 kDa human CD45 isoform also had PTPase activity; recombinant CD45 cytoplasmic domain activity with Tyr(P) raytide > Tyr(P) angiotensin II	606
(LCA)	Recombinant constructs of cytoplasmic domain expressed in *E. coli* extracts	Substrate dephosphorylation	More active *vs.* Tyr(P) raytide > MBP; demonstrated that a critical cysteine residue in proximal domain (Cys-828) is essential for activity; first domain is active by itself; when first domain is inactivated (Cys-828 → Ser), second domains lack intrinsic PTPase activity; catalytic activity of the proximal domain is especially influenced by interdomain sequences (AA 969–987)	343
cdc25	Recombinant human $p54^{CDC25H}$ protein, *E. coli*	Activation of $p34^{cdc2}$/cyclin B complex *in vitro*	Inactive $p34^{cdc2}$/cyclin B complex from G2-arrested starfish oocytes is dephosphorylated by the recombinant human cdc25 protein which activates its H1 histone kinase activity	342

Table 7 Continued

PTPase	Protein source	Parameter	Data	Refs
cdc25	Recombinant *Drosophila* cdc25 protein, *E. coli; in vitro* translation product from reticulocyte lysates	Dephosphorylation of $p34^{cdc2}$ *in vitro*	Relatively specific interaction with $p34^{cdc2}$ leading to dephosphorylation of Tyr-15, less active towards other Tyr(P) substrates; highly specific for Tyr-15 in p34	262
	HeLa cell extract; *in vitro* translation	Molecular mass	Two proteins of M_r 55 and 52 kDa	303
	Recombinant GST-fusion protein of C-terminal 23 kDa segment cdc25 from *S. pombe* (p80cdc25)	Substrate dephosphorylation	*In vivo* active domain localized to the C-terminal 23 kDa; bacterially expressed fusion protein of this domain dephosphorylated and activated $p34^{cdc2}$, hydrolyses *p*NPP, as well as p-ser casein	304
	Recombinant GST-fusion protein of human cdc25 (p80cdc25Hs)	Substrate dephosphorylation	Induced meiotic maturation of *Xenopus* oocytes, activated histone H1 kinase activity in oocyte extracts, activated $p34^{cdc2}$/cyclin B complexes *in vitro* by stimulating its dephosphorylation; also hydrolyses *p*NPP; catalytic activity resides in the C-terminal domain	296
	Recombinant GST-fusion protein of human cdc25	Substrate dephosphorylation	Depohosphorylates and activates cdk2 isolated from interphase cells	261
	Recombinant C-terminal catalytic domain of *Drosophila* enzyme, *E. coli*	Substrate dephosphorylation	Dephosphorylates *p*NPP; inhibited by vanadate at IC_{50} of 10 μM with total inhibition at 100 μM; also inhibited by Zn sulfate and molybdate, but not EDTA, $MgCl_2$ or poly (Glu/Tyr); $V_{max} = 56$ nmol/min/mg and $K_m = 0.50$ mM; turnover number ~ 2 for *p*NPP; also active towards Tyr(P) cdc2 peptide or angiotensin	259
	Recombinant C-terminal catalytic domain of *Drosophila* enzyme, *E. coli*	Substrate dephosphorylation; activa-tion of MPF in *Xenopus* oocyte extracts	cdc25 elicits the mitotic state by inducing tyrosine dephosphorylation of cdc2	291
cdc25A	Western blot of protein in rat fibroblast NRK cells	Molecular mass	$M_r = 65$ kDa	161
	Immunoprecipitation of metabolically labelled cells; *in vitro* translation	Molecular mass	$M_r = 75$ kDa	503
cdc25B	Recombinant human enzyme, *E. coli*	*In vitro* dephosphorylation of human $p34^{cdc2}$ kinase on both Thr-14 and Tyr-15	cdc25B phosphatase dephosphorylates both Thr-14 and Tyr-15 but not Thr-161	275
cdc25M1	Recombinant catalytic domain as GST-fusion protein, *E. coli*	Substrate dephosphorylation	Dephosphorylates *p*NPP; inhibited by vanadate at IC_{50} of 25 μM; also dephosphorylated Tyr(P) peptide corresponding to $p34^{cdc2}$ Tyr-15 domain, as well as p-ser casein; also activates *in vitro* the histone kinase activity of $p34^{cdc2}$ by tyrosine dephosphorylation	569

cdc25M2	Recombinant GST-fusion protein, *E. coli*	Activation of CDK2 kinase by dephosphorylation of Thr-14 and Tyr-15.	CDK2 is phosphorylated on Thr-14 and Tyr-15; treatment of cyclin A or cyclin E immunoprecipitates with bacterially expressed cdc25M2 (the mouse homologue of human CDC25B) ↑ the histone H1 kinase activity of these immune complexes 5- to 10-fold; tryptic peptide mapping demonstrated that cdc25M2 treatment of cyclin A or cyclin B1 immune complexes resulted in the specific dephosphorylation of Thr-14 and Tyr-15 on CDK2 or CDC2, respectively. Thus, cdc25 family members comprise a class of dual-specificity phosphatases	334
	Recombinant GST-fusion protein with catalytic domain (AA 278–576), *E. coli*	Substrate dephosphorylation	Hydrolyses *p*NPP	535
Cdi1	Immunoprecipitated from metabolically labelled HeLa cell extracts	Molecular mass	$M_r = 21\,000$ kDa	515
	Recombinant GST-fusion protein, *E. coli*	Substrate dephosphorylation	Efficiently dephosphorylated *p*NPP; also active against Tyr(P) raytide, but did not dephosphorylate p-ser/thr H1 histone phosphorylated by cdc2, suggesting specificity for Tyr(P); catalysis dependent on Cys-140; inhibited by vanadate	515
ChPTPλ **ChPTPα** **LAR**	Recombinant proteins, *E. coli* and transfected COS cells	Comparison of PTPase activity of ChPTP-λ, ChPTP-α, and hLAR with v-Src and c-Src	*In vitro,* all three enzymes dephosphorylated v-Src, but only ChPTP-λ dephosphorylated c-Src; activities were also compared in Cos cells by expressing cytoplasmic domains and recombinant cytoplasmic domain constructs that were myristylated and associated with plasma membrane; c-Src Tyr-527 was dephosphorylated and the kinase was activated 3–4-fold when coexpressed with myristoylated ChPTP-λ, less obvious with the other enzymes; thus, dephosphorylation of Src Tyr-416 may have less specific recognition than Src Tyr-527 by PTPases	260
ChPTPλ	Western blot	Molecular mass	Several bands between 170 and 210 kDa in various cultured haematopoietic cell lines	500
	Immunoprecipitated from DT40 cells and recombinant intracellular domain, *E. coli*	Substrate dephosphorylation	PTPase specific for phosphotyrosine, inhibited >70% by 1 mM orthovanadate, 0.1 mM molybdate, 1 mM iodoacetate, 10 mM Zn^{2+}	500
CL100	Recombinant protein with N-terminal poly-histidine tag, *E. coli*	Molecular mass, substrate dephosphorylation	$M_r = 44\,000$ kDa; rapidly hydrolyses *p*NPP, inhibited by vanadate and Zn^{2+}, activity dependes on presence of reducing agents	537

Table 7 Continued

PTPase	Protein source	Parameter	Data	Refs
CL100 (Cont.)	Recombinant protein, *E. coli*	Substrate dephosphorylation	CL100 rapidly and potently inactivates recombinant MAP kinase *in vitro* by the concomitant dephosphorylation of both p-thr and p-tyr residues; CL100 also suppresses the ras-induced activation of MAP kinase in a cell-free system from *Xenopus* oocytes; both activities are dependent on catalytic activity of the CL100; CL100 shows no measurable catalytic activity towards a number of other substrate proteins modified on serine, threonine or tyrosine residues	982a
	Recombinant protein, *E. coli*	Substrate dephosphorylation	Able to dephosphorylate both tyrosine and threonine residues of activated p42 MAP kinase *in vitro*	1108
DdPTP1	Recombinant GST-fusion protein, *E. coli*	Substrate dephosphorylation	Hydrolysed *p*NPP and dephosphorylated Tyr(P) $pp60^{src}$; also has a low level of activity towards dephosphorylation of p-ser (casein and regulatory subunit of PKA)	277
DdPTP2 (DdPTPa)	Recombinant enzyme expressed full length in *E. coli*	Molecular mass, substrate dephosphorylation	$M_r = \sim 45$ kDa; dephosphorylates *p*NPP and Tyr(P) RCM lysozyme; *p*NPPase activity inhibited 50% by 1mM vanadate; also inhibited poorly by $ZnCl_2$	588
DEP-1	Transfected COS cells	Molecular mass; substrate dephosphorylation	Expressed in COS cells as 180 kDa glycoprotein (tunicamycin treatment of cells ↓ M_r to 140 kDa; PTPase activity detected in immunoprecipitates; also detected in WI-38 human embryonic lung fibroblast cells	576
DLAR	Western blotting with monoclonal antibodies	Molecular mass	Major band of 210 kDa	614
DPTP	Recombinant cytoplasmic domains, *E. coli*	Substrate dephosphorylation	Cytoplasmic domains each have PTPase activity, with Tyr(P) raytide > Tyr(P) angiotensin II	606
DPTP10D	Western blotting with monoclonal antibodies	Molecular mass	A single 190 kDa species	614
DPTP4E	Recombinant cytoplasmic domains, *E. coli*	Substrate dephosphorylation	Cytoplasmic domain demonstrated to have PTPase activity	575
DPTP61F	Recombinant GST-fusion protein, *E. coli*	Substrate dephosphorylation	PTPase activity was similar with *p*NPP; K_m of p61/62m form was 4.8 mM with a k_{cat} of 0.9 s^{-1}, values with p61/62n form were 6.3 mM and 0.8 s^{-1}, respectively; activity specific for Tyr(P), does not dephosphorylate p-ser casein; inhibited by vanadate	558
DPTP99A	Western blotting with monoclonal antibodies	Molecular mass	Antibody to the extracellular domains blots to a 160 kDa protein band, while antibody to the cytoplasmic domain also recognizes a 100 kDa band in addition to the 160 kDa band; this protein may correspond to the 4.8 kb mRNA, and may be ubiquitously expressed as noted by diffuse immunostaining in early embryos	614

ERP	*In vitro* translation, without and with microsomes	Molecular mass	M_r = 40 000 kDa as predicted from sequence, no effect of cotranslational incubation with microsomes	571
	Recombinant GST-fusion protein, *E. coli*	Substrate dephosphorylation	Compared with GST-VH1 recombinant fusion protein, both have *p*NPPase activity, inhibited by vanadate, GST-ERP only 50–80% of activity of GST-VH1; GST-ERP dephosphorylated p-ser casein, also 5-fold lower activity than GST-VH1	571
GLEPP1	Western blot	Molecular mass	235 kDa glomerular protein in non-reducing and 205 kDa in reducing gels (?heterodimer structure), also probably glycosylated	613
	Cell lysate protein	Molecular mass	Apparent molecular weight of approximately 200 kDa	1130
HA1, HA2	Cultured rat adipocytes and preadipocytes	Dephosphorylation of Tyr-19 of the 15 kDa adipocyte fatty acid binding protein *in vitro*	Two enzymes, both inhibited by PAO; HA1 is ~60 kDa and HA2 ~38 kDa; activity is specific for Tyr(P); HA2 is expressed in 3T3-L1 preadipocytes and adipocytes; HA1 is expressed only in adipocytes	297
HePTP	Recombinant GST-fusion protein, *E. coli*	Molecular mass, substrate dephosphorylation	M_r = 68 kDa (fusion protein); active in *p*NPP hydrolysis	630
	Rat mast cells	Molecular mass, p*I*	By 2D electrophoresis, the protein was 40 kDa and had a p*I* of 6.9 as predicted from the deduced amino acid sequence	923
hVH-1	Recombinant GST-fusion protein, *E. coli*	Substrate dephosphorylation	Specifically dephosphorylated Tyr(P) and p-thr of MEK-activated ERK-1 but not p-ser residues; did not dephosphorylate Tyr(P) or p-ser residues of casein, autophosphorylated v-abl kinase or GST-MEK2; ERK-1 inactivated by hVH1 can be reactivated by MEK kinase	384
hVH-2		Substrate dephosphorylation	Recombinant hVH-2 exhibited a high substrate specificity toward activated ERK and dephosphorylated both threo9 and tyrosine residues of activated ERK1 and ERK2	1100
	Recombinant protein, *E. coli*	Substrate dephosphorylation	Dephosphorylated both Tyr and Thr on phosphorylated ERK-1, demonstrating dual specificity; pH optimum *vs.* *p*NPP was 5.2	950
	GST-fusion protein, *E. Coli*	Substrate dephosphorylation	Hydrolyses *p*NPP and inactivates MAP kinase	1110
IA-2	*In vitro* translation	Molecular mass	M_r = 106 kDa	543
	Recombinant GST-fusion protein, *E. coli*	Substrate dephosphorylaiton	No activity with conventional peptide and protein substrates or tyr(P) cellular proteins, suggests very narrow substrate specificity	548
IphP	Expression of *N. commune* genomic DNA in plasmid construct in *E. coli* host	Molecular mass, substrate dephosphorylation	M_r = 30 000; has PTPase activity towards RCM-lysozyme with pH optimum of 5.0; also dephosphorylates p-ser groups on casein phosphorylated with cAMP-dependent protein kinase	584
KAP	Recombinant GST-fusion protein, *E. coli*	Substrate dephosphorylation	Sharp pH optimum at 6.5 with no activity above 7.0 or below 6.0; dephosphorylated Tyr(P) and p-ser substrates, inhibited by orthovanadate, iodoacetic acid and NEM, but not okadaic, tartrate, F^-, or tetramizole	518

Table 7 Continued

PTPase	Protein source	Parameter	Data	Refs
KAP (Cont.)		Substrate dephosphorylation	Human KAP was unable to dephosphorylate Tyr-15 and only dephosphorylated Thr-160 in native monomeric cdk2; the binding of cyclin A to cdk2 inhibited the dephosphorylation of Thr(160) by KAP but did not preclude the binding of KAP to the cyclin A–cdk2 complex; suggest that KAP binds to cdk2 and dephosphorylates Thr-160 when the associated cyclin subunit is degraded or dissociates	1028
LAR	Transfected HeLa cells	Molecular mass	M_r =∼ 200 kDa (proprotein)	110
	Transfected COS cells and rat 208F fibroblasts	Molecular mass	M_r = 190 kDa (proprotein)	116
	Recombinant 40 kDa fragment corresponding to D1 PTPase domain	Regional dephosphorylation of a phosphotyrosine peptide derived from the IR kinase domain (pY positions 5, 9 and 10)	NMR analysis was performed to distinguish fingerprint regions for each of the seven phosphotyrosyl peptides (mono, di and tri); LAR showed a strong preference for dephosphorylation of the pY5 position in mono-, di- and tri-phosphotyrosyl peptides; following pY5 dephosphorylation, there is equal dephosphorylation of pY9 or pY10, followed by full dephosphorylation	295
	Recombinant LAR and CD45 cytoplasmic domains, *E. coli*	Substrate dephosphorylation	LAR D1-D2 (615 AA) showed almost identical specific activity and high catalytic efficiency as the 40 kDa LAR D1, consistent with a single active site in the 70 kDa LAR cytoplasmic domain; 90 kDa fragment of CD45 D1-D2 had similar catalytic activity towards a Tyr(P) undecapeptide	244
	Immunoprecipitated from transfected mouse B cells (300-19) transfected with the human LAR cDNA; recombinant cytoplasmic domains, *E. coli*	Substrate dephosphorylation	LAR immunoprecipitated with a monoclonal antibody directed at an epitope tag that was added to the extreme 5′ end of the LAR molecule had PTPase activity with Tyr(P) raytide; recombinant LAR cytoplasmic domain activity with Tyr(P) raytide > Tyr(P) angiotensin II	606
	Recombinant constructs of cytoplasmic domain expressed in *E. coli* extracts	Substrate dephosphorylation	LAR cytoplasmic domain rapidly inactivates the IR kinase by preferentially dephosphorylating the receptor kinase regulatory domain compared to PTPase1B and the cytoplasmic domain of RPTP-α	273
	Recombinant constructs of cytoplasmic domain expressed in *E. coli* extracts	Substrate dephosphorylation	Dephosphorylated both intact IR and EGF-R *in vitro* without preference; did not dephosphorylate p-ser/thr residues of casein phosphorylated with cAMP-dependent protein kinase; inhibited by vanadate, molybdate; effects of spermine and EDTA depended on substrate (RCM-lysozyme or MBP); with RCM-lysozyme, K_m = 7.9 μM and turnover number = 150 s^{-1}; with MBP, K_m = 4.4 μM and turnover number = 7 s^{-1}	631
	Recombinant constructs of cytoplasmic domain expressed in *E. coli* extracts	Substrate dephosphorylation	More active *vs.* Tyr(P) MBP > raytide; demonstrated that a critical cysteine residue in proximal domain (Cys-1522) is essential for activity; first domain is active by itself, second domains lack intrinsic PTPase activity but influences the catalytic activity of the proximal domain	343

LAR	Recombinant rat cytoplasmic domain expressed in *E. coli* extracts	Substrate dephosphorylation	With *p*NPP, pH optimum = 5.0, K_m = 420 μM, k_{cat} = 6.1 s^{-1}; with angiotensin I, pH optimum = 6.0, K_m = 118 μM, k_{cat} = 3.1 s^{-1}; vanadate (100 mM) inhibited by 50%, this is similar to CD45 inhibition but 10-fold higher than than required for PTPase1B; PAO did *not* inhibit rLAR, even at 10 times concentration used to inhibit CD45; LAR was not inhibited by Zn ion; iodoacetamide was a poor inhibitor compared to iodoacetate suggesting interaction with positive charge near active site	326
	Recombinant PTPase domains (D1 and D1/D2) purified from *E. coli*	Substrate dephosphorylation	In most assays, D1 and D1/D2 constructs were indistinguishable; polycationic polypeptides strongly stimulated the PTPase activity of D1/D2 but not D1 alone, suggesting that D2 may have a regulatory function	280
	Recombinant proximal PTPase domain (D1) purified from *E. coli*	Substrate dephosphorylation	Studied kinetics on 6–12 residue synthetic, non-radioactive Tyr(P) phosphopeptides corresponding to autophosphorylation sites of two major types of kinases, including IR peptide, EGF-R peptide and $p60^{src}$ and $p56^{lck}$ domains; K_m varied from 27μM to 4.1 mM and k_{cat} values were 20–70 s^{-1}; lowest K_m was with the IR Tyr(P)-1146 peptide	245
LAR-PTP2	Primary cultures of fetal rat lung	Molecular mass	Immunoblotting with an antibody to the extracellular domain demonstrated a weak band at ~190 kDa, corresponding to the full-length protein, and a prominent ~100 kDa band likely to represent the proteolytically cleaved extracellular domain	193
	Recombinant constructs of cytoplasmic domain expressed in *E. coli* extracts	Substrate dephosphorylation	Dephosphorylated both intact IR and EGF-R *in vitro* without preference; did not dephosphorylate p-ser/thr residues of casein phosphorylated with cAMP-dependent protein kinase; inhibited by vanadate, molybdate (more sensitive than LAR); effects of spermine and EDTA depended on substrate (RCM-lysozyme or MBP); with RCM-lysozyme, K_m = 3.3 μM and turnover number = 78 s^{-1}; with MBP, K_m = 2.9 μM and turnover number =4 s^{-1}	631
Low M_r PTPase (AcP1, AcP2)	Purified protein, rat liver	Substrate dephosphorylation	With *p*NPP, pH optimum 4.5–5.5; P1 and P2 differ in substrate affinity (similar K_m of 0.31 and 0.27 mM, but differ in V_{max}: (AcP1 = 98.9, AcP2 = 69.3) and sensitivity to activators and inhibitors (cGMP ↑ activity of AcP2 by 3.2-fold; but had no effect on AcP1)	550
(HAAP-β)	Recombinant human adipocyte enzyme, GST-fusion protein, *E. coli*	Substrate dephosphorylation	HAAP-β dephosphorylated *p*NPP and Tyr(P)-ALBP; was inactivated by PAO and inhibited by vanadate at K_i = 17 μM; also efficiently dephosphorylated the cytoplasmic IR kinase domain *in vitro*	600
	Purified protein from bovine liver	Cysteine oxidation state	All eight half-cystine residues are in the free thiol form; two half-cystines at or near the active site were identified through the reaction of the enzyme with [^{14}C] iodoacetate in the presence or in the absence of a competitive inhibitor (i.e. inorganic phosphate)	482

Table 7 Continued

PTPase	Protein source	Parameter	Data	Refs
Low M_r PTPase (Cont.)	Recombinant bovine heart enzyme, *E. coli*	Determined pK_a values for histidines near active site by Inactivation and kinetic studies	His-66 titrates with a pK_a of 8.4 while for His-72 the pK_a is 9.2	254
	Human	Identification and pK_a determination of the histidine residues using an MLEV-17 spectral editing scheme	All three or four of the histidine ^{1}H-NMR signals of two human low M_r PTPases (HCPTP-A or -B) were readily detected; peak assignments were accomplished through the use of His → Ala of HCPTP-A and -B and a homologous bovine enzyme; titration of HCPTP-A and -B with vanadate, a strongly bound competitive inhibitor, caused the His-72 peak to appear as two signals at nearly equimolar concentrations of protein and vanadate, while the other histidine peaks were not affected; indicating that His-72 is at the enzyme active site	385
	Bovine heart	Leaving group dependence and proton inventory studies	Determined k_{cat} and K_m values for the bovine heart low M_r PTPase catalysed hydrolysis of a series of aryl phosphate monoesters and alkyl phosphate monoesters; the values of k_{cat} are effectively constant for the aryl phosphate monoesters, consistent with the catalysis being nucleophilic in nature, with the existence of a common covalent phosphoenzyme intermediate, and with the breakdown of this intermediate being rate-limiting; k_{cat} for the alkyl phosphate monoesters is much smaller and the rate-limiting step for these substrates is interpreted to be the phosphorylation of the enzyme; the D_2O solvent isotope effect and proton inventory experiments indicate that only one proton is 'in flight' in the transition state of the phosphorylation process	378
	Purified protein	Molecular mass, substrate dephosphorylation	Purified low M_r PTPase from bovine brain, had M_r of 17 kDa; enzyme dephosphorylated M_r 170 kDa protein identified as EGF receptor in brain extracts using specific antibody	336
(CPTP-A, CPTP-B)	Recombinant human enzymes, *E. coli*	Molecular mass, substrate dephosphorylation	M_r = 18 kDa; highly active towards *p*NPP and phosphotyrosine, but not phosphothreonine or phosphoserine; for *p*NP CPTP-A K_m = 0.18–0.30 mM; for CPTP-B K_m = 0.49–1.24 mM	619
	Bovine heart	Molecular mass substrate dephosphorylation	A monomer with M_r = 18 kDa, isoelectric point = 7.0; absorption coefficient, $E_{1\%}/1$ cm was 9.65 at 280 nm; pH optima were 5.3 and 6.0 with the substrates *p*NPP and tyrosine phosphate, respectively; at pH 5, K_m = 0.38 and 14 mM for *p* NPP and tyr-*O*-phosphate, respectively; molybdate and vanadate were potent inhibitors with K_i values of 37 and 29 μM, respectively; seven or eight accessible cysteines were detected on the monomeric protein and at least one was essential for enzyme activity; the enzyme also had phosphotransferase activity, and was able to transfer phosphate from *p* NPP to a wide variety of alcohol acceptors	377

Low M_r PTPase	Recombinant protein, *E. coli.*	Molecular mass; substrate dephosphorylation	$M_r = 18$ kDa; a kinetically competent phosphoenzyme intermediate was trapped from a phosphatase-catalysed reaction; using ^{31}P NMR, the covalent intermediate was identified as a cysteinyl phosphate; suggested use of term cysteine phosphatases for these enzymes	365
	Bovine heart	Pre-steady-state and steady-state reaction kinetics with *p*NPP	A transient presteady-state 'burst' of *p*-nitrophenol was formed with a rate constant of 48 s^{-1}; the burst was effectively stoichiometric and corresponded to a single enzyme active site/molecule; this was followed by a slow steady-state turnover of the phosphoenzyme intermediate with a rate constant of 1.2 s^{-1}; an enzyme-catalysed ^{18}O exchange between inorganic phosphate and water was detected and occurred with $k_{cat} = 4.47 \times 10^{-3}\ s^{-1}$; these results were all consistent with the existence of a phosphoenzyme intermediate in the catalytic pathway and with the breakdown of the intermediate being the rate-limiting step. The true Michaelis binding constant $K_s = 6.0$ mM, and the apparent $K_m = 0.38$ mM	379
	Bovine liver	PTPase activity with EGF receptor as substrate	The activity was significantly inhibited by orthovanadate and *p*-hydroxymercuribenzoate; indicating that free sulfhydryl groups are required for PTPase activity; pH optimum was broad, between pH 5.5 and 7.5; apparent K_m for ^{32}P-EGF receptor dephosphorylation was 4 nM; enzyme not ^{32}P-Ser-casein, p-ser or p-thr	328
	Purified from rat liver	Substrate dephosphorylation	Tested a series of phosphotyrosyl peptides for dephosphorylation by two isoforms of the low M_r PTPase (AcP1 and AcP2), affinity varied as well as reaction rate	341
		Substrate dephosphorylation	Cyclic GMP and guanosine activate the isoenzyme AcP2 $\gg$ AcP1; cyclic GMP activates the AcP2 isoenzyme by increasing the rate of the step that leads to the hydrolysis of the covalent enzyme–substrate phosphorylated complex formed during the catalytic process	989
LTP1	Recombinant protein, *E. coli*	Substrate dephosphorylation	Efficiently hydrolyses phosphotyrosine and p-Tyr peptide related to Fyn, but it shows low activity toward p-Ser and p-Thr, catalytic activity toward a number of substrates was approximately 30-fold lower than the corresponding values measured for the bovine low M_r PTPase; yeast enzyme was markedly activated by adenine (30-fold) and cAMP and CGMP	1116
MAP kinase PTPase	Purified protein from *Xenopus laevis* eggs	Molecular mass, substrate dephosphorylation	$M_r = 47$ kDa; absolute specificity for Tyr(P) residues and not p-thr or p-ser residues of MAP kinase phosphorylated by MEK; pH optimum 7.0 with K_m of 9.0 μM, inhibited by 2 μM molybdate, 250 μM vanadate, mM levels of Mn^{2+}, Zn^{2+} and *p*NPP, but not okadaic acid	333

Table 7 Continued

PTPase	Protein source	Parameter	Data	Refs
MKP-1 (3CH134)	Recombinant protein with N-terminal 16 AA extension, *E. coli* (labile to proteolysis during isolation)	Substrate dephosphorylation	pH optimum 7.5, sensitive to vanadate and sulfhydryl modifying agents, insensitive to serine/threonine phosphatase inhibitors; activity is highly selective and in the order of $p42^{MAPK}$ dephosphorylated 15-times more rapidly than RCM-lysozyme > MBP > EDNDYINASL peptide; no activity against autophosphorylated IR kinase or p-thr or p-ser residues of casein or RCM-lysozyme phosphorylated by PKA; $p42^{MAPK}$ was dephosphorylated very slowly by PTP1B	238
	Epitope-tagged, recombinant protein, *E. coli*	Substrate dephosphorylation	Dephosphorylated both Thr-183 and Tyr-185 in $p42^{MAPK}$ and abolished its kinase activity *in vitro*	345
MKP-2	Recombinant protein, *E. coli*	Substrate dephosphorylation	MKP-2 displayed vanadate-sensitive phosphatase activity against MAP kinase *in vitro*	1111
MSG5	Recombinant GST-fusion protein, *E. coli*	Substrate dephosphorylation	Hydrolyses *p*NPP; inhibited by vanadate; also inactivated autophosphorylated (Ser/Tyr) FUS3 kinase; dephosphorylated FUS3 only in its native conformation	497
OST-PTP	Recombinant GST-fusion protein of cytoplasmic domain, *E. coli*	Substrate dephosphorylation	pH optimum was 5.6, wih *p*NPP, K_m was 0.52 mM and k_{cat} estimated to be 41 s^{-1}; dephosphorylation specific for Tyr(P) and enzyme did not dephosphorylate p-ser kemptide	557
PAC-1	*In vitro* translation; immunoprecipitation from metabolically labelled T cells	Molecular mass	$M_r = 32\,000$ kDa	589
PAC1	Recombinant GST-fusion protein, *E. coli*	Substrate dephosphorylation	Remarkably specific for ERK2 (MAP kinase), did not dephosphorylate Tyr(P) raytide, angiotensin, casein, v-Abl or p-ser/p-thr casein or MEK; time course showed Tyr(P) dephosphorylated before p-thr on ERK2; both events inhibited by vanadate	361
PC-PTP1	Rat PC12 cells	Molecular mass	Western blot analysis using a polyclonal Ab (antibody) raised against the cytoplasmic region of PC-PTP1 detected two products, a major 65 kDa and minor 42 kDa protein that correspond to the products translated from the second and fifth methionine of PC-PTP1	1124
PC12-PTP1	*In vitro* translation	Molecular mass	*In vitro* translation product of ~39 kDa, smaller than predicted from largest ORF	1121
PCS phosphatase		Dephosphorylation of a Tyr(P) EGF receptor and a Tyr(P) RSV peptide	PTPase activity of the PCS phosphatases is increased by direct effects of ATP or PP_i on the enzyme; ↑ PTPase activity is associated with ↓ phosphorylase phosphatase activity, demonstrating a change in substrate specificity; this PTPase activity is a substantial portion of the PTPase activity in the cytosol of *Xenopus laevis* oocytes	266

PEP	70Z/3 pre-B cells and transfected S194 plasmacytoma cells	Half-life	Pulse-chase analysis demonstrated half-life of > 5 h, raising a query about significance of 'PEST' motif	146
	In vitro translation	Molecular mass	89 and 97 kDa products	553
PP2A (PTPase activator, PTPA)	Purified from rabbit skeletal muscle	Induction of PTPase activity of PP2A	PTPA could only be detected in cytosolic fractions; interaction of PTPA with PP2A in a 1:1 ratio induces a low ($k_{cat=3}$ min^{-1}) ATPase activity that is inhibited by okadaic acid, ADP and non-hydrolysable ATP analogues	1039
PRL-1	*In vitro* translation	Molecular mass	M_r = 20 000 kDa	493
	Recombinant His-tagged fusion protein, *E. coli*	Substrate dephosphorylation	With *p*NPP, has V_{max} of 2.2 nmol/min/mg, inhibited by orthovanadate (60% at 1 mM), and requires reducing agents for activity; pH optimum is 7.3–7.5; did not dephosphorylate p-ser casein; when phosphorylated itself by immunoprecipitated Src kinase, PRL-1 rapidly autodephosphorylates; Tyr(P) casein and angiotensin were not dephosphorylated by PRL-1	493
PTP-5	Bovine brain	Substrate dephosphorylation	Dephosphorylation of model substrates by bovine brain PTP-5 using sequences of casein and RCM-lysozyme and compared to recombinant rat PTP1B; K_m for hydrolysis of simple aromatic phosphate esters was ~5000-fold higher than Tyr(P) peptides and proteins, indicating that the surrounding sequences contribute to PTPase substrate recognition; substitutions indicated that Asp or Glu within the 1st five residues on the N-terminal side of Tyr(P) ↑ the peptide reactivity with both PTPs; Asn was only weakly able to substitute for Asp, indicating the importance of acidic residues at this position	274
PTP-BAS (PTPL1)	Transiently transfected COS-1 cells, metabolic labelling	Molecular mass; substrate dephosphorylation	M_r = 250 kDa; immunoprecipitated PTPL1 from transfected COS-1 cells dephosphorylated Tyr(P) MBP	596
(PTP1E)	Recombinant catalytic domain expressed in *E. coli*	Substrate dephosphorylation	Active towards *p*NPP and Tyr(P) raytide, inhibited by vanadate	475
PTP-PEST	Recombinant GST fusion protein, *E. coli*	Substrate dephosphorylation	Construct used residues 1–492 including catalytic domain; specific for Tyr(P); pH optimum 6–7; active against IR kinase domain, RCM-lysozyme > MBP; effect of various agents depended on substrate	622
PTP1B	Partially purified placental enzyme	Dephosphorylation of Tyr(P) peptide substrates corresponding to IR kinase domain	Dephosphorylated each site with similar affinity (K_m = 1.3–2.5 μM)	395
	In vitro translation	Molecular mass	M_r = 56 000 kDa	487

Table 7 Continued

PTPase	Protein source	Parameter	Data	Refs
PTP1B (Cont.)	Recombinant GST fusion protein, *E. coli*	Molecular mass, substrate dephosphorylation	49 kDa form detected in membranes; *in vitro*, cellular substrate proteins for the EGFR kinase were dephosphorylated by recombinant PTP1B fusion protein	268
PTP1B, TCPTP, LAR (cytoplasmic domain), CD45 (full length)	Recombinant or purified protein	Order of dephosphorylation of a triply tyrosine-phosphorylated peptide (positions 5, 9 and 10) derived from the kinase domain of the insulin receptor	All four PTPases dephosphorylated the tris-Tyr(P) peptide to various Tyr(P) forms; PTP1B preferentially dephosphorylated the two Tyr(P) residues at positions 9 and 10, while CD45 and LAR preferentially dephosphorylated Tyr(P) at position 5; TCPTP did not exhibit any regiospecificity for dephosphorylation; thus, primary structure surrounding the Tyr(P) sites may contribute to substrate specificity	327
	Recombinant GST-fusion protein of C-terminal truncated rat PTP1B, *E. coli*	Substrate dephosphorylation	Enzyme displayed classical Michaelis–Menten kinetics, with K_m of 68 μM and 42 μM and k_{cat}/K_m values of 4.9 to 6.9 × 10^5 s^{-1} M^{-1} for phosphorylated peptides of lengths four and 10 residues, respectively	399
	Recombinant protein, *E. coli*, also immunoprecipitated proteins from 293 cell transfections as full-length and truncated forms Δ321 and Δ299	Substrate dephosphorylation	Kinetic characterization for *p*NPP; from *E. coli*, Δ321 and Δ299 forms had K_m values of 0.08 and 0.11 mM; wild-type protein isolated from 293 cells had K_m of 0.38 mM, Δ321 form from 293 cells had K_m ↓ to 0.1 mM	634
	Human placenta (soluble 37 kDa form)	Substrate dephosphorylation	RCM lysozyme K_m = 280 nM, myelin basic protein K_m = 250 nM; activated by EDTA, spermine, and myelin basic protein; inhibited by vanadate, molybdate, Zn^{2+}, Mn^{2+}, heparin and poly(Glu/Tyr); other substrates – IR	352
	Human placenta (~46 kDa membrane-associated form)	Substrate dephosphorylation	RCM lysozyme or src peptide K_m = 2 μM; inhibited by heparin and poly(Glu/Tyr); Other substrates – *src* peptide, EGFR kinase, $p56^{lck}$	318
	Recombinant protein, *E.coli*	Substrate dephosphorylation	Similar rate of dephosphorylation of EGFR and IR *in vitro*	790
	Recombinant protein, *E. coli*	Substrate dephosphorylation kinetics	Product inhibition and ^{18}O exchange data are consistent with the reaction proceeding through two steps, formation and breakdown of a covalent phosphoenzyme intermediate, a reaction mechanism is also deduced from these and additional detailed kinetic studied; compared with the *Yersinia* PTPase, the data suggest that the mechanism is conserved from bacteria to mammals	1044
	Catalytically inactivate (Cys-215→Ser) recombinant protein, *E. Coli*	Substrate recognition	Binds with high affinity (100 nm) to cognate phosphotyrosylpeptide from EGF receptor (pY-992); Ala scanning mutagenesis around pY site altered binding affinity in pattern distinct from PLCγ SH2 domain binding	1071
PTPD1	Transient expression in 293 embryonic fibroblasts	Molecular mass, substrate dephosphorylation	M_r = 130 kDa on Western blot; immunoprecipitated enzyme hydrolyses *p*NPP poorly, but able to hydrolyse poly(Glu : Tyr)	564

PTPH1	*In vitro* translation	Molecular mass	$M_r \sim 120$ kDa	624
	Baculovirus vector system	Molecular mass, substrate dephosphorylation	Purified protein had apparent M_r of 120 000 Da on SDS gels; native enzyme dephosphorylated MBP five-fold more actively than RCML, with K_m of 1.45 μM and 1.6 μM, respectively; phosphorylation of PTPH1 by PKC *in vitro* decreased the K_m but not V_{max}; removal of the N-terminal band 4.1 homology domain of PTPH1 by limited trypsin cleavage stimulated dephosphorylation of RCML but inhibited its activity toward MBP, suggesting that in addition to a potential role in controlling subcellular localization, the N-terminal band 4.1 homology domain of PTPH1 may exert a direct effect on catalytic function	1043
PTP-SL	*In vitro* translation	Molecular mass	*In vitro* translation product of ~60 kDa as predicted	1103
PTPX1	Recombinant protein, Sf9 cells	Molecular mass, substrate dephosphorylation	$M_r = 79$ kDa; inhibited by 100 μM vanadate and molybdate 10 μM, but not okadaic acid; HA-tagged PTPX1 isolated from the membrane fraction of *Xenopus* oocytes is ~4-times more active than PTPX1 from the soluble fraction	490
PTPX10	Recombinant protein, Sf9 cells	Molecular mass	$M_r = 69$ kDa	490
pyp1$^+$	Immunoprecipitation from transfected, metabolically labellled cells	Molecular mass	$M_r \sim 63$ kDa; protein was not detected in unsynchronized cells suggesting that it is not abundant	577
	Recombinant protein, *E. coli*	Substrate dephosphorylation	An active phosphatase with Tyr(P) raytide; labelled p107^{wee1} is also a substrate *in vitro*	270
pyp2$^+$	Recombinant GST-fusion protein of C-terminal catalytic domain, *E. coli*	Substrate dephosphorylation	Dephosphorylates *p*NPP and Tyr(P) raytide, but not p-ser or p-thr H2 or H6 peptides; with *p*NPP, $V_{max} = 0.4$ μM/min/mg and $K_m = 15.8$ mM; inhibited by vanadate ($IC_{50} = 90$ nM) and $ZnCl_2$ ($IC_{50} = 30$ μM), but not by NaF	561
	Recombinant protein, *E. coli*	Substrate dephosphorylation	An active phosphatase with Tyr(P) raytide	270
pyp3$^+$	Recombinant GST-fusion protein, *E. coli*	Substrate dephosphorylation	Cleaves *p*NPP at 17.6 μmol/min/mg, also a Tyr(P) peptide substrate; inhibited by vanadate; also activates p34^{cdc2}/cyclin activity by dephosphorylation of Tyr-15	560
RKPTP	Recombinant protein, *E. coli*	Substrate dephosphorylation	Active *in vitro* against MBP and p43$^{v\text{-}abl}$	566
RPTP-α	Recombinant protein, Sf9 insect cells	Molecular mass, substrate dephosphorylation	Two N-linked glycosylation variants observed of 98 and 114 kDa; pH optimum six with high K_m (μM range) towards MBP and RCM-lysozyme; effect of reagents depended on substrates and assay pH; kinetic analysis suggested enzyme could exist in low and high substrate affinity states	252

Table 7 Continued

PTPase	Protein source	Parameter	Data	Refs
RPTP-α, RPTP-αi	Transient transfection in COS-1 cells	Molecular mass, substrate dephosphorylation	No distinction found between isoforms in apparent size, *in vitro* enzymatic activity or glycosylation pattern; both isoforms generate glycosylated species of 100 and 130 kDa on Western blot analysis; emerge as 200 and 340 kDa proteins on Superose 12 molecular sieving, suggesting presence of dimers	84, 595
	Recombinant cytoplasmic D1 and D2 domains, *E. coli*	Substrate dephosphorylation	D1 and D2 are enzymatically active but with different specificities: D1 prefers MBP ~ RR-src > *p*NPP, D2 favors *p*NPP >>> RR-src and is inactive towards MBP; regions N-terminal to each domain contribute to catalytic activity; other sequences contribute to the functional interactions between D1 and D2 in the native protein which is more highly active than either individual domain	359
(LRP)	Recombinant protein, *E. coli*	Substrate dephosphorylation	*In vitro*, dephosphorylates intact insulin receptors more rapidly than a cognate peptide from the IR kinase domain; also dephosphorylated EGF receptors more rapidly than insulin receptors	790
RPTP-β	Recombinant catalytic domain expressed in *E. coli*	Substrate dephosphorylation	Cytoplasmic domain exhibits high activity toward all substrates tested and is potently inhibited by zinc, vanadate and polyanions; *in vitro* phosphorylation by PKC did not affect catalytic activity	360
	Recombinant epitope-tagged fusion protein of intracellular domain of RPTP-β (44 kDa), *E. coli*	Substrate dephosphorylation using series of phosphopeptides and 96-well plate assay	Strongly inhibited by vanadate, molybdate, heparin, poly (Glu/Tyr) and Zn^{2+}; for *p*NPP, $K_m = 2.5$ mM; $V_{max} = 103$ μmol/min/mg and $k_{cat} = 76\ s^{-1}$; using Tyr(P) phosphopeptides (13 residues) corresponding to candidate physiological substrates of RPTP-β, found that k_{cat} values were between 120 and 258 s^{-1} in preference of src (Tyr-527) > PDGF-R (Tyr-740) > ERK1 (Tyr-204 $\gg$ CSF-1R (Tyr 708), with K_m values ranging from 140 μM to >10 mM; different fom ref. 280, they found that spermine was inhibitory at high concentrations ($\geq$2 mM)	271
RPTP-β, LAR, RPTP-α, RPTP-δ	Recombinant proteins expressed in *E. coli*	Substrate dephosphorylation	Specific activity of PTP-β was significantly higher than the other enzymes tested; comparing relative activities with MBP versus raytide, CD45 had previously been shown to be more active wih raytide than MBP, in contrast, LAR is more active with MBP, the enzymes exhibited a 100-fold difference in relative activity with the two substrates, PTP-α, β and δ were similar in relative activity to LAR	540
RPTP-δ	Western blotting in mouse tissue lysates	Molecular mass	Antibody to extracellular domain identified a 210 kDa protein in brain and kidney lysates	563

RPTP-γ	Sf9 cells	Molecular mass, substrate dephosphorylation	M_r = 185 000 Da protein accompanied by a protein with $M_r \sim$ 120 000 Da as a putative cleavage product; modified by *N*-linked glycosylation and constitutive phosphorylation of serine residues; dephosphorylated MBP at a pH optimum of 7.5 and a K_m of 12.6 μM and RCM-lysozyme at a pH optimum of 6.0 and a K_m of 12 μM and *p*NPP with a pH optimum of 5.5 and a K_m of 3.5 mM; inhibited by $ZnCl_2$ and sodium orthovanadate, Mg^{2+}, Mn^{2+} and Ca^{2+} were ineffective; only in a partially purified form, the enzyme was allosterically activated by triphosphorylated nucleosides with a preference for purines; the enzyme protein was specifically bound by an ATP–agarose matrix through its intracellular domain	1035
RPTP-κ	Transient transfection in HeLa cells	Molecular mass	210 kDa precursor form, processing into smaller forms observed in COS cells	529
RPTP-μ	Transient transfection in COS cells	Molecular mass	M_r = 195 kDa by immunoblotting, suggests glycosylation of core protein	504
	Purified recombinant full-length and cytoplasmic domain, Sf9 cells	Molecular mass, substrate dephosphorylation	M_r = 200 kDa full-length, 80 kDa cytoplasmic domain; full-length protein was membrane-associated, cytoplasmic domain soluble; cytoplasmic domain was 2-fold more active than full-length protein and exhibited a preference for RCM-lysozyme (V_{max} = 725 nmol/min/mg; K_m = 400 nM) over MBP or a Tyr(P) nonapeptide; pH optimum 7.5; vanadate and molybdate abolished enzyme activity; also inhibited by Zn^{2+}, Mn^{2+}, polylysine, poly(Glu/Tyr) and spermine; trypsin inhibited enzyme activity	236
	Purified recombinant cytoplasmic domain, Sf9 cells; additional mutations expressed as GST-fusion proteins in *E. coli*	Substrate dephosphorylation	High activity towards RCM-lysozyme (V_{max} = 4500 nmol/min/mg; K_m = 1000 nM) and MBP (V_{max} = 8500 nmol/min/mg; K_m = 1550 nM)), but negligible activity towards Tyr(P) angiotensin or a nonapeptide that is readily dephosphorylated by PTP1B; inhibited by Zn^{2+}, Mn^{2+}, vanadate, phenylarsine poxide and heparin; proximal domain is only 10–20% as active as tandem domain construct	263
RPTP-η	Transfected cells	Molecular mass	220–250 kDa protein identified by immunoblotting	1105
RPTP-σ	GST-fusion protein. *E. coli*	Substrate dephosphorylation	Activity *vs.* MBP and raytide compared to that of LAR, RPTP-δ and CD45; RPTP-σ is similar to LAR and RPTP-δ, with MBP/raytide ratios of 14.9, 24.4 and 14, respectively; CD45 has ratio of 0.4	1117
(LAR-PTP2B)	Transient expression in human embryonic kidney 293 cells	Molecular mass	M_r of full-length protein = 200 kDa; immunoprecipitated protein had active PTPase	621
RPTP-σ (PTP-P1 PTP-PS, LAR-PTP2B)	Recombinant cytoplasmic domains expressed in *E. coli* cell lysates	Molecular mass; substrate dephosphorylation	PTP-P1 expressed as 68 kDa cytoplasmic domain, PTP-PS is 40 kDa; both cytoplasmic domains can dephosphorylate *p*NPP, a Tyr(P) peptide related to cdc2, and Tyr(P) raytide with similar activity	579

Table 7 Continued

PTPase	Protein source	Parameter	Data	Refs
RPTP-σ (PTP-NE3 LAR-PTP2B)	Recombinant cytoplasmic domain cleaved from GST-fusion protein, *E. coli*	Substrate dephosphorylation	pH optimum 5.6; with *p*NPP; inhibited by vanadate; no activity towards p-ser casein; with *p*NPP, $K_m = 1.5$ mM and $k_{cat} = 1.2\ s^{-1}$	216
RPTP-ζ/β	Immunoprecipitation from a human neuroblastoma cell line	Molecular mass	Mass of the glycosylated form = 300 kDa	546
	Recombinant constructs of D1 and D2 cytoplasmic domain expressed in *E. coli* extracts	Substrate dephosphorylation	Intact cytoplasmic domain was more active towards Tyr(P) raytide than MBP, as found with LCA (CD45), but other PTPases (LAR, RPTP-α, RPTP-β, RPTP-δ) have demonstrated the opposite ratio; isolated proximal PTPase (D1) domain was also more active on raytide > MBP	539
SAP-1	Colon carcinoma WiDr cells and transiently tranfected COS cells	Molecular mass	$M_r = 200$ kDa by immunoblotting, suggesting glycosylation	552
	Recombinant GST-fusion protein with cytoplasmic domain, *E. coli*	Substrate dephosphorylation	Hydrolyses *p*NPP and dephosphorylates Tyr(P) raytide	552
SH-PTP1 (PTP1C)	Recombinant proteins, *E. coli;* full-length and with the N-terminal SH2 domain deleted (AA 108–597) or the catalytic domain alone (AA 209–597)	Dephosphorylation of various Tyr(P) forms of an IR kinase domain peptide	No selectivity for dephosphorylation of Tyr(P) sites of the triphosphorylated IR peptide; kinetic data (including K_m, V_{max}, and K_{cat}/K_m) is given for the various Tyr(P) forms of the IR peptide with the catalytic domain; best substrate for the catalytic domain alone was the triphosphorylated peptide $K_m = 1.6\ \mu$M; in contrast, the full-length enzyme had a 22-fold lower affinity for the triphosphorylated IR peptide; deletion of only the N-terminal SH2 domain increased the affinity of PTP1C to a level similar to the catalytic domain itself; results suggest the SH2 domain exerts an inhibitory effect on the PTPase activity	355
	Immunoprecipitation from various cell lines	Molecular mass	Antibody to PTP1C directed at C-terminal 18 AA, recognized 66 kDa (called PTP1C-α) and 62 kDa (called PTP1C-β) forms of the protein which were differentially expressed in different cell types HL-60 (only 66 kDa α-form), HeLa (only 62 kDa β-form), in gastric cancer cell lines (MKN-45 and KATO-III) both were expressed, and in ZR-75-1, from which PTP1C was originally isolated, α-form > β-form	212
	Western blot of insulin-stimulated IM-9 human lymphoblasts	Molecular mass	$M_r = 65\,000$ kDa	113
(SHP)	*In vitro* translation, immunoprecipitation from LSTRA cells	Molecular mass	67.5–68 kDa	553

(HCP)	Immunoprecipitated from murine myeloid cell line (32Dcl); *in vitro* translation	Molecular mass, substrate dephosphorylation	M_r = 68 000 kDa; activity specific for Tyr(P)	627
(PTP1C)	Recombinant protein, from 293 cells infected with adenovirus expression system	Molecular mass, substrate dephosphorylation	M_r = 68 000 kDa; specific for Tyr(P); pH optimum ~7.0 for RCM-lysozyme and MBP, but < 5.0 for *p*NPP; V_{max} *vs.* protein substrates is < 1% that of other PTPases (including PTP1B, TCPTP, RPTP-α and CD45), but activity is comparable with *p*NPP; with *p*NPP, K_m = 1.5 mM, V_{max} = 33 000 nmol/min/mg, k_{cat} = 37 s^{-1}; effect of activators and inhibitors varied with the substrate tested; limited trypsinolysis of C-terminal segment generated a 63 kDa fragment with 20- and 10-fold ↑ activity towards RCM-lysozyme and MBP, respectively	381
	Recombinant protein, full-length and with truncation of both SH2 domains, *E. coli*	Molecular mass, substrate dephosphorylation	M_r = 63 000 for full-length; activity specific for *p*NPP and Tyr(P); pH optimum 5.5 for full-length and 6.3 for catalytic domain construct; optimal [NaCl] = 250–300 mM; for the catalytic domain, K_m was high for *p*NPP (24–38 mM), V_{max} 148 μmol/min/mg, k_{cat} = 110 s^{-1} and k_{cat}/K_m = 2.9–4.6 × 10^3 M^{-1} s^{-1}; for the full-length enzyme K_m was very high for *p*NPP (148 mM), V_{max} 147 μmol/min/mg and k_{cat} = 155 s^{-1}; classical saturation kinetics were not observed for any phosphotyrosyl peptide substrates tested	319
	Recombinant GST fusion protein, *E.coli*	Substrate dephosphorylation	Specific phosphatase activity against Tyr(P) raytide and MBP, inhibited by orthovanadate; no catalytic activity towards p-ser kemptide	582
(PTP1C)	Recombinant protein, adenovirus expression system in 293 cells	Activation of PTPase activity by phospholipids	Some phospholipids activated PTP1C > 1000-fold with specific protein substrates including MBP or ERK2; low PTPase activity towards EGFR, PLC-γ, and enolase was unchanged; activity towards *p*NPP, a Tyr(P) peptide, RCM-lysozyme and the IR kinase domain was actually inhibited by phospholipids; phospholipid vesicles were much more effective than detergent micelles, Triton X-100 almost abolished the activity; effects were not observed with other PTPases including CD45, RPTP-α, TCPTP; phospholipid treatment of PTP1C and trypsin led to rapid and extensive degradation of the PTPase suggesting that the phospholipid induced significant conformational changes in the enzyme; truncated enzyme was also strongly stimulated by phospholipids (although at a 5× higher concentration), indicating that the C terminus does not play a role in the activation process	382
(SHP), SH-PTP2	Recombinant protein, *E. coli*	Substrate dephosphorylation	Catalytic activities were regulated by the presence of the linked SH2 domains; both SHP and SH-PTP2 demonstrated a similar specificity pattern; both failed to elicit detectable phosphate release from several phosphopeptide substrates, while displaying catalytic efficiencies that ranged over 40–1.6 × 10^3 M^{-1} s^{-1} towards other substrates; in contrast, PTP1B did not exhibit substantial substrate specificity; suggests that physiological specificity of the SH2 domain-containing PTPases is at least in part contained in the SH2 domains	991

Table 7 Continued

PTPase	Protein source	Parameter	Data	Refs
SH-PTP2 (Syp)	*In vitro* translation; Western blot of cultured cells	Molecular mass	M_r = 65 000 kDa	501
	Recombinant protein, *E. coli*	Molecular mass, substrate dephosphorylation	M_r = 68 000 kDa; pH optimum ~7 for pY peptides, but 5.6 for *p*NPP; for RCM-lysozyme, K_m (1.7 μM) is 2000-fold lower and k_{cat} (0.11 s^{-1}) is 2.4-fold higher than with *p*NPP; among candidate pY peptides, high activity exhibited against Tyr(P)-992 domain of EGF-R and Tyr(P)-1009 and Tyr(P)-1021 of PDGF-R; Tyr(P)-1009 data suggest substrate and allosteric effects of the peptide	344
SH-PTP2 (PTP-L1)	Rat liver	Molecular mass, substrate dephosphorylation	M_r = 67 000 kDa; Tyr(P) poly(Glu:Tyr) dephosphorylated at 6500 nmol P_i released min^{-1} mg^{-1}; only mildly sensitive to inhibition by Zn^{2+} and vanadate, activated by poly(Glu:Tyr)	522
	Cleaved recombinant GST-fusion proteins, *E. coli*	Substrate dephosphorylation	Studied full-length protein as well as construct lacking SH2 domains (ΔSH2); demonstrated marked substrate preference among Tyr(P) peptides studied from no dephosphorylation to a k_{cat}/K_m of 1.1×10^5 with a PDGF-R pY-1021 peptide; pY-1009, a binding site for SH-PTP2 was also a good substrate *in vitro*; ΔSH2 form had higher catalytic efficiency, soluble SH2 domain were found also to inhibit the enzyme, suggesting a functional association between the two domains	256
SH-PTP2 (Syp)	Recombinant GST-fusion protein, *E. coli*	Substrate dephosphorylation	P210bcr-abl tyrosine kinase dephosphorylated *in vitro*	457
SH-PTP2 (PTP1D PTP1Di)	Recombinant GST-fusion protein, *E. coli*	Substrate dephosphorylation	V_{max} of PTP1Di relative to PTP1D was 8- to 20-fold lower towards *p*NPP, AChR and MBP; K_m also lower for PTP1Di for AChR and MBP; PTP1Di more sensitive to inhibition by orthovanadate, molybdate and spermidine than PTP1D (MBP substrate)	301
SH-PTP2 (PTP1D)	Western blot from cultured cells	Molecular mass	M_r = 68 000 kDa	616
SH-PTP2 (PTP2C)	Recombinant protein, *E. coli*	Substrate dephosphorylation	pH optimum ~7 for pY proteins, but <5.5 for pY peptides and *p*NPP; specific activity <0.1% of other PTPs towards pY protein substrates; activated by anionic phospholipids and inhibited by vanadate, molybdate and Zn^{2+}	383
	Recombinant protein, *E. coli*	Substrate dephosphorylation	Tyr phosphorylated IRS-1 was rapidly dephosphorylated by SH-PTP2 *in vitro*, enzyme less active towards IR kinase domain	289
	Recombinant protein, *E. coli*, lacking SH2 domains	Substrate dephosphorylation	Dephosphorylated intracellular domains of EGF-R, IR and lipocortin I; activity against RCM-lysozyme and MBP considerably lower than PTP1B or TCPTP	472

STEP	*In vitro* translation	Molecular mass	$M_r = 46$ kDa	547
	Rat brain proteins; recombinant GST-fusion protein, *E. coli*	Molecular mass, substrate dephosphorylation	Western blot shows a triplet of polypeptides with M_r of 46, 37 and 33 kDa enriched in the striatum; detergent phase separation showed that the proteins were in soluble and not membrane fractions; affinity purified GST-STEP fusion protein was an active PTPase (with *p*NPP); mutated form of STEP Cys-300-Ser had no detectable PTPase activity; activity inhibited by vanadate and molybdate but not poly(Gly/Tyr) or heparin	169
	Adult rat brain	Molecular mass	Immunoblot analysis revealed multiple polypeptides of 33, 37 and 46 kDa enriched in cytosol and a 64–66 kDa doublet enriched in membrane fractions (see Table 5)	1095
stp1$^+$	Recombinant protein	Substrate dephosphorylation, kinetic analysis	Possesses intrinsic phosphatase activity toward both aryl phosphates (pY) and alkyl phosphates (p-ser) as well as phosphotyrosyl peptide/protein substrates; yeast enzyme was six-fold slower than the mammalian enzymes; burst kinetics suggested that the rate-limiting step corresponds to the decomposition of the phosphoenzyme intermediate	1132
TCPTP	Recombinant full-length protein and TCΔC11PTP (lacking the C-terminal 11 kDa), *E. coli*	Continuous spectrophotometric assay	For full-length TCPTP, $K_m = 304\ \mu$M, $V_{max} = 62\,000$ U/mg; for TCΔC11PTP , $K_m = 194\ \mu$M, $V_{max} = 73\,000$ U/mg; D- and L-forms of Tyr(P) were equally effective as substrates; optimum pH is 4.75 for both enzymes	407
TCPTP (MPTP)	*In vitro* translation	Molecular mass	$M_r = 45\,000$ kDa	567
TCPTP	*In vitro* translation	Molecular mass	$M_r = 48\,000$ kDa	488
TCPTP (PTP-S)	Rat fibroblasts, lymphocytes	Molecular mass	$M_r = 42\,000$–$44\,000$ kDa by immunoblotting	187, 188
TCPTP	Purified from porcine spleen, immunoreactive with antibody to TCPTP	Molecular mass, catalytic activity and inhibition by phospholipids and inositol trisphosphate	$M_r = 53\,000$; SA ~ 75 nmol min^{-1} mg^{-1} against poly(Tyr:4Glu) as substrate; no serine phosphatase activity; RCM-lysozyme dephosphorylation was 5-fold > MBP dephosphorylation; strongly inhibited by phosphatidyl inositol ($IC_{50} = 6\ \mu$M), phosphatidyl serine ($IC_{50} = 3.7\ \mu$M), *myo*-inositol 1,4,5-trisphosphate ($IC_{50} \sim 2\ \mu$M) as well as vanadate ($IC_{50} = 5\ \mu$M)	340
TCPTP (MPTP)	Recombinant, cleaved GST-fusion protein, *E. coli*	Molecular mass, substrate dephosphorylation	$M_r = 40{,}000$; purified protein had V_{max} for *p*NPP of 44.4 μmol/min/mg protein and K_m of 0.476 mM	207

Table 7 Continued

PTPase	Protein source	Parameter	Data	Refs
TCPTP	Recombinant protein, full length and with deleted C-terminal region, *E. coli*	Substrate dephosphorylation	Truncated TCPTP readily dephosphorylates sub-μM concentrations of a variety of peptides but with some exhibits either very slow or undetectable activity; amino acid substitutions around Tyr(P) in the peptides strongly influences the catalytic efficiency: acidic residues on the N-terminal side of Tyr(P), especially position 3, play a crucial role in substrate recognition; basic residues at the same position act as negative determinants; with the full-length T cell-PTPase, the dephosphorylation efficiencies of all peptides tested are dramatically impaired with an ↑ in K_m values and a ± decrease in V_{max} values, suggesting that the C-terminal region alters the affinity of the enzyme for its substrates	331
	Recombinant protein, full length (48 kDa) and with deleted 11 kDa C-terminal region, BHK cells	Substrate dephosphorylation	Full-length enzyme (extracted from the particulate fraction) is essentially inactive until subjected to limited trypsinization; truncated form is fully active	251
	Recombinant protein, full length (48 kDa) and with deleted 11 kDa C-terminal region, baculovirus	Substrate dephosphorylation	Full-length enzyme (extracted from the particulate fraction) exhibited low level of activity for RCM-lysozyme, and was 12 times more active towards MBP; C-terminally truncated form was found in the soluble fraction and displayed five times higher activity towards RCM-lysozyme > MBP, suggesting that the C-terminal segment plays a role in regulation of the catalytic activity; both forms inhibited by μM vanadate, molybdate and Zn^{2+}, activation by polycationic molecules varied with substrate but was more pronounced with the full-length enzyme, suggesting that they interact with the C-terminal region	369
***T. californica* PTPase**	Purified protein	Substrate dephosphorylation	Dephosphorylated the tyrosine phosphorylated nicotinic acetylcholine receptor much faster than the insulin receptor and RCM-lysozyme, but did not dephosphorylate pp15, a fatty acid binding protein from adipocytes, suggesting relative specificity	1160
VH1	Recombinant GST-fusion protein, *E. coli,* cleaved with thrombin	Substrate dephosphorylation	Highly active towards *p*NPP, Tyr(P) raytide and MBP as well as p-ser casein; not inhibited by okadaic acid (1 μM), but completely blocked by 1 mM vanadate; unlike Type 2 Ser/Thr phosphatases have residual PTPase activity which is dependent on divalent cations	511
VHR	Recombinant GST-fusion protein, *E. coli*	Substrate dephosphorylation	Active with *p*NPP, with K_m in low μM range, ever after cleavage from GST fusion; dephosphorylated several autophosphorylated growth factor receptor (including PDGF, EGF, insulin and KGF); also dephosphorylated p-ser casein	524

VHR	Recombinant protein, *E. coli*	Molecular mass, substrate dephosphorylation	Protein is monomeric with mass of 20 500 Da by mass spectroscopy; preferred diphosphorylated peptide substrates over singly phosphorylated substrates by three to eight-fold; demonstrated rapid dephosphorylation of pY in peptide with slower dephosphorylation of pT, suggesting ordered reaction	995, 996
Yeast PTP1	Recombinant GST-fusion protein, *E. coli,* cleaved with thrombin	Molecular mass, substrate dephosphorylation	$M_r \sim 39$ kDa; hydrolysed Tyr(P) raytide 1000 times faster than p-ser casein substrate; specific activity for *p*NPP was 27 μM/mg/min; MBP 0.14 μM/mg/min; raytide 0.043 μM/mg/min	509
***Yersinia* PTPase**	Recombinant proteins, *E. coli*	*p*NPP hydrolysis	Kinetic constants at pH 5.0 – $k_{cat} = 1230\,s^{-1}$ with $V_{max} = 2.55$ mM	380
***Yersinia* PTPase, PTP1B**	Recombinant proteins, *E. coli*; PTP1B truncated form	Substrate dephosphorylation	Used a peptide corresponding to the autophosphorylation site of the EGF receptor (AA 988–998); Ala-scan was done by changing each residue of the peptide to Ala and analysed by kinetic constants with each PTPase; sequence for PTPase recognition from wild-type EGFR (DADEpYLIPQQG) demonstrated to be (DADEpYAAPA); presence of acidic residues proximate to the N-terminal side of phosphorylation site is critical for high-affinity binding and catalysis; K_m value for the peptide decreased as the pH increased suggesting that the phosphate dianion is favored for substrate binding; demonstrated that chemical features in the primary structure surrounding the pY in the substrate contribute to PTPase substrate specificity	376
	Recombinant proteins, *E. coli*; PTP1B truncated form	Substrate dephosphorylation	Used various-sized synthetic phosphotyrosine-containing peptides corresponding to the EGFR autophosphorylation site at Tyr-992; for both enzymes, efficient binding and catalysis required six residues (4 AA N-terminal to the pY and one residue C-terminal to the pY; dephosphorylated peptides do not bind to the PTPases	374
	Recombinant Yop51* and C-terminal truncated PTP1B	Substrate dephosphorylation by continuous spectrophotometric and fluorometric assay	K_{cat} values were relatively constant using a series of phosphotyrosyl peptides but K_m varied with changes in the AA sequence surrounding the tyrosine residue; Yop51* carries a Cys-235 → Arg mutation.	406
	Recombinant protein, *E. coli*	Dephosphorylation of alkyl phosphate substrates	k_{cat} values for alkyl phosphates (pyridoxal 5′-phosphate, D-glucose 6-phosphate, *O*-phospho-L-serine, *O*-phospho-L-theonine) are orders of magnitude slower than those for aryl phosphates (*p*NNP and *O*-phospho-L-tyrosine) and are similar to the k_{cat} values for the PTPase-catalysed ^{18}O-exchange reaction between inorganic phosphate and water; suggests that the rate-limiting step for the hydrolysis of alkyl phosphates is the formation of the phosphoenzyme intermediate	1045

Table 7 Continued

PTPase	Protein source	Parameter	Data	Refs
Yop51	Recombinant protein, *E. coli*		Specific for Tyr(P) residues, active on angiotensin, raytide and a Src-related peptide; no activity on p-ser/thr casein or histone H2a; rat insulin receptor is also a substrate for Yop51	513
	Recombinant protein, *E. coli*	Kinetic characterization of catalytic mechanism	Phosphate monoester hydrolysis proceeds by two-step formation and breakdown of a covalent phosphoenzyme intermediate; under acidic conditions, the rate-determining step is the breakdown of the intermediate; under more alkaline conditions, substrate effects also contribute to the reaction rate	375
	Recombinant protein, *E. coli*	Substrate dephosphorylation; physicochemical parameters	Yop51 is monomeric in solution; 31% α-helix, turnover number of $1200\,s^{-1}$ with *p*NPP; *Yersinia* PTPases are most highly reactive PTPases known; kinetic parameters are sensitive to the ionic strength of the medium	371
YVH1	Expressed as GST-fusion protein, *E. coli*	Substrate dephosphorylation	Active with *p*NPP and Tyr(P) casein, but not with p-ser casein or p-ser histone; thus, PTPase activity is limited to Tyr(P), unlike *Vaccinia* virus homologue VH1	510

Table 8 General methods used for assay, purification and cloning of PTPases

PTPase	Method	Source	Comments	Refs
Assays			Nonapeptide ENDYINASL derived from the TCPTP sequence phosphorylated by the EGF receptor kinase provides a sensitive substrate for PTPases	388
	Two non-radioactive assays for PTPases with IR as substrate		Assays use either HPLC to separate dephosphorylated peptide products or substrates immobilized on microtitre plates, with analysis by ELISA with antiphosphotyrosine antibodies	395
	96-well plate assay for Tyr(P) phosphopeptide dephosphorylation using malachite green reagent			271
	Automated, non-isotopic assay		Uses antiphosphotyrosine antibodies in a particle concentration immunoassay technique	386
	Colorimetric plate assay			390
	Continuous and discontinuous assays using Tyr(P) peptide substrates		Substrates included synthetic peptide Glu-Glu-Tyr(P)-Ala-Ala and the Fmoc derivative; hydrolysis of Tyr(P) was followed either spectrophotometrically or fluorometrically or of the Fmoc-derivative by discontinuous HPLC fluorescence assay	398
	Continuous spectrophotometric and fluorometric assay using Tyr(P) containing peptides		Dephosphorylation causes ↑ absorbance at 282 nm or ↑ fluorescence at 305 nm which can be followed continuously and analysed for kinetic parameters using the integrated form of the Michaelis–Menten equations; results were similar to those obtained with initial rates from inorganic phosphate assay	406
	Continuous spectrophotometric assay		Analysis employs changes in UV absorption and fluorescence of phosphotyrosine	407
	Immobilon-based solid phase assay	Human platelet membranes	Membrane proteins are resolved by PAGE and then blotted to a polyvinylidene difluoride membrane surface-labelled with [^{32}Tyr(P)](Glu$_4$:Tyr$_1$); PTPase activity appears as clear areas in the autoradiograph of the renatured ^{32}P Western blot; used to identify a 53 kDa membrane PTPase	389
	Microtitre enzyme-linked immunosorbent assay		Uses immobilized phosphotyrosine with assay by ELISA with antiphosphotyrosine antibody	397
	Phosphorylated synthetic peptides as substrates for PTPases			400
	Synthesis of Tyr(P) containing peptides used as substrates for PTPases		Synthesis of phosphopeptides using direct incorporation of N-α-(9-fluorenylmethyloxycarbonyl)-*O*-phospho-L-tyrosine (unprotected side chain); kinetic analysis revealed that both the rat brain PTPase and the human adipocyte acid phosphatase catalysed peptide dephosphorylation but with different rates and affinities	399

Table 8 Continued

PTPase	Method	Source	Comments	Refs
Assays (Cont.)	Use of a synthetic IR kinase domain peptide as substrate		Peptide is phosphorylated by the IR kinase	393
	Continuous spectrophotometric assay		Incorporates a coupled enzyme system that uses purine nucleoside phosphorylase and the chromophoric substrate 7-methyl-6-thioguanosine for the quantitation of inorganic phosphate	1052
	Direct and non-radioactive localization of phosphocysteine		Phosphopeptide first identified by liquid chromatography–electrospray mass spectrometry; following chemical modification with alkaline ethanethiol, *S*-ethylcysteine was identified during Edman degradation, demonstrating that phosphocysteine reacts like phosphoserine	1040
	Use of 2-methoxybenzoyl phosphate		Useful for high sensitivity continuous fluorometric and UV absorption spectrophotometric assays	1054
	Characterization of PTPase activity in intact cells		By using specific inhibitors of tyrosine kinases which enter the cells rapidly and do not affect the activity of PTPases the decay of phosphotyrosine in activated growth factor receptors can be monitored by immunoblotting with antiphosphotyrosine antibodies to study specific PTPases in intact cells	1050
CD45	Non-radioactive assay for CD45 PTPase activity isolated directly from cells		Uses immunoprecipitated CD45 in a colorimetric assay performed in microtitre plates with malachite green; sensitive to 100 pmol of free phosphate released	1053
	Purification and assay	Human spleen		405
Low M_r PTPase	Affinity chromatography	Bovine brain	Use of *p*-aminobenzylphosphonic acid–agarose for affinity chromatography	332
	Chemical gene synthesis and expression in *E. coli*		Recombinant protein shows reduced specific activity but retains affinity for substrates and inhibitors	190
Multiple	Affinity chromatography		Use of Zn^{2+}-iminodiacetate agarose provides quantitative recoveries of Zn^{2+}-inhibited protein PTPases; gradients of competing Zn^{2+} ligands, such as imidazole, provide the best purification	402
	Analysis of multigene PTPase family by DNA fingerprinting of amplified conserved enzyme domains		Detected at least 20 different mouse PTPases during development in specific patterns	731
	Assessment of PTPase expression levels by RT-PCR with degenerate primers			391

	Enzyme purification scheme	Bovine brain	Seven PTPase activities were isolated from bovine brain using phosphotyrosyl casein as a substrate; M_r values ranged from 24 kDa to 104 kDa; PTP-5 was the major activity accounting for 26% of the total; PTP-5 had a neutral pH optimum, and using Tyr(P)-casein as substrate it had a K_m of 130 nM and a V_{max} of 10 μmol P_i released min^{-1} mg^{-1}	392, 802
	Enzyme purification scheme	Human placenta		404
	Methods to distinguish various types of protein phosphatase activity		PTPases are specifically inhibited by μM Zn^{2+} or vanadate, and show maximal activity in the presence of EDTA; other cellular phosphatases, specific for protein Ser(P) and Thr(P) residues, are inhibited by fluoride and EDTA	387
	Use of vanadate as a PTPase inhibitor			673
PP2A	Purification of PTPase activator of PP2A	Rabbit skeletal muscle	Present in all tissues examined with highest abundance in brain	1039
PTP1B	Enzyme purification scheme	Human placenta	Use of thiophosphorylated, derivatized lysozyme as affinity ligand	403
	Enzyme purification scheme	Human placenta	Use of novel phosphotyrosine analogue affinity ligand, L-histidyldiazobenzylphosphonic acid–agarose	318
Various	Molecular, immunohistochemical and *in situ* hybridization	Brain	Review of various techniques for studying PTPases in brain tissue	1051

Table 9 Summary of PTPase mutations and their biochemical effects

PTPase	Mutation(s)	Expression system	Results	Comments	Refs
BVP	Cys-119 → Ser	Recombinant enzyme, *E. coli*	Abolished catalytic activity	Identifies catalytic site Cys residue	335
CD45	D1 alone; C-terminal Δ 78 AA; C-terminal Δ 91 AA; D2 Δ 21 acidic AA insert; various substitutions around HCSAG and GXGXXG motifs; Tyr-671, 729, 756, 1181, individually → Phe	*In vitro* transcription/translation system	D1 was not active alone; Δ C-terminal 78 did not affect activity but Δ additional 13 AA from D2 abolished activity; Δ acidic insert in D2 ↓ activity 4-fold; Tyr-729 in D1 is required for activity	Both PTPase domains and the 77 AA juxtamembrane residues were required for activity	281
	Wild-type and inactivated by mutagenesis of catalytic Cys residues	Transfection into CD45-deficient host cells	Catalytic activity of the membrane proximal domain is necessary and sufficient for restoration of TCR-mediated signalling events; putative catalytic activity of the 2nd domain is not required for signalling		257
(LCA)	Cys-379 → Ser Cys-828 → Ser Cys-1047 → Ser Cys-1144 → Ser	Recombinant cytoplasmic domain, *E. coli*	Only mutation in second Cys of the proximal domain (D1 Cys-828 → Ser) affected and completely abolished PTPase activity towards Tyr(P) raytide or angiotensin II		606
	Glu-180 → Gly in D2 domain	Recombinant enzyme, *E. coli*	Abolished PTPase activity against a phosphorylated Fyn peptide	Suggests importance of D2 domain in catalysis	1021
	Δ AA 876–931, region linking the D1 and D2 domains	Recombinant enzyme, *E. coli*	Abolished PTPase activity against a phosphorylated Fyn peptide	Suggests importance of D1–D2 intervening segment in catalysis	1021
cdc25	Arg-385 → Lys/Met	*Drosophila* cdc25	Abolished ability to dephosphorylate $p34^{cdc2}$ *in vitro* and to promote *Xenopus* oocyte maturation	Evidence that this Arg is involved in catalysis	262
	Arg-486 → Lys	*S. pombe* cdc25	Abolished ability to rescue a conditional cdc25 mutant in *S. pombe*	Evidence that this Arg is involved in catalysis	262
	Cys-379 → Ser	Recombinant C-terminal catalytic domain of *Drosophila* cdc25, *E. coli*	Abolished catalytic activity towards *p*NPP, or Tyr(P) cdc2 peptide or angiotensin		259
	Cys-379 → Ser/Ala	*Drosophila* cdc25	Abolished ability to dephosphorylate $p34^{cdc2}$ *in vitro* and to promote *Xenopus* oocyte maturation	Evidence that this is the catalytic Cys in the enzyme active site	262

(p80cdc25)	Cys-480 → Ser	*S. pombe* cdc25 expressed as GST-fusion protein of C-terminal catalytic domain in *E. coli*	Abolished the catalytic activity *in vitro*		304
	Cys-480 → Ser/Ala	*S. pombe* cdc25	Abolished ability to rescue a conditional cdc25 mutant in *S. pombe*	Evidence that this is the catalytic Cys in the enzyme active site	262
	His-335 → Ala	*Drosophila* cdc25	Retained ability to dephosphorylate $p34^{cdc2}$ *in vitro* and to promote *Xenopus* oocyte maturation	Not a catalytic site residue	262
	His-378 → Ala	*Drosophila* cdc25	Retained ability to dephosphorylate $p34^{cdc2}$ *in vitro* and to promote *Xenopus* oocyte maturation	Not a catalytic site residue	262
	His-479 → Ala	*S. pombe* cdc25	Retained activity towards rescuing a conditional cdc25 mutant in *S. pombe*	Not a catalytic site residue	262
	Thr-48 → Val Thr-67 → Val Thr-138 → Val Ser-205 → Ala Ser-285 → Ala	*Xenopus* cdc25 expressed as GST-fusion protein in Sf9 cells	The triple Thr mutation and the quintuple mutation abolish the *in vitro* phosphorylation of cdc25 by $p34^{cdc2}$ and inhibit the activation of cdc25 by this kinase by 70 and 90%, respectively; also in cdc25-depleted oocyte extracts, the mutant cdc25 proteins cannot activate $p34^{cdc2}$	Suggests a positive feedback loop between cdc2 and cdc25 is necessary for the full activation of cyclin B/$p34^{cdc2}$ that induces mitosis	91
cdi1	Cys-140 → Ser	Recombinant GST-fusion protein, *E. coli*	PTPase activity abolished		515
ERP (3CH134)	Δ257-264	Recombinant GST-fusion protein, *E.coli*	*In vitro*, deletion of six amino acids within the PTPase catalytic domain (HCXAGXXR)	Mutated protein lacked detectable *p*NPPase activity	571
KAP	Cys-139 → Ser	Recombinant, *E.coli*	*In vitro* phosphatase activity towards both Tyr(P) and p-ser abolished	Suggests common catalytic mechanism	518
LAR	Δ1275-1529 (most of first PTPase domain deleted)	Recombinant cytoplasmic domain expressed in *E. coli* extract	Catalytic activity abolished		343
	Δ1614-1881 (second PTPase domain deleted)	Recombinant cytoplasmic domain expressed in *E. coli* extract	Reversal of MBraytide activity such that raytide > MBP; ↓ MBP activity 8.5-fold		343
	Cys-1522 → Ser	Recombinant cytoplasmic domain (human) expressed in *E. coli* extract	Catalytic activity abolished		343

Table 9 Continued

PTPase	Mutation (s)	Expression system	Results	Comments	Refs
LAR (Cont.)	Cys-1522 → Ser	Recombinant cytoplasmic domain (rat) expressed in *E. coli* extract	Catalytic activity abolished		326
	R1149 → A	Transfection of full-length cDNA into HeLa cells	Proteolytic processing into extracellular and TM/cytoplasmic domain subunits is blocked		110
	Random mutation in D1 between positions 1329 and 1407	Recombinant proximal PTPase domain (D1) purified from *E. coli*	23 mis-sense mutations chemically induced; eight were temperature sensitive and localized to region between position 1329 and 1407; a second-site mutation at Cys-1446 → Tyr suppressed several temperature-sensitive mutations and enhanced the folding of LAR protein expressed in *E. coli*		356
	Various AA substitutions flanking catalytic Cys-1522	Recombinant cytoplasmic domain expressed in *E. coli* extract	At most positions, any substitution severely reduced enzyme activity		343
LCA (CD45)	Δ924-1281 (second PTPase domain deleted)	Recombinant cytoplasmic domain expressed in *E. coli* extract	Catalytic activity abolished		343
	Δ969-987 (intercatalytic domain deleted)	Recombinant cytoplasmic domain expressed in *E. coli* extract	Reversal of MBraytide activity such that MBP > raytide; ↑ MBP activity 6.5-fold		343
	Cys-828 → Ser	Recombinant cytoplasmic domain expressed in *E. coli* extract	Catalytic activity abolished		343
Low M_r PTPase	Asp-129 → Ala Asp-56 → Ala Asp-56 → Asn Asp-92 → Ala	Recombinant bovine enzyme, *E. coli*	Asp-129 → Ala caused a >2000-fold ↓ in V_{max} with *p*NPP, due to a change in the rate-limiting step of the catalytic reaction; Asp-129 is the proton donor for the leaving group; Asp-56 → Ala, Asp-56 → Asn, Asp-92 → Ala did not change V_{max} but did alter the K_m up to 7-fold		370

	Asp-129 → Ala		Only 0.04% specific activity of wild-type	Trapping experiments showed it retained ability to form covalent E–P complex; data suggest the Asp-129 is involved in both 1st step and the rate-limiting step of catalytic mechanism (nucleophilic attack)	347
(bovine heart)	Cys-12 → Ala/Ser Cys-17 → Ala/Ser Cys-62 → Ser Cys-90 → Ser Cys-109 → Ser Cys-145 → Ser Cys-148 → Ser Cys-149 → Ser Arg-18 → Ala	Recombinant protein, *E. coli*	CD spectra for each single mutation were essentially unchanged; only Cys-12, Cys-17 and Arg-18 had significantly altered catalytic activity towards *p*NPP, with Cys-12 and Arg-18 completely inactive; the Cys-17 mutant has a 4-fold higher phosphate K_i and slightly higher *p*NPP K_m values, suggesting that Cys-17 may serve to position the substrate phosphate moiety	Suggested Cys-12 is the catalytic nucleophile	253
	Cys-12 → Ser Cys-17 → Ser Arg-18 → Lys or Met	Recombinant protein, *E. coli*	Identified Cys-12 as the active site residue that is involved in nucleophilic attack and the formation of the phosphoenzyme intermediate; Cys-17 and Arg-18 are involved in substrate binding	Also demonstrated that Cys-62 and Cys-145 are not involved in the active site or formation of the covalent intermediate	249
(bovine liver)	Cys-17 → Ser Cys-62 → Ser Cys-145 → Ser Cys-12 → Ser or Ala	Synthetic gene expressed as fusion with maltose binding protein in *E. coli*	Kinetic properties studied with *p*NPP assay; Cys-12 mutants were both completely inactive; Cys-17 mutant was 200-fold decreased; Cys-62 mutant was 2.5-fold decreased; Cys-145 mutant was unchanged	Confirm the involvement of Cys-17 and Cys-62 in active site and suggest that Cys-62 and Cys-145 may destabilise the enzyme structure	241
	His-66 → Ala/Asn His-72 → Ala/Asn His-66/His-72 → Ala	Recombinant protein, *E. coli*	In mutant enzymes, ^{1}H-NMR spectra revealed no tertiary structure alterations; kinetic studies showed neither His residue was essential for catalysis; His-66 mutants had unchanged catalytic properties compared to wild-type; His-72 mutants had reduced specific activity and higher phosphatase K_i and lower K_m values at pH 5 and above	Suggesting His-72 has a significant role at the enzyme active site	254

Table 9 Continued

PTPase	Mutation (s)	Expression system	Results	Comments	Refs
Low M_r PTPase (bovine liver)	His-66 → Gln His-72 → Gln His-66/His-72 → Gln	Synthetic gene expressed as fusion with maltose binding protein in *E. coli*	Both single and double mutants showed ↓ k_{cat} values (30% and 7% of control for His-66 and His-72, resp); rate-determining step of His-66–Gln enzyme was same as wild-type, but for the His-72–Gln enzyme, the formation of the phosphoenzyme covalent intermediate and the dephosphorylation of the covalent enzyme intermediate contributed to the k_{cat} value; double mutant was completely inactive	Suggested that both His residues play a role in the enzyme active site	240
MPK-1 (3CH134)	Cys-258 → Ser	Epitope-tagged recombinant protein, *E. coli*	*In vitro,* abolished dephosphorylating activity towards Tyr(P) and p-thr of $p42^{MAPK}$; *in vivo* in transfected cells, augments MAP kinase phosphorylation and allows the formation of a complex with $p42^{MAPK}$	Suggests MPK-1 is a physiological MAP kinase phosphatase	345
MSG5	Cys-319 → Ala	GST-fusion protein, *E. coli*	Catalytically inactive against *p*NPP; also lost ability to promote recovery from pheromone-induced cell cycle arrest *in vivo*		497
PAC1	Cys-257 → Ser	Recombinant GST-fusion protein, *E. coli*	Phosphatase activity abolished towards ERK2 (MAP kinase) *in vitro* and *in vivo* in transfected cells		361
PRL-1	Cys-104 → Ser	Recombinant His-tagged fusion protein, *E. coli*	PTPase activity abolished	Catalytically inactive mutant	493
PTP1B	Cys-121 → Ser Cys-215 → Ser His-214 → Gln	Recombinant, *E. coli* Recombinant, *E. coli* Recombinant, *E. coli*	PTPase activity ↓ to 44% of control Complete loss of PTPase activity PTPase activity ↓ to 5% of control		89

RPTP-β	Deletion of the juxtamembrane segment, and successive deletion of N-terminal sequence	Recombinant cytoplasmic domain, *E. coli*	Deletion of the juxtamembrane segment (residues 1622–1639) can increase activity 5-fold; successive deletion of N-terminal sequence prior to residue 1684 had little effect on substrate affinity and, at most, reduced activity about 6-fold; further removal of residues 1684–1686 resulted in a marked 50–500-fold drop in activity, and loss of N-terminal sequence prior to residue 1690 abolished activity	Suggests that the juxtamembrane domain potentially functions as a negative regulatory sequence; also, identified a highly conserved motif (E/q) (F/y)XX(L/i), corresponding to positions 1684–1688 of RPTP-β	360
RPTP-μ	Cys-1095 → Ser	Recombinant GST fusion protein with cytoplasmic domain, *E. coli;* also immunoprecipitated protein expressed in COS cells	PTPase activity abolished	Catalytically inactive mutant	263
	Deletion of 53 AA in juxtamembrane region	Recombinant GST fusion protein of cytoplasmic domain, *E. coli*	Reduction in PTPase activity by ~2-fold		263
RPTP-ζ/β	Cys-1913 → Ser	Recombinant cytoplasmic domain expressed in *E. coli* extract	Catalytic activity abolished, suggesting that only the proximal PTPase domain is catalytically active	Consistent with natural replacement of the obligatory catalytic cysteine residue in D2 with aspartate; suggests that role of D2 is regulatory and not catalytic	539
SH-PTP1 (PTP1C)	ALLQ inserted between Asn-402 and Glu-403	Recombinant GST fusion protein, *E. coli*	PTPase activity of PTP1Ci (with insertion) ↓ to 11%–24% of wild-type control	Insertion into PTP1C analogous to sequence difference between PTP1D and PTP1Di splice variants	301
SH-PTP1 (PTP1C)	Cys-455 → Ser	Transfected 293 cells	↑ phosphorylation in response to PDGF due to loss of autodephosphorylation activity	Catalytically inactive mutant	79
SH-PTP1 (PTP1C)	Tyr-538 → Phe	Transfected 293 cells	Loss of Tyr phosphorylation in response to growth factor stimulation	Tyr-538 identified as site of Tyr(P) *in vitro* by IR kinase and *in vivo* by PDGF stimulation of transfected 293 cells	79
SH-PTP1	Deletion of both N-terminal SH2 domains	Recombinant enzyme, *E. coli*	Activated PTPase enzyme 30-fold	Additional data suggest SH2 domains autoinhibit the catalytic activity of the PTPase domain in a pY-independent fashion	1027

Table 9 Continued

PTPase	Mutation (s)	Expression system	Results	Comments	Refs
SH-PTP1	Deletion of 35 AA at C terminus	Recombinant enzyme, *E. coli*	Activated PTPase enxyme 24- to 30-fold; truncation of 60 AA at C terminus gave no activation over full-length enzyme	Suggests C-terminus also has inhibitory role to modulate PTPase activity	1027
SH-PTP2	Arg-138 → Lys	Recombinant, *E. coli*	Reduced allosteric activation of PTPase by pY peptides from IRS-1; affects additive in double Arg → Lys mutant in both SH2 domains	Mutation in C-SH2 domain	453
	Arg-32 → Lys	Recombinant, *E. coli*	Reduced allosteric activation of PTPase by pY peptides from IRS-1	Mutation in N-SH2 domain	453
	Tyr-580 → Phe	*In vitro* phosphorylation by PDGF-R	Analysis of tryptic phosphopeptides shows this to also be phosphorylated by PDGF-R, and is a consensus site for Grb2 binding	Syp variant of SH-PTP2 lacks the Tyr-580 site in the C-terminal segment which may impact on signalling by Syp *vs.* other SH-PTP2 forms	78
	Tyr-542 → Phe	Transfection of ATWT cells	Analysis of loss of tryptic phosphopeptides by mutation of Tyr-542 establishes this as major *in vivo* phosphorylation site by PDGF-R		78
(PTP2C)	Deletion of both SH2 domains	Recombinant, *E. coli*	PTPase activity ↑ 12- to 45-fold after SH2 deletion, depending on substrate		383
(PTP2C)	Deletion of both SH2 domains	Recombinant, *E. coli*	*In vitro* dephosphorylation of IRS-1 was ↓ to 33% of control after deletion of SH2 domain; activity against IR kinase domain and other substrates ↑ > 10-fold after SH2 domain deletion	Suggests SH-PTP2 is activated by binding to IRS-1 and may dephosphorylate it *in vivo*	289
STEP	Cys-300 → Ser	Recombinant GST-fusion protein, *E.coli*	PTPase activity abolished		169
VH1	Cys-110 → Ser	Recombinant GST-fusion protein, *E. coli,* cleaved with thrombin	Abolishes catalytic activity towards both Tyr(P) and p-ser substrates		511
VHR	Cys-124 → Ser	Recombinant GST-fusion protein, *E. coli*	Abolished activity towards *p*NPP, as well as both Tyr(P) and p-ser phosphatase activities	Cys-124 proved to be catalytic cysteine in HCXXXXXR motif	524

	Ser-131 → Ala	Recombinant enzyme, *E. coli*	The k_{cat} value for the S131A mutant is 100-fold lower than that for the native enzyme, and the reaction became independent of pH over the range 4.5–9.0	Role of the conserved hydroxyl in the active-site sequence motif HCXXGXXRS(T) was investigated; data suggest that the S131A mutation alters the rate-limiting step in the catalytic mechanism	993
	Asp-92 → Asn Glu-6 → Gln Glu-32 → Gln Asp-14 → Asn Asp-110 → Asn		Asp-92 → Asn enzyme was 100-fold less active than the native enzyme and exhibited the loss of the basic limb in the pH profile, suggesting that in the native enzyme D92 must be protonated for activity; other mutations had less than a two-fold effect on the kinetic parameters compared to native enzyme	The D92 residue is conserved throughout the entire family of dual-specific phosphatases	994
	Cys-124 → Ser	Recombinant enzyme, *E. coli*	Abolished formation of phosphoenzyme intermediate and catalytic activity	Identifies catalytic site Cys residue and shows that the dual-specificity phosphatases have similar mechanism to tyrosine specific PTPases	1048
***Yersinia* PTPase**	Glu and Asp residues, including: Glu-276 → Gln Glu-290 → Gln Asp-356 → Asn Glu-224 → Gln Asp-243 → Asn Glu-363 → Gln Glu-459 → Gln		Asp-356 and Glu-290 are the general acid and base catalysts responsible for PTPase-catalysed phosphate ester hydrolysis; double Glu-290 → Gln and Asp-356 → Asn mutant showed no pH dependence for catalysis, but did exhibit a rate enhancement of 2.6×10^6 over non-catalysed *p*NPP hydrolysis by water probaby via a transition-state stabilisation (wild-type enzyme has a rate enhancement of 10^{11})	Suggests that PTPases use a common mechanism that depends on the formation of a thiol phosphate intermediate and use general acid–base catalysis; kinetic constants at pH 5.0 – $k_{cat} = 1230\ s^{-1}$ with $V_{max} = 2.55$ mM	380
	Cys-403 → Ser	Recombinant enzyme, *E. coli*	Abolished catalytic activity and blocked formation of covalent phosphoenzyne intermediate	Identified catalytic site Cys residue by site-directed mutagenesis as well as chemical modification which suggests	1046
	Arg-409 → Lys/Ala	Recombinant enzyme, *E. coli*	Significantly reduced catalytic activity and altered enzyme affinity for arsenate	critical role in PTPase reaction and importance of the Cys(X)$_5$-Arg catalytic motif	1046

Table 9 Continued

PTPase	Mutation (s)	Expression system	Results	Comments	Refs
Yop51	Cys-235 → Arg	Recombinant protein, *E. coli*	↑ Yield of recombinant enzyme	Structure and kinetics identical to wild-type Yop51	371
	Cys-403 → Ala or Ser	Recombinant protein, *E. coli*	Abolished enzymatic activity		513
	His-402 → Ala	Recombinant protein, *E. coli*	Altered active site thiol pK_a from 4.67 to 7.35	Similar effects as His-402 → Asn mutation	373
	His-402 → Asn	Recombinant protein, *E. coli*	Altered active site thiol pK_a from 4.67 to 5.99	Also exhibited enhanced reactivity towards iodoacetate; suggests involvement of His-402 in stabilising the catalytic thiolate anion	373

Table 10 Modification of PTPases by toxins, chemical agents, proteolysis and PTPase inhibitors

PTPase	AA/region modified	Reagent/manipulation	Protein source/form	Functional parameter measured	Comments	Refs
		Thiophosphotyrosylated RCM-lysozyme (TRCML)		Microinjection into sea urchin eggs, studied development	At least one site of action is late in the first cell cycle near the G2/M boundary; TRCML did no inhibit MPF activation	1143
cdc25B		Benzoquinoid antitumour compounds, dnacin A1 and dnacin B1	Recombinant GST-fusion protein with the catalytic domain of cdc25B		Using *p*NPP, dnacin A1 and dnacin B1 inhibited phosphatase activity in a non-competitive manner	1144
CD45		Suramin			Alters Tyr(P) content of proteins in many cancer cell lines; strongly inhibits activity of CD45 in a non-competitive, irreversible reaction that is complete within 10 min	672
		Aporphine alkaloids, annonaine, nornuciferine and roemerine, metabolites of *Rollinia ulei*		High-throughput screening bioassay for CD45		1147
Endogenous and PTP1B		Synthetic tris-sulfotyrosyl dodecatpeptide analog of the insulin receptor kinase domain		Inhibition of IR dephosphorylation *in situ* in permeabilized cells and by PTP1B *in vitro*		693
LAR	Catalytic domain			Exchange of oxygen from phosphate to water	Demonstrated enzymatic exchange of ^{18}O from $P^{18}O_4$ inorganic phosphate into $H_2{}^{16}O$ at $1 \times 10^{-2}\ s^{-1}$; this exchange appears to proceed by a phosphoenzyme intermediate which was detected after incubation of enzymes with a ^{32}Tyr(P) peptide substrate	244

Table 10 Continued

PTPase	AA/region modified	Reagent/manipulation	Protein source/form	Functional parameter measured	Comments	Refs
LAR (Cont.)	Cys-1522	[^{14}C]iodoacetate labelling	Recombinant cytoplasmic domain, *E. coli*	Active site labelling	LAR is inactivated by iodoacetate and not by iodoacetamide; iodoacetate binds at the active site of the enzyme with a stoichiometry of 0.8 mol of iodoacetate bound per mol of LAR; a [^{14}C]iodoacetate-labelled peptide was isolated from LAR, and Cys-1522 was found to contain the radiolabel; iodoacetate reacts only with the first domain of this double domain PTPase; this is the highly reactive cysteine residue at the active site of LAR that participates as a nucleophile in the catalytic reactions	325
	Cys-1522	[^{32}P]–tyr-angiotensin I	Recombinant cytoplasmic domain, *E. coli*	Trapping of ^{32}P-phosphoenzyme intermediate	Not visualized with Cys-1522 → Ser catalytic site mutation	326
	Cys-1522	Rapid chemical quenching	Recombinant 40 kDa fragment of LAR D1, *E. coli*	Trapping of a covalent phosphocysteine enzyme intermediate	Rapid chemical quench studies identified a phosphocysteine intermediate as a covalent phosphoryl compound; rates of formation ($1200\,s^{-1}$) and decay ($80\,s^{-1}$) of the intermediate were faster than the steady-state turnover rate ($24\,s^{-1}$) indicating that product release is most likely rate-limiting	242
Low M_r PTPase		Pyridoxal 5′-phosphate		PTPase assay	The type two isoform is strongly inhibited by pyridoxal 5′-phosphate; using a series of analogues, showed that pyridoxal 5′-phosphate interacts with the enzyme in both the phosphate and aldehyde groups; by site-directed mutagenesis, showed that the sites of pyridoxal 5′-phosphate binding involves Cys-17, which binds the phosphate moiety; Cys-12 does not participate in binding	248

Low M_r PTPase		Pyridoxal 5′-phosphate	Purified protein, porcine liver		Strongly inhibited by pyridoxal 5′-phosphate ($K_i = 21\ \mu M$) as seen with the bovine liver enzyme, rat liver (AcP2 isoenzyme), and human erythrocyte (B-slow isoenzyme)	483
	Cys-?			Trapping of a covalent phosphocysteine enzyme intermediate, identified by ^{31}P NMR		253
	(Cys-17 → Ala mutant)	Diethyl pyrocarbonate, phenylglyoxal, cyclohexanedione, iodoacetate, iodoacetamide, phenylarsine oxide, and certain epoxides; phosphomycin is simply a competitive inhibitor of the enzyme, but 1,2-epoxy-3-(*p*-nitrophenoxy)propane (EPNP) and (*R*)- and (*S*)-benzylglycidol act as irreversible covalent inactivators, consistent with the importance of a hydrophobic moiety on the substrate in controlling substrate specificity; phenylarsine oxide acts as a very slow, tight-binding inhibitor of the enzyme			Results are interpreted in terms of an active site model that incorporates a histidine–cysteine ion pair, similar to that present in papain	372
	Cys-62 and Cys-145 in active site	EPNP (1,2-epoxy-3-(*p*-nitrophenoxy)propane), irreversible inhibitor		Tryptic digests of enzymes labelled with the EPNP showed involvement of Cys-62 and Cys-145 in catalysis;	Suggested a model that included a histidine–cysteine ion pair at the enzyme active site	372

Table 10 Continued

PTPase	AA/region modified	Reagent/manipulation	Protein source/form	Functional parameter measured	Comments	Refs
Low M_r PTPase (Cont.)		Iodoacetate and 1,2-cycloexanedione; pyridoxal 5′-phosphate and analogues pyridoxamine 5′-phosphate and pyridoxal	Purified from bovine brain		Inactivated by iodoacetate and 1,2-cycloexanedione; P_i, a competitive inhibitor, protected the enzyme from inactivation and demonstrated the involvement of both cysteine(s) and arginine(s) at the active site; the strong inhibition exerted by pyridoxal 5′-phosphate and the low inhibitory capacity possessed by the pyridoxal 5′-phosphate analogues pyridoxamine 5′-phosphate and pyridoxal, indicate that at least one lysine residue is present at the active site	332
	Cys-12 and Cys-17	Nitric oxide-generating compounds	Recombinant proteins	PTPase activity	Both enzymes were inactivated by nitric oxide-generating compounds; and inorganic phosphate, as a competitive inhibitor, protected the enzymes from inactivation; inhibition results from oxidation of Cys-12 and Cys-17 to form S–S bond; enzyme was reactivated with thiol-containing reagents; also studies *Yersinia enterocolitica* PTPase	82
PTP1B	Cys-215	^{32}Tyr(P)-raytide or ^{32}P-*p*NPP	Recombinant, *E. coli*	Trapping of ^{32}P as cysteine–phosphate intermediate	Suggests catalytic mechanism	89
		Phosphonmethylphenyl-alanine (Pmp) and phosphonodifluoro-methyl-phenylalanine as non-hydrolysable Tyr(P) mimetics in hexameric Pmp-peptide (Ac-D-A-D-E-Pmp-L-amide)		Dephosphorylation of the IR	K_i of 200 μM for (D/L) Pmp or 100 nM for L-F_2-Pmp using PTP1B dephosphorylation of the IR; use of phosphonodifluoromethyl phenylalanine gives three orders of enhancement of affinity relative to Pmp	658

	Catalytic domain	SH-modifying agents (e.g. NEM, IA)	Purified from human placenta/C-terminal truncated	PTPase activity	Inactivated by SH-blocking agents; can be reactivated after oxidation by DTT > β-ME	352
		pTyr mimetic, L-*O*-malonyltyrosine (L-OMT) incorporated into peptides		Inhibition of PTP1B enzyme activity	The non-phosphorus containing p-tyr L-OMT is superior to phosphonomethyl phenylalanine (Pmp) as a pTyr mimetic when incorporated into the hexamer peptide Ac-D-A-D-E-X-L-amide X = D,L-Pmp, IC_{50} = 200 μM; X = L-OMT, IX_{50} = 10 μM)	1145
SH-PTP1 (PTP1C)	C terminus (~5 kDa segment deleted)	Limited tryptic cleavage	Recombinant protein, from 293 cells infected with adenovirus expression system	PTPase activity	Limited trypsinolysis of C-terminal segment generated a 63 kDa fragment with 20- and 10-fold ↑ activity towards RCM-lysozyme and MBP, respectively	381
(SHP)		Suicide inactivation agent, 4-difluoromethylphenyl bis(cyclohexylammonium) phosphate	Recombinant protein, *E. coli*		4-Difluoromethylphenyl bis(cyclohexylammonium) phosphate shown to be a time-dependent suicide inactivator of the SHP PTPase; inactivation of SHP followed pseudo-first-order kinetics, with a $t_{1/2}$ ~ 15 min in the presence of 8.2 mM inhibitor; mechanism of inactivation probably involves the enzymatic release of difluoromethyl phenol which rapidly eliminates fluoride, generating a quinone methide. This potent electrophile then reacts with residues at the active site of the enzyme	720
SH-PTP2	C terminus (at Lys536, Arg-537 or Lys-538)	Limited tryptic cleavage	Recombinant, *E. coli*	PTPase activity	Tryptic cleavage ↑ PTPase activity 27-fold, but further activation by pY peptides is lost; larger or smaller truncations retain negative autoregulation	453

Table 10 Continued

PTPase	AA/region modified	Reagent/manipulation	Protein source/form	Functional parameter measured	Comments	Refs
SH-PTP2 (PTP2C)	C terminus (~4 kDa segment deleted)	Limited tryptic cleavage	Recombinant, *E. coli*	PTPase activity	Tryptic cleavage of wild-type enzyme ↑ PTPase activity 2- to 5-fold, but had no effect on form with truncation of SH2 domains	383
TCPTP	C terminus (11 kDa segment deleted)	Limited tryptic cleavage	Recombinant protein, from BHK cell expression system	PTPase activity	Full-length enzyme (extracted from the particulate fraction) is essentially inactive until subjected to limited trypsinization; truncated form is fully active	251
Various		(Phosphonomethyl)-phenyl-alanine (Pmp), a phosphonate-based mimetic of *p*Tyr as a non-hydrolysable phosphotyrosyl mimetic			Useful as a phosphatase-resistant ligand to interfere with SH2 domain interactions; pmp analogues bearing fluorine or hydroxyl substituents on the phosphonate alpha-methylene carbon demonstrate a binding potency in the order HPmp < Pmp < FPmp < F2Pmp = pTyr	659, 662
		Two heat-stable PTPase inhibitors from bovine brain			Two PTPase inhibitors, inhibitor H (M_r > 500 kDa) and inhibitor L (M_r 38 kDa) were partially purified; both are proteins, as judged by their inactivation by proteinase K, and they exhibited remarkable stability during incubation at 95°C	686
		Dephostatin			Produced by *Streptomyces;* inhibited PTPase prepared from a human neoplastic T-cell line with an IC_{50} at 7.7 μM; the inhibitory pattern was competitive against the substrate; also inhibited the growth of Jurkat cells; structure was elucidated to be 2-(N-methyl-N-nitroso)hydroquinone by spectral and chemical analyses	685, 689

		Thiophosphoryl protein and peptide substrate analogues		Substrate dephosphorylation	Inhibition was rapid and reversible and competitive in nature; in general K_i was parallel to the corresponding K_i for the phosphorylated substrate	681
		Vanadate				673
VHR		RK-682 (3-hexadecanoyl-5-hydroxymethyl-tetronic acid) isolated from microbial metabolites			In vitro, RK-682 inhibited dephosphorylation activity of CD45 and VHR with IC_{50} of 54 and 2.0 µm, respectively; *in situ*, sodium orthovanadate and RK-682 enhanced the phosphotyrosine level of Ball-1 cells; sodium orthovanadate inhibited the cell cycle progression at G_2/M boundary phase, on the other hand, RK-682 inhibited the G_1/S transition	1141
***Yersinia* PTPase, PTP1B**		Thiophosphoryl peptide analogues and phosphonomethylphenyl-alanine peptide analogues	Recombinant protein, *E. coli*	Modified phosphotyrosyl peptide dephosphorylation	Thiophosphoryl analogues are dephosphorylated by the PTPases; a phosphonomethyl phenylalanine analog is a competitive and nonhydrolysable inhibitor with K_i values of 18.6 and 10.2 μM for *Yersinia* enzyme and PTP1B, respectively	374
Yop51	Cys-403	Iodoacetate	Recombinant protein, *E. coli*	PTPase activity	Alkylating agent iodoacetate inactivated the Yop51 enzyme and Cys-403 was labelled by iodoacetate indicating its involvement in catalytic mechanism; also favored iodoacetate over iodoacetamide (940-fold); pK_a of reactive thiol group = 4.67	373

Covalent modification of PTPases

Post-translational modifications of PTPases include proteolytic processing of enzyme proproteins, glycosylation, phosphorylation and myristoylation (Table 11).

Extracellular domains

The large, transmembrane PTPases have been shown in several cases to undergo proteolytic processing of a proprotein that produces protein subunits, including LAR, RPTP-κ and RPTP-σ. The site and determinants of LAR proprotein cleavage have recently been evaluated [105, 110, 116]. LAR is synthesized initially as a ~200 kDa proprotein that is processed into a complex of two non-covalently associated subunits. The E-subunit (150 kDa) contains the cell adhesion molecule domains, the P-subunit (85 kDa) contains a short segment of extracellular region and the transmembrane and cytoplasmic domains. Cleavage occurs at a pentabasic (RRRRR) motif at residues 1148–1152, and is catalysed by a subtilisin-like endogenous enzyme. The unprocessed enzyme can also be detected at the cell surface. Studies in transfected HeLa cells have shown that the E-subunit is shed during cell growth; whether this is a mechanism for regulating PTPase function remains to be determined [110].

Like many receptor-like cell-surface molecules, the extracellular domain of most of the transmembrane PTPases is glycosylated. The RPTP-ζ/β homologues is more extensively processed to a chondroitin sulfate proteoglycan which forms a complex with the extracellular matrix protein tenascin [77, 106, 556]. This processed extracellular domain may be involved in neurodevelopmental processes involving axonal migration.

Phosphorylation of PTPases

Several PTPases are modified by phosphorylation on serine or threonine residues; tyrosine phosphorylation of PTPases has also been observed but the effects on PTPase activity have been difficult to evaluate because of autodephosphorylation and the requirement for PTPase inhibitors in these studies. SH-PTP2 is phosphorylated on tyrosine by the PDGF receptor kinase *in vivo* and *in vitro* [78, 435]. This elicits the association of SH-PTP2 with the GRB-2 adapter protein and may have physiological significance in downstream signalling pathways. In the C-terminal region, the alternatively spliced 'Syp' form of the enzyme lacks one of the potential tyrosine phosphorylation sites for the PDGF receptor kinase (Tyr-580), which may influence the signalling potential of this PTPase isoform. PTP1B is phosphorylated on Ser residues in a pattern that accompanies progression or arrest of the cell cycle [86, 104]. Other important examples of phosphorylation of PTPases are summarized in Table 11.

Table 11 Post-translational modification of PTPases

Protein modified	Site of modification	Protein source/ form	Modification (in vivo/in vitro)	Enzyme	Functional parameter measured	Comments	Refs
CD45	Extracellular domain	K562 erythroleukaemia cells	Glycosylation	Endogenous		Treatment with tunicamycin to inhibit protein N-glycosylation produced CD45 polypeptides of 130 and 140 kDa; tunicamycin ↓ cell surface expression of CD45 and associated ↓ in PTPase activity; intact glycosylation is required for stability and cell surface transport of CD45	102
			Glycosylation	Endogenous	Glycosylation carbohydrate structure	Asn-linked CHO chains comprised six mol per mol of CD45 protein; were primarily sialyl derivatives as α-2,6-linked oligosaccharides as bi-, tri- and tetra-antennary complexes; presence of additional sugar groups was also characterized	103
			Glycosylation	Endogenous	Isoform-dependent glycosylation pattern	Amino acids at the junction of exons 3 and 7 are responsible for generating glycosylation-related CD45 isoforms that are expressed in a manner related to cell lineage and dependent on cell activation (identified by monoclonal antibodies)	586
			Glycosylation	Endogenous		Exponentially growing cells expressed a change in N-linked saccharide structure compared to resting cells; occurred at late stage of post-translational processing of CD45	100
		YAC-1 cultured T lymphocytes	Myristoylation	Endogenous		The labelling of CD45 was resistant to mild alkaline methanolysis and was found in fatty acid and sphingosine, indicating a novel chemical attachment to the protein	58, 112

Table 11 Continued

Protein modified	Site of modification	Protein source/ form	Modification (in vivo/in vitro)	Enzyme	Functional parameter measured	Comments	Refs
CD45	Alternatively spliced extracellular domain	Lymphocytes	*O*-linked glycosylation	Endogenous		Amino acid sequencing detected extensive *O*-linked glycosylation	93
			Ser/Tyr phosphorylation (*in vivo*/*in vitro*)	Endogenous	PTPase activity	PTPase activity of CD45 is enhanced when phosphorylation of tyrosine precedes that of serine but phosphorylation in the reverse order yields no activation; any of four protein-tyrosine kinases tested, in combination with the protein-serine/threonine kinase, casein kinase II, was capable of mediating this activation *in vitro*	108
		IL-4 dependent cell line	Ser phosphorylation	Endogenous	PTPase activity	Phosphorylated in response to IL-2 stimulation; 3- to 5-fold ↑ in Ser phosphorylation in time- and concentration-dependent manner on different Ser residues than that observed in resting cells; pattern was different from that seen with treatment of cells with PMA; no change in amount of cell-surface CD45 seen in IL-2 treated cells; inc. in p-Ser content did not affect PTPase activity towards RCM-lysozyme	114
	Cytoplasmic domain	Peripheral T cells	Ser/Thr phosphorylation	Endogenous		Stimulation with phorbol 12,13-dibutyrate markedly ↑ phosphorylation of CD45 within 30 s	75
	Cytoplasmic domain	Human T cells	Ser/Thr phosphorylation	Endogenous PKC	PTPase activity	Stimulation with PMA resulted in Ser/Thr phosphorylation of various CD45 isoforms relative to their surface expression; phosphorylation significantly downregulated the CD45 PTPase activity	223
		Jurkat T-cell leukaemia cell line	Transient Tyr phosphorylation *in vivo*	Endogenous		Stimulation with PHA or anti-CD3 antibodies induced transient Tyr(P) phosphorylation (40 min duration), only under conditions of PTPase inhibition	107

	Numerous		Sulfation *in vivo*	Endogenous		CD45 is a sulfated molecule with numerous potential sulfation sites	917
	Extracellular domain	Exons expressed in *E. coli*	*O*-glycosylation	Endogenous	Binding of monoclonal antibodies	Assessed contribution to antigenicity of polypeptide backbone and *O*-linked glycosylation of CD45; 14/17 mAbs for human and 4/6 mAbs for rat CD45R reacted with polypeptides containing exons 4–6 of CD45; carbohydrate was shown to affect the kinetics of binding to the polypeptide backbone	935
cdc25			Serine and threonine phosphorylation *in vivo* and *in vitro*	Cyclin A/cdk2 and E/cdk2	Enzyme activation	Cdc25 is activated at the G2/M transition by phosphorylation on Ser and Thr residues; cdc35 was found to be phosphorylated and activated by cyclin A/cdk2 and cyclin E/cdk2 *in vitro*; in interphase *Xenopus* egg extracts with no detectable cdc2 and cdk2, treatment with the phosphatase inhibitor microcystin activated a distinct kinase that could phosphorylate and activate cdc25	919
	Thr-48, Thr-67, Thr-138	*Xenopus* cdc25 expressed as GST-fusion protein in Sf9 cells	*In vitro*	Cyclin A/$p34^{cdc2}$ and cyclin B/$p34^{cdc2}$	$P34^{cdc2}$ kinase activating ability; mobility shift on electrophoresis	Phosphorylation of these sites activates the ability of cdc25 to increase the p34 kinase activity; mutagenesis of these residues abolishes this activating function of cdc25	91
		Xenopus cdc25	Ser/Thr phosphorylation *in vivo*	Endogenous	Phosphorylation and PTPase activity	cdc25 activity oscillates in meiotic and mitotic cell cycles, ↓ in interphase and ↑ in M phase; ↑ activity accompanied by ↑ phosphorylation and an ↑ of mobility from an apparent M_r of 76 to 92 kDa; reversed by treatment with PP-1 or PP-2A	92

Table 11 Continued

Protein modified	Site of modification	Protein source/ form	Modification (in vivo/in vitro)	Enzyme	Functional parameter measured	Comments	Refs
cdc25C (Cont.)						Phosphorylation of cdc25C in mitotic HeLa cell extracts or *in vitro* by cdc2-cyclin B ↑ its catalytic activity; only after cdc25C is stably thiophosphorylated can it activate cyclin B1-cdc2 and induce *Xenopus* oocyte maturation	90
			In vivo and *in vitro*	$p34^{cdc2}$	cdc25C (Ser/Thr) phosphorylation; activation of $p34^{cdc2}$ and mitosis	Phosphorylation of cdc25C ↑ during G_2/M transition and it is phosphorylated by $p34^{cdc2}$ on five sites both *in vivo* and *in vitro*; phosphorylation of cdc26C ↑ its PTPase activity by 2- to 3-fold; microinjection studies show that only the phosphorylated form of cdc25C is effective in activating G_2 cells into premature prophase	109
	Ser 216	HeLa cells	Serine phosphorylation	Partially purified		Ser-216 is the major site of cdc25C phosphorylation; isolated a protein kinase that binds to cdc25C and phosphorylates serine 216; the cdc25C-associating kinase was purified over 8000-fold from rat liver as a 36–38 kDa doublet of proteins	922
DPTP69D, DPTP10D, DPTP99A	Extracellular domains	*Drosophila* embryos	Glycosylation resulting in the 'HRP' epitope	Endogenous		Carbohydrate determinant identified with antibody recognizing horse-radish peroxidase (HRP) epitope	1099
HePTP		Rat mast cells	Tyrosine phosphorylation	Endogenous		Aggregation of high affinity IgE receptors induced tyrosine phosphorylation of HePTP; effect was mimicked by stimulation with calcium ionophore A23187 and was dependent on a rise in intracellular Ca^{2+}; suggest that HePTP is involved in the IgE receptor signalling cascade	923

IphP	? site of prokaryotic signal sequence	Expression in *E. coli* host	Secretion, proteolytic processing	*E. coli* enzyme		In cells, IphP has $M_r = 30\,000$ kDa; when secreted into medium, is 1 kDa smaller, suggesting proteolytic processing during secretion	584
LAR	Extracellular domain	Transfected murine pre-B lymphocyte cell line 300-19	Proprotein proteolytic processing and shedding of extracellular domain	Endogenous		Using site-directed mutagenesis, scanned residues in the small ectodomain of the P subunit (includes the TM and cytoplasmic domains) and identified three residues that are essential for proprotein cleavage, two are in the penta-arginine sequence and one C-terminal to the cleavage site; also, several non-contiguous residues were identified that are important for subunit association; LAR E-domain shedding is a consequence of proteolytic cleavage at a 2nd site within the P-subunit ectodomain near the TM peptide; P-subunit N-terminal sequence of transfected enzyme after aminopeptidase treatment was AERLKPY, indicting that the cleavage site is between AA 1152 and 1153; three regions play a role in subunit association, 1172–1173, 1183–1200, 1210–1211	105
		Transfected HeLa cells	Proteolytic cleavage	Endogenous (subtilisin-like)	Appearance of LAR subunits and shedding of extracellular domain	LAR cDNA expressed initially as a ~200 kDa proprotein that is processed into aa complex of two non-covalently associated subunits, E-subunit (150 kDa) contains the cell adhesion molecule domains, the P-subunit (85 kDa) contains a short segment of extracellular region and the TM and cytoplasmic domains; cleavage occurs at a pentabasic site RRRRR (AA 1148–1152), as shown by mutational analysis; unprocessed LAR can be expressed at the cell surface; uncleaved LAR has comparable PTPase activity to the wild-type enzyme; E-subunit apparently shed during cell growth and may be mechanism for regulating PTPase function	110

Table 11 Continued

Protein modified	Site of modification	Protein source/ form	Modification (in vivo/in vitro)	Enzyme	Functional parameter measured	Comments	Refs
LAR (Cont.)	Extracellular domain	Transfected rat fibroblast 208F and COS cells	Proteolytic cleavage; glycosylation	Endogenous (subtilisin-like)	Appearance of LAR subunits	Expressed as 190 kDa precursor that is subsequently cleaved into two fragments that remain non-covalently associated: 145 kDa (glycosylated, fully extracellular) and 85 kDa (TM and cytoplasmic domains)	116
	Cytoplasmic domain	Recombinant cytoplasmic domain, *E. coli*	Ser/Thr phosphorylation (*in vitro*)	Protein kinase C		Phosphorylated by PKC to 0.6 mol phosphate/mol. of LAR; without apparent effect on PTPase activity, LAR not a substrate *in vitro* for cAMP-dependent protein kinase or casein kinase II	326
		Recombinant cytoplasmic domain, *E. coli*	Tyr phosphorylation (*in vitro*)	$P43^{v-abl}$ kinase		Catalytically inactive (Cys-1522 → Ser) mutation of LAR was phosphorylated on tyrosine by $p43^{v\text{-}abl}$ kinase to 0.4 mol phosphate/mol. of LAR; autodephos-phorylation of native enzyme prevented assessment of effect of Tyr(P) phosphorylation on PTPase enzyme activity	326
PTP-PEST	Ser-39 and Ser-435	Recombinant protein expressed in baculovirus vector and purified to homogeneity	Ser phosphorylation (*in vitro*)	PKA and PKC	PTPase activity	*In vitro*, phosphorylated by PKA and PKC at Ser-39 and Ser-435; p-ser also at these sites *in vivo* after treatment of HeLa cells with TPA, forskolin or IBMX; *in vitro*, also PTP-PEST phosphorylation ↓ PTPase activity by ↓ affinity for substrate; Ser-39 found to be the PTPase regulatory site by examining truncated protein constructs	88
PTP1B	N-terminal Met	Human placenta	N-acetylation	Endogenous	–	–	484
	C-terminal ER targeting sequence	Human platelets	Proteolytic cleavage (*in vivo*)	Calpain (endogenous)	Relocalization from ER to cytosol	Induced by platelet agonists; requires engagement on platelet surface; cleavage ↑ PTP1B PTPase activity by 2-fold	87

	Segment 283–364 (?Ser-352)	HeLa cells	Ser phosphorylation (*in vivo*)	Endogenous	PTPase activity (↑ 4–5-fold in cell membranes)	Phosphorylation induced by activation of cAMP -dependent or Ca^{+2}/phospholipid-dependent kinases	81
	Ser-352	HeLa cells	Ser phosphorylation (*in vivo*)	Endogenous (kinase unknown)	PTPase activity (70% of control)	Increased p-ser accompanies $G_{two} \rightarrow M$ phase of cell cycle	86
	Ser-378	HeLa cells	Ser phosphorylation (*in vivo*)	Endogenous (PKC)	PTPase activity (unchanged)	Induced by TPA stimulation; same site phosphorylated by PKC *in vitro*	86
	Ser-386	HeLa cells	Ser phosphorylation (*in vivo*)	Endogenous ($p34^{cdc2}$ kinase)	–	Also induced by $p34^{cdc2}$ *in vitro*	86
	–	HeLa cells	Ser phosphorylation (*in vivo*)	Endogenous	PTPase activity (unchanged)	Induced by mitotic arrest of cells	104
RPTP-α	?	Rat L6 myoblasts	Tyr phosphorylation (*in vivo*)	Endogenous	Complex formation with Grb2	Constitutive Tyr phosphorylation in stable and transient transfection systems confers association with Grb2	111
	Tyr-789	NIH 3T3 cells	Tyr phosphorylation (*in vivo*)	Endogenous (?c-src)	Complex formation with Grb2	Constitutively phosphorylated in Tyr at C-terminal region Tyr-789, increased by cotransfection of c-src, evidence of autodephosphorylation in a non-functional mutant of RPTP-α	85
RPTP-α, RPTP-αi	?	COS-1 cells (transient expression)	Glycosylation	Endogenous	PTPase activity	Each isoform +i and −i gave rise to a 100 kDa form which had only N-linked CHO and was a precursor form to a 130 kDa form which had N- and *O*-linked CHO; enzymatic activity was unchanged	84
RPTP-α	Ser-180 and Ser-204	NIH 3T3 cells	Serine phosphorylation	Protein kinase C		RPTP-α is constitutively phosphorylated in NIH 3T3 cells, predominantly on two serines, Ser-180 and Ser-204, in the juxtamembrane domain; additional studies showed that RPTP-α is a direct substrate for protein kinase C	924

Table 11 Continued

Protein modified	Site of modification	Protein source/ form	Modification (in vivo/in vitro)	Enzyme	Functional parameter measured	Comments	Refs
RPTP-α (Cont.)	Multiple sites	Transfected cells overexpressing RPTP-α	Serine phosphorulation	Protein kinase C (activated by phorbol ester)	PTPase activity	PTPase activity ↑ due to a 2- to 3-fold ↑ in substrate affinity	922
RPTP-δ	Extracellular domain	Transfected COS-7 cells	Proteolytic cleavage of extracellular domain *in vivo*	Endogenous furin-like endoprotease	Subunit structure	Two subunits are non-covalently associated as a complex; the ectodomain is shed from the cell surface	959
RPTP-κ	RTKR proteolytic processing site (640–643)	Transfected COS cells	Proteolytic processing	Endogeneous (furin-like)	Production of subunits	210 kDa precursor is cleaved into 100 kDa N-terminal and 100 kDa TM and cytoplasmic domains; mutation of RTKR (640–643) to LTNR blocked cleavage; some non-covalent association between cleaved subunits occurs	529
RPTP-μ	Extracellular domain	Recombinant human RPTP-μ, Sf9 cells	*In vivo*	Endogenous	Molecular mass	M_r = 195 kDa, as a glycoprotein, size reduced to ~160 by endoglycosidase F treatment	421
RPTP-σ		Transient expression in human embryonic kidney 293 cells	Proteolytic processing	Endogenous		Antibody to extracellular Fn-III domain of RPTP-σ immunoprecipitated only a protein of 200 kDa; immunoblot analysis revealed 200 and 100 kDa proteins that apparently result from proprotein cleavage	621
RPTP-ζ/β	Extracellular domain	Transfected human 293 fibroblasts	CSPG	Endogenous	Metabolic labelling with [^{35}S]sulfate, and digestion with chondroitinase		77

		Bovine and human brain	CSPG	Endogenous	Susceptibility to chondroitinase digestion	Diffuse band (350–500 kDa) on immunoblotting human brain PG-enriched fraction with antibodies to RPTP-ζ/β, after chondroitinase digestion collapses to a 310/300 kDa doublet representing core protein; further reduced by 10 kDa after N-glycanase digestion	106
RPTP-ζ/β (Phosphacan)	Extracellular domain	Rat brain	CSPG	Endogenous	cDNA homology to known CSPG from rat brain		556
RPTP-ζ/β	Extracellular domain	Immuno-precipitation from a human neuro-blastoma cell line	Glycosylation	Endogenous	Change in molecular mass after tunicamycin treatment of cells	In cells treated with the glycosylation inhibitor tunicamycin, the mass of the core protein shifts from 300 → 250 kDa	546
	Extracellular domain	Rat brain	CSPG	Endogenous	Change in M_r	First isolated membrane-bound proteoglycan fractions from 8-day-old rat brains and demonstrated high PTPase activities; chondroitin sulfate proteoglycan with 380 and 170 kDa core proteins were associated with the PTPase activity; the 380 kDa core protein was identified as RPTPβ/ζ bearing HNK-1 carbohydrate	97
SH-PTP1 (PTP1C)	?	BAC1.2F5 macrophage cell line	*In vivo*	? CSF-1R or endogenous tyrosine kinase	–	64 kDa protein was rapidly phosphorylated in response to CSF-1; purified and identified as PTP1C by protein sequence analysis; also identified by immunoreactivity with PTP1C antibody blotting of Tyr(P) reactive proteins from CSF-stimulated cells	115

Table 11 Continued

Protein modified	Site of modification	Protein source/ form	Modification (in vivo/in vitro)	Enzyme	Functional parameter measured	Comments	Refs
SH-PTP1	Tyr-536 and Tyr-564	T cells, thymocytes, LSTRA cells	Tyr and Ser phosphorylation (*in vivo*); Tyr phosphorylation (*in vitro*)	Lck (tyr-p), endogenous (ser-p)	–	In resting T cells, SH-PTP1 is phosphorylated on Ser; after stimulation of CD4 or CD8 in T cell hybridoma cells or in primary thymocytes, SH-PTP1 becomes phosphorylated on Tyr; in Lck-overexpressing LSTRA lymphoma cells, SH-PTP1 is constitutively phosphorylated *in* vivo on tyr; *in vitro*, SH-PTP1 is also phosphorylated on Tyr by recombinant Lck; Tyr(P) SH-PTP1 from LSTRA cells autodephosphorylates; two sites identified as sites of direct Tyr-P by Lck *in vitro* (Tyr-536 and Tyr-564) are identical to sites of Tyr-P in LSTRA cells; after activation of Lck in T cells, one site (Tyr-564) becomes phosphorylated, and may play a role in early T cell signalling	96
(PTP1C)	Tyr-538	CHO cells overexpressing IR, IM-9 human lymphoblasts, rat H35 hepatoma cells	Tyr phosphorylation (*in vivo* and *in vitro*)	IR kinase	PTPase activity (↑ 3–4-fold *in vitro*)	Insulin stimulates Tyr(P) on PTP1C in cultured cell lines *in vivo;* IR also phosphorylates PTP1C *in vitro;* Tyr-538 located in the C terminus; *in vitro,* tyr-p stimulates PTPase activity of PTP1C by 3–4-fold towards Tyr(P)-raytide	113
	?	v-src transformed SR-3Y1 rat fibroblasts	Tyr phosphorylation (*in vivo* and *in vitro*)	v-src		Phosphorylation does not require SH2 domains	99
	?	Transfected 293 cells	Ser phosphorylation (*in vivo* and *in vitro*	Endogenous	–	Phosphorylated *in vitro* by PKA and PKC; also phosphorylated on serine in ^{32}P-labelled transfected 293 cells; no threonine phosphorylation detected	79

?	HL-60 cells	Ser phosphorylation (*in vivo*)	Endogenous PKC	Translocation from cytosol to plasma membrane	With PMA-induced differentiation of HL-60 cells into macrophages, PTP1C becomes phosphorylated on serine residues and protein expression is induced	226
Tyr-538	Endogenous enzyme and recombinant protein from transfected 293 cells	Tyr phosphorylation (*in vivo* and *in vitro*	Growth factor receptors	–	Phosphorylated on Tyr by EGFR and IR *in vitro*; autodephosphorylates when kinase reaction is blocked; also phosphorylated transiently on Tyr by EGF in A431 cells and by PDGF in 293 cells overexpressing PTP1C; catalytically inactive mutant (Cys-455 → Ser) with autodephosphorylation blocked shows ↑ Tyr(P); Tyr-538 identified as site of Tyr(P) *in vitro* by IR kinase and *in vivo* by PDGF stimulation of 293 cells; mutagenesis of Tyr-538 → Phe abolished phosphorylation; sequence surrounding Tyr-538 is consensus for GRB2 binding (pYXNX)	79
		Serine and tyrosine phosphorylation *in vivo*	Endogenous	PTPase activity	Occurs in response to thrombin or PMA; PTPase activity is increased by 40 and 40%, respectively; phosphorylation of SH-PTP1 could be provoked in permeabilized platelets by thrombin or GTPγS; data identify SH-PTP1 as a substrate of a putative protein tyrosine kinase linked to the thrombin receptor by a Gi protein	920
	Rat PC12 cells	Tyrosine phosphorylation		PTPase activity	SH-PTP1, but not SH-PTP2, becomes tyrosine phosphorylated following NGF, but not EGF; enzymatic activity of SH-PTP1 toward an exogenous substrate following NGF treatment is increased two-fold	925

Table 11 Continued

Protein modified	Site of modification	Protein source/ form	Modification (in vivo/in vitro)	Enzyme	Functional parameter measured	Comments	Refs
SH-PTP1, SH-PTP2			Tyrosine phosphorylation *in vivo*	Endogenous		Tyrosine phosphorylation of both PTP1C and PTP1D was increased in response to insulin, PDGF and α-thrombin	961
SH-PTP2	SH2 domains	Recombinant fusion protein with GST, *E. coli*	*In vitro*	IR kinase	PTPase activity against the IR	SH-PTP2 was phosphorylated by the IR kinase; and physically associated with the insulin receptor *in vitro*. The N-terminal SH2 domain was more phosphorylated than the other SH2 domain of SH-PTP2. However, both SH2 domains of SH-PTP2 were necessary for association with the IR; Tyr(P) of the SH2 domains of SH-PTP2 resulted in decreased PTPase activities toward the phosphorylated IR; these results suggest that the IR can negatively regulate SH-PTP2 activity by means of phosphorylating the SH2 domains	98
	?	Cell lines over-expressing EGF-R or PDGF-R	Tyr and Ser phosphorylation (*in vivo*)	Endogenous or transfected receptor kinase	–	Ser phosphorylated in resting A431 cells, ↑ Tyr phosphorylation or SH-PTP2 after EGF stimulation, also ↑ Tyr phosphorylation after PDGF stimulation	431
(Syp)		M07e cells	Tyr phosphorylation (*in vivo*)	c-kit kinase	Complex formation with Grb2	Ligand activation of c-kit leads to Tyr(P) of Syp and complex formation with Grb-2, which activates Ras and Raf	456
	Tyr-304 and/or Tyr-542	FDMAC11/4.5 IL-4 dependent murine myeloid cells	Tyr phosphorylation	Endogenous	Complex formation with Grb2 and PI 3′-kinase; PTPase activity	Cellular binding of IL-3 or GM-CSF induces Tyr(P) of SH-PTP2 and complex formation with Grb2 and PI 3′-kinase; also ↑ PTPase activity by 2-3 fold; IL-4 was without effect	462

	SH2 domains	–	Tyr phosphorylation (*in vitro*)	IR	–	GST fusion protein with SH2 domains of SH-PTP2, but not SH-PTP1 (PTP1C), is phosphorylated by IR	438
	Tyr-542	PDGF-R overexpressing ATWT cells	Tyr phosphorylation (*in vivo* and *in vitro*)	PDGF-R	–	After PDGF stimulation, SH-PTP2 binds to PDGF-R and is phosphorylated on Tyr-542 *in vivo*; pY^{542}TNI motif of SH-PTP2 binds to Grb2, linking PDGF-R to Ras signalling pathway via Grb2/Sos complex	78
(PTP1D)	?	Human 293 cells	Tyr phosphorylation (*in vivo*)	Transfected chimeric PDGF-Rβ	PTPase activity of ↑ by 1.5–2.5 fold	Suggested that differences in activation of PTP1D and Syp by phosphorylation due to various spliced forms at C terminus	616
	?	Transient expression in human 293 embryonic fibroblasts	Tyr phosphorylation (*in vivo*)	Cotransfected v-src or c-src	Complex formation with $pp60^{c-src}$	Heavily tyrosine phosphorylated in cells also cotransfected with v-src or c-src; has putative tyrosyl phosphorylation and SH2 binding motifs (Y^{158}ESQ and Y^{217}GEE)	564
(PTP2C)	?	PC12 cells	Ser phosphorylation (*in vivo* and *in vitro*)	? p44 MAP kinase	PTPase activity	*In vivo,* PTP2C constitutively phosphorylated on Ser, EGF treatment ↑ pThr content and ↓ PTPase activity; *in vitro,* p44 MAP kinase phosphorylated PTP2C on Ser and Thr and ↓ PTPase activity	101
	?	–	Ser, Thr and Tyr phosphorylation (*in vitro*)	See comments	PTPase activity	Phosphorylated by MAP kinase on Thr, PKC on Ser, and by c-src and IR kinase on Tyr; no effect on PTPase activity was detected	383
(Syp)	?	A31, Rat-2 and Rat-1 cells	Tyr phosphorylation (*in vivo*)	Endogenous	–	Tyrosine phosphorylation of (Syp) induced by stimulation of cells with EGF, PDGF or in v-src transformed cells; no change in (Syp) PTPase activity observed	501
	?	Murine cell lines	Tyr phosphorylation (*in vivo*)	Transfected p210bcl-abl oncoprotein	–	Tyrosyl phosphorylation constitutive in transfected cells	457

Table 11 Continued

Protein modified	Site of modification	Protein source/ form	Modification (in vivo/in vitro)	Enzyme	Functional parameter measured	Comments	Refs
	Tyr-546 and/or 584	Transfection of BHK cells overexpressing the insulin receptor	Tyrosine phosphorylation (*in vivo*)	Endogenous (possibly the insulin receptor kinase)	Association with Grb2	Insulin stimulation caused tyrosine phosphorylation of the transfected catalytically inactive mutant of PTP1D, and it coprecipitated with Grb2; tyrosine phosphorylation of the mutant PTP1D is significantly reduced when wild-type enzyme is coexpressed suggesting autodephosphorylation; substitution of Tyr-546 and 584 with Phe abrogates tyrosine phosphorylation of the catalytically inactive mutant and abolishes its interaction with Grb2	1036
	Tyr-546 and/or Tyr-584	BHK cells overexpressing the insulin receptor	Tyrosine phosphorylation	Insulin receptor kinase	Complex with Grb2	A catalytically inactive mutant of PTP1D, Cys-463 → A, becomes tyrosine-phosphorylated and coprecipitates with Grb2; tyrosine phosphorylation of this mutant is significantly reduced when wild-type PTP1D is coexpressed; substitution of tyrosine residues 546 and 584 with phenylalanine abrogates tyrosine phosphorylation of the catalytically inactive mutant and abolishes its interaction with Grb2	916

Modification of PTPase function by chemical agents and proteolysis

Work in this area has provided insight into structure–function relationships in the PTPases as well as potential agents that might be used for pharmaceutical intervention in modulating PTPase function in clinical situations (Table 10). Proteolysis of PTPases in vitro *has provided evidence for autoinhibitory domains in PTP1B, TCPTP and the SH2-domain-containing PTPases. It will be of interest to compare these findings to the results of the three-dimensional studies of PTPases to examine the changes in enzyme folding and substrate access to the catalytic centre which may occur with proteolytic cleavage on non-catalytic regions.*

Studies with chemical agents that inhibit the PTPase activity have also provided some early insight into the nature of the chemical cleavage reaction by derivatizing the nucleophilic cysteine with labelled iodoacetate and trapping of covalent reaction intermediates with the catalytic cysteine moiety [244, 325, 387]. Vanadate has been used in many studies as a PTPase inhibitor both *in vitro* and in intact cells; its use is summarized in the review by Gordon [673]. More recently, Posner and his colleagues developed peroxovanadium compounds which are PTPase inhibitors that may have a role in enhancing insulin signalling [702]. The mechanism of action of vanadate and related compounds is unclear but may involve an augmentation of the phosphorylation state of the insulin receptor or receptor substrates. A number of studies have shown that oral vanadate is effective in reducing hyperglycemia in diabetic animal models, and clinical trials in humans are currently underway to explore the potential effectiveness of vanadate as an antidiabetic agent. Because of the relevance of vanadate to PTPases, general references in which vanadate has been used in various studies on cellular metabolism and in disease states in animals are included in the bibliography.

In attempts to develop more specific PTPase inhibitors, several groups have synthesized modified phosphotyrosyl peptide analogues that use thiophosphoryl groups, which are hydrolysed slowly, or phosphonomethylphenylalanine as a Tyr(P) analogue, which is non-hydrolysable [374, 658, 659, 662, 681, 693]. These agents can be incorporated into peptides that mimic potential physiological substrates in order to identify PTPases, or for the purposes of developing pharmaceutical agents that can interfere with individual PTPases in the cell. As the signalling networks involving phosphotyrosine-containing proteins, their PTPases and potential ligands with SH2 domains become elucidated, the role of these interactions in cellular physiology as well as in disease states will be further evaluated by the use of these types of specific inhibitors.

Interactions of PTPases with other proteins, peptides and macromolecules

Some of the most exciting work in cell signalling has been in the characterization of protein–protein intermolecular reactions. PTPases have been shown to be involved in a variety of interactions that hold promise in identifying associated cellular proteins, potential substrate targets and eventually to help elucidate the physiological roles of the various PTPase homologues (Table 12). The initial sequence characterization of several transmembrane PTPases revealed structural similarity in the extracellular domains to the N-CAM family of cell adhesion molecules and to repeats of fibronectin type III (Fn-III) motifs [756, 891] suggesting that these domains might be regulated by homotypic cell–cell interactions [605, 901]. This hypothesis has been demonstrated to be true at least in the case of two of the receptor-type PTPases, RPTP-μ and RPTP-κ, where recombinant domains on inert spheres as well as in the surface of overexpressing eukaryotic cells have been shown to engage in homotypic aggregation [412, 421, 447, 1089].

Numerous studies have been published on proteins that associate with CD45 or on the effects of 'ligating' CD45 to itself or other lymphocyte surface proteins by antibody binding. This work has provided details regarding the central role of CD45 in propagating a number of lymphocyte responses to antigen stimulation or T cell receptor activation by other means (summarized in Table 12). The SH2 domain-containing PTPases have also been demonstrated to interact with autophosphorylated growth factor receptors or their phosphorylated substrates; for example, SH-PTP1 binds to c-Kit and to the IL-3 receptor β chain [463, 464] and SH-PTP2 binds to the autophosphorylated EGF and PDGF receptors as well as the insulin receptor substrate IRS-1 [430, 431, 432, 501].

Identifying interactions between PTPases and other proteins has been successfully performed by several novel approaches. The yeast 'two hybrid' system has been used to discover an interaction between PTP-PEST and cellular Shc homologues [424]. One successful approach has also used site-directed mutagenesis of the catalytic cysteine residue to eradicate the activity of the PTPase domain, which generates an enzyme that can still bind to potential substrates without hydrolysing the phosphotyrosine moiety. This technique has been used to characterize an association between PTP1B and the EGF receptor [440] and recently to identify a specific interaction between CD45 and the CD3 ζ chain in T lymphocytes [420].

Table 12 Interactions between PTPases and other proteins, peptides and macromolecules

PTPase	Protein interactions	*In vivo* or *in vitro*	Comments	Refs
CD45	180 kDa isoform coisolates with fodrin and spectrin from mouse T-lymphoma cells	*In vivo* and *in vitro*	*In vitro* binds to fodrin with K_d of 1.1 nM or spectrin with K_d of 3.2 nM; this interaction ↑ the V_{max} of CD45 PTPase activity by 7.5 and 3.2-fold, respectively, without changing the K_m	436
	Activation of human PBL T cells with anti-CD3 mAb or during MLR culture elicits association of CD4 or CD8 with CD45 on the cell surface	*In vivo*	Association is not elicited by mitogen stimulation, only TCR-CD3 mediated signalling; maximal association at 72–96 h after anti CD3 exposure or after six days of MLR culture	441
	Anti CD26 (dipeptidyl peptidase IV) coprecipitated CD45 from T cell lysates	*In vivo*	Binding of anti-CD26 mAb lead to enhanced Tyr phosphorylation of CD3ζ and ↑ CD4-associated $p56^{lck}$ Tyr kinase activity, providing evidence of association between CD45 and CD26	458
	Anti CD45 antibodies coimmunoprecipitated a surface glycoprotein of 275 kDa from intraepithelial lymphocytes	*In vivo*		429
	Antibody binding to CD45 in human neutrophils	*In vivo*	Some CD45 antibodies inhibited PMN chemotaxis but only when induced by leukotriene B4 or C5a; binding of anti-CD45 did not affect chemotaxis to IL-8 and superoxide production was not blocked	272
	Antibody binding to CD45 in mouse T cells	*In vivo*	Binding of CD3 induces early signals that are inhibited by cross-linking with antibody to CD45 including an increase in inositol phosphates and cytoplasmic Ca^{2+}, indicating that CD45 is involved in the regulation of PL-C	827
	Association with 32–33 kDa p-ser phosphoproteins	*In vivo*	Observed association between CD45 and several proteins in murine B and T lymphocytes by chemical cross-linking; prominent ones were 32–33 kDa p-ser phosphoproteins	410
	Association with LPAP in T lymphocytes	*In vivo*	LPAP (lymphocyte phosphatase associated protein), a 32 kDa phosphoprotein that interacts with CD45; associates with CD45 in human T lymphocytes and T-lymphoma cell lines; GenBank accession X81422; expression restricted to B and T lymphocytes, most likely represents a novel substrate for CD45 since its phosphorylation is affected by CD45 activity; in the absence of CD45 expression, it is degraded	451
	Association with tyrosine-phosphorylated CD3 ζ chain	*In vitro*	Used recombinant GST-CD45 cytoplasmic domain construct incubated with tyrosine phosphorylated T cell proteins; enzymatically inactive CD45 cytoplasmic domain, with a Cys-828 → Ser mutation bound tightly to the phosphorylated CD3 ζ chain; binding was specific for the CD45 PTPase, since LAR or hybrid recombinant constructs with either D1 or D2 of LAR swapped with the corresponding CD45 domain did not bind; also, the CD3 ζ chain was preferentially dephosphorylated by CD45 *in vitro;* these studies suggest that CD3 ζ is a high-affinity substrate for CD45 and may be involved in termination of the T cell response	420

Table 12 Continued

PTPase	Protein interactions	*In vivo* or *in vitro*	Comments	Refs
CD45 (Cont.)	B cell antigen receptor transduced a Ca^{2+} mobilization signal only if cells expressed CD45; cells treated with anti-CD45 antibody lost surface expression of CD45 as well as membrane IgM, suggesting formation of a protein complex	*In vivo*	Also, CD45 dephosphorylated a complex of membrane IgM-associated proteins involved in signalling through the antigen receptor	282
	Binding mAb to CD45 inhibited IgE-mediated histamine release from human basophils	*In vivo*	Did not block histamine release elicited by A23187 ionophore treatment or PMA	276
	Binding of a combination of two CD45 antibodies to Jurkat cells	*In vivo*	Increased $p56^{lck}$ kinase activity and internalization of the kinase; use of various antibodies suggested complex formation between CD45, CD2 and p56lck	439
	Binding of a mAb to the 190 kDa isoform of CD45 augmented CD2 and CD3-induced T cell proliferation and Ca^{2+} influx	*In vivo*	Antibody reacted only to a subset of mature thymocytes and peripheral T cells suggesting that the 190 kDa isoform of CD45 on human CD4 cells is heterogeneous	354
	Binding of CD45 antibodies	*In vivo*	Cross-linking of CD45 antibodies to resting T cells and quiescent T lymphoblasts costimulated the CD3-induced proliferation of lymphocytes including IL-2 production	313
	Binding of CD45 antibodies	*In vivo*	CD45-ligation induced aggregation of peripheral blood mononuclear cells by LFA-1 and ICAM-1 interactions required presence of both T cells and monocytes, CD45 ligation on T cells initiates clustering to monocytes aggregation is independent of PKC and appears to involve cAMP/cGMP-dependent protein kinases	298
	Binding of CD45 antibodies	*In vivo*	Tyrosine phosphorylation of ezrin was markedly enhanced by ligation of either CD3 or CD4 antigen and peaked between 1 and two min; CD45R ligation modulated the ezrin tyrosine phosphorylation	349
	Binding of CD45 antibodies	*In vivo*	Antibody engagement of CD45 external domains enhances lck tyrosine kinase activity in CD4 + CD8 + thymocytes, inhibits TCR expression, and inhibits differentiation of immature $CD4^+/CD8^+$ thymocytes into mature T cells	230
	Binding of CD45 antibodies	*In vivo*	Induced homotypic adhesion of activated T lymphocytes but not resting cells	452
	Binding of CD45 antibodies	*In vivo*	Inhibition of degranulation of wild-type mast cells	122
	Binding of CD45 antibodies	*In vivo*	Binding of monoclonal CD45 antibodies inhibited the ICAM-3-induced aggregation of JM and HAFSA cells	229
	Binding of CD45 antibodies	*In vivo*	Inhibited T cell proliferation induced by PHA involving both IL-2 production and IL-2 receptor expression	232
	Binding of CD45 antibodies	*In vivo*	Synergizes with PHA/PMA to stimulate expression of IL-2 and IL-2 mRNA; only occurs with CD45R-specific antibody and not antibody to common determinants	255

Binding of CD45 antibodies inhibited B cell proliferative response to various activators	*In vivo*	Block appeared at $G_0(G_1)$ to S phase transition and also affected Ig synthesis by the B cells	306
Binding of CD45 antibodies inhibited non-MHC restricted lysis of of murine mastocytoma P815 cells mediated by CD3–CD16+ NK cells	*In vivo*	Also inhibited Ca^{2+} mobilization induced by binding of antibodies to LFA, suggesting that CD45 may regulate the activation state and function of the LFA molecule	323
Binding of CD45 antibodies to CD4(–) and CD4 transfected Jurkat T cells	*In vivo*	CD45 ligation inhibited TCR-induced calcium influx; effect depended on presence of intact CD4 receptor protein; CD45 cross-linking dephosphorylated lck and reduced its capacity to phosphorylate enolase autophosphorylation and activity of fyn was not affected	305
Binding of CD45 antibodies to human thymocytes and immature T cell lines	*In vivo*	Causes rapid aggregation of T cells partially dependent on a LFA-1/ICAM-3 dependent pathway	411
Binding of CD45 antibodies to murine B-lymphoma cells	*In vivo*	Inhibited Ag-induced p21 ras activation; also Ag-induced ras activation required CD45 expression and was blocked in CD45(–) B cells	283
Binding of CD45 antibodies to T cells regulates signal transduction through both the IL-2 receptor and the CD3/Ti TCR complex	*In vivo*	Proliferative response of T cells to both anti-CD3 or IL2 was inhibited by ligating CD45; CD45 downregulated CD3-induced T cell activation when the CD45 and CD3 molecules were ligated simultaneously with immobilized antibodies	265
Binding of CD45 to cell-surface molecules	*In vivo*	CD45 on the A4 cell surface binds 3T3 cells; various glycosaminoglycans, particularly heparin and heparan sulfate, blocked binding of A4 cell surface molecules to the 3T3 cells; the data suggest that heparan sulfate on the 3T3 cell surface is a ligand for CD45	417
Binding of monoclonal antibody to epitope common to all isoforms of CD45 inhibits cell–cell adhesion mediated by homotypic binding of MHC class I or II, CD19, CD20, CD21, CD40 and Leu-13	*In vivo*		461
CD45 and $p59^{fyn}$ codistribute in functional T lymphocytes by capping experiments using double indirect immunofluorescence	*In vivo*	*In vitro*, fyn is a substrate for CD45 with a rapid dephosphorylation of the regulatory Tyr-531 on fyn causing a several-fold ↑ in the kinase activity of fyn; CD45-mediated dephosphorylation of fyn occurs at a slow basal rate in resting T cells	309
CD45 associates with lck independently of both CD4 and CD8 in a CD4+ T cell line and peripheral T cells; interaction between CD45, lck and a 32–34 kDa protein is stable after T cell stimulation	*In vivo*		445
CD45 cross-linked to CD2 with mAbs, inhibited the activation of MAP kinase in response to interaction of CD2 with stimulatory anti-CD2 mAb	*In vivo*		310

Table 12 Continued

PTPase	Protein interactions	*In vivo* or *in vitro*	Comments	Refs
CD45 (Cont.)	CD45 ligation	*In vivo*	In normal T cells and Jurkat cells, T cell receptor complex (CD3/Ti)-mediated PI production was inhibited by CD45 ligation, associated with suppressed Tyr(P) of specific substrates in response to TCR stimulation	433
	CD45-associated 30 kDa protein was isolated and cloned; it is a leucocyte-specific protein with a unique sequence		*In vitro* translation product binds to CD45 *in vitro*	454
	CD45/p56^{lck}-associated 32 kDa phosphoprotein is composed of two distinct proteins of M_r 29 and 32 kDa in both resting T lymphocytes and in proliferating T-lymphoma lines	*In vivo*	Activation of lymphocytes with CD2 or CD3 mAb or PMA resulted in loss of pp29 and pp32 from the CD45/p56^{lck} complex and association of phosphoproteins of different M_r (30 and 31 kDa); antiserum to GTP-binding consensus sequence suggested that the 29–32 kDa associated proteins might have GTP-binding properties; V8 digestion suggested that they differ in low M_r fragments	450
	Coaggregation of CD45 with CD2	*In vivo*	Inhibits inracellular Ca^{2+} rise and T cell proliferation in response to CD2 ligation alone, also ↓ PI hydrolysis and Tyr(P) of a 100 kDa endogenous protein	446
	Co-clustering of CD45 with CD4 or CD8 inhibited the anti-CD4-induced phosphorylation of p56^{lck} on Tyr and the concomitant ↑ in Tyr kinase activity	*In vivo*	Suggesting that p56^{lck} is a substrate for CD45 *in vivo*	443
	Coimmunoprecipitation of CD45 with Thy-1 or TCR	*In vivo*	In T cells, a substantial fraction of Thy-1 and TCR can be cross-linked to CD45	460
	Coimmunoprecipitation with the kinase, lyn as well as multiple components of the B cell antigen receptor complex including the MB-1/B29 heterodimer	*In vivo*		413
	Coprecipitation of membrane IgM with anti-CD45 antibody	*In vivo*	CD45 also dephosphorylated a complex of membrane Ig-associated proteins that may function in the receptor signal transduction	252
	Covalently binding of CD45 to a 80 kDa polypeptide	*In vivo*	In B cell chronic lymphocytic leukaemia cells, identified a disulfide-linked 80 kDa protein to a high M_r isoform of CD45	442
	Cross-linking with CD45 antibodies	*In vivo*	Cross-linking of CD3, CD2 or CD28 on surface of T lymphocytes with CD45 abolished the increase in intracellular Ca^{2+} observed when homoaggregates of these receptors are formed; in contrast, linking CD45 to CD4 amplified the signal indicated by CD4 homoaggregation	434
	Demonstrated an association between CD45, CD16 and an unidentified 33 kDa molecule in human NK cells and an association between zeta (CD3ζ) with CD16, CD45 and an unidentified molecular of ~150 kDa	*In vivo*		409

Demonstrated that CD45 forms homodimers using cleavable and bifunctional chemical cross-linking agents	*In vivo*	Also noted association of a protein of 30 kDa associated with both the CD45 monomer and dimer that was phosphorylated and not labeled by surface iodination	455
Demonstration of a functional complex between CD45, $p56^{lck}$ and a 32 kDa phosphoprotein in immunoprecipitates of human T lymphocytes	*In vivo* and *in vitro*	p32 is phosphorylated by $p56^{lck}$ and dephosphorylated by CD45 *in vitro*	448
Engagement of CD45 with mAb on the monocyte surface triggers TNF-α and IL-1 β release	*In vivo*		364
In activated human tumour infiltrating lymphocytes, CD45RO isoform coimmunoprecipitates with CD28	*In vivo*		467
In human T lymphocytes, a specific codistribution observed for $p56^{lck}$ and CD45, but not other proteins on the cell surface	*In vivo*	After capping induced by CD4 antibody, $p56^{lck}$ and CD45 were colocalized and ↑ Tyr(P) noted at CD4 cap sites; coclustering of CD4 and CD45 did not result in ↑ Tyr(P) at cap sites	423
In Jurkat T cells identified a trimolecular complex containing CD45, lck and a 34 kDa protein	*In vivo*	Kinase activity associated with CD45 requires expression of lck in Jurkat cell clones as well as the *in vivo* phosphorylation of p34; association between CD45 and lck does not require the presence or activation of the TCR; also, association between p34 and CD45 does not require the presence of the lck kinase	285
In monocytic leukaemia cell line THP-1, cocross-linking of an F(ab′)2 fragment of a monoclonal CD45 antibody with antibodies to cell surface proteins mediating signalling through Fc gamma RI (CD64) or Fc gamma RII(CDw32) inhibited Ca^{2+} mobilization and tyrosine phosphorylation of specific proteins	*In vivo*	Suggests that CD45 regulates signalling through the Fc gamma receptor	329
In vitro, CD45 was phosphorylated by $p50^{csk}$ on two tyrosines, one was Tyr-1193, which was not phosphorylated by $p56^{lck}$ or by $p59^{fyn}$; Tyr-1193 was also phosphorylated *in vivo* in T cells	*In vivo* and in vitro	Cotransfection of CD45 and csk into COS-1 cells caused Tyr phosphorylation of CD45; Tyr(P) CD45 bound $p56^{lck}$ via its SH2 domains and Tyr phosphorylation of CD45 ↑ its PTPase activity several-fold	76
Interaction between CD45 and the CD3-TCR complex	*In vivo*	CD45 antibody blocks activation of peripheral blood T cells and downregulation of the CD3-TCR complex induced by an anti-TCR antibody	428
mAb to CD45 are strongly comitogenic with CD2 antibodies but not CD3 antibodies	*In vivo*	Demonstrated that CD45 is physically associated with CD2 on the T cell surface by chemical cross-linking techniques	449

Table 12 Continued

PTPase	Protein interactions	*In vivo* or *in vitro*	Comments	Refs
CD45 (Cont.)	mAb to mouse CD2 precipitated a high M_r glycoprotein from various T cells identified as CD45	*In vivo*	Specific for CD2, since CD3 mAb did not precipitate CD45	408
	Mapping of fodrin binding domain	*In vitro*	Cytoskeletal binding site is in cytoplasmic domain; deleted various portions of CD45 cytoplasmic domain and expressed the truncated recombinant proteins using *in vitro* transcription/translation system; identified fodrin binding domain between AA 825 and 939; expression of this domain alone as a fusion protein is sufficient to bind fodrin, with a critical role played by the sequence ^{930}EENKKKNRN^{939}S; this sequence is also involved in the fodrin-induced up-regulation of CD45 PTPase activity	425
	Modulates binding of lck to a phosphopeptide encompassing the negative regulatory tyrosine of lck	*In vitro*	CD45 can modulate the binding of the lck to an 11 amino acid tyrosine phosphorylated peptide containing the C terminus of lck (lckP); CD45 did not influence the binding of Fyn, PLC-γ 1, GAP and Vav to the same phosphopeptide. Lck protein which bound the peptide was dephosphorylated on Tyr-505 and consisted of only 5–10% of the total cellular lck	338
	Monoclonal CD45 antibody effects on PTPase activity in murine T cell hybridoma cells	*In vivo*	Two antibodies gave rise to markedly different effect one elicited IL-2 production, ↑ intracellular Ca^{+2}, and ↑ Tyr(P) off specific proteins; the other inhibited biological responses in a PTPase-dependent manner; suggests that certain interactions of the extracellular domain of CD45 can activate its PTPase activity	422
	Thy-1 and CD45 associate in thymocytes	*In vivo*	Monoclonal anti-Thy-1 antibody also reduced the PTPase activity of isolated membranes, demonstrating a functional coupling between Thy-1 and membrane PTPases	437
	Using heterologous receptor aggregation, CD45 shown to inhibit induction of MAP kinase activity in Jurkat cells during CD3 or CD3/CD4 stimulation	*In vivo* and *in vitro*	Purified CD45 also dephosphorylated MAP kinase isolated from lymphoid cells	324
	Ligation of CD4 with antibodies	*In vivo*	Pretreatment of primary CD4+, but not CD8+ T cells with anti-CD45 inhibits activation signals induced through the T cell receptor for antigen (TCR$\alpha\beta$), involving phosphorylation of phospholipase Cγ1, Ca^{2+} mobilization and DNA synthesis; these changes were not accompanied by changes in the level of CD45 expression, its membrane distribution or changes in its intrinsic phosphatase activity; also anti-CD3ε-mediated activation is unaffected, suggesting that the antigen receptor complex can be functionally uncoupled	1015

	Co-cross-linking with either Fc γ RI or Fc γ RII	*In vivo*	Co-cross-linking prevented Fc γ RI RII-mediated calcium mobilization and protein tyrosine phosphorylation; cross-linking of the CD45 itself was sufficient to induce IL-6 production	1061
	Interaction between the extracellular domain of CD45 and CD22	*In vivo*	Engagement of CD45 by soluble CD22 can modulate early T-cell signals in antigen receptor/CD3-mediated stimulation	1079
	Interaction with p56(lck)	*In vivo*	In the antigen-specific T-cell line (BI-141), unique sequences in the amino-terminal domain of p56lck regulates its interactions with PTPases, notably CD45	1065
	Interactions between CD3, CD4 and CD5 in T cells	*In vivo*	Simultaneous cross-linking of CD45, CD3 and CD4 abrogates the sustained increase in calcium seen following CD3 and CD4 cross-linking	1011
	Interactions between cross-linked Fc γ receptors and CD45 on polymorphonuclear leucocytes	*In vivo*	Treatment with a mAb F(ab′)(2) fragment recognizing CD45 phosphatase suppressed Fc γ R-induced calcium mobilization in a dose-dependent manner, was immediately terminated when activation was followed by co-cross-linking Fc γ R and CD45, suggesting that the initial steps of Fc γ R signal transduction pathways are influenced by the state of tyrosine phosphorylation	1005
	Binding to CD45-AP	*In vivo*	30 kDa phosphorylated proteins, CD45-AP, specifically associates with CD45 by the potential transmembrane segment of CD45-AP; propose that CD45-AP interacts with CD45 at the plasma membrane and that the bulk of CD45-AP located in the cytoplasm act as an adapter	1067
	Homophilic binding between B cells involving CD45 and LFA-1/ICAM-1 and LFA-1/ICAM-3 adhesion pathways	*In vivo*	Three anti-CD45 monoclonal Abs to epitopes on the alternative spliced exon-A-encoded region of CD45, induced B-cell aggregation; CD45-mediated cell aggregation was also inhibited by preincubation with some conventional anti-CD45 mAbs; CD45-mediated intercellular adhesion was abrogated upon incubation with anti-LFA-1, ICAM-1, and ICAM-3 Abs	1088
	Association with a 116 kDa glycoprotein	*In vivo* and *in vitro*	116 kDa tyrosine phosphorylated glycoprotein identified in immunoprecipitates with CD45; association not dependent on cytoplasmic domain	1056
	Chemical cross-linking to Thy-1	*In vivo*	Method using surface biotinylation and cross-linking with DTSSP	1049
	Chemical cross-linking to $p56^{lck}$	*In vivo*	Method using lysolecithin permeabilization of T-cell lines, followed by biotinylation and immunoprecipitation	1049
	Cross-linking with anti-CD45 antibodies	*In vivo*	Stimulates IFN-γ production in NK3.3 cell line	1033
cdc25	Coimmunoprecipitation with $p34^{cdc2}$	*In vivo*	2–8% of cdc25 injected into oocytes is coimmunoprecipitated with antibody to $p34^{cdc2}$, association was higher in M phase than in interphase	262
	Complex formation and activation of cdc25 by cyclin B-cdc2		A ‘P box’ domain on cyclin B contributes to its interaction with cdc2 and its ability to activate cdc25	466
	Complex formation between $p72^{cdc25}$ and cdc2-cyclin B	*In vivo*	In *Xenopus* embryos, cdc25 associates with cdc2-cyclin B in a cell cycle-dependent manner, reaching a peak at M phase	426
	Association with raf-1	*In vivo*	cdc25 associates with raf-1 in mammalian cells and in meiotic frog oocytes; cdc25 can be activated *in vitro* in a raf-1-dependent manner	1064
	Association with 14-3-3 proteins	*In vivo* and *in vitro*	The 14-3-3 ε protein and 14-3-3 β interact with human cdc25A and cdc25B; 14-3-3 protein does not affect the phosphatase activity of cdc25A; 14-3-3 may facilitate the association of cdc25 with raf-1 *in vivo*, linking mitogenic signalling with the cell cycle machinery	1060

Table 12 Continued

PTPase	Protein interactions	*In vivo* or *in vitro*	Comments	Refs
cdc25A	Complex formation between cdc25A and B-type cyclins	*In vivo*	In HeLa cells, cdc25 associates with cyclin B1/ cdc2 by coimmunoprecipitation and Western blotting; PTPase activity of both cdc25A and cdc25B is ↑ 4- to 5-fold by addition of cyclin B1 or B2 but not cyclin A or D; suggests activation of cdc25 by its physiological substrate	503
cdc25C	Dephosphorylation by a Type-2A protein phosphatase	*In vitro*	cdc25C is maintained in a low activity, dephosphorylated state by a Type 2A protein phosphatase	83
cdi1	Complex formation with cdks	*In vivo*	Interacts in yeast two-hybrid system with human cdc2 > cdk3 > cdk2 > (*S. cerevisiae*) cdc28; not cdk4 or other cdc2 family proteins; also coimmunoprecipitated with cdk2 from HeLa cells	515
DPTP10D	Binding to gp150, a transmembrane glycoprotein	*In vivo* and *in vitro*	Binding is dependent on catalytic activity of DPTP10D; tyrosine phosphorylated gp150 is efficiently dephosphorylated *in vitro* vy DPTP10D	1081
FAP-1 (PTP-BAS)	Association with cell surface receptor fas	*In vivo*	Identified by yeast interaction trap; FAP-1 is mouse homologue of human PTP-BAS; interacts with the C-terminal 15 AA residues of the cytoplasmic domain of Fas	1118
IA-2	Binding to autoantibodies in patients with insulin-dependent diabetes mellitus	*In vivo*	Screened a cDNA expression library from rat insulinoma cells with a serum from an IDDM patient which precipitated the 37/40 kDa antigen; identified one cDNA clone whose open reading frame encodes a deduced amio acid sequence with high homology to IA-2	1026
KAP	Complex formation with cdk2 and cdc2	*In vivo*	Associates in yeast two-hybrid system with cdk2 and cdc2; also can coimmunoprecipitate HA-tagged KAP with cdc2 > cdk2 in transiently transfected COS cells	518
LAR	Binding between LAR D2 domain and LIP.1	*In vivo* and *in vitro*	LIP.1 is a 160 kDa phosphoserine protein, interaction with LAR identified by yeast interaction trap (GenBank No. U22815, U22816); LIP.1 may function to localize LAR to focal adhesions that may be undergoing disassembly	1078
MKP-1 (3CH134)	Complex formation with $p42^{MAPK}$	*In vivo*	Transfection of Cys-258 → Ser, catalytically inactive mutant allows coimmunoprecipitation of complex between MPK-1 and endogenous $p42^{MAPK}$	345
PCS phosphatase	PCS phosphatase association with a ‘PTPase activator’, PTPA from *Xenopus* oocytes	*In vitro*	PTPA, a PTPase activator of protein phosphatases known as PCS_{H2} and PCS_L (polycation stimulated protein phosphatases) is 40 kDa; when activated these enzymes represent up to 50% of the PTPase activity in *Xenopus laevis* oocytes	415
(PP2A)	Molecular cloning and characterization of PTPA, the PTPase activator of protein phosphatase 2A	*In vitro*	Sequence from purified protein used to clone cDNA from rabbit muscle and human heart cDNA libraries; protein encoded was 323 amino acids; expressed in various tissues; GenBank accession numbers X73478 and X73479	416

PP2A	Association with polyomavirus small-T or middle-T antigen as an okadaic acid-sensitive PTPase	*In vivo*	PTPase activity in immunoprecipitates of small or middle-T from polyoma virus-transformed or polyoma virus-infected mouse 3T3 fibroblasts had characteristics of PP2A as well as an okadaic acid-sensitive PTPase activity	743
PTP-PEST	Complex formation with $p52^{shc}$ and $p66^{shc}$	*In vivo* and *in vitro*	Using $p52^{shc}$ as bait, isolated C-terminal region of PTP-PEST in yeast two-hybrid system; p52 and p66 forms of Shc, but not p42 form interact with PTP-PEST; interaction with N-terminal half of Shc proteins; in HeLa cell lysates, also coprecipitated p52 and p66 forms of Shc with PTP-PEST antiserum; complex formation was ↑ 6–8-fold by activators of PKC	424
PTP1B	Binding of labelled PTP1B fusion protein to immunoprecipitated EGFR; apparent $K_d = 100\,nM$	*In vitro*	PTP1B made catalytically inactive by mutating Cys-215 → Ser; binding enhanced by EGFR autophosphorylation; site specificity for EGFR Tyr(P)-992 and Tyr(P)-1148; binding inhibited by SH2 domain-containing proteins including PLC-γ and *ras*-GAP	440
PTPD1	Complex formation with $pp60^{c-src}$	*In vivo*	Is heavily tyrosyl phosphorlyated in 293 cells cotransfected with c-src or v-src; coimmunoprecipitates with c-src in doubly transfected cell system; has potential SH2 domain binding motif and putative SH3 binding domains (residues 334–343 and 565–574)	564
RPTP-α	Complex formation between Tyr-phosphorylated RPTP-α and Grb2	*In vivo*	In cultured L6 and 293 cells, RPTP-α is phosphorylated on Tyr and binds the adaptor protein Grb2 via SH2 domains of Grb2	111
	Complex formation between Tyr-phosphorylated RPTP-α and Grb2	*In vivo*	In NIH 3T3 cells, Tyr-phosphorylated RPTP-α binds the adaptor protein Grb2 at Tyr(P)-789; complex did not contain Sos, suggesting role for attenuation of Grb2 signalling	85
RPTP-δ	Binding between RPTP-δ D2 domain and LIP.1	*In vivo* and *in vitro*	Demonstrated by yeast interaction trap assay and by immunoprecipitation from LIP.1-transfected COS cells; similar to LAR; no binding to CD45	1117
RPTP-κ	Homophilic binding	*In vivo* and *in vitro*	Aggregation induced between transfected *Drosophila* S2 cells overexpressing RPTP-κ and between synthetic beads coated with recombinant extracellular domain of RPTP-κ; association is independent of Ca^{2+}, does not require proteolytic processing of the extracellular domain or PTPase activity; no change in PTPase activity detected after homophilic binding	447
RPTP-μ	Homophilic binding between extracellular domains	*In vivo* and *in vitro*	Human RPTP-μ expressed with or without the PTPase domains in Sf9 cells, leads to aggregation via a homophilic mechanism; interaction also demonstrated between RPTP-μ coated fluorescent beads (Covaspheres) and endogenously expressed RPTP-μ on mink lung MvLu cells, as well as by binding of coated Covaspheres to Petri plates adsorbed with a GST-RPTP-μ extracellular domain fusion protein; no gross changes in PTPase activity with cell aggregation were noted	412

Table 12 Continued

PTPase	Protein interactions	*In vivo* or *in vitro*	Comments	Refs
RPTP-μ (Cont.)	Homophilic binding between extracellular domains	*In vivo*	Human RPTP-μ expressed in Sf9 cells; promoted cell–cell adhesion by homophilic interactions in a Ca^{2+}-independent manner requiring pH ≥ 6.3; interaction dependent on expression of the extracellular domain of RPTP-μ and not on the PTPase catalytic activity, as demonstrated with a catalytically-inactive point mutation (Cys-1095 $\rightarrow$ Ser) or a recombinant construct lacking the entire cytoplasmic domain; also transfection of cells did not affect adhesion to substratum on dishes coated with laminin, fibronectin, or vitronectin; unable to demonstrate an effect of aggregation on catalytic activity of RPTP-μ	421
	Binding to cadherins and catenins	*In vivo* and *in vitro*	RPTP-μ associates with a complex containing cadherins, α- and β-catenin in mink lung (MvLu) cells, and in rat heart, lung, and brain tissues; greater than 80% of the cadherin in the cell is cleared from lysates of MvLu cells after immunoprecipitation with antibodies to PTP-μ; *in vitro*, the intracellular segment of PTP-μ binds directly to the intracellular domain of E-cadherin, but not α- or β-catenin; RPTP-μ, cadherins, and catenins all localized to points of cell–cell contact in MvLu cells	1057
	Homophilic extracellular domain binding	*In vivo*	Extracellular Ig domain is the determinant of homophilic binding; neither the MAM domain nor the Fn-III repeats were capable of homophilic interaction	986
	Homophilic interactions via the MAM domain	*In vivo*	Despite having MAM domains, RPTP-μ and RPTP-κ fail to interact in a heterophilic manner, RPTP-μ lacking the MAM domain fails to promote cell–cell adhesion; homophilic cell adhesion is restored in a chimeric RPTP-μ molecule containing the MAM domain of RPTP-κ; however, this chimeric RPTP-μ does not interact with either native RPTP-μ or RPTP-κ	1089, 1090
RPTP-σ	Binding between RPTP-σ D2 domain and LIP.1	*In vivo* and *in vitro*	Demonstrated by yeast interaction trap assay and by immunoprecipitation from LIP.1-transfected COS cells; similar to LAR; no binding to CD45	1117
RPTP-ζ/β (phosphacan)	Complex formation with tenascin	*In vitro*	Covaspheres coated with RPTP-ζ / β (phosphacan; 3F8 PG) specifically co-aggregate with covaspheres coated with tenascin, an extracellular matrix protein, and not Ng-CAM (neuron-glia cell adhesion molecule) or aggrecan (a cartilage CSPG); the 3F8-tenascin interaction is blocked by excess 3F8 protein or Fab$'$ fragments of 3F8 monoclonal antibody	77
	Binding between extracellular domain (Phosphacan CSPG) and N-CAMs	*In vivo* and *in vitro*	Reversible, saturable binding between Phosphacan and N-CAM or Ng-CAM on microbeads at high affinity ($K_d \simeq 0.1$ nM); also binds to cultured neurones	1072
	Specific binding between the CAH domain and contactin	*In vivo* and *in vitro*	Contactin is a 140 kDa neuronal cell recognition molecule that is a GPI membrane-anchored protein; the CAH domain of RPTP-ζ/β induced cell adhesion and neurite growth of primary tectal neurones, and differentiation of neuroblastoma cells; these responses were blocked by antibodies against contactin	1075

	Binding of extracellular domain (Phosphacan) to NCAM/L1 and N-CAM and to tenascin	*In vivo*	Analysis of tryptic digests of labelled Phosphacan and glycopeptide structure demonstrated that the interactions of Phosphacan/RPTP-ζ/β with neural cell adhesion molecule and tenascin are mediated by asparagine-linked oligosaccharides present in their carbonic anhydrase and fibronectin type III-like domains	1073
SH-PTP1 (PTP1C)	Associates with autophosphorylated IR *in vitro* by C terminus of PTP1C (not by SH2 domains)	*In vitro*	Does not bind to unphosphorylated IR; C-terminal 38 AA of PTP1C are necessary for binding to IR as well as for phosphorylation of Tyr-538, leading to ↑ PTPase activity; suggests that PTP1C is a target of the IR kinase that may function both in the regulation of PTPase activity and in the association of PTP1C with the IR	113
SH-PTP1	Asssociates with somatostatin receptors	*In vivo*	In rat pancreatic plasma membranes, SH-PTP1 copurified and coimmunoprecipiated with somatostatin receptors	465
	Coimmunoprecipitates with a 160 kDa phosphotyrosyl protein in LSTRA cells	*In vivo*	160 kDa protein did not react with a panel of antibodies to candidate proteins	96
(PTP1C)	Complex formation with autophosphorylated EGFR	*In vivo* and *in vitro*	Recombinant SH2 domains of PTP-1C expressed as fusion protein binds to *in vitro* autophosphorylated EGF-receptor, also 110, 70 and 58–60 kDa pY proteins present in A431 cell lysate phosphorylated *in vitro*; associated primarily with EGFR in *in vivo* phosphorylated proteins from EGF-stimulated A431 cells	602
(HCP)	Complex formation with c-kit	*In vivo* and *in vitro*	HCP transiently associates with ligand-activated c-kit but not c-Fms in Mo73 cells; *in vitro*, interaction shown to occur through SH2 domains; *in vitro*, GST-HCP can dephosphorylate c-kit and c-Fms	463
(HCP)	Complex formation with IL-3R β chain	*In vivo* and *in vitro*	In IL-3 dependent myeloid cell line, DA-3, transfected with recombinant GST-fusion constructs of full length and individual domains of PTP1C; associates via the N-terminal SH2 domain to the β chain of the IL-3 receptor both *in vivo* and *in vitro* and can dephosphorylate it	464
(PTP1C)	Binding to autophosphorylated growth factor receptors	*In vivo*	Coimmunoprecipitates only with HER2-*neu* kinase domain, not with PDGF-Rβ, p145$^{\text{c-kit}}$ kinase or the EGF-R	616
(PTP1C)	Coimmunoprecipitated with a 25 kDa protein in metabolically labelled HL-60 cells	*In vivo*	May complex with PTP1C; precipitation not affected by excess peptide used to make PTP1C antibodies	212
(SHP)	Complex formation with $M_r = 15$ kDa protein in a lysate from LSTRA cells	*In vitro*	Association with SH2 domains of SHP and LSTRA cell 15 kDa protein	553
(PTP1C)	Association with the resting BCR complex	*In vivo*	Data include coprecipitation of PTP1C protein and phosphatase activity with BCR components and the depletion of BCR-associated tyrosine phosphatase activity by anti-PTP1C antibodies; PTP1C specifically induced dephosphorylation of a 35 kDa BCR-associated protein likely representing Ig-α; following membrane Ig cross-linking, tyrosine phosphorylation of PTP1C was increased and it was no longer detected in the BCR complex, suggesting that PTP1C dissociates from the activated receptor complex	1074

Table 12 Continued

PTPase	Protein interactions	*In vivo* or *in vitro*	Comments	Refs
(SHP)	Binding of SHP to CD22 and its activation	*In vivo*	CD22 is tyrosine phosphorylated when Ig is ligated; tyrosine-phosphorylated CD22 binds and activates SHP; ligation of CD22 to prevent its coaggregation with mIg lowers the threshold at which mIg activates the B cell by a factor of 100; CD22 is thus a molecular switch for SHP	997
(PTP1C)	Complex formation with tyrosine phosphorylated CD22	*In vivo*	A 60 kDa protein that coprecipitates with CD22 after anti-Ig stimulation which was identified with molecule PTP1C; the association is dependent upon tyrosine phosphorylation of CD22	1058
	Recruitment to Fc γRIIB1	*In vivo*	The Fc γRIIB1 recruits PTP1C after BCR coligation; the association is mediated by the binding of a 13-amino acid tyrosine-phosphorylated sequence to the carboxyl-terminal src homology 2 domain of PTP1C and activates PTP1C; PTP1C is an effector of BCR-Fc γRIIB1 negative signal cooperativity	1062
	Complexed with Epo-R in stimulated cells and regulation of Jak2 activation	*In vivo* and *in vitro*	SH-PTP1 associates via its SH2 domains with the tyrosine-phosphorylated EPO-R at Tyr-429 mutant EPO-Rs lacking Y429 do not bind SH-PTP1; cells expressing such mutants are hypersensitive to EPO and display prolonged EPO-induced autophosphorylation of JAK2	1069
	Binding to and dephosphorylation of activated EGF receptors	*In vivo* and *in vitro*	PTP1C and PTP1D were both found associated with epidermal growth factor (EGF) receptor which was purified from A431 cell membranes and when purified EGF receptors are added to lysates of MCF7 mammary carcinoma cells; receptor dephosphorylation was initiated by treatment of the receptor PTPase complex with phosphatidic acid; the association of PTP1C but not of PTP1D was enhanced in the presence of PA; data suggest that PTP1C but not PTP1D participates in dephosphorylation of activated EGF receptors	1082
	Binding of a 44 kDa phosphoprotein	*In vivo* and *in vitro*	PTP1C undergoes rapid tyrosine phosphorylation following mIgM or mIgD cross-linking and associates with a number of other phosphoproteins in stimulated W delta cells; a 44 kDa coprecipitated with PTP1C in mIgM-ligated cells but was not detected in PTP1C immunoprecipitates from mIgD-ligated cells	1083
(HCP)	Binding to CD22 in activated B cells	*In vivo*	Ligation of the B cell antigen receptor induces tyrosine phosphorylation of HCP and the formation of a multimolecular complex containing additional 68–70 and 135 kDa phosphoproteins (CD22), suggesting that *in vivo* B cell antigen receptor constituents or associated molecules may serve as substrates for its catalytic activity	1070
(HCP)	Stable binding between HCP and Tyk-2	*In vivo*	The tyrosine kinase Tyk-2 forms stable complexes with HCP in several haematopoietic cell lines; the IFN α-induced tyrosine-phosphorylated form of Tyk-2 is a substrate for the phosphatase activity of HCP in *in vitro* assays	1086
(HCP)	Binding between HCP and tyrosine phosphorylated EPO receptor	*In vivo*	Binds through amino-terminal SH2 domain of HCP; potential binding sites to the C-terminal region of the EPO receptor are identified	1087

SH-PTP2 (Syp)	Associates with ligand-activated c-kit receptor; also becomes phosphorylated on Tyr and binds to Grb2	*In vivo* and *in vitro*	Associate *in vivo* in M07e cells; *in vitro,* both N-terminal and C-terminal SH2 domains of Syp were able to interact with c-kit; also, Syp became phosphorylated on Tyr after c-kit activation, and Tyr(P) Syp complexed to Grb-2 in M073 cells as well as *in vitro;* this interaction lead to Ras activation and binding of Ras and Raf	456
	Binding and activation by pY peptides derived from IRS-1	*In vitro*	Peptides from IRS-1 pY895, pY1172 and pY1222 ↑ activity of SH-PTP2 up to 50-fold; analysis of N-SH2 and C-SH2 showed differences in binding specificity	453
	Binding to pY peptides from PDGF-R and complex formation with PDGF-R	*In vitro and in vivo*	By peptide competition assay, Tyr(P)-1009 domain is the major binding site for SH-PTP2 on PDGF-R; complex formation of SH-PTP2 with activated PDGF-R from ATWT cells demonstrated; Tyr(P)-1009 domain peptide activates PTPase activity 5- to 10-fold	432
	Complex formation between PDGF-R, Grb2-Sos and	*In vivo* and *in vitro*	SH-PTP2 binds to activated PDGFR at Tyr-1009 and forms complex linking Grb2-Sos with PDGFR; Grb2 binds directly to Tyr(P) SH-PTP2 *in vitro*	435
	Complex formation with EGF-R and PDGF-R	*In vivo* and *in vitro*	Growth factor activation of various cultured cells elicits coimmunoprecipitation of SH-PTP2 and EGF-R or PDGF-R; *in vitro* complex formation can arise *via* N-SH2 domain with much weaker binding activity of C-SH2 domain	431
	Complex formation with Grb2 and PI 3′-kinase	*In vivo*	In FDMAC11/4.5 IL-4 dependent murine myeloid cells, binding of IL-3 or GM-CSF induces Tyr(P) of SH-PTP2 and complex formation with Grb2 and PI 3′-kinase	462
	Complex formation with IR	*In vitro*	Both SH2 domains of SH-PTP2 were necessary for association with the IR	98
	GST fusion protein with SH2 domains of SH-PTP2 binding to IR and IRS-1	*In vitro*	Binding of SH-PTP2 SH2 domains to IR abolished by truncating IR C terminus or mutating C-terminal Tyr → Phe; SH2 binds to C-terminal sites of IRS-1	459
	GST fusion protein with SH2 domains of SH-PTP2 or SH-PTP1 (PTP1C) binding to IR	*In vitro*	SH2 domain of SH-PTP2, but not SH-PTP1, bound to autophosphorylated C terminus of IR	438
(N-SH2 domain)	Binding to candidate pY peptides from growth factor PDGF-R and EGF-R and IRS-1	*In vitro* (binding to GST-fusion protein with SH2 domains)	Highest binding to Tyr-1009 domain of candidate PDGFR peptides, Tyr-954 of EGFR and to Tyr-546, 895 and 1172 of IRS-1 sequences; consensus N-SH2 domain binding specificity: pY-(Val,Ile,Thr)-X-(Val,Leu,Ile)	414
(PTP1D)	Binding to autophosphorylated growth factor receptors	*In vivo*	By co-immunoprecipitatoin, associated strongly with HER2-*neu* kinase domain and with PDGF-Rβ, less affinity to p145$^{c\text{-}kit}$ kinase and EGF-R. M_r ↑ in association with PDGF-Rβ, suggesting phosphorylation (see below)	616
(Syp)	Binding of SH-PTP2 (Syp) to autophosphorylated EGF and PDGF receptors by its SH2 domains	*In vivo*	In mouse A31 fibroblasts (PDGF stimulation) and Rat-1 cells overexpressing the human EGFR (EGF stimulation)	501
(Syp)	Binding to autophosphorylated growth factor receptors	*In vitro*	Either N- or C-terminal SH2 domain of Syp can mediate binding to activated PDGF-Rβ or EGFR with synergistic binding for tandem SH2 domains; site identified as Tyr-1009 on PDGF-R	419

Table 12 Continued

PTPase	Protein interactions	*In vivo* or *in vitro*	Comments	Refs
(Syp)	Binding to p210bcr-abl oncoprotein	*In vivo*	Protein complex formation between p210bcr-abl, Syp and Grb2 in p210bcr-abl transfected mouse cells; mediated by N-SH2 domain of Syp	457
(Syp)	Complex formation between IRS-1 and	*In vivo*	Complex immunoprecipitated from insulin-stimulated 3T3 L1 adipocytes, binding occurs via SH2 domains demonstrated with a recombinant GST-SH2 domain fusion construct; IR did not bind SH-PTP2 and SH-PTP2 was not phosphorylated by insulin stimulation	430
	Binding to and dephosphorylation of specific PDGF-β receptor sites	*In vivo* and *in vitro*	Syp displayed a clear preference for certain receptor phosphorylation sites; the most efficiently dephosphorylated sites were phosphotyrosines pY-771 and pY-751, followed by pY-740, while pY-1021 and pY-1009 were very poor substrates; in contrast, PTP1B displayed no particular selectivity for the various β-PDGFR phosphorylation sites; a β-PDGFR mutant (F1009) that associates poorly with syp, had a much slower *in vivo* rate of receptor dephosphorylation than the wild-type receptor; syp is substrate-selective and both the catalytic domain and the SH2 domains contribute to syp's ability to choose substrates	1068
	Complex formation wih a 100 kDa protein in rat brain membranes		SH-PTP2 was coimmunoprecipitated with a 100 kDa tyrosine-phosphorylated membrane protein, which may couple SH-PTP2 to brain membranes	968
	Complex formation with a 115 kDa protein in EGF-stimulated cells and with a 105 kDa protein in activated T cells	*In vivo*	EGF stimulation of HepG2 and NIH 3T3 cells expressing high levels of the human EGF receptor resulted in the tyrosine phosphorylation of a 115 kDa protein that was coimmunoprecipitated with SH-PTP2; only trace amounts of the total EGF receptor pool were associated with SH-PTP2; activation of Jurkat cells with a T-cell receptor agonist monoclonal antibody resulted in the coimmunoprecipitation of a 105 kDa protein with SH-PTP2	1084
(syp)	Complexed with Grb2 in Epo-stimulated cells and binding directly to tyrosine-phosphorylated EpoR	*In vivo* and *in vitro*	Syp became phosphorylated on tyrosine after stimulation with Epo in M07ER cells overexpressing human EpoR; syp complexed with Grb2 in the cells and binding between syp and Grb2 was observed *in vitro*; syp also bound directly to tyrosine-phosphorylated EpoR on M07ER cells and *in vitro* to the EpoR by both N- and C-terminal SH2 domains	1080
(syp)	Association between syp and gp130 and JAK2 induced by IL-11 treatment	*In vivo* and *in vitro*	IL-11 treatment of 3T3-L1 cells increases the tyrosine phosphorylation of syp; in cell lysates, the C-terminal SH2 domain of syp associated with several proteins of 70, 130, 150 and 200 kDa that were tyrosine phosphorylated in response to Il-11; syp was inducibly associated with both gp130 and Janus kinase 2 (JAK2)	1063
	Binding between SH-PTP2 and a 115 kDa protein tyrosine phosphorylated by the insulin receptor	*In vivo*	Use of a myc epitope-tagged wild-type SH-PTP2 and a catalytically inactive point mutant demonstrated an insulin-dependent association of SH-PTP2 with tyrosine-phosphorylated pp115 which was markedly enhanced with the mutant enzyme; pp115 is the predominant *in vivo* SH-PTP2 binding protein	1085

	Analysis of binding specificity between phosphotyrosol peptide and tandem SH2 domains	*In vitro*	Used cognate peptide from PGDF receptor Tyr1009 domain and SPR analysis; minimum peptide with high binding affinity was VL(pY)TAV; Ala scan showed critical residues to be pY + 1 and pY + 3 as well as pY − 2; also examined cognate peptides from IRS-1	1066
	Binding between cognate insulin receptor phosphopeptides and the SH2 domains of SH-PTP2	*In vitro*	SH-PTP2 binds to insulin receptor C terminus at Tyr-1322 by its SH2 domains	1079a
	Binding between cognate IRS-1 phosphopeptides and the SH2 domains of SH-PTP2	*In vitro*	Simultaneous occupancy of both SH2 domains by tethered pY-peptides potently stimulates PTPase 37-fold; diphosphorylated peptide binds with higher affinity than monophosphorylated peptide sequence, suggests potent regulation by substrates with tandem pY sites	1076
(syp)	Binding to the IGF-1 receptor	*In vivo* and *in vitro*	GST-fusion protein containing both SH2 domains of syp coprecipitate with the IGF-1 receptor from purified receptor preparations and from whole cell lysates; binding of syp-GST was inhibited by a C-terminal peptide containing pY1316	1077
TCPTP (PTP-S)	Binding to DNA by basic domains at C terminus	*In vitro*	Expressed as recombinant full-length and GST-fusion constructs in *E. coli*, C-terminal 57 AA are sufficient for binding to calf thymus DNA, analysed by affinity chromatography, Southwestern blotting and gel retardation assays; binding did not occur for rat PTP1B	444

The physiology of PTPases

Studies using overexpression, microinjection and gene disruption

Many of the physiological effects of PTPases have either been discovered or substantiated in studies employing a variety of current molecular and cellular techniques in whole animal or cultured cell models. For the purposes of organizing the available data, I have separated these studies into two tables. Table 13 incorporates experiments using either transfection of recombinant PTPase cDNA constructs or microinjection of enzymes or antibodies that can perturb the activity of certain enzymes *in situ*. Table 14 compiles studies using those methodologies that eradicate the expression of specific PTPases either in transgenic animal models, known mutations in PTPases in animal strains, or expression of inactivated or mutated PTPases. Although there is some overlap in the published literature with some studies employing a combination of these techniques, this separation provides a useful way of presenting these data. This is one of the most active areas of current investigation, aimed at deciphering the role of specific PTPase homologues in growth, development and the function of various cells and organ systems. Insight into the available data for certain enzymes can be gleaned from the tabulated summaries.

As might be expected, evaluation of the physiological substrates and cellular role for PTPases has been challenging and is an exciting area of current work. For example, many PTPases have been found to be active against particular substrates such as the autophosphorylated insulin receptor *in vitro*, including CD45 [353]; however, in intact cells, high-level coexpression of recombinant CD45 and insulin receptors showed little dephosphorylation effect, while PTPase activity against other tyrosine phosphorylated receptor substrates in the transfected cells was observed [293]. The restricted expression of some PTPases to specific cell types is an essential factor in determining their roles in cellular function. For lymphocytes, CD45 has been shown to activate $p56^{lck}$ by the dephosphorylation of a negative regulatory residue (Tyr-505) near the C terminus of lck. This effect is essential for the role of CD45 in T lymphocyte responsiveness to antigen activation, which probably involves additional protein interactions and dephosphorylation events [8, 68]. Similarly, the transforming ability of RPTP-α in fibroblast cells involves the dephosphorylation of the autoregulatory Tyr-527 at the C terminus of c-*src*, which activates the kinase and can lead to cell proliferation [228]. MKP-1 and related PTPases have been shown to have particular specificity towards the dephosphorylation and deactivation of MAP kinase both *in vivo* and *in vitro* [345].

Two additional examples of PTPases that participate in the activation of cellular processes by interacting with specific substrates include cdc25 and the *corkscrew* gene product. The activity of the $p34^{cdc2}$ kinase, which is required for cellular entry into the $G_2 \rightarrow M$ phase of the cell cycle, is tightly regulated by reversible phosphorylation of Tyr-15. The phosphorylation state of this Tyr residue is balanced by the action of cdc25 in mammalian cells or homologues of the cdc25 PTPase in diverse organisms including *Drosophila* and related proteins in yeast [3]. The *Drosophila* corkscrew protein has been shown to potentiate the action of the *Drosophila* c-*raf* homologue to transmit signals positively downstream of the *torso* receptor tyrosine kinase and is involved in developmental morphogenesis [579, 580]. SH-PTP2, which appears to be the mammalian homologue of corkscrew, has been shown to have a similar positive role in growth factor signalling in mammalian cells. Studies employing expression of catalytically inactive recombinant SH-PTP2 constructs or by microinjection of blocking antibodies, SH2 domains of SH-PTP2, or phosphopeptides known to bind to its SH2 domains, all block insulin-stimulated activation of MAP kinase and mitogenic signalling [302, 367].

Table 13 Effects of PTPase expression/microinjection on cellular processes

PTPase	Transfection (T)/microinjection (M) construct	Host cell type	Functional parameter measured	Comments	Refs
CD45	(T) CD45	Murine cell line C127	↓ Ligand-activated autophosphorylation of IGF-1 and PDGF receptors as well as phosphorylation of endogenous sunstrates for these receptor kinases and mitogenic signalling	Basal (growth-factor independent) tyrosine phosphorylation of cellular proteins was unaffected by CD45 transfection	176
	(T) CD45		PDGF and IGF-1 signalling	CD45 expression ↓ activation of PI-3 kinase, phosphorylation of PL-C γ-1 and GAP by PDGF stimulation as well as ↓ activation of PI-3 kinase, tyrosine phosphorylation of IRS-1 and IRS-1 association of PI-3 kinase by IGF-1 stimulation	362
	(T) CD45	Plasmacytoma cell line J558Lμm3	B-cell antigen receptor	Receptor transduced a Ca^{2+}-mobilizing signal only if cells expressed CD45	252
	(T) CD45 RA + B + C isoform	CD45(−) HPB-ALL T cells	Kinase activity of $p59^{fyn}$ was ↓ in CD45(−) compared to CD45(+) cells, but $p56^{lck}$ activity was unchanged	In CD45(−) cells, multiple TCR-coupled signalling pathways were disrupted that were restored by CD45 transfection and correlated to increased $p59^{fyn}$ activity	337
	(T) CD45RABC isoform *vs.* CD45RO form	T cells	Thymic T cell responses	CD45RABC increased T cell proliferation in MLC and following TCR antibody binding as well as enhancing Ca^{2+} mobilization and Tyr(P) accumulation; CD45RO did not enhance these responses	246
	(T) chimeric cDNA encoding the intracellular domain of murine CD45 preceded by a short amino-terminal sequence from $p60^{c\text{-}src}$	CD45(−) T cells	TCR-mediated signalling	Expression of this chimeric protein corrected most of the TCR signalling abnormalities observed in the absence of CD45, including TCR-mediated enhancement of tyrosine kinase activity and Ca^{2+} flux; thus, the enzymatically active intracellular portion of CD45 is sufficient to allow TCR TM signalling	104
	(T) chimeric cDNA with intracellular PTPase domains of CD45 and the extracellular and TM domains of an MHC protein	CD45(−) T cells	TCR-mediated signalling	Expression of the enzymatically active intracellular portion of CD45 is sufficient to restore TCR TM signalling, extracellular domains not necessary to couple to signalling machinery	156

Table 13 Continued

PTPase	Transfection (T)/microinjection (M) construct	Host cell type	Functional parameter measured	Comments	Refs
CD45 (Cont.)	(T) chimeric protein consisting of EGF receptor extracellular and TM domains and the CD45 cytoplasmic domain		Expression of chimera is sufficient to restore TCR-mediated signalling in a CD45(−) cell line	Indicates that the cytoplasmic domain mediates TCR signal transduction; also, binding of EGF ligands functionally inactivated the chimeric protein in a manner dependent on dimerization	418
	(T) chimeric protein consisting of extracellular and TM domains of MHC class I molecule and the catalytic domain of yeast PTP1	J45.01 CD45(−) variant Jurkat T cell	Rescue of TCR-mediated signalling	Expression of the single domain yeast PTPase at the cell surface restored proximal and distal T cell signalling events	177
	(T) chimeric protein containing CD45 cytoplasmic domain and extracellular and TM domains of MHC class I A2 allele	J45.01 CD45(−) variant Jurkat T cell	Rescue of TCR-mediated signalling	Transfection of the chimeric CD45 protein rescued TCR- and CD2-mediated increases in protein tyrosine kinase activity and PI turnover; the cytoplasmic domain of CD45 is sufficient for mediating signalling through TCR and CD2 receptor pathways	141
	(T) full-length CD45	C127 mouse mammary tumour cells	c-src activation and dephosphorylation	Expression of CD45 ↓ PDGF-dependent receptor autophosphorylation as well as the activation of pp60$^{c\text{-}src}$, but did not affect basal c-src activity; *in vitro*, CD45 dephosphorylated and activated the c-src kinase; results suggested that the PDGF receptor was an *in vivo* substrate for CD45, whereas pp60$^{c\text{-}src}$ was not, perhaps due to compartmentalization or other factors regulating intracellular PTPase–substrate interactions	217
	(T) transient co-overexpression of PTPases with a panel of protein tyrosine kinases	293 embryonic fibroblast cells	Receptor kinase dephosphorylation	*In vivo*, inactive against HER1-2 receptor, little effect on ligand-activated IR and IGF-1R, but exhibited stronger dephosphorylation of stimulated and unstimulated α and β-PDGF receptors, EGFR, CSF-1R and highly effective at dephosphorylation of the chimeric c-kit kinase domain	293
	Thymocyte-specific overexpression of CD45 in transgenic mice using an lck promoter	Thymic T cells	TCR-mediated apoptosis and negative selection of TCRs	↑ Thymic expression of CD45RO ↑ the efficacy of TCR-mediated apoptosis and MHC-restricted negative selection of HY TCRs *in vivo*; augmented thymic expression of CD45 also resulted in activation of endogenous p56lck tyrosine kinase in CD4 + CD8 + thymocytes	312

	(T) CD45 isoforms	T-cell clones	Stimulation through the T-cell antigen receptor	CD45 isoforms differ in their ability to participate in antigen recognition; the null isoform that is predominantly found on memory CD4 T cells is the most effective; the ability of the CD45 ectodomain to differentially affect sensitivity to specific ligands represents a novel way of regulating the efficacy of signalling through a receptor without altering its specificity	1023
	CD45 (+) and (−) cell clones	Jurkat cells	Fc ϵ RI, high affinity IgE receptor activation	CD45 is necessary for the initiation of calcium flux through the transfected Fc ϵ RI; the tyrosine phosphorylation levels of β and γ after aggregation of Fc ϵ RI are reduced in the CD45-deficient cells; Fx ϵ RI aggregation induces an increase in lck enzymatic activity only in wild-type Jurkat and the CD45-deficient Jurkat reconstituted with chimeric CD45 suggesting that CD45 controls aggregation-induced receptor phosphorylation by regulation of src-family tyrosine kinase	981
	(T) CD45 isoforms	CD45-antisense transfected Jurkat T-cell clones that lack CD45	IL-2 production and tyrosine phosphorylation of cellular proteins after anti-CD3 stimulation	Transfected cells express either the smallest, CD45(O), or the largest, CD45(ABC), isoform exhibited isoform-dependent differences in IL-2 production and tyrosine phosphorylation of cellular proteins, including Vav after anti-CD3 stimulation; results demonstrate that the distinct CD45 extracellular domains differentially regulate T cell receptor-mediated signalling	1016
cdc25	(M) cdc25 antibodies	HeLa cells	Cell cycle progression	Inhibited entry into mitosis	303
	(M) full-length and truncated forms of *Drosophila* enzyme	*Xenopus laevis* oocytes	Induction of oocyte maturation	Injection of mRNA or protein for cdc25 rapidly induced maturation, effect was dependent on the presence of the C-terminal catalytic domain	262
	(T)	*Schizosaccharomyces pombe*	Cell cycle progression	Rescues defect in a temperature-sensitive yeast cdc25+ mutant that is unable to initiate mitosis	195
	(T)	*Schizosaccharomyces pombe*	Cell cycle progression	Overexpression causes mitosis to initiate at a reduced cell size; functions as a dose-dependent inducer of mitotic control	591
	In vitro transcription assay	*Xenopus laevis* oocyte extracts	Transcription of the $tRNA^{tyrC}$	Recombinant cdc25 selectively inhibited $tRNA^{tyrC}$ gene transcription	330

Table 13 Continued

PTPase	Transfection (T)/microinjection (M) construct	Host cell type	Functional parameter measured	Comments	Refs
cdc25A	(M) blocking antibody	HeLa cells	Cell cycle progression	Causes mitotic arrest	503
	(M) blocking antibody	Rat fibroblast NRK cells	Cell cycle progression	Blocked progression from G_1 to S phase	161
	(M) immunodepleting antibody		Cell cycle progression	Immunodepletion of cdc25A in rat cells by microinjection of a specific antibody effectively blocks their cell cycle progression from $G_1 \rightarrow S$ phase, indicating that cdc25A is not a mitotic regulator but a novel phosphatase that plays a crucial role in the start of the cell cycle; cdc25 expression is ↑ in G_1 phase	161
	(M) recombinant bacterially expressed human protein	*Xenopus* oocytes	Activation of MPF and arrest of meiosis at metaphase I	cdc25A protein into *Xenopus* prophase oocytes provokes the activation of p34cdc2; cdc25 induces the assembly of a metaphase I spindle which is abnormally located in the deep cytoplasm and oocytes arrest at the metaphase I stage	1030
	(M) inhibitory antibodies	HeLa cells	$G_1 \rightarrow S$ phase transition	Microinjection of anti-cdc25 antibodies into G_1 cells blocks entry into S phase	906
cdc25C	(M) blocking antibody	Hamster BHK cells	Chromosome condensation	Antibody injection inhibited chromosome condensation induced by a tsBN2 mutation	599
cdi1	(T)	*S. cerevisiae*/HeLa cells	Cell cycle progression	Overexpression delays progression through the cell cycle in both cell models; effect dependent on PTPase activity, catalytically inactive Cys-140 → Ser mutant was without effect	515
DdPTP1	(T)	*Dictyostelium discoideum*	Growth and development	Overexpression of DdPTP1 to a high level causes failure of cell aggregatation; overexpression to a moderate degree from the DdPTP1 promoter itself leads to severe morphological defects following aggregation including multiply tipped aggregates and aberrant fruiting bodies; also, in mutant strains, specific changes are seen in Tyr(P) proteins in the host cells by Western blot	277

	(T)	*Dictyostelium discoideum*	Actin tyrosine phosphorylation; cell shape	Cell lines overexpressing PTP1 had diminished amplitude and duration of actin tyrosine phosphorylation and exhibited diminished cell-shape change	278
DdPTP2	(T)	*Dictyostelium discoideum*	Growth and development	Overexpression of PTP2 to a high level causes slow growth on bacterial lawns and small cells in axenic medium; initiation of development in high expressing cells leads to aggregation but a cessation of morphogenesis at 6–8 h after which time a variable fraction of cells produce diminutive fruiting bodies with normal timing; the severity of the growth rate and developmental abnormalities corresponds to the level of overexpression; a unique pattern of alteration in Tyr(P) proteins in the cells is noted that is distinct from DdPTP1 overexpressing and null strains	523
(DdPTPa)	(T)	*Dictyostelium discoideum*	Growth and development	Transformed into *D. discoideum* using a non-cell-type specific promoter (actin 6), organism does not develop normally: fails to aggregate, and after 24 h, gives rise to only a few abnormal slugs and small fruiting bodies	588
ERP (3CH134)	(T)	Murine NIH 3T3 cells	Ability to isolate neomycin-resistant transfectant colonies; cell morphology; cell growth rate	Increased level of expression of ERP correlated with decreased appearance of neomycin-resistant transfectant colonies (100 : 1), ΔERP mutation with deletion in catalytic domain had same transfection ability as control indicating effect is due to active phosphatase domain; ERP-transfected cells have altered morphology: larger and generally multinucleated, and also have a slower growth rate than the parental cell line	571
FAP-1 (PTP-BAS)	(T) FAP-1	T-cell lines	Fas-induced apoptosis	Elevation in FAP-1 abundance partially abolished Fas-induced apoptosis	1118

Table 13 Continued

PTPase	Transfection (T)/microinjection (M) construct	Host cell type	Functional parameter measured	Comments	Refs
HA2	(T) wild-type enzyme	3T3-L1 preadipocytes	Adipocyte differentiation	Constitutive expression of PTPase HA2 in 3T3-L1 preadipocytes prevents adipocyte gene expression and differentiation; appropriately timed exposure of transfected preadipocytes to vanadate as clonal expansion ceases restores their capacity to differentiate, indicating that a critical tyrosine dephosphorylation event occurs during clonal expansion	1012
HePTP	(T)	NIH 3T3 fibroblasts	Cell morphology and growth	Transfected cells exhibited altered cell morphology, disorganized growth, anchorage independent colony formation and changes in phosphorylation pattern of Tyr(P) proteins	629
hVH-2	(T)	NIH 3T3 cells	Inhibition of map kinase signalling	Inhibited the v-src and MEK-induced transcriptional activation of a serum-responsive element containing promoter, consistent with the notion that hVH-2 promotes the inactivation of MAP kinase	1100
Low M_r PTPase (bovine liver)	(T)	NIH 3T3 cells	Cell proliferation (thymidine incorporation)	Overexpression ↓ mitogenic response to PDGF up to 90% also had a pronounced inhibitory effect on serum stimulation of thymidine incorporation; hormone-dependent stimulation of PDGF receptor autophosphorylation was significantly reduced	123
Low M_r PTPase	(T) synthetic full-length gene	Normal and v-erbB, v-src and v-raf transformed NIH 3T3 fibroblasts	Inhibition of proliferation	Inhibition correlated with level of transfected PTPase; also suppressed growth in soft agar, but not turnover of inositol lipids stimulated by PDGF	194
	(T) synthetic full-length gene	Normal and v-erbB-transformed NIH 3T3 cells	Cell proliferation; colony formation in soft agar	In transfected cells, dephosphorylation of a number of cellular proteins was observed; cell proliferation (thymidine incorporation) was reduced in both normal and v-erbB- transfected cells overexpressing recombinant CPTP; also ability of transformed cells to grow in soft agar was markedly decreased by overexpression of CPTP; may be involved in control of mitogenic signalling and the control of cell proliferation	189

(bovine liver)	(T) synthetic full-length gene	Normal 3T3 cells and 3T3 cells transformed by v-erbB, src or raf	Cell proliferation and GSH levels	Overexpression of the low M_r PTPase ↓ the proliferation of normal and transformed NIH 3T3 cells; also decreases GSH levels in normal and in v-erbB-transformed fibroblasts and increased GSH levels in src and raf-transformants; suggested a role for tyrosine dephosphorylation in GSH metabolism in normal and transformed cells	172
MKP-1	(T) wild-type and catalytically inactive enzymes	CCL39 fibroblasts	MAP kinase activation; subcellular localization	Transient transfection with epitope-tagged MKP-1 showed the protein to be entirely nuclear; a catalytically inactive mutant (Cys → Ser) was predominantly cytoplasmic; expression of either MKP-1 did not alter the cytosolic localization of MAP kinase in quiescent cells nor the ability of MAP kinase to translocate to the nucleus following mitogen where G_1-specific gene transcription and S-phase entry in fibroblasts was blocked	987
MKP-1 (3CH134)	Both (M) and (T) cell systems used for expression of plasmids	Rat embryonic fibroblast REF-52 cells	Inhibition of ras-induced DNA synthesis	Expression in REF-52 cells blocked to induction of DNA synthesis in quiescent cells by microinjection of an activated mutant of ras, V^{12}ras; PTP1B was without effect	346
	(T)	COS cells	Dephosphorylation and inactivation of MAP kinase	Expression in COS cells caused selective dephosphorylation of $p42^{MAPK}$ and blocked phosphorylation and activation of $p42^{MAPK}$ by serum, or cotransfected oncogenic ras or activated raf	345
MKP-2	(T) wild-type nzyme	PC12 cells		Overexpression of MKP-2 *in vivo* inhibited MAP kinase-dependent gene transcription	1111
MSG5	(T)	*Saccharomyces cerevisiae*	Recovery from pheromone-induced cell cycle arrest	Overexpression of MSG5 dramatically promoted recovery from pheromone-induced arrest in G_1 phase; conversely, loss of MSG5 function leads to a diminished adaptive response to pheromone; acts at stage where STE7 and FUS3 kinases function to transmit the pheromone signal	497

Table 13 Continued

PTPase	Transfection (T)/microinjection (M) construct	Host cell type	Functional parameter measured	Comments	Refs
PAC1	(T) wild-type and Cys-257 → Ser mutant	Several cell systems	Inhibition of MAP kinase (ERK2) activation	*In vivo* expression inhibits MAP kinase stimulation by EGF, PMA or T-cell cross-linking; also blocks MAP kinase regulated reporter gene expression (c-Fos SRE); effects abolished by catalytic site mutation	361
PRL-1	(T) stable overexpression of full-length cDNA	NIH 3T3 cells	Viability of transfected cells; morphology; growth rate; colony formation in soft agar	PRL-1 cDNA ↓ frequency of viable transfected colonies by 90%; in stable clones overexpressing 3- to 5-fold higher levels of PRL-1 compared to controls (primarily nuclear in a speckled pattern); some of the transfected cells were grossly abnormal including multinucleated giant cells; all selected clones had rapid growth rates (20 h) compared to control (24 h), as well as ↑ cell saturation density and ↑ ability to form colonies in soft agar (rate comparable to v-src transfected cells)	493
PTP-α	(T)	Rat embryo fibroblasts	Activation of c-Jun and MAP kinase; c-Src/c-Jun complex formation	PTP-α transfected cells have enhanced Tyr(P) phosphorylation of p39 which was identified as c-Jun; c-Jun forms complex with c-src; in transfected cells, c-Jun DNA binding activity and c-Jun mediated transcription is ↑; also MAP kinase is activated in transfected cells and translocated to the nucleus	227
PTP1, PTP1B, pyp1$^+$, pyp2$^+$	(T) *Saccharomyces cerevisiae* PTP1 or human PTP1B	*Schizosaccharomyces pombe*	Complementation of a cdc25 mutant	In a cdc25-22 or wee1-50 genetic background, established that either *Saccharomyces cerevisiae* PTP1 or human PTP1B can complement the cdc25 mutant, opposing the *wee1$^+$/mik1$^+$* pathway, in contrast to pyp1$^+$ or pyp2$^+$, which were ineffective	270
PTP1	(T) of PTP1 and PTP1 mutant cells	*Saccharomyces cerevisiae*	Dephosphorylation of pY-protein substrates	Effects of Ptp1 expression or mutant cells showed that it can dephosphorylate a broad range of substrates *in vivo*; one phosphotyrosyl protein (p70) was identified as FPR3 gene product, a nucleolarly localized proline rotamase	1041

PTP1B	(M) neutralizing antibodies	Rat hepatoma cells	Insulin receptor activation and signalling	In cells loaded with PTP1B antibody, insulin signalling was enhanced and insulin-stimulated receptor autophosphorylation and kinase activity was increased, suggesting that PTP1B has a role in the negative regulation of insulin signalling acting at least in part at the level of the receptor	982
	(M) purified, truncated cytoplasmic form of PTP1B	*Xenopus* oocytes	Induction of meiotic cell division	Microinjection markedly retarded (by up to 5 h) maturation induced by insulin; maturation was significantly retarded even when the PTPase was injected 2 to 4 h after exposure of the cells to insulin; PTP1B also retarded maturation induced by progesterone and MPF, pointing to a second site of action of the PTPase in meiotic cell division, downstream of the insulin receptor	351
	(M) purified, truncated cytoplasmic form of PTP1B	*Xenopus* oocytes	Inhibition of insulin action	Microinjection blocked the insulin-stimulated phosphorylation of tyrosyl residues on endogenous proteins, including a protein likely to be the β-subunit of the IR or IGF-1 R; also PTP1B blocked the activation of an S6 peptide kinase known to be activated early in response to insulin	247
	(T)		Inhibition of transcription	Transient transfection specifically blocks transctiptional activation by AP-1 or CREB, but not by estrogen receptor, GAL4, GAL4-VP16 fusion protein or activity of SV40 early promoter or MMLV-LTR; effect also seen with TCPTP	132
	(T) full-length human cDNA	Normal and *neu* oncogene transformed NIH 3T3 cells	Cell transformation	Overexpression of PTP1B in NIH 3T3 cells suppressed subsequent transformation by expression of *neu* oncogene, involving focus formation, anchorage-independent growth and tumourigenicity	125
	(T) full-length rat cDNA	Normal and v-src-transformed mouse NIH 3T3 fibroblasts	Cell transformation	Overexpressed 2- to 25-fold; enzyme associated with ER; changes noted in morphology with ↑ bi- and multinucleate cells; no change in pY content of $pp60^{v\text{-}src}$ or in cell proliferation	222
	(T) transient co-overexpression of PTPases with a panel of protein tyrosine kinases	293 embryonic fibroblast cells	Receptor kinase dephosphorylation	*In vivo* able to dephosphorylate a wide range of receptors including EGFR, a chimeric EGF-c-*erb*B2 receptor (HER1-2), IR, IGF-1R, PDGFR (α and β), a chimeric EGF/c-kit receptor, and CSF-1R	293

Table 13 Continued

PTPase	Transfection (T)/microinjection (M) construct	Host cell type	Functional parameter measured	Comments	Refs
pyp1$^+$	(T)	*Schizosaccharomyces pombe*	Cell growth	Overexpression of pyp1$^+$ leads to a delay in mitosis	578
pyp1$^+$, pyp2$^+$		*Schizosaccharomyces pombe*	Dephosphorylation of Sty1	Pyp2 associates with and tyrosine dephosphorylates Sty1 *in vitro*; induction of pyp2 mRNA is responsible for tyrosine dephosphorylation of Sty1 *in vivo* on prolonged exposure to osmotic stress; pyp1 and pyp2 are tyrosine-specific MAP kinase phosphatases that inactivate an osmoregulated MAP kinase, Sty1, which acts to control cell size at division in fission yeast	1018
pyp2$^+$	(T)	*Schizosaccharomyces pombe*	Cell growth	Overexpression of pyp2$^+$ leads to a delay in mitosis	578
	(T)	*Schizosaccharomyces pombe*	Cell growth	Overexpression of pyp2$^+$ delays the onset of mitosis	561
pyp3$^+$	(T)	*Schizosaccharomyces pombe*	Cell growth	Overexpression of pyp3$^+$ causes an advancement of mitosis, cells divide at one-half the normal size; different from pyp1$^+$ or pyp2$^+$ overexpression which delays onset of mitosis; pyp3$^+$ acts cooperatively with p80cdc25 to activate p34^{cdc2}	560
RPTP-α	(T) full-length human RPTP-α	Fischer rat embryo fibroblasts	Cell transformation and activation of pp60$^{c\text{-}src}$	Overexpression results in persistent activation of pp60$^{c\text{-}src}$ kinase, cell transformation and tumourigenesis; c-src activation is accompanied by dephosphorylation of c-src Tyr-527 *in vivo*, a site which is susceptible to dephosphorylation by RPTP-α *in vitro;* thus, RPTP-α may have oncogenic potential when overexpressed	228
	(T) full-length murine RPTP-α	Mouse P19 embryonal carcinoma cells	Activation of pp60$^{c\text{-}src}$ and neuronal differentiation	Treatment of P19 cells overexpressing RPTP-α with retinoic acid induces neuronal differentiation while in native cells the same treatment leads to endo- and meso-dermal differentiation; expression of RPTP-α results in persistent activation of pp60$^{c\text{-}src}$ kinase, accompanied by dephosphorylation of c-src Tyr-527 *in vivo;* RPTP-α also dephosphorylated this site *in vitro*	139

	(T) human RPTP-α	BHK cells	Blockade of insulin-induced cell detachment	Identified as a negative regulator of the insulin receptor tyrosine kinase	1019
	(T) human RPTP-α	REF cells	Transcription factor activity	Overexpression of RPTP-α resulted in a constitutive nuclear NF-κ B-like DNA-binding activity	1025
RPTP-ε	(T) human RPTP-ϵ	BHK cells	Blockade of insulin-induced cell detachment	Identified as a negative regulator of the insulin receptor tyrosine kinase	1019
SH-PTP1 (HCP)	(T) inducible expression of sense and antisense cDNAs of HCP	IL-3 dependent myeloid cell line, DA-3	IL-3R β chain tyrosine phosphorylation and cell proliferation	↑ Levels of HCP decreased IL-3R Tyr(P) content and suppressed cell growth; conversely ↓ HCP by antisense increased IL-3R phosphorylation and slightly increased cell growth rate	464
(PTP1C)	(T) full-length human cDNA	Human 293 fibroblasts	Growth factor receptor dephosphorylation	Cotransfection with growth factor receptor cDNAs; after ligand stimulation, SH-PTP1 led to complete or partial dephosphorylation of receptors for PDGF(α), PDGF(β), EGF, insulin, IGF-1 or chimeric HER2-*neu* and p145$^{c\text{-}kit}$ kinases	616
(PTP1C)	(T) full-length and construct lacking SH2 domains	v-src transformed SR-3Y1 rat fibroblasts	Cell growth and anchorage-independent colony formation	Cell growth rate was decreased in low serum medium, also SH-PTP1 suppressed anchorage-independent colony formation; presence of the SH2 domains enhances these two effects	99
(HCP)	(T) HCP	CEM-6, MOLT-4 cells and other lymphoid cell lines	Involvent in Fas-mediated apoptosis	Tyr(P) dephosphorylation after Fas cross-linking occurred in Fas apoptosis-sensitive CEM-6 cells but not in Fas apoptosis-resistant MOLT-4 cells; the dephosphorylation correlated with increased HCP inactivity; HCP was corrlated with Fas apoptosis function in 11 lymphoid cell lines; expression of recombinant HCP in the MOLT-4 cell line converted this FAS apoptosis-resistant cell line to Fas apoptosis sensitive; HCP-mutant mev/mev mice exhibited decreased Fas-mediated apoptosis in lymphoid organs	1037
SH-PTP2 (PTP1D)	(T) full-length human cDNA	Human 293 fibroblasts	Growth factor receptor dephosphorylation	Cotransfection with growth factor receptor cDNAs; no effect on ligand-stimulated autophosphorylation of receptors for PDGF(α), PDGF(β), EGF, insulin, IGF-1, and a chimeric HER2-*neu* kinase; p145$^{c\text{-}kit}$ kinase ↓ 40%	616

Table 13 Continued

PTPase	Transfection (T)/microinjection (M) construct	Host cell type	Functional parameter measured	Comments	Refs
SH-PTP2 (Syp)	(M) Syp SH2 domain-GST fusion protein	Rat-1 fibroblasts overexpressing the human IR	Growth factor-induced BrdUrd incorporation	Microinjection of Syp SH2-GST fusion protein, anti-Syp Aby, or phosphonopeptides derived from Syp binding sites to IRS-1 blocked stimulation of BrdUrd incorporation by insulin, IGF-1 and EGF but not serum	367
(Syp)	(T) full-length SH-PTP2 from human fetal liver, PTPase inactivated by Cys-459→Ser mutation	NIH 3T3 cells overexpressing the human IR	Activation of mitogenesis and MAP kinase by insulin	Expression of mutant SH-PTP2 blocked stimulation of mitogenesis and phosphorylation and activation of MAP kinase kinase and MAP kinase by insulin without affecting insulin-induced tyrosine phosphorylation of the IR, IRS-1 or Shc; identified a 120 kDa insulin-stimulated pY protein that binds to the SH2 domains of SH-PTP2	302
(PTP2C)	(T) wild-type and catalytically inactive enzymes	293 kidney cell line	Insulin-elicited protein-tyrosine phosphorylation	Insulin caused the rapid tyrosine phosphorylation of a 160 kDa protein, possibly IRS-2, as well as a 100 kDa polypeptide, probably a mixture of the insulin and IGF-1 receptor β-subunits; there was no difference in the extent of tyrosine phosphorylation or in the rate of reversal upon insulin withdrawal, suggesting that PTP2C does not dephosphorylate these proteins *in vivo*	1008
(PTP2C)	(T) wild-type and catalytically inactive enzymes	293 embryonic kidney cells	EGF-stimulated MAP kinase activation; substrate phosphorylation	Native PTP2C enhanced EGF-stimulated MAP kinase activity by 30%, whereas expression of the inactive mutant reduced the stimulated activity by 50%; native PTP2C was located entirely in the cytosol, while the inactive mutant was nearly equally distributed in cytosolic and membrane fractions; expression of inactive PTP2C caused hyperphosphorylation on tyrosine of a 43 kDa protein	1047
SH-PTP2, SH-PTP1	(T) wild-type and catalytically inactive enzymes	Fibroblasts	DNA synthesis	Overexpression of PTP1C or PTP1D had no effect on DNA synthesis stimulated by different growth factors; a mutated inactive form of PTP1D strongly inhibited the stimulatory effects of both PDGF and α-thrombin on early gene transcription and DNA synthesis, suggesting that it is a positive mediator of mitogenic signals induced by both tyrosine kinase receptors and G protein-coupled receptors	961

TCPTP	(T)	NIH 3T3 fibroblasts transformed by v-*erb*B or v-fms	Suppresion of transformation	Cell transformation by v-erbB or v-fms can be suppressed by expression of TCPTP only with a deregulated, truncated form lacking the C-terminal localization signal for the ER	293
	(T)		Inhibition of transcription	Transient transfection specifically blocks transcriptional activation by AP-1 or CREB, but not by estrogen receptor, GAL4, GAL4-VP16 fusion protein or activity of SV40 early promoter or MMLV-LTR; effect also seen with PTP1B	132
	(T) full-length and C-terminal truncated (37 kDa) forms	Baby hamster kidney (BHK) cells	Cell growth and phosphorylation of cellular proteins	Expression of the enzymatically active, truncated form reduced the cellular growth rate to ~50% of control, but expression of *both* forms reduced level of pY phosphorylation in several proteins (but not the PDGF receptor itself) after 2–5 min of PDGF stimulation indicating they are both active *in vivo*	251
	(T) full-length and C-terminal truncated (37 kDa) forms	Baby hamster kidney (BHK) cells	Cell morphology and nuclear division	Stable overexpression of truncated form leads to cytokinetic failure and asynchronous division in BHK cells leading to multinucleated cells exhibiting synchronous mitosis of the syncytial nuclei; cells overexpressing the full-length enzyme are morphologically similar to controls	250
	(T) full-length and C-terminal truncated (37 kDa) forms	Rat-2 cells	Suppression of v-fms - induced cell transformation	In host cells transformed with v-fms, overexpression of full-length TCPTP had no effect, while cells overexpressing the truncated form exhibited changes in cell morphology, loss of anchorage-independent growth in soft agar and and reduced tumour formation in nude mice; both increases and decreases in the Tyr(P) content of cell proteins were detected in the transfected cells	368
	(T) full-length cDNA	*Schizosaccharomyces pombe*	Dephosphorylation of Tyr-15 on pp34 *in vivo* and *in vitro*	Dephosphorylation of pp34 by TCPTP triggers activation of pp34–cyclin complex in *S. pombe;* this can complement the activity of the mitotic activator, $p80^{cdc25}$	267

Table 13 Continued

PTPase	Transfection (T)/microinjection (M) construct	Host cell type	Functional parameter measured	Comments	Refs
	(T) transient co-overexpression of PTPases with a panel of protein tyrosine kinases	293 embryonic fibroblast cells	Receptor kinase dephosphorylation	*In vivo* able to dephosphorylate a wide range of receptors including EGFR, a chimeric EGF-c-erbB2 receptor (HER1-2), IR, IGF-1R, PDGFR (α and β), a chimeric EGF/c-kit receptor and CSF-1R; truncated form had higher level of dephosphorylation of all substrates than full-length; catalytically inactive mutant of TCPTP (Cys-216 → Ser) had no effect on receptor tyrosine phosphorylation	293
VHR	(M) wild-type and catalytically inactive enzymes	*Xenopus* oocytes	GVBD, maturation	VHR induced GVBD while the inactive (Cys → Ser) mutant lacked activity in injected oocytes; VHR may be promoting G2/M transition by weakly mimicking the action of cdc25, the dual specificity phosphatase that physiologically activates the MPF, and VHR itself may be responsible for activation of cdc–cyclin complexes	983
YopH	Cells infected with *Yersinia pseudotuberculosis* carrying wild-type YopH or a Cys-403 → Ala mutant which is catalytically inactive	Murine macrophage-like cell line J774A.1	Dephosphorylation of host cell proteins	YopH specifically and rapidly dephosphorylated a macrophage protein of 120 kDa; this protein and a previously detected 5-kDa substrate of YopH coprecipitated with the catalytically inactive mutant; the 120- and 55-kDa proteins were tyrosine kinases *in vitro* suggesting that YopH dephosphorylates activated tyrosine kinases *in vivo*	235
	Y. pseudotuberculosis carrying plasmid pIB1 encoding YopH, or an inactivated YopH PTPase (Cys-403 → Ala mutant)	Intact mice and cell line J774A.1	Infectivity in mice; Tyr(P) protein dephosphorylation in cultured cells	Intrinsic PTPase activity of YopH is essential for causing disease in the intact animals; in cultured cells, dephosphorylation of specific host proteins of 120 and 60 kDa was due to YopH PTPase activity; phagocytosis of bacteria was not required for dephosphorylation of host cell proteins, suggesting that functional YopH is expressed by extracellular bacteria	234

(T) wild-type and catalytically inactive enzymes	Murine bone marrow-derived macrophages	Effects of YopH on tyrosine phosphorylation and respiratory burst activity	YopH suppressed both tyrosine phosphorylation and respiratory burst activity in response to zymosan; mutational inactivation of YopH reversed the suppressive effect, but failed to reverse the inhibitory effect of YopH on the zymosan-triggered respiratory burst	1001
Macrophage infection with *Y. pseudotuberculosis* carrying YopH plasmid	J774A.1 macrophage cells	Oxidative burst to secondary bacterial infection	Infection of macrophages with plasmid-containing *Y. pseudotuberculosis* inhibited the oxidative burst triggered by secondary infection with opsonized bacteria; PTPase activity of YopH was necessary for this inhibition, indicating that YopH inhibits Fc receptor-mediated signalling in macrophages	985

Table 14 Effect of PTPase mutations/deletions/gene disruption in cell and animal models

PTPase	Animal/cell model	Mutation	Phenotypic parameters	Comments	Refs
CD45	CD45(−) CD8+ T cell clone L3M-93	CD45(−) clones	↑ Tyr(P) content of $p56^{lck}$ and $p59^{fyn}$ and ↓ kinase activity; phosphorylation of C-terminal negative regulatory sites ↑ 8-fold in $p56^{lck}$ and 2-fold in $p59^{fyn}$	CD45 dephosphorylates the negative regulatory sites of multiple Src family kinases and this correlates with the ability to respond efficiently to antigen stimulation	300
	CD45(−) T cell clones		Failure of anti-Thy-1 to induce stimulation of proliferation	Also, anti-Thy-1 fails to induce $p56^{lck}$ activity in CD45(−) cells; cells retain responsiveness to lectins (Con A and PHA); PMA together with ionomycin also elicits proliferation in CD45(−) cells; data suggest that CD45 is required for the activation of Tyr kinase activity at a proximal step	321
	CD45-deficient mice		Cross-linking of surface IgE does not induce mast cell degranulation as in wild-type; CD45-deficient mice are resistant to IgE-dependent systemic anaphylaxis		231
	Human CD45(−) T cell acute leukaemia cells	CD45(−) clones	Cell surface $p56^{lck}$ activity was 78% lower in CD45(−) cells and TCR-associated kinase activity was 84% lower in CD45(−) cells	Demonstrated coupling of CD45 with receptor-associated kinase pools	233
	Human T cell leukaemia cell line	Defective in CD45 expression	Stimulation of TCR failed to ↑ PI derived second messengers	Reconstitution of CD45 restored early TCR signalling; CD45 is essential for coupling TCR with PI signalling pathway	287
	Immature B-cell line	CD45(−) clones	In the CD45(−) cells, tyrosine phosphorylation was constitutively induced by anti-IgM stimulation and anti-IgM-induced Ca^{2+} flux was prolonged	The degree of growth arrest and DNA fragmentation induced by anti-IgM antibody was more evident in CD45 clones	311
	Jurkat cell mutants	CD45(−) clones	↓ Ca^{2+} influx following TCR-CD3 stimulation; also had ↓ secretion of various lymphokines after stimulation with anti-CD3 and PMA	Stimulated lymphokine secretion reappeared in a CD45(+) revertant subclone	320

Jurkat cell mutants	CD45(−) clones	CD45 expression is required for activation of a TCR-associated tyrosine kinase as well as the PI pathway; also, activation of T cells via the CD2 accessory molecule requires CD45; activation by CD28 was not impaired in the CD45(−) cells	Expression of CD45 is essential for TCR coupling to PI second messenger pathways	286
Leukemic T cell	CD45(−) clones	Phosphorylation of $p56^{lck}$ at the inhibitory Tyr-505 position was ↑ by 2- to 8-fold in 3 different cell lines; phosphorylation of fyn at Tyr-531 was increased 2.5-fold in 2/3 cell lines; phosphorylation of p60c-src was unchanged	Effect of CD45 expression is much greater on lck than fyn, and that CD45 exhibits substrate specificity *in vivo;* hyperphosphorylation of lck may be cause of unresponsiveness of CD45(−) lymphoid cells to antigenic stimulation	279
Mouse CD8+ cytolytic T-cell clones	Specific defect in the expression of CD45	↓ CD45 mRNA expression	Cells were greatly diminished in their ability to respond to antigen, affecting cytolysis of targets, proliferation and cytokine production; functions were restored in a CD45(+) revertant	363
Mouse T-cell clones	Defective in CD45 expression	Unable to proliferate in response to antigen or to CD3 cross-linking	Response to IL-2 was preserved; a CD45(+) revertant responded normally to antigen and CD3 cross-linking	322
Murine and human CD45(−) T-cell lines	CD45(−) clones	In CD45(−) cells, $p56^{lck}$ and $p59^{fyn}$ were hyperphosphorylated (lck on Tyr-505), but had paradoxically increased tyrosine kinase activity		237
Murine lymphoma cell line	Defective in CD45 expression	Loss of CD45 increases Tyr(P) in $p56^{lck}$ on Tyr-505	Tyr-505 is a putative negative regulatory site of $p56^{lck}$	314
T cell variants expressing progressively lower amounts of CD45	Reduced CD45 expression	CD45 abundance was inversely related to Tyr(P) content of several cellular proteins including TCR-ζ chain and was directly correlated to TCR-mediated PI hydrolysis; CD45(−) cells also exhibited a delayed ↑ in intracellular Ca^{2+}	CD45(+) revertants had normal signalling properties	357

Table 14 Continued

PTPase	Animal/cell model	Mutation	Phenotypic parameters	Comments	Refs
CD45 (Cont.)	T cells	CD45(−) clones	Fail to signal via the TCR pathway to PLC-γ-1 and ↑ PI hydrolysis	Reconstitution of CD45 into a deficient Jurkat T-cell mutant restored the signalling potential of the TCR	164
	Transgenic mouse	Homozygous exon 6 (−/−)	Thymocyte maturation was blocked at transitional stage from immature CD4+/CD8+ to mature CD4+ and CD8+ cells; only a few T cells detected in peripheral lymphoid organs; B cell development appeared normal	In homozygous mice, B cells and most T cells did not express CD45	284
	LEC rats	CD4+ cells which are deficient in expression of the CD45RC isoform	IL-2 production in LEC rat peripheral CD4+ T cells	LEC rat CD4+ cells are deficient in expression of the CD45RC isoform, but not of CD45 molecules; CD45RC low cells exhibit a defect in IL-2 production; PTPase activity in the membrane fraction of LEC rat CD4+ cells was threefold higher than that of normal rat CD4+ cells; suggest that an elevated CD45 PTPase activity is reponsible for a defect in IL-2 production in LEC rat peripheral CD4+ T cells	964
	T cells	CD45 (−) clones	Regulation of the Syk-family kinase p70zap	In CD45 (−) T cells, p70zap was constitutively phosphorylated on tyrosine and coimmuniprecipitated with the TcRζ chain; in resting wild-type CD45-positive cells, p70zap was mainly unphosphorylated and was rapidly phosphorylated on tyrosine upon treatment of the cells with anti-CD3 or PTPase inhibitors;; tyrosine phosphorylated p70zap was also dephosphorylated by CD45 *in vitro*	1020
	Jurkat cells	CD45 (−) clones	Oscillations of free intracellular Ca^{2+} in response to a magnetic field	CD45-deficient Jurkat cell line was unable to respond to magnetic field stimulation	1013
	BAL-17 cells (B-lymphocyte line)	CD45 (−) clones	Growth inhibition	B cell receptor (BcR) stimulation led to growth inhibition in the parental cells, but signals for growth inhibition were completely blocked in the CD45-negative clones	1025
	CD45 knockout mice		Development of Vγ3 dendritic epidermal T cells	Skin contained reduced numbers of dendritic AT cells in CD45-deficient mice; maturation was blocked at immature stage in thymus	1007

	CD45 (−) clones of immature B-cell line WEHI-231		Effects on protein-tyrosine kinase activation	Lyn is hyperphosphorylated and activated in CD45 (−) clones (no change in Syk or Lck), suggesting Lyn as *in vivo* substrate for CD45 in B cells	1006
	CD45 (−) T cells	Mutation of catalytic Cys or expression of CD45 in cell cytoplasm	Restoration of TCR-mediating signalling	Used a chimeric CD45 containing the myristoylation sequence of Src that can restore TCR-mediated signalling; mutation of catalytic cysteine in 1st PTPase domain blocked this effect of CD45; mutation destroying the myristoylation site caused expression of active CD45 in the cytoplasm but failed to restore signalling, indicating membrane localization is critical	1022
	Human lymphocytes	Heterozygosity for a point mutation at nucleotide position 77 of exon A, leading to a C → G transition	Associated with continuous expression of CD45RA molecules on activated and memory T cells but no change in T-cell reactivity in the heterozygous state	Does not change the protein sequence of the CD45RA isoform; appears to be part of a motif necessary for splicing of exon A, which might prevent binding of a *trans*-acting splice factor	973
cdc25 (string)	*Drosophila melanogaster*	*string* cdc25Dm locus	Cell cycle arrest at interphase 14 transition	Suggests the *stg* is required for the initiation of mitosis	498
	Drosophila melanogaster	mat(2)synHB5 flies harbour a point mutation Pro-295 → Leu in *twine* protein	Complete block of meiosis in males and severe meiotic defects in females with sterility	Pro-295 is highly conserved among cdc25 homologues	489
cdc25 (twine)	*Drosophila melanogaster*	Homozygous *twine* mutant	Meiosis does not occur in homozygous *twine* mutant males which produce abnormal spermatids	Gene required for oogenesis, during syncytial embryonic development and for male meiosis	474
CSW	*Drosophila melanogaster*	Gene disruption	Maternally required for cell fate determination at the termini of the embryo	Signalling pathway mediated by a tyrosine kinase (*torso*), a serine/threonine kinase (D-*raf*), and *tailless*; analysis of double mutants demonstrates that *csw* acts downstream of *torso* and in concert with D-*raf* to positively transduce the *torso* signal via *tailless* to downstream terminal genes; *csw* is required for zygotic viability; embryos derived from females carrying *csw* mutations are twisted or U-shaped; mutant embryos show abnormal development of terminal structures	581, 582
DdPTP1	*Dictyostelium discoideum*	Gene disruption	Cell growth and development	Causes an acceleration of development	277

Table 14 Continued

PTPase	Animal/cell model	Mutation	Phenotypic parameters	Comments	Refs
DdPTP1 (Cont.)	*Dictyostelium discoideum*	Gene disruption	Tyrosine phosphorylation of actin (primarily of a minor acidic isoform) was rapid and more prolonged	Also exhibited accelerated kinetics of cell rounding	278
DdPTP2	*Dictyostelium discoideum*	Gene disruption	Cell growth and development	DdPTP1 and DdPTP2 possess distinct regulatory functions in controlling growth and development of *D. discoideum;* not detectably altered in growth or temporal pattern of development; *Ddptp2* null slugs are significantly larger than wild-type suggesting a possible role in regulating multicellular structures	523
DPTP99A	*Drosophila*	Null mutation	No obvious phenotypic changes noted	Flies are viable and fertile with no detectable alterations in embryonic nervous system	1002
LAR	McA-RH7777 rat hepatoma cells	Expression of antisense RNA to reduce LAR protein abundance	IR phosphorylation and kinase activation	↓ of LAR protein by 63% lead to a 150% increase in IR autophosphorylation, 35% ↑ in IR kinase activity and 350% ↑ in insulin-stimulated PI 3′-kinase activity	290
	Transgenic mice	Gene knockout	Development	mRNA expression was eradicated in homozygous mice carrying insertions in the gene, but embryonic development was normal	1055
	Rat McARH-7777 hepatoma cells	Antisense expression of LAR mRNA	Ligand dependent activation of multiple receptor tyrosine kinases	*In vivo* insulin-dependent tyrosine phosphorylation of both IRS-1 and Shc was increased by a similar three-fold with LAR suppression, and were paralleled by increases in insulin-dependent PI 3-kinase association with IRS-1 and activation of the MAP kinase pathway; reduced LAR levels also resulted in increases in EGF- and hepatocyte growth factor-dependent receptor autophosphorylation, as well as increases in substrate tyrosine phosphorylation	1009
LTP1	*S. cerevisiae*	Disruption of *LTP1* gene	No apparent phenotype	Neither the disruption of the *LTP1* gene nor an approximately 10-fold overexpression of LTP1 in *S. cerevisiae* caused any apparent phenotypic changes; no proteins related to LTP1 could be detected in extracts of the null mutant, suggesting that LTP1 is the only low M_r PTPase in *S. cerevisiae*	1116
MKP-1	Vascular smooth muscle cells	Use of antisense oligonucleotides	MAP kinase activation by angiotensin II	Inhibition of MKP-1 expression caused prolonged activation of the p42 and p44 MAP kinases by angiotensin II	998

PTP2	*Saccharomyces cerevisiae*	Double mutant of PTP2 and a protein serine/threonine phosphatase gene (PTC1)	Synthetic growth defect	PTP2 mutation has no obvious phenotype by itself, but has a profound effect on cell growth when combined with mutations in a novel protein phosphatase gene, PTC1, a homologue of the mammalian protein serine/threonine phosphatase 2C (PP2C); disruption of PTC1 itself showed that the PTC1 function is non-essential; however, PTC1/PTP2 double mutants showed a marked growth defect	299
pyp1$^+$, pyp2$^+$	*Schizosaccharomyces pombe*	pyp1$^+$ Cys-470 → Ser, pyp2$^+$ Cys-630 → Ser (catalytically inactive), or overexpression of a non-catalytic N-terminal domain	Cell cycle progression	In a cdc25-22 background, overexpression of either N-terminal domain of pyp1$^+$ or a catalytically inactive mutant Cys-470 → Ser causes cell cycle arrest, this phenotype reverses the suppression of a cdc25 temperature-sensitive mutation caused by pyp1 disruption; pyp1$^+$(Cys-470 → Ser) or a catalytically inactive form of pyp2(Cys-630 → Ser) induce mitotic delay as do the wild-type enzymes, further analysis showed that the *in vitro* catalytic activity as well as their biological activity depends on the presence of N-terminal sequences not normally considered part of the catalytic domains	269
pyp1$^+$	*Schizosaccharomyces pombe*	Gene disruption	Cell growth	Resulted in viable cells	577
				Disruption of pyp1$^+$ produces cells that divide at a smaller size and an advancement of mitosis similar to inactivation of the wee1 mitotic inhibitor	561
pyp1$^+$/ pyp2$^+$ double knockout	*Schizosaccharomyces pombe*	Gene disruption	Cell growth	Double disruption of pyp1$^+$ and pyp2$^+$ causes synthetic lethality; simultaneous manipulation of multiple genes suggests that pyp1$^+$ and pyp2$^+$ act as negative regulators of mitosis upstream of the wee1$^+$/mik1$^+$pathway	578
		Gene disruption	Cell growth	Double disruption of pyp1$^+$ and pyp2$^+$ is lethal; simultaneous manipulation of multiple genes suggests that pyp1$^+$ and pyp2$^+$ act as negative regulators of mitosis upstream of the wee1$^+$/mik1$^+$pathway	561
pyp2$^+$	*Schizosaccharomyces pombe*	Gene disruption	Cell growth	Disruption of pyp2$^+$, like pyp1$^+$, results in viable cells	578
				pyp2$^+$disruption produces cells with normal growth rate and morphology, but a slightly smaller cell division size, suggesting that pyp2$^+$ has a weak role as a mitotic inhibitor, while pyp1$^+$ is dominant	561
pyp3$^+$	*Schizosaccharomyces pombe*	Gene disruption	Cell growth	Causes mitotic delay that is greatly exacerbated in cells with partial cdc25 defect	560
RPTP-γ	Mouse L-cells	Allelic deletions	Cell transformation; sarcomatous growth	L-cells have homozygous, intragenic deletions in the PTPRG gene that spans the CA domain from the 2nd intron to the 4th intron	618

Table 14 Continued

PTPase	Animal/cell model	Mutation	Phenotypic parameters	Comments	Refs
RPTP-κ	Transgenic mice	Gene knockout	Development	mRNA expression was eradicated in homozygous mice carrying insertions in the gene, but embryonic development was normal	1055
SH-PTP1 (HCP)	Moth-eaten (*me*/*me*) mouse	Point mutation leads to aberrant mRNA splicing that predicts a 100 bp deletion in the 1st SH2 domain and alters the reading frame; no HCP mRNA is observed in tissues		Macrophage proteins showed no immunoreactive material by Western blot for HCP with a relative increase in tyrosine phosphorylation of several proteins in the cells carrying the HCP mutation; Northen blots showed no HCP mRNA from bone marrow; PCR analysis of HCP cDNA showed that the 1st SH2 domain has a 100 bp deletion which alters the reading frame and predicts a truncated HCP of 102 AA; sequence analysis of cDNA and genomic clones showed that both *me* and me^v are due to point mutations that result in aberrant splicing of the HCP transcript	603
(HCP)	Viable moth-eaten (me^v/me^v) mouse	Point mutation leads to aberrant mRNA splicing that results in two types of altered transcripts affecting the sequence of the PTPase catalytic domain, one is a 69 bp insertion, the other is a 15 bp deletion, both maintaining the reading frame		Western blot of macrophage proteins showed abnormally migrating HCP protein at 67 and 71 kDa (compared to 68 kDa in wild-type) with a relative increase in tyrosine phosphorylation of several proteins in the cells carrying the HCP mutation; Northern blots showed normal levels of HCP mRNA in bone marrow cells; PCR analysis of HCP cDNA showed that two types of altered transcripts are found with different abnormalities in the PTPase catalytic domain, one is a 69 bp insertion, the other is a 15 bp deletion, both maintaining the reading frame; sequence analysis of cDNA and genomic clones showed that both *me* and me^v are due to point mutations that result in aberrant splicing of the HCP transcript	603
(PTP1C)	Moth-eaten (*me*/*me*) mouse	Premature termination of PTP1C polypeptide in 1st SH2 domain	Immune system devopment	No PTP1C protein or mRNA detected in tissues; PTP1C PTPase activity absent from cells homozygous for *me* mutation; autoimmune and immunodeficiency condition with death by age 3 or 9 weeks; severe haematopoietic dysregulation with overexpansion of $CD5^+$ peripheral B cells, impaired T and NK cell function, and increased tissue accumulation of granulocytes and macrophages	288

(PTP1C)	Viable moth-eaten (*me*^{v}/*me*^{v}) mouse	Insertion or deletion in phosphatase domain	Same as *me* phenotype with ↑ viability	Homozygous mice expressed levels of full-length PTP1C protein comparable to wild type, and recombinant SH2 domains of *me*^{v} /*me*^{v} mice bind normally to Tyr(P) ligands *in vitro* (activated EGF and PDGF receptors),but enzyme has marked reduction of PTPase activity	288
	B cells from moth-eaten viable (*mev*) mice	SH-PTP1 deficiency	Elevation of intracellular Ca^{2+} to antigen stimulation	Antigen triggered a greater and more rapid elevation of intracellular calcium in *mev* B cells, indicating that SH-PTP1 negatively regulates immunoglobulin signalling	990
SH-PTP2	CHO cells overexpressing the IR	Cys-459 → Ser(catalytically inert)	Attenuated the insulin stimulation of Ras activation	Also inhibited MAP kinase activation by insulin but not by TPA; in contrast, mutant SH-PTP2 did not affect insulin-stimulated PI kinase association with IRS-1, binding of Grb2 to IRS-1, the Tyr phosphorylation of Shc or the association of Shc with Grb2; ↑ SH-PTP2 association with IRS-1 was observed; *in vitro*, SH-PTP2 dephosphorylated IRS-1 Tyr(P) peptide corresponding to Tyr-1172, the putative SH-PTP2 binding site, suggesting that SH-PTP2 may bind to IRS-1 via its SH2 domains in response to insulin, and in turn, dephosphorylate its binding site and self-regulate its association with IRS-1; also, SH-PTP2 appears to regulate an upstream pathway necessary for Ras activation and this may be required for both the Grb2 and Shc-mediated pathways	178
	Human glioma cell line SNB19	Catalytically inactive	SNB19 cells lack the ability to proliferate in response to EGF; stable overexpression of an interfering SH-PTP2 mutant can restore the ability of these cells to proliferate in response to EGF	Suggests that SH-PTP2 can also function to negatively regulate EGF-mediated signal transduction	1029
	CHO cells overexpressing the insulin receptor	Catalytically inactive SH-PTP2 or its isolated SH2 domains	Insulin stimulation of c-*fos* reporter gene expression and MAP kinase activation	Dominant interfering mutants of SH-PTP2 inhibited insulin signal transduction suggesting role as positive mediator of insulin action	1042
	Microinjected *Xenopus* oocytes	Internal PTPase domain deletion	Oocyte development	Mutant SH-PTP2 acts as a dominant negative causing severe posterior truncations; also blocks FGF and activin-mediated induction of mesoderm and MAP kinase activation	1038
	3T3 fibroplasts overexpressing the insulin receptor and 3T3-L1 adipocytes	Injection of SH2 domains as GST-fusion proteins or anti-SH-PTP2 antibodies	Insulin signal transduction	Microinjection of either SH2 domains or antibodies blocked insulin-induced DNA synthesis in fibroblasts and insulin-stimulated GLUT1 expression in adipocytes, but did not block insulin-stimulated translocation of GLUT4 to cell surface, indicating role in mitogenesis	1004

Table 14 Continued

PTPase	Animal/cell model	Mutation	Phenotypic parameters	Comments	Refs
SH-PTP2 (syp)		Catalytically inactive SH-PTP2	Acrivation of MEK and raf-1 kinase by insulin	Expression of dominant negative syp blocked the activation of MEK and raf-1 kinase in reponse to insulin and had no detectable effect on insulin-induced activation of p21(ras). These data suggest that the target of the syp phosphatase may reside in proteins immediately downstream of p21(ras)	1032
stp1$^+$	*S. pombe*	Disruption of *stp1*$^+$ gene	No obvious phenotypic changes noted.		1112
twine	*Drosophila* testes	Twine and twine, Dmcdc2(ts) double mutant	Meiotic spindles and metaphase plates were never formed, but chromosomes condensed in late spermatocytes and spermatid differentiation continued	Suggest that the cdc2 kinase activity required for meiotic divisions is activated by the twine; partial rescue is observed by ectopic expression of string	1034
Yeast PTP1	*Saccharomyces cerevisiae*	Gene disruption	Cell growth	No effect on vegetative growth	509
Yeast PTP1 or PTP2	*Saccharomyces cerevisiae*	Gene disruption	Cell growth	Neither gene is essential either singly or in combination; neither deletion nor overexpression results in any strong phenotypes in a number of assays	528
Yeast PTP2	*Saccharomyces cerevisiae*	Gene disruption	Cell growth	PTP2 knockout alone or PTP2 / PTP1 combined null mutant showed that neither was essential for growth; in the cells disrupted for both PTP1 and PTP2, no differences were seen in sensitivity to growth temperature, UV light or heat shock	512
				Null *ptp2* mutants grow slowly, are hypersensitive to heat and are viable in the presence or absence of the N-end rule pathway	316
YopH	J774 macrophage cells	YopH negative mutants in *Yersinia pseudotuberculosis*	Antiphagocytic effect of the bacteria	YopH-negative mutants did not induce antiphagocytosis, but were readily ingested, almost to the same extent as that of the translocation mutants YopB and YopD and the plasmid-cured strain	999
YVH1	*Saccharomyces cerevisiae*	Gene disruption	Cell growth	Inactivation of YVH1 causes decreased growth rate, with a doubling time ↑ from 100 to 300 min; however, cells do not accumulate in specific cell cycle stage	510

Bibliography

References for first edition

Reviews

1*. **Alexander, D.** 1990. The role of phosphatases in signal transduction. *New Biologist* **2**: 1049–1062.

2. **Alexander, D., M. Shiroo, A. Robinson, M. Biffen, and E. Shivnan.** 1992. The role of CD45 in T-cell activation – Resolving the paradoxes. *Immunol. Today* **13**: 477–481.

3*. **Atherton-Fessler, S., G. Hannig, and H. Piwnica-Worms.** 1994. Reversible tyrosine phosphorylation and cell cycle control. *Sem. Cell Biol.* **4**: 433–442.

4*. **Bignon, J. S. and K. A. Siminovitch.** 1994. Identification of PTP1C mutation as the genetic defect in motheaten and viable motheaten mice: a step toward defining the roles of protein tyrosine phosphatases in the regulation of hemopoietic cell differentiation and function. *Clin. Immunol. Immunopathol.* **73**: 168–179.

5. **Brautigan, D. L.** 1992. Great expectations – protein tyrosine phosphatases in cell regulation. *Biochim. Biophys. Acta* **1114**: 63–77.

6*. **Brautigan, D. L.** 1994. Protein phosphatases. *Recent Prog. Horm. Res.* **49**: 197–214.

7*. **Cayla, X., J. Goris, J. Hermann, C. Jessus, P. Hendrix, and W. Merlevede.** 1990. Phosphotyrosyl phosphatase activity of the polycation-stimulated protein phosphatases and involvement of dephosphorylation in cell cycle regulation. *Adv. Enz. Regul.* **30**: 265–285.

8. **Chan, A. C., D. M. Desai, and A. Weiss.** 1994. The role of protein tyrosine kinases and protein tyrosine phosphatases in T cell antigen receptor signal transduction. *Ann. Rev. Immunol.* **12**: 555–592.

9*. **Charbonneau, H. and N. K. Tonks.** 1992. 1002 Protein phosphatases. *Annu. Rev. Cell Biol.* **8**: 463–493.

10. **Clark, E. A. and J. A. Ledbetter.** 1989. Leukocyte cell surface enzymology: CD45 (LCA,T200) is a protein tyrosine phosphatase. *Immunol. Today* **10**: 225–228.

11. **Cohen, P.** 1992. Signal integration at the level of protein kinases, protein phosphatases and their substrates. *Trends Biochem. Sci.* **17**: 408–413.

12*. **Cool, D. E. and E. H. Fischer.** 1994. Protein tyrosine phosphatases in cell transformation. *Sem. Cell Biol.* **4**: 443–453.

13. **Fearon, D. T.** 1993. The CD19-CR2-TAPA-1 complex, CD45 and signaling by the antigen receptor of B lymphocytes. *Curr. Opin. Immunol.* **5**: 341–348.

14. **Feng, G. S. and T. Pawson.** 1994. Phosphotyrosine phosphatases with SH2 domains – regulators of signal transduction. *Trends Genet.* **10**: 54–58.

15. **Fischer, E. H., H. Charbonneau, D. E. Cool, and N. K. Tonks.** 1992. Tyrosine phosphatases and their possible interplay with tyrosine kinases. *Ciba Found. Symp.* **164**: 132–140.

16*. **Fischer, E. H., H. Charbonneau, and N. K. Tonks.** 1991. Protein tyrosine phosphatases – a diverse family of intracellular and transmembrane enzymes. *Science* **253**: 401–406.

17. **Fischer, E. H., N. K. Tonks, H. Charbonneau, M. F. Cicirelli, D. E. Cool, C. D. Diltz, E. G. Krebs, and K. A. Walsh.** 1990. Protein tyrosine phosphatases: a novel family of enzymes involved in transmembrane signalling. *Adv. Second Messenger Phosphoprotein Res.* **24**: 273–279.

18. **Goldstein, B. J.** 1992. Protein-tyrosine phosphatases and the regulation of insulin action. *J. Cell. Biochem.* **48**: 33–42.

19*. **Goldstein, B. J.** 1993. Regulation of insulin receptor signalling by protein-tyrosine dephosphorylation. *Receptor* **3**: 1–15.

20. **Goldstein, B.J.** 1996. Protein-tyrosine phosphatases and the regulation of insulin action. In *Diabetes Mellitus: A Fundamental and Clinical Text.* D. LeRoith, J.M. Olefsky, and S.I. Taylor, eds. Lippincott, Philadelphia. pp. 174–186.

21. **Goldstein, B. J., J. Meyerovitch, W. R. Zhang, J. M. Backer, P. Csermely, N. Hashimoto, and C. R. Kahn.** 1991. Hepatic protein-tyrosine phosphatases and their regulation in diabetes. *Adv. Prot. Phosphatases* **6**: 1–17.

22. **Goldstein, B. J., W. R. Zhang, N. Hashimoto, and C. R. Kahn.** 1992. Approaches to the molecular cloning of protein-tyrosine phosphatases in insulin-sensitive tissues. *Mol. Cell. Biochem.* **109**: 107–113.

23. **Guan, K., D. Hakes, J. E. Dixon, H. D. Park, and T. G. Cooper.** 1993. The yeast open reading frame encoding a dual specificity phosphatase. *Trends Biochem. Sci.* **18**: 6.

24*. **Guan, K. L. and J. E. Dixon.** 1993. Bacterial and viral protein tyrosine phosphatases. *Sem. Cell Biol.* **4**: 389–396.

25. **Hathcock, K. S., H. Hirano, and R. J. Hodes.** 1993. CD45 expression by murine B cells and T cells: alteration of CD45 isoforms in subpopulations of activated B cells. *Immunol. Res.* **12**: 21–36.

26*. **Hunter, T.** 1989. Protein-tyrosine phosphatases: the other side of the coin. *Cell* **58**: 1013–1016.

27. **Janeway, C. A., Jr.** 1992. The T cell receptor as a multicomponent signalling machine: CD4/CD8 coreceptors and CD45 in T cell activation. *Annu. Rev. Immunol.* **10**: 645–674.

28. **Justement, L. B., V. K. Brown, and J. J. Lin.** 1994. Regulation of B-cell activation by CD45: a question of mechanism. *Immunol. Today* **15**: 399–406.

29. **Keyse, S. M. and M. Ginsburg.** 1993. Amino acid sequence similarity between CL100, a dual-specificity MAP kinase phosphatase and cdc25. *Trends Biochem. Sci.* **18**: 377–378.

30*. **Koretzky, G. A.** 1993. Role of the CD45 Tyrosine phosphatase in signal transduction in the immune system. *FASEB J.* **7**: 420–426.

31. **Kumagai, A. and W. G. Dunphy.** 1991. Molecular mechanism of the final steps in the activation of MPF. *Cold Spring Harb. Symp. Quant. Biol.* **56**: 585–589.

32. **Lau, K. H. and D. J. Baylink.** 1993. Phosphotyrosyl protein phosphatases: potential regulators of cell proliferation and differentiation. *Crit. Rev. Oncog.* **4**: 451–471.

33*. **Lau, K. H., J. R. Farley, and D. J. Baylink.** 1989. Phosphotyrosyl protein phosphatases. *Biochem. J.* **257**: 23–36.

* Key references.

34. **Ledbetter, J. A., J. P. Deans, A. Aruffo, L. S. Grosmaire, S. B. Kanner, J. B. Bolen, and G. L. Schieven.** 1993. CD4, CD8 and the role of CD45 in T-cell activation. *Curr. Opin. Immunol.* **5**: 334–340.

35. **Mauro, L. J. and J. E. Dixon.** 1994. Zip codes direct intracellular protein tyrosine phosphatases to the correct cellular address. *Trends Biochem. Sci.* **19**: 151–155.

36*. **Mcfarland, E. C., E. Flores, R. J. Matthews, and M. L. Thomas.** 1994. Protein tyrosine phosphatases involved in lymphocyte signal transduction. *Chemical Immunology* **59**: 40–61.

37. **Millar, J., C. McGowan, R. Jones, K. Sadhu, A. Bueno, H. Richardson, and P. Russell.** 1991. cdc25 M-phase inducer. *Cold Spring Harb. Symp. Quant. Biol.* **56**: 577–584.

38. **Millar, J. B., G. Lenaers, C. McGowan, and P. Russell.** 1992. Activation of MPF in fission yeast. *Ciba Found. Symp.* **170**: 50–58.

39*. **Millar, J. B. and P. Russell.** 1992. The cdc25 M-phase inducer: an unconventional protein phosphatase. *Cell* **68**: 407–410.

40. **Moreno, S. and P. Nurse.** 1991. Clues to action of Cdc25 protein. *Nature* **351**: 194.

41. **Mourey, R. J. and J. E. Dixon.** 1994. Protein tyrosine phosphatases: characterization of extracellular and intracellular domains. *Curr. Opin. Genet. Dev.* **4**: 31–39.

42*. **Neel, B. G.** 1993. Structure and function of SH2-domain containing tyrosine phosphatases. *Sem. Cell Biol.* **4**: 419–432.

43*. **Pallen, C. J.** 1994. The receptor-like protein tyrosine phosphatase α: a role in cell proliferation and oncogenesis. *Sem. Cell Biol.* **4**: 403–408.

44. **Penninger, J. M., V. A. Wallace, K. Kishihara, and T. W. Mak.** 1993. The role of $p56^{lck}$ and $p59^{fyn}$ tyrosine kinases and CD45 protein tyrosine phosphatase in T-cell development and clonal selection. *Immunol. Rev.* **135**: 183–214.

45. **Pilarski, L. M.** 1993. Adhesive interactions in thymic development: does selective expression of CD45 isoforms promote stage-specific microclustering in the assembly of functional adhesive complexes on differentiating T lineage lymphocytes? *Immunology & Cell Biology* **71**: 59–69.

46. **Pilarski, L. M. and J. P. Deans.** 1989. Selective expression of CD45 isoforms and of maturation antigens during human thymocyte differentiation: observations and hypothesis. *Immunol. Lett.* **21**: 187–198.

47. **Pot, D. A. and J. E. Dixon.** 1992. A 1000 and 2 protein tyrosine phosphatases. *Biochim. Biophys. Acta* **1136**: 35–43.

48*. **Saito, H.** 1994. Structural diversity of eukaryotic protein tyrosine phosphatases: functional and evolutionary implications. *Sem. Cell Biol.* **4**: 379–387.

49*. **Saito, H. and M. Streuli.** 1991. Molecular characterization of protein tyrosine phosphatases. *Cell Growth Differ.* **2**: 59–65.

50. **Saito, H., M. Streuli, N. X. Krueger, M. Itoh, and A. Y. Tsai.** 1992. CD45 and a family of receptor-linked protein tyrosine phosphatases. *Biochem. Soc. Trans.* **20**: 165–169.

51. **Sale, G. J.** 1991. Insulin receptor phosphotyrosyl protein phosphatases and the regulation of insulin receptor tyrosine kinase action. *Adv. Prot. Phosphatases* **6**: 159–186.

52. **Sale, G. J.** 1992. Serine/threonine kinases and tyrosine phosphatases that act on the insulin receptor. *Biochem Soc. Trans.* **20**: 664–670.

53. **Sefton, B. M. and M. A. Campbell.** 1991. The role of tyrosine protein phosphorylation in lymphocyte activation. *Annu. Rev. Cell Biol.* **7**: 257–274.

54. **Shaw, A. and M. L. Thomas.** 1991. Coordinate interactions of protein tyrosine kinases and protein tyrosine phosphatases in T-cell receptor-mediated signalling. *Curr. Opin. Cell Biol.* **3**: 862–868.

55. **Shechter, Y.** 1990. Insulin-mimetic effects of vanadate: possible implications for future treatment of diabetes. *Diabetes* **39**: 1–5.

56*. **Stone, R. L. and J. E. Dixon.** 1994. Protein-tyrosine phosphatases. *J. Biol. Chem.* **269**: 31323–31326.

57. **Sun, H. and N. K. Tonks.** 1994. The coordinated action of protein tyrosine phosphatases and kinases in cell signaling. *Trends Biochem. Sci.* **19**: 480–485.

58. **Takeda, A.** 1993. Sphingolipid-like molecule linked to CD45, a protein tyrosine phosphatase. *Adv. Lipid Res.* **26**: 293–317.

59*. **Thomas, M. L.** 1989. The leukocyte common antigen family. *Annu. Rev. Immunol.* **7**: 339–369.

60. **Thomas, M. L.** 1994. The regulation of B- and T-lymphocyte activation by the transmembrane protein tyrosine phosphatase CD45. *Curr. Opin. Cell Biol.* **6**: 247–252.

61. **Tonks, N. K.** 1990. Protein phosphatases: key players in the regulation of cell function. *Curr. Opin. Cell Biol.* **2**: 1114–1124.

62. **Tonks, N. K.** 1993. Protein tyrosine phosphatases. *Sem. Cell Biol.* **4**: 373–377.

63. **Tonks, N. K. and H. Charbonneau.** 1989. Protein tyrosine dephosphorylation and signal transduction. *Trends Biochem. Sci.* **14**: 497–500.

64*. **Tonks, N. K., A. J. Flint, M. F. Gebbink, H. Sun, and Q. Yang.** 1993. Signal transduction and protein tyrosine dephosphorylation. *Adv. Second Messenger Phosphoprotein Res.* **28**: 203–210.

65. **Tonks, N. K., Q. Yang, and P. J. Guida.** 1991. Structure, regulation, and function of protein tyrosine phosphatases. *Cold Spring Harb. Symp. Quant. Biol.* **56**: 265–273.

66. **Trowbridge, I. S.** 1991. CD45 – a prototype for transmembrane protein tyrosine phosphatases. *J. Biol. Chem.* **266**: 23517–23520.

67. **Trowbridge, I. S., H. L. Ostergaard, and P. Johnson.** 1991. CD45 – a leukocyte-specific member of the protein tyrosine phosphatase family. *Biochim. Biophys. Acta* **1095**: 46–56.

68*. **Trowbridge, I. S. and M. L. Thomas.** 1994. CD45: an emerging role as a protein tyrosine phosphatase required for lymphocyte activation and development. *Annu. Rev. Immunol.* **12**: 85–116.

69*. **Tsui, F. W. L. and H. W. Tsui.** 1994. Molecular basis of the motheaten phenotype. *Immunol. Rev* **138**: 185–206.

70*. **Walton, K. M. and J. E. Dixon.** 1993. Protein tyrosine phosphatases. *Annu. Rev. Biochem.* **62**: 101–120.

71. **Weaver, C. T., J. T. Pingel, J. O. Nelson, and M. L. Thomas.** 1992. CD45: a transmembrane protein tyrosine phosphatase involved in the transduction of antigenic signals. *Biochem. Soc. Trans.* **20**: 169–174.

72. **Woodford-Thomas, T. and M. L. Thomas.** 1993. The leukocyte common antigen, CD45 and other protein tyrosine phosphatases in hematopoietic cells. *Sem. Cell Biol.* **4**: 409–418.

73*. **Zhang, Z. Y. and J. E. Dixon.** 1994. Protein tyrosine phosphatases: mechanism of catalysis and substrate specificity. *Advances in Enzymology and Related Areas of Molecular Biology* **68**: 1–36.

74*. **Zinn, K.** 1993. *Drosophila* protein tyrosine phosphatases. *Sem. Cell Biol.* **4**: 397–401.

Derivatization

75. **Autero, M. and C. G. Gahmberg.** 1987. Phorbol diesters increase the phosphorylation of the leukocyte common antigen CD45 in human T cells. *Eur. J. Immunol.* **17**: 1503–1506.

76. **Autero, M., J. Saharinen, T. Pessa-Morikawa, M. Soula-Rothhut, C. Oetken, M. Gassmann, M. Bergman, K. Alitalo, P. Burn, and C. G. Gahmberg.** 1994. Tyrosine phosphorylation of CD45 phosphotyrosine phosphatase by p50csk kinase creates a binding site for p56lck tyrosine kinase and activates the phosphatase. *Mol. Cell. Biol.* **14**: 1308–1321.

77*. **Barnea, G., M. Grumet, P. Milev, O. Silvennoinen, J. B. Levy, J. Sap, and J. Schlessinger.** 1994. Receptor tyrosine phosphatase β is expressed in the form of proteoglycan and binds to the extracellular matrix protein tenascin. *J. Biol. Chem.* **269**: 14349–14352.

78*. **Bennett, A. M., T. L. Tang, S. Sugimoto, C. T. Walsh, and B. G. Neel.** 1994. Protein-tyrosine-phosphatase SHPTP2 couples platelet-derived growth factor receptor β to Ras. *Proc. Natl Acad. Sci. USA* **91**: 7335–7339.

79. **Bouchard, P., Z. Z. Zhao, D. Banville, F. Dumas, E. H. Fischer, and S. H. Shen.** 1994. Phosphorylation and identification of a major tyrosine phosphorylation site in protein tyrosine phosphatase 1C. *J. Biol. Chem.* **269**: 19585–19589.

80. **Bourgoin, S. and S. Grinstein.** 1992. Peroxides of vanadate induce activation of phospholipase D in HL-60 cells. Role of tyrosine phosphorylation. *J. Biol. Chem.* **267**: 11908–11916.

81. **Brautigan, D. L. and F. M. Pinault.** 1993. Serine phosphorylation of protein tyrosine phosphatase (PTP1B) in HeLa cells in response to analogues of cAMP or diacylglycerol plus okadaic acid. *Mol. Cell. Biochem.* **127/128**: 121–129.

82. **Caselli, A., G. Camici, G. Manao, G. Moneti, L. Pazzagli, G. Cappugi, and G. Ramponi.** 1994. Nitric oxide causes inactivation of the low molecular weight phosphotyrosine protein phosphatase. *J. Biol. Chem.* **269**: 24878–24882.

83. **Clarke, P. R., I. Hoffmann, G. Draetta, and E. Karsenti.** 1993. Dephosphorylation of cdc25-C by a type-2A protein phosphatase: specific regulation during the cell cycle in *Xenopus* egg extracts. *Mol. Biol. Cell* **4**: 397–411.

84. **Daum, G., S. Regenass, J. Sap, J. Schlessinger, and E. H. Fischer.** 1994. Multiple forms of the human tyrosine phosphatase RPTPα: isozymes and differences in glycosylation. *J. Biol. Chem.* **269**: 10524–10528.

85*. **den Hertog, J., S. Tracy, and T. Hunter.** 1994. Phosphorylation of receptor protein-tyrosine phosphatase a on Tyr789, a binding site for the SH3-SH2-SH3 adaptor protein GRB-2 *in vivo*. *EMBO J.* **13**: 3020–3032.

86*. **Flint, A. J., M. F. B. G. Gebbink, B. R. Franza, D. E. Hill, and N. K. Tonks.** 1993. Multi-site phosphorylation of the protein tyrosine phosphatase, PTP1B – identification of cell cycle regulated and phorbol ester stimulated sites of phosphorylation. *EMBO J.* **12**: 1937–1946.

87*. **Frangioni, J. V., A. Oda, M. Smith, E. W. Salzman, and B. G. Neel.** 1993. Calpain-catalyzed cleavage and subcellular relocation of protein phosphotyrosine phosphatase-1B (PTP-1B) in human platelets. *EMBO J.* **12**: 4843–4856.

88*. **Garton, A. J. and N. K. Tonks.** 1994. PTP-PEST: a protein tyrosine phosphatase regulated by serine phosphorylation. *EMBO J.* **13**: 3763–3771.

89*. **Guan, K. L. and J. E. Dixon.** 1991. Evidence for protein-tyrosine-phosphatase catalysis proceeding via a cysteine-phosphate intermediate. *J. Biol. Chem.* **266**: 17026–17030.

90. **Hoffmann, I., P. R. Clarke, M. J. Marcote, E. Karsenti, and G. Draetta.** 1993. Phosphorylation and activation of human cdc25-C by cdc2–cyclin B and its involvement in the self-amplification of MPF at mitosis. *EMBO J.* **12**: 53–63.

91. **Izumi, T. and J. L. Maller.** 1993. Elimination of Cdc2 phosphorylation sites in the Cdc25 phosphatase blocks initiation of M-phase. *Mol. Biol. Cell* **4**: 1337–1350.

92. **Izumi, T., D. H. Walker, and J. L. Maller.** 1992. Periodic changes in phosphorylation of the *Xenopus* cdc25 phosphatase regulate its activity. *Mol. Biol. Cell* **3**: 927–939.

93. **Jackson, D. I. and A. N. Barclay.** 1989. The extra segments of sequence in rat leucocyte common antigen (L-CA) are derived by alternative splicing of only three exons and show extensive O-linked glycosylation. *Immunogenetics* **29**: 281–287.

94. **Kiyomoto, H., B. Fouqueray, H. E. Abboud, and G. G. Choudhury.** 1994. Phorbol 12-myristate 13–acetic acid inhibits PTP1B activity in human mesangial cells – A possible mechanism of enhanced tyrosine phosphorylation. *FEBS Lett.* **353**: 217–220.

95. **Lefebvre, J. C., V. Giordanengo, A. Doglio, L. Cagnon, J. P. Breittmayer, J. F. Peyron, and J. Lesimple.** 1994. Altered sialylation of CD45 in HIV-1-infected T lymphocytes. *Virology* **199**: 265–274.

96. **Lorenz, U., K. S. Ravichandran, D. Pei, C. T. Walsh, S. J. Burakoff, and B. G. Neel.** 1994. Lck-dependent tyrosyl phosphorylation of the phosphotyrosine phosphatase SH-PTP1 in murine T cells. *Mol. Cell. Biol.* **14**: 1824–1834.

97. **Maeda, N., H. Hamanaka, T. Shintani, T. Nishiwaki, and M. Noda.** 1994. Multiple receptor-like protein tyrosine phosphatases in the form of chondroitin sulfate proteoglycan. *FEBS Lett.* **354**: 67–70.

98. **Maegawa, H., S. Ugi, M. Adachi, Y. Hinoda, R. Kikkawa, A. Yachi, Y. Shigeta, and A. Kashiwagi.** 1994. Insulin receptor kinase phosphorylates protein tyrosine phosphatase containing SRC homology 2 regions and modulates its PTPase activity *in vitro*. *Biochem. Biophys. Res. Commun.* **199**: 780–785.

99. **Matozaki, T., T. Uchida, Y. Fujioka, and M. Kasuga.** 1994. Src kinase tyrosine phosphorylates PTP1C, a protein tyrosine phosphatase containing Src homology-2 domains that down-regulates cell proliferation. *Biochem. Biophys. Res. Commun.* **204**: 874–881.

100. **Ohta, T., K. Kitamura, A. L. Maizel, and A. Takeda.** 1994. Alterations in CD45 glycosylation pattern accompanying different cell proliferation states. *Biochem. Biophys. Res. Commun.* **200**: 1283–1289.

101. **Peraldi, P., Z. Z. Zhao, C. Filloux, E. H. Fischer, and E. Vanobberghen.** 1994. Protein-tyrosine-phosphatase 2C is phosphorylated and inhibited by 44-kDa mitogen-activated protein kinase. *Proc. Natl Acad. Sci. USA* **91**: 5002–5006.

102. **Pulido, R. and F. Sanchez-Madrid.** 1992. Glycosylation of CD**45**: carbohydrate processing through Golgi apparatus is required for cell surface expression and protein stability. *Eur. J. Immunol.* **22**: 463–468.

103. **Sato, T., K. Furukawa, M. Autero, C. G. Gahmberg, and A. Kobata.** 1993. Structural study of the sugar chains of human leukocyte common antigen CD45. *Biochemistry* **32**: 12694–12704.

104*. **Schievella, A. R., L. A. Paige, K. A. Johnson, D. E. Hill, and R. L. Erikson.** 1993. Protein tyrosine phosphatase-1B undergoes mitosis-specific phosphorylation on serine. *Cell Growth Differ.* **4**: 239–246.

105*. **Serra-Pages, C., H. Saito, and M. Streuli.** 1994. Mutational analysis of proprotein processing, subunit association, and shedding of the LAR transmembrane protein tyrosine phosphatase. *J. Biol. Chem.* **269**: 23632–23641.

106*. **Shitara, K., H. Yamada, K. Watanabe, M. Shimonaka, and Y. Yamaguchi.** 1994. Brain-specific receptor-type protein-tyrosine phosphatase RPTP β is a chondroitin sulfate proteoglycan *in vivo*. *J. Biol. Chem.* **269**: 20189–20193.

107. **Stover, D. R., H. Charbonneau, N. K. Tonks, and K. A. Walsh.** 1991. Protein-tyrosine-phosphatase-CD45 is phosphorylated transiently on tyrosine upon activation of Jurkat T-cells. *Proc. Natl Acad. Sci. USA* **88**: 7704–7707.

108. **Stover, D. R. and K. A. Walsh.** 1994. Protein-tyrosine phosphatase activity of CD45 is activated by sequential phosphorylation by two kinases. *Mol. Cell. Biol.* **14**: 5523–5532.

109. **Strausfeld, U., A. Fernandez, J. P. Capony, F. Girard, N. Lautredou, J. Derancourt, J. C. Labbe, and N. J. Lamb.** 1994. Activation of p34cdc2 protein kinase by microinjection of human cdc25C into mammalian cells. Requirement for prior phosphorylation of cdc25C by p34cdc2 on sites phosphorylated at mitosis. *J. Biol. Chem.* **269**: 5989–6000.

110*. **Streuli, M., N. X. Krueger, P. D. Ariniello, M. Tang, J. M. Munro, W. A. Blattler, D. A. Adler, C. M. Disteche, and H. Saito.** 1992. Expression of the receptor-linked protein tyrosine phosphatase LAR: proteolytic cleavage and shedding of the CAM-like extracellular region. *EMBO J.* **11**: 897–907.

111*. **Su, J., A. Batzer, and J. Sap.** 1994. Receptor tyrosine phosphatase R-PTP-α is tyrosine-phosphorylated and associated with the adaptor protein Grb2. *J. Biol. Chem.* **269**: 18731–18734.

112. **Takeda, A. and A. L. Maizel.** 1990. An unusual form of lipid linkage to the CD45 peptide. *Science* **250**: 676–679.

113. **Uchida, T., T. Matozaki, T. Noguchi, T. Yamao, K. Horita, T. Suzuki, Y. Fujioka, C. Sakamoto, and M. Kasuga.** 1994. Insulin stimulates the phosphorylation of Tyr-538 and the catalytic activity of Ptp1C, a protein tyrosine phosphatase with Src homology-2 domains. *J. Biol. Chem.* **269**: 12220–12228.

114. **Valentine, M. A., M. B. Widmer, J. A. Ledbetter, F. Pinault, R. Voice, E. A. Clark, B. Gallis, and D. L. Brautigan.** 1991. Interleukin 2 stimulates serine phosphorylation of CD45 in CTLL-2.4 cells. *Eur. J. Immunol.* **21**: 913–919.

115. **Yeung, Y. G., K. L. Berg, F. J. Pixley, R. H. Angeletti, and E. R. Stanley.** 1992. Protein tyrosine phosphatase-1C is rapidly phosphorylated in tyrosine in macrophages in response to colony stimulating factor-1. *J. Biol. Chem.* **267**: 23447–23450.

116*. **Yu, Q., T. Lenardo, and R. A. Weinberg.** 1992. The N-terminal and C-terminal domains of a receptor tyrosine phosphatase are associated by non-covalent linkage. *Oncogene* **7**: 1051–1058.

Expression

117. **Adachi, M., M. Sekiya, M. Ishino, H. Sasaki, Y. Hinoda, K. Imai, and A. Yachi.** 1994. Induction of protein-tyrosine phosphatase LC-PTP by IL-2 in human T cells – LC-PTP is an early response gene. *FEBS Lett.* **338**: 47–52.

118. **Ahmad, F. and B. J. Goldstein.** 1995. Alterations in specific protein-tyrosine phosphatases accompany the insulin resistance of streptozotocin-diabetes. *Am. J. Physiol.* **268**: E932–E940.

119*. **Ahmad, F. and B. J. Goldstein.** 1995. Purification, identification and subcellular distribution of three predominant protein-tyrosine phosphatase enzymes in skeletal muscle tissue. *Biochim. Biophys. Acta* **1248**: 57–69.

120. **Ahmad, F. and B. J. Goldstein.** 1995. Increased abundance of specific skeletal muscle protein-tyrosine phosphatases in a genetic model of obesity and insulin resistance. *Metabolism* **44**: 1175–1184.

121. **Athanasou, N. A., J. Quinn, and J. O. McGee.** 1987. Leucocyte common antigen is present on osteoclasts. *J. Pathol.* **153**: 121–126.

122. **Berger, S. A., T. W. Mak, and C. J. Paige.** 1994. Leukocyte common antigen (CD45) is required for immunoglobulin E-mediated degranulation of mast cells. *J. Exp. Med.* **180**: 471–476.

123. **Berti, A., S. Rigacci, G. Raugei, D. Deglinnocenti, and G. Ramponi.** 1994. Inhibition of cellular response to platelet-derived growth factor by low M_r phosphotyrosine protein phosphatase overexpression. *FEBS Lett.* **349**: 7–12.

124. **Birkeland, M. L., P. Johnson, I. S. Trowbridge, and E. Pure.** 1989. Changes in CD45 isoform expression accompany antigen-induced murine T-cell activation. *Proc. Natl Acad. Sci. USA* **86**: 6734–6738.

125*. **Brown-Shimer, S., K. A. Johnson, D. E. Hill, and A. M. Bruskin.** 1992. Effect of protein tyrosine phosphatase-1B expression on transformation by the human Neu oncogene. *Cancer Res.* **52**: 478–482.

126. **Buzzi, M., L. Lu, A. J. Lombardi, M. R. Posner, D. L. Brautigan, L. D. Fast, and A. R. Frackelton.** 1992. Differentiation-induced changes in protein-tyrosine phosphatase activity and commensurate expression of CD45 in human leukemia cell lines. *Cancer Res.* **52**: 4027–4035.

127. **Caldwell, C. W., W. P. Patterson, and Y. W. Yesus.** 1991. Translocation of CD45RA in neutrophils. *J. Leukocyte Biol.* **49**: 317–328.

128. **Caldwell, C. W. and W. P. Patterson.** 1991. Relationship between CD45 antigen expression and putative stages of differentiation in B-cell malignancies. *Am. J. Hematol.* **36**: 111–115.

129. **Caldwell, C. W., W. P. Patterson, B. D. Toalson, and Y. W. Yesus.** 1991. Surface and cytoplasmic expression of CD45 antigen isoforms in normal and malignant myeloid cell differentiation. *Am. J. Clin. Pathol.* **95**: 180–187.

130. **Canoll, P. D., G. Barnea, J. B. Levy, J. Sap, M. Ehrlich, O. Silvennoinen, J. Schlessinger, and J. M. Musacchio.** 1993. The expression of a novel receptor-type tyrosine phosphatase suggests a role in morphogenesis and plasticity of the nervous system. *Brain Res. Dev. Brain Res.* **75**: 293–298.

131. **Carrasco, D. and R. Bravo.** 1993. Expression of the nontransmembrane tyrosine phosphatase gene erp during mouse organogenesis. *Cell Growth Differ.* **4**: 849–859.

132. **Champion-Arnaud, P., M. C. Gesnel, N. Foulkes, C. Ronsin, P. Sassone Corsi, and R. Breathnach.** 1991. Activation of transcription via AP-1 or CREB regulatory sites is blocked by protein tyrosine phosphatases. *Oncogene* **6**: 1203–1209.

133. **Chang, H. L., L. Lefrancois, M. H. Zaroukian, and W. J. Esselman.** 1991. Developmental expression of CD45 alternate exons in murine T cells. Evidence of additional alternate exon use. *J. Immunol.* **147**: 1687–1693.

134. **Craig, W., S. Poppema, M. T. Little, W. Dragowska, and P. M. Lansdorp.** 1994. CD45 isoform expression on human haemopoietic cells at different stages of development. *Br. J. Haematol.* **88**: 24–30.

135. **Cui, Y., K. Harvey, L. Akard, J. Jansen, C. Hughes, R. A. Siddiqui, and D. English.** 1994. Regulation of neutrophil responses by phosphotyrosine phosphatase. *J. Immunol.* **152**: 5420–5428.

136. **Deans, J. P., A. W. Boyd, and L. M. Pilarski.** 1989. Transitions from high to low molecular weight isoforms of CD45 (T200) involve rapid activation of alternate mRNA splicing and slow turnover of surface CD45R. *J. Immunol.* **143**: 1233–1238.

137. **Deans, J. P., H. M. Serra, J. Shaw, Y. J. Shen, R. M. Torres, and L. M. Pilarski.** 1992. Transient accumulation and subsequent rapid loss of messenger RNA encoding high molecular mass CD45 isoforms after T cell activation. *J. Immunol.* **148**: 1898–1905.

138. **den Hertog, J., C. E. G. M. Pals, L. J. C. Jonk, and W. Kruijer.** 1992. Differential expression of a novel murine non-receptor protein-tyrosine phosphatase during differentiation of P19 embryonal carcinoma cells. *Biochem. Biophys. Res. Commun.* **184**: 1241–1249.

139*. **den Hertog, J., C. E. G. M. Pals, M. P. Peppelenbosch, L. G. J. Tertoolen, S. W. Delaat, and W. Kruijer.** 1993. Receptor protein tyrosine phosphatase-a activates pp60(c-src) and is involved in neuronal differentiation. *EMBO J.* **12**: 3789–3798.

140. **Ding, W., W. R. Zhang, K. Sullivan, N. Hashimoto, and B. J. Goldstein.** 1994. Identification of protein-tyrosine phosphatases prevalent in adipocytes by molecular cloning. *Biochem. Biophys. Res. Commun.* **202**: 902–907.

141. **Donovan, J. A., F. D. Goldman, and G. A. Koretzky.** 1994. Restoration of CD2-mediated signaling by a chimeric membrane protein including the cytoplasmic sequence of CD45. *Hum. Immunol.* **40**: 123–130.

142. **Ducommun, B., G. Draetta, P. Young, and D. Beach.** 1990. Fission yeast cdc25 is a cell-cycle regulated protein. *Biochem. Biophys. Res. Commun.* **167**: 301–309.

143. **Duff, J. L., M. B. Marrero, W. G. Paxton, C. H. Charles, L. F. Lau, K. E. Bernstein, and B. C. Berk.** 1993. Angiotensin-II induces 3CH134, a protein-tyrosine phosphatase, in vascular smooth muscle cells. *J. Biol. Chem.* **268**: 26037–26040.

144. **Faure, R. and B. I. Posner.** 1993. Differential intracellular compartmentalization of phosphotyrosine phosphatases in a glial cell line – TC-PTP versus PTP-1B. *Glia* **9**: 311–314.

145. **Fernandez-Luna, J. L., R. J. Matthews, B. H. Brownstein, R. D. Schreiber, and M. L. Thomas.** 1991. Characterization and expression of the human leukocyte-common antigen (CD45) gene contained in yeast artificial chromosomes. *Genomics* **10**: 756–764.

146. **Flores, E., G. Roy, D. Patel, A. Shaw, and M. L. Thomas.** 1994. Nuclear localization of the PEP protein tyrosine phosphatase. *Mol. Cell. Biol.* **14**: 4938–4946.

147. **Forsyth, K. D., K. Y. Chua, V. Talbot, and W. R. Thomas.** 1993. Expression of the leukocyte common antigen CD45 by endothelium. *J. Immunol.* **150**: 3471–3477.

148*. **Frangioni, J. V., P. H. Beahm, V. Shifrin, C. A. Jost, and B. G. Neel.** 1992. The nontransmembrane tyrosine phosphatase PTP-1B localizes to the endoplasmic reticulum via 35 amino acid C-terminal sequence. *Cell* **68**: 545–560.

149. **Freiss, G. and F. Vignon.** 1994. Antiestrogens increase protein tyrosine phosphatase activity in human breast cancer cells. *Mol. Endocrinol.* **8**: 1389–1396.

150. **Fujii, Y., M. Okumura, K. Inada, K. Nakahara, and H. Matsuda.** 1992. CD45 isoform expression during T cell development in the thymus. *Eur. J. Immunol.* **22**: 1843–1850.

151. **Gillitzer, R. and L. M. Pilarski.** 1990. In situ localization of CD45 isoforms in the human thymus indicates a medullary location for the thymic generative lineage. *J. Immunol.* **144**: 66–74.

152. **Girard, F., U. Strausfeld, J. C. Cavadore, P. Russell, A. Fernandez, and N. J. Lamb.** 1992. cdc25 is a nuclear protein expressed constitutively throughout the cell cycle in nontransformed mammalian cells. *J. Cell Biol.* **118**: 785–794.

153. **Hashimoto, N. and B. J. Goldstein.** 1992. Differential regulation of mRNAs encoding three protein-tyrosine phosphatases by insulin and activation of protein kinase C. *Biochem. Biophys. Res. Commun.* **188**: 1305–1311.

154. **Hathcock, K. S., H. Hirano, S. Murakami, and R. J. Hodes.** 1992. CD45 expression by B cells. Expression of different CD45 isoforms by subpopulations of activated B cells. *J. Immunol.* **149**: 2286–2294.

155. **Heald, R., M. McLoughlin, and F. McKeon.** 1993. Human wee1 maintains mitotic timing by protecting the nucleus from cytoplasmically activated Cdc2 kinase. *Cell* **74**: 463–474.

156*. **Hovis, R. R., J. A. Donovan, M. A. Musci, D. G. Motto, F. D. Goldman, S. E. Ross, and G. A. Koretzky.** 1993. Rescue of signaling by a chimeric protein containing the cytoplasmic domain of CD45. *Science* **260**: 544–546.

157. **Huelin, C., M. Gonzalez, S. Pedrinaci, B. de la Higuera, M. A. Piris, J. San Miguel, F. Ruiz-Cabello, and F. Garrido.** 1988. Distribution of the CD45R antigen in the maturation of lymphoid and myeloid series: the CD45R negative phenotype is a constant finding in T CD4 positive lymphoproliferative disorders. *Br. J. Haematol.* **69**: 173–179.

158. **Ide, R., H. Maegawa, R. Kikkawa, Y. Shigeta, and A. Kashiwagi.** 1994. High glucose condition activates protein tyrosine phosphatases and deactivates insulin receptor function in insulin sensitive rat 1 fibroblasts. *Biochem. Biophys. Res. Commun.* **201**: 71–77.

159. **Iivanainen, A. V., C. Lindqvist, T. Mustelin, and L. C. Andersson.** 1990. Phosphotyrosine phosphatases are involved in reversion of T lymphoblastic proliferation. *Eur. J. Immunol.* **20**: 2509–2512.

160. **Jensen, G. S., S. Poppema, M. J. Mant, and L. M. Pilarski.** 1989. Transition in CD45 isoform expression during differentiation of normal and abnormal B cells. *Int. Immunol.* **1**: 229–236.

161*. **Jinno, S., K. Suto, A. Nagata, M. Igarashi, Y. Kanaoka, H. Nojima, and H. Okayama.** 1994. Cdc25A is a novel phosphatase functioning early in the cell cycle. *EMBO J.* **13**: 1549–1556.

162. **Kaneko, Y., S. Takano, K. Okumura, J. Takenawa, H. Higashituji, M. Fukumoto, H. Nakayama, and J. Fujita.** 1993. Identification of protein tyrosine phosphatases expressed in murine male germ cells. *Biochem. Biophys. Res. Commun.* **197**: 625–631.

163. **Kenner, K. A., D. E. Hill, J. M. Olefsky, and J. Kusari.** 1993. Regulation of protein tyrosine phosphatases by insulin and insulin-like growth factor-I. *J. Biol. Chem.* **268**: 25455–25462.

164. **Koretzky, G. A., M. A. Kohmetscher, T. Kadleck, and A. Weiss.** 1992. Restoration of T-cell receptor-mediated signal transduction by transfection of CD45 cDNA into a CD45-deficient variant of the Jurkat T-cell line. *J. Immunol.* **149**: 1138–1142.

165. **Kume, T., K. Tsuneizumi, T. Watanabe, M. L. Thomas, and M. Oishi.** 1994. Induction of specific protein tyrosine phosphatase transcripts during differentiation of mouse erythroleukemia cells. *J. Biol. Chem.* **269**: 4709–4712.

166. **Lacal, P., R. Pulido, F. Sanchez-Madrid, and F. Mollinedo.** 1988. Intracellular location of T200 and Mo1 glycoproteins in human neutrophils. *J. Biol. Chem.* **263**: 9946–9951.

167. **LaSalle, J. M. and D. A. Hafler.** 1991. The coexpression of CD45RA and CD45RO isoforms on T cells during the S/G2/M stages of cell cycle. *Cell. Immunol.* **138**: 197–206.

168. **Li, R. Y., F. Gaits, A. Ragab, J. M. F. Ragabthomas, and H. Chap.** 1994. Translocation of an SH2-containing protein tyrosine phosphatase (SH-PTP1) to the cytoskeleton of thrombin-activated platelets. *FEBS Lett.* **343**: 89–93.

169. **Lombroso, P. J., J. R. Naegele, E. Sharma, and M. Lerner.** 1993. A protein tyrosine phosphatase expressed within dopaminoceptive neurons of the basal ganglia and related structures. *J. Neurosci.* **13**: 3064–3074.

170. **Longo, F. M., J. A. Martignetti, J. M. Le Beau, J. S. Zhang, J. P. Barnes, and J. Brosius.** 1993. Leukocyte common antigen-related receptor-linked tyrosine phosphatase. Regulation of mRNA expression. *J. Biol. Chem.* **268**: 26503–26511.

171. **Mackay, C. R., J. F. Maddox, and M. R. Brandon.** 1987. A monoclonal antibody to the p220 component of sheep LCA identifies B cells and a unique lymphocyte subset. *Cell. Immunol.* **110**: 46–55.

172. **Marraccini, P., T. Iantomasi, S. Rigacci, S. Pacini, M. Ruggiero, M. T. Vincenzini, and G. Ramponi.** 1994. Effect of phosphotyrosine phosphatase over-expression on glutathione metabolism in normal and oncogene-transformed cells. *FEBS Lett.* **344**: 157–160.

173. **Marty, L. M., C. W. Caldwell, and T. L. Feldbush.** 1992. Expression of CD45 isoforms by Epstein–Barr virus-transformed human B lymphocytes. *Clin. Immunol. Immunopathol.* **62**: 8–15.

174. **Melkerson-Watson, L. J., M. E. Waldmann, A. D. Gunter, M. H. Zaroukian, and W. J. Esselman.** 1994. Elevation of lymphocyte CD45 protein tyrosine phosphatase activity during mitosis. *J. Immunol.* **153**: 2004–2013.

175. **Minami, Y., F. J. Stafford, J. Lippincott-Schwartz, L. C. Yuan, and R. D. Klausner.** 1991. Novel redistribution of an intracellular pool of CD45 accompanies T-cell activation. *J. Biol. Chem.* **266**: 9222–9230.

176*. **Mooney, R. A., G. G. Freund, B. A. Way, and K. L. Bordwell.** 1992. Expression of a transmembrane phosphotyrosine phosphatase inhibits cellular response to platelet-derived growth factor and insulin-like growth factor-1. *J. Biol. Chem.* **267**: 23443–23446.

177. **Motto, D. G., M. A. Musci, and G. A. Koretzky.** 1994. Surface expression of a heterologous phosphatase complements CD45 deficiency in a T cell clone. *J. Exp. Med.* **180**: 1359–1366.

178*. **Noguchi, T., T. Matozaki, K. Horita, Y. Fujioka, and M. Kasuga.** 1994. Role of SH-PTP2, a protein-tyrosine phosphatase with Src homology 2 domains, in insulin-stimulated ras activation. *Mol. Cell. Biol.* **14**: 6674–6682.

179. **Ogata, M., M. Sawada, A. Kosugi, and T. Hamaoka.** 1994. Developmentally regulated expression of a murine receptor- type protein tyrosine phosphatase in the thymus. *J. Immunol.* **153**: 4478–4487.

180. **Ogimoto, M., T. Katagiri, K. Hasegawa, K. Mizuno, and H. Yakura.** 1993. Induction of CD45 isoform switch in murine B cells by antigen receptor stimulation and by phorbol myristate acetate and ionomycin. *Cell. Immunol.* **151**: 97–109.

181. **Paramithiotis, E., L. Tkalec, and M. J. Ratcliffe.** 1991. High levels of CD45 are coordinately expressed with CD4 and CD8 on avian thymocytes. *J. Immunol.* **147**: 3710–3717.

182. **Pei, J. J., E. Sersen, K. Iqbal, and I. Grundkeiqbal.** 1994. Expression of protein phosphatases (PP-1, PP-2A, PP-2B and PTP-1B) and protein kinases (MAP kinase and P34(cdc2)) in the hippocampus of patients with Alzheimer disease and normal aged individuals. *Brain Res.* **655**: 70–76.

183. **Pilarski, L. M., R. Gillitzer, H. Zola, K. Shortman, and R. Scollay.** 1989. Definition of the thymic generative lineage by selective expression of high molecular weight isoforms of CD45 (T200). *Eur. J. Immunol.* **19**: 589–597.

184. **Pulido, R., M. Cebrian, A. Acevedo, M. O. de Landazuri, and F. Sanchez-Madrid.** 1988. Comparative biochemical and tissue distribution study of four distinct CD45 antigen specificities. *J. Immunol.* **140**: 3851–3857.

185. **Pulido, R., P. Lacal, F. Mollinedo, and F. Sanchez-Madrid.** 1989. Biochemical and antigenic characterization of CD45 polypeptides expressed on plasma membrane and internal granules of human neutrophils. *FEBS Lett.* **249**: 337–342.

186. **Purushotham, K. R., G. A. Paul, P. Wang, and M. G. Humphreys-Beher.** 1994. Characterization of an SH2 containing protein tyrosine phosphatase in rat parotid gland acinar cells. *Life Sci.* **54**: 1185–1194.

187. **Radha, V., S. Nambirajan, and G. Swarup.** 1994. Subcellular localization of a protein-tyrosine phosphatase – evidence for association with chromatin. *Biochem J.* **299**: 41–47.

188. **Rajendrakumar, G. V., V. Radha, and G. Swarup.** 1993. Stabilization of a protein-tyrosine phosphatase messenger RNA upon mitogenic stimulation of T-lymphocytes. *Biochim. Biophys. Acta* **1216**: 205–212.

189. **Ramponi, G., M. Ruggiero, G. Raugei, A. Berti, A. Modesti, D. Degl'Innocenti, L. Magnelli, C. Pazzagli, V. P. Chiarugi, and G. Camici.** 1992. Overexpression of a synthetic phosphotyrosine protein phosphatase gene inhibits normal and transformed cell growth. *Int. J. Cancer* **51**: 652–656.

190. **Raugei, G., R. Marzocchini, A. Modesti, G. Ratti, G. Cappugi, G. Camici, G. Manao, and G. Ramponi.** 1991. Chemical synthesis and expression of a gene coding for bovine liver phosphotyrosine-protein phosphatase. *Biochem. Int.* **23**: 317–326.

191. **Rothstein, D. M., H. Saito, M. Streuli, S. F. Schlossman, and C. Morimoto.** 1992. The alternative splicing of the CD45 tyrosine phosphatase is controlled by negative regulatory trans-acting splicing factors. *J. Biol. Chem.* **267**: 7139–7147.

192. **Rothstein, D. M., A. Yamada, S. F. Schlossman, and C. Morimoto.** 1991. Cyclic regulation of CD45 isoform expression in a long term human CD4+CD45RA+ T cell line. *J. Immunol.* **146**: 1175–1183.

193. **Rotin, D., B. J. Goldstein, and C. A. Fladd.** 1994. Expression of the protein-tyrosine phosphatase LAR-PTP2 is developmentally regulated in lung epithelia. *Am. J. Physiol.* **267**: L263–270.

194. **Ruggiero, M., C. Pazzagli, S. Rigacci, L. Magnelli, G. Raugei, A. Berti, V. P. Chiarugi, J. H. Pierce, G. Camici, and G. Ramponi.** 1993. Negative growth control by a novel low M_r phosphotyrosine protein phosphatase in normal and transformed Cells. *FEBS Lett.* **326**: 294–298.

195. **Sadhu, K., S. I. Reed, H. Richardson, and P. Russell.** 1990. Human homolog of fission yeast cdc25 mitotic inducer is predominantly expressed in G2. *Proc. Natl Acad. Sci. USA* **87**: 5139–5143.

196. **Saga, Y., J. S. Tung, F. W. Shen, and E. A. Boyse.** 1987. Alternative use of 5′ exons in the specification of Ly-5 isoforms distinguishing hematopoietic cell lineages. *Proc. Natl Acad. Sci. USA* **84**: 5364–5368.

197. **Sahin, M. and S. Hockfield.** 1993. Protein tyrosine phosphatases expressed in the developing rat brain. *J. Neurosci.* **13**: 4968–4978.

198. **Seimiya, H. and T. Tsuruo.** 1993. Differential expression of protein tyrosine phosphatase genes during phorbol ester-induced differentiation of human leukemia U937 cells. *Cell Growth Differ.* **4**: 1033–1039.

199. **Serra, H. M., J. F. Krowka, J. A. Ledbetter, and L. M. Pilarski.** 1988. Loss of CD45R (Lp220) represents a post-thymic T cell differentiation event. *J. Immunol.* **140**: 1435–1441.

200. **Shah, V. O., C. I. Civin, and M. R. Loken.** 1988. Flow cytometric analysis of human bone marrow: differential quantitative expression of T-200 common leukocyte antigen during normal hemopoiesis. *J. Immunol.* **140**: 1861–1867.

201*. **Shifrin, V. I. and B. G. Neel.** 1993. Growth factor-inducible alternative splicing of nontransmembrane phosphotyrosine phosphatase PTP-1B pre-messenger RNA. *J. Biol. Chem.* **268**: 25376–25384.

202. **Shimohama, S., S. Fujimoto, T. Taniguchi, M. Kameyama, and J. Kimura.** 1993. Reduction of low-molecular-weight acid phosphatase activity in Alzheimer brains. *Ann. Neurol.* **33**: 616–621.

203. **Sridhar, T. S., G. Swarup, and A. Khar.** 1993. Downregulation of phospho-tyrosine phosphatases in a macrophage tumor. *FEBS Lett.* **326**: 75–79.

204. **Streuli, M., L. R. Hall, Y. Saga, S. F. Schlossman, and H. Saito.** 1987. Differential usage of three exons generates at least five different mRNAs encoding human leukocyte common antigens. *J. Exp. Med.* **166**: 1548–1566.

205. **Streuli, M. and H. Saito.** 1989. Regulation of tissue-specific alternative splicing: exon-specific cis-elements govern the splicing of leukocyte common antigen pre-mRNA. *EMBO J.* **8**: 787–796.

206. **Taetle, R., H. Ostergaard, M. Smedsrud, and I. Trowbridge.** 1991. Regulation of CD45 expression in human leukemia cells. *Leukemia* **5**: 309–314.

207. **Tillmann, U., J. Wagner, D. Boerboom, H. Westphal, and M. L. Tremblay.** 1994. Nuclear localization and cell cycle regulation of a murine protein tyrosine phosphatase. *Mol. Cell. Biol.* **14**: 3030–3040.

208. **Tsai, A. Y., M. Streuli, and H. Saito.** 1989. Integrity of the exon 6 sequence is essential for tissue-specific alternative splicing of human leukocyte common antigen pre-mRNA. *Mol. Cell. Biol.* **9**: 4550–4555.

209. **Tsukamoto, T., T. Takahashi, R. Ueda, K. Hibi, and H. Saito.** 1992. Molecular analysis of the protein tyrosine phosphatase gamma gene in human lung cancer cell lines. *Cancer Res.* **52**: 3506–3509.

210. **Tsuneizumi, K., T. Kume, T. Watanabe, M. F. B. G. Gebbink, M. L. Thomas, and M. Oishi.** 1994. Induction of specific protein tyrosine phosphatase transcripts during differentiation of mouse embryonal carcinoma (F9) cells [correction]. *FEBS Lett.* **351**: 295.

211. **Tsuneizumi, K., T. Kume, T. Watanabe, M. F. B. G. Gebbink, M. L. Thomas, and M. Oishi.** 1994. Induction of specific protein tyrosine phosphatase transcripts during differentiation of mouse embryonal carcinoma (F9) cells. *FEBS Lett.* **347**: 9–12.

212. **Uchida, T., T. Matozaki, K. Matsuda, T. Suzuki, S. Matozaki, O. Nakano, K. Wada, Y. Konda, C. Sakamoto, and M. Kasuga.** 1993. Phorbol ester stimulates the activity of a protein tyrosine phosphatase containing SH2 domains (PTP1C) in HL-60 leukemia cells by increasing gene expression. *J. Biol. Chem.* **268**: 11845–11850.

213. **Villa-Moruzzi, E.** 1993. Activation of the cdc25C phosphatase in mitotic HeLa cells. *Biochem. Biophys. Res. Commun.* **196**: 1248–1254.

214. **Visser, L., R. Lai, and S. Poppema.** 1993. Patterns of leucocyte common antigen expression in peripheral blood T cell populations. *Cell. Immunol.* **151**: 218–224.

215. **Volarevic, S., B. B. Niklinska, C. M. Burns, C. H. June, A. M. Weissman, and J. D. Ashwell.** 1993. Regulation of TCR signaling by CD45 lacking transmembrane and extracellular domains. *Science* **260**: 541–544.

216*. **Walton, K. M., K. J. Martell, S. P. Kwak, J. E. Dixon, and B. L. Largent.** 1993. A novel receptor-type protein tyrosine phosphatase is expressed during neurogenesis in the olfactory neuroepithelium. *Neuron* **11**: 387–400.

217. **Way, B. A. and R. A. Mooney.** 1994. Differential effects of phosphotyrosine phosphatase expression on hormone-dependent and independent pp60(c- src) activity. *Mol. Cell. Biochem.* **139**: 167–175.

218. **Werfel, T., G. Sonntag, M. H. Weber, and O. Gotze.** 1991. Rapid increases in the membrane expression of neutral endopeptidase (CD10), aminopeptidase N (CD13), tyrosine phosphatase (CD45), and Fc gamma-RIII (CD16) upon stimulation of human peripheral leukocytes with human C5a. *J. Immunol.* **147**: 3909–3914.

219. **Wiener, J. R., J. A. Hurteau, B. J. M. Kerns, R. S. Whitaker, M. R. Conaway, A. Berchuck, and R. C. Bast.** 1994. Overexpression of the tyrosine phosphatase Ptp1B is associated with human ovarian carcinomas. *Am. J. Obstet. Gynecol.* **170**: 1177–1183.

220. **Wiener, J. R., B. J. M. Kerns, E. L. Harvey, M. R. Conaway, J. D. Iglehart, A. Berchuck, and R. C. Bast.** 1994. Overexpression of the protein tyrosine phosphatase PTP1B in human breast cancer – association with p185(c-erbB-2) protein expression. *J. Natl Cancer Inst.* **86**: 372–378.

221. **Woodford, T. A., K. L. Guan, and J. E. Dixon.** 1991. Expression of Rat PTP1 in normal and transformed cells. *Adv. Prot. Phosphatases* **6**: 503–524.

222. **Woodford-Thomas, T. A., J. D. Rhodes, and J. E. Dixon.** 1992. Expression of a protein tyrosine phosphatase in normal and v-src-transformed mouse 3T3 fibroblasts. *J. Cell Biol.* **117**: 401–414.

223. **Yamada, A., M. Streuli, H. Saito, D. M. Rothstein, S. F. Schlossman, and C. Morimoto.** 1990. Effect of activation of protein kinase C on CD45 isoform expression and CD45 protein tyrosine phosphatase activity in T cells. *Eur. J. Immunol.* **20**: 1655–1660.

224. **Zhai, Y. F., H. Beittenmiller, B. Wang, M. N. Gould, C. Oakley, W. J. Esselman, and C. W. Welsch.** 1993. Increased expression of specific protein tyrosine phosphatases in human breast epithelial cells neoplastically transformed by the neu oncogene. *Cancer Res.* **53**: 2272–2278.

225. **Zhang, W. R. and B. J. Goldstein.** 1991. Identification of skeletal muscle protein-tyrosine phosphatases by amplification of conserved cDNA sequences. *Biochem. Biophys. Res. Commun.* **178**: 1291–1297.

226. **Zhao, Z. Z., S. H. Shen, and E. H. Fischer.** 1994. Phorbol ester-induced expression, phosphorylation, and translocation of protein-tyrosine-phosphatase 1C in HL-60 cells. *Proc. Natl Acad. Sci. USA* **91**: 5007–5011.

227. **Zheng, X. M. and C. J. Pallen.** 1994. Expression of receptor-like protein tyrosine phosphatase a in rat embryo fibroblasts activates mitogen-activated protein kinase and c-Jun. *J. Biol. Chem.* **269**: 23302–23309.

228*. **Zheng, X. M., Y. Wang, and C. J. Pallen.** 1992. Cell transformation and activation of pp60c-src by overexpression and activation of a protein tyrosine phosphatase. *Nature* **359**: 336–339.

Function

229. **Arroyo, A. G., M. R. Campanero, P. Sanchez-Mateos, J. M. Zapata, M. A. Ursa, M. A. del Pozo, and F. Sanchez-Madrid.** 1994. Induction of tyrosine phosphorylation during ICAM-3 and LFA-1-mediated intercellular adhesion, and its regulation by the CD45 tyrosine phosphatase. *J. Cell Biol.* **126**: 1277–1286.

230. **Benveniste, P., Y. Takahama, D. L. Wiest, T. Nakayama, S. O. Sharrow, and A. Singer.** 1994. Engagement of the external domains of CD45 tyrosine phosphatase can regulate the differentiation of immature CD4+CD8+ thymocytes into mature T cells. *Proc. Natl Acad. Sci. USA* **91**: 6933–6937.

231. **Berger, S. A., T. W. Mak, and C. J. Paige.** 1994. Leukocyte common antigen (CD45) is required for immunoglobulin E-mediated degranulation of mast cells. *J. Exp. Med.* **180**: 471–476.

232. **Bernabeu, C., A. C. Carrera, M. O. de Landazuri, and F. Sanchez-Madrid.** 1987. Interaction between the CD45 antigen and phytohemagglutinin. Inhibitory effect on the lectin-induced T cell proliferation by anti-CD45 monoclonal antibody. *Eur. J. Immunol.* **17**: 1461–1466.

233. **Biffen, M., D. McMichael-Phillips, T. Larson, A. Venkitaraman, and D. Alexander.** 1994. The CD45 tyrosine phosphatase regulates specific pools of antigen receptor-associated p59fyn and CD4–associated p56lck tyrosine in human T-cells. *EMBO J.* **13**: 1920–1929.

234*. **Bliska, J. B., K. L. Guan, J. E. Dixon, and S. Falkow.** 1991. Tyrosine phosphate hydrolysis of host proteins by an essential *Yersinia*-virulence determinant. *Proc. Natl Acad. Sci. USA* **88**: 1187–1191.

235. **Bliska, J. B., J. C. Clemens, J. E. Dixon, and S. Falkow.** 1992. The *Yersinia* tyrosine phosphatase – specificity of a bacterial virulence determinant for phosphoproteins in the J774A.1 macrophage. *J. Exp. Med.* **176**: 1625–1630.

236. **Brady-Kalnay, S. M. and N. K. Tonks.** 1993. Purification and characterization of the human protein tyrosine phosphatase, PTPμ, from a baculovirus expression system. *Mol. Cell. Biochem.* 128131–141.

237. **Burns, C. M., K. Sakaguchi, E. Appella, and J. D. Ashwell.** 1994. CD45 regulation of tyrosine phosphorylation and enzyme activity of src family kinases. *J. Biol. Chem.* **269**: 13594–13600.

238*. **Charles, C. H., H. Sun, L. F. Lau, and N. K. Tonks.** 1993. The growth factor-inducible immediate-early gene 3CH134 encodes a protein-tyrosine-phosphatase. *Proc. Natl Acad. Sci. USA* **90**: 5292–5296.

239. **Chernoff, J., H. C. Li, Y. S. Cheng, and L. B. Chen.** 1983. Characterization of a phosphotyrosyl protein phosphatase activity associated with a phosphoseryl protein phosphatase of $M_r = 95\,000$ from bovine heart. *J. Biol. Chem.* **258**: 7852–7857.

240. **Chiarugi, P., P. Cirri, G. Camici, G. Manao, T. Fiaschi, G. Raugei, G. Cappugi, and G. Ramponi.** 1994. The role of His-66 and His-72 in the reaction mechanism of bovine liver low M_r phosphotyrosine protein phosphates. *Biochem J.* **298**: 427–433.

241. **Chiarugi, P., R. Marzocchini, G. Raugei, C. Pazzagli, A. Berti, G. Camici, G. Manao, G. Cappugi, and G. Ramponi.** 1992. Differential role of 4 cysteines on the activity of a low M_r phosphotyrosine protein phosphatase. *FEBS Lett.* **310**: 9–12.

242. **Cho, H., R. Krishnaraj, E. Kitas, W. Bannwarth, C. T. Walsh, and K. S. Anderson.** 1992. Isolation and structural elucidation of a novel phosphocysteine intermediate in the LAR protein tyrosine phosphatase enzymatic pathway. *J. Am. Chem. Soc.* **114**: 7296–7298.

243. **Cho, H. J., R. Krishnaraj, M. Itoh, E. Kitas, W. Bannwarth, H. Saito, and C. T. Walsh.** 1993. Substrate specificities of catalytic fragments of protein tyrosine phosphatases (HPTPb, LAR and CD45) toward phosphotyrosylpeptide substrates and thiophosphorylated peptides as inhibitors. *Prot. Sci.* **2**: 977–984.

244*. **Cho, H. J., S. E. Ramer, M. Itoh, E. Kitas, W. Bannwarth, P. Burn, H. Saito, and C. T. Walsh.** 1992. Catalytic domains of the LAR and CD45 protein tyrosine phosphatases from *Escherichia coli* expression systems – purification and characterization for specificity and mechanism. *Biochemistry* **31**: 133–138.

245. **Cho, H. J., S. E. Ramer, M. Itoh, D. G. Winkler, E. Kitas, W. Bannwarth, P. Burn, H. Saito, and C. T. Walsh.** 1991. Purification and characterization of a soluble catalytic fragment of the human transmembrane leukocyte antigen related (LAR) protein tyrosine phosphatase from an *Escherichia coli* expression system. *Biochemistry* **30**: 6210–6216.

246. **Chui, D., C. J. Ong, P. Johnson, H. S. Teh, and J. D. Marth.** 1994. Specific CD45 isoforms differentially regulate T cell receptor signaling. *EMBO J.* **13**: 798–807.

247. **Cicirelli, M. F., N. K. Tonks, C. D. Diltz, J. E. Weiel, E. H. Fischer, and E. G. Krebs.** 1990. Microinjection of a protein-tyrosine-phosphatase inhibits insulin action in *Xenopus* oocytes. *Proc. Natl Acad. Sci. USA* **87**: 5514–5518.

248. **Cirri, P., P. Chiarugi, G. Camici, G. Manao, L. Pazzagli, A. Caselli, I. Barghini, G. Cappugi, G. Raugei, and G. Ramponi.** 1993. The role of Cys-17 in the pyridoxal 5′-phosphate inhibition of the bovine liver low M_r phosphotyrosine protein phosphatase. *Biochim. Biophys. Acta* **1161**: 216–222.

249. **Cirri, P., P. Chiarugi, G. Camici, G. Manao, G. Raugei, G. Cappugi, and G. Ramponi.** 1993. The Role of Cys12, Cys17 and Arg18 in the catalytic mechanism of low-M_r cytosolic phosphotyrosine protein phosphatase. *Eur. J. Biochem.* **214**: 647–657.

250. **Cool, D. E., P. R. Andreassen, N. K. Tonks, E. G. Krebs, E. H. Fischer, and R. L. Margolis.** 1992. Cytokinetic failure and asynchronous nuclear division in BHK cells overexpression a truncated protein-tyrosine phosphatase. *Proc. Natl Acad. Sci. USA* **89**: 5422–5426.

251. **Cool, D. E., N. K. Tonks, H. Charbonneau, E. H. Fischer, and E. G. Krebs.** 1990. Expression of a human T-cell protein-tyrosine phosphatase in baby hamster kidney cells. *Proc. Natl Acad. Sci. USA* **87**: 7280–7284.

252. **Daum, G., N. F. Zander, B. Morse, D. Hurwitz, J. Schlessinger, and E. H. Fischer.** 1991. Characterization of a human recombinant receptor-linked protein tyrosine phosphatase. *J. Biol. Chem.* **266**: 12211–12215.

253. **Davis, J. P., M. M. Zhou, and R. L. Van Etten.** 1994. Kinetic and site-directed mutagenesis studies of the cysteine residues of bovine low molecular weight phosphotyrosyl protein phosphatase. *J. Biol. Chem.* **269**: 8734–8740.

254. **Davis, J. P., M. M. Zhou, and R. L. Van Etten.** 1994. Spectroscopic and kinetic studies of the histidine residues of bovine low-molecular-weight phosphotyrosyl protein phosphatase. *Biochemistry* **33**: 1278–1286.

255. **Deans, J. P., J. Shaw, M. J. Pearse, and L. M. Pilarski.** 1989. CD45R as a primary signal transducer stimulating IL-2 and IL-2R mRNA synthesis by CD3-4-8-thymocytes. *J. Immunol.* **143**: 2425–2430.

256. **Dechert, U., M. Adam, K. W. Harder, I. Clarklewis, and F. Jirik.** 1994. Characterization of protein tyrosine phosphatase SH-PTP2 – study of phosphopeptide substrates and possible regulatory role of SH2 domains. *J. Biol. Chem.* **269**: 5602–5611.

257. **Desai, D. M., J. Sap, O. Silvennoinen, J. Schlessinger, and A. Weiss.** 1994. The catalytic activity of the CD45 membrane-proximal phosphatase domain is required for TCR signaling and regulation. *EMBO J.* **13**: 4002–4010.

258. **Dissing, J., B. Rangaard, and U. Christensen.** 1993. Activity modulation of the fast and slow isozymes of human cytosolic low-molecular-weight acid phosphatase (ACP1) by purines. *Biochim. Biophys. Acta* **1162**: 275–282.

259*. **Dunphy, W. G. and A. Kumagai.** 1991. The cdc25 protein contains an intrinsic phosphatase activity. *Cell* **67**: 189–196.

260. **Fang, K. S., H. Sabe, H. Saito, and H. Hanafusa.** 1994. Comparative study of three protein-tyrosine phosphatases – chicken protein-tyrosine phosphatase λ dephosphorylates c-Src tyrosine 527. *J. Biol. Chem.* **269**: 20194–20200.

261*. **Gabrielli, B. G., M. S. Lee, D. H. Walker, H. Piwnica-Worms, and J. L. Maller.** 1992. Cdc25 regulates the phosphorylation and activity of the *Xenopus* cdk2 protein kinase complex. *J. Biol. Chem.* **267**: 18040–18046.

262*. **Gautier, J., M. J. Solomon, R. N. Booher, J. F. Bazan, and M. W. Kirschner.** 1991. cdc25 is a specific tyrosine phosphatase that directly activates p34cdc2. *Cell* **67**: 197–211.

263. **Gebbink, M. F. B. G., M. H. G. Verheijen, G. C. M. Zondag, I. Van Etten, and W. H. Moolenaar.** 1993. Purification and characterization of the cytoplasmic domain of human receptor-like protein-tyrosine phosphatse RPTPμ. *Biochemistry* **32**: 13516–13522.

264. **George, R. J. and C. W. Parker.** 1990. Preliminary characterization of phosphotyrosine phosphatase activities in human peripheral blood lymphocytes: identification of CD45 as a phosphotyrosine phosphatase. *J. Cell. Biochem.* **42**: 71–81.

265. **Gilliland, L. K., G. L. Schieven, L. S. Grosmaire, N. K. Damle, and J. A. Ledbetter.** 1990. CD45 ligation in T cells regulates signal transduction through both the interleukin-2 receptor and the CD3/Ti T-cell receptor complex. *Tissue Antigens* **35**: 128–135.

266. **Goris, J., C. J. Pallen, P. J. Parker, J. Hermann, M. D. Waterfield, and W. Merlevede.** 1988. Conversion of a phosphoseryl/threonyl phosphatase into a phosphotyrosyl phosphatase. *Biochem. J.* **256**: 1029–1034.

267. **Gould, K. L., S. Moreno, N. K. Tonks, and P. Nurse.** 1990. Complementation of the mitotic activator, $p80^{cdc25}$, by a human protein-tyrosine phosphatase. *Science* **250**: 1573–1576.

268. **Gunaratne, P., C. Stoscheck, R. E. Gates, L. Y. Li, L. B. Nanney, and L. E. King.** 1994. Protein tyrosyl phosphatase-1B is expressed by normal human epidermis, keratinocytes, and A-431 cells and dephosphorylates substrates of the epidermal growth factor receptor. *J. Invest. Dermatol.* **103**: 701–706.

269. **Hannig, G., S. Ottilie, and R. L. Erikson.** 1994. Negative regulation of mitosis in fission yeast by catalytically inactive pyp1 and pyp2 mutants. *Proc. Natl Acad. Sci. USA* **91**: 10084–10088.

270*. **Hannig, G., S. Ottilie, A. R. Schievella, and R. L. Erikson.** 1993. Comparison of the biochemical and biological functions of tyrosine phosphatases from fission yeast, budding yeast and animal cells. *Yeast* **9**: 1039–1052.

271. **Harder, K. W., P. Owen, L. K. H. Wong, R. Aebersold, I. Clarklewis, and F. R. Jirik.** 1994. Characterization and kinetic analysis of the intracellular domain of human protein tyrosine phosphatase (HPTP-β) using synthetic phosphopeptides. *Biochem. J.* **298**: 395–401.

272. **Harvath, L., J. A. Balke, N. P. Christiansen, A. A. Russell, and K. M. Skubitz.** 1991. Selected antibodies to leukocyte common antigen (CD45) inhibit human neutrophil chemotaxis. *J. Immunol.* **146**: 949–957.

273*. **Hashimoto, N., E. P. Feener, W. R. Zhang, and B. J. Goldstein.** 1992. Insulin receptor protein-tyrosine phosphatases – leukocyte common antigen-related phosphatase rapidly deactivates the insulin receptor kinase by preferential dephosphorylation of the receptor regulatory domain. *J. Biol. Chem.* **267**: 13811–13814.

274. **Hippen, K. L., S. Jakes, J. Richards, B. P. Jena, B. L. Beck, L. B. Tabatabai, and T. S. Ingebritsen.** 1993. Acidic residues are involved in substrate recognition by two soluble protein tyrosine phosphatases, PTP-5 and rrbPTP-1. *Biochemistry* **32**: 12405–12412.

275. **Honda, R., Y. Ohba, A. Nagata, H. Okayama, and H. Yasuda.** 1993. Dephosphorylation of human p34(cdc2) kinase on both Thr-14 and Tyr-15 by human cdc25B phosphatase. *FEBS Lett.* **318**: 331–334.

276. **Hook, W. A., E. H. Berenstein, F. U. Zinsser, C. Fischler, and R. P. Siraganian.** 1991. Monoclonal antibodies to the leukocyte common antigen (CD45) inhibit IgE-mediated histamine release from human basophils. *J. Immunol.* **147**: 2670–2676.

277*. **Howard, P. K., B. M. Sefton, and R. A. Firtel.** 1992. Analysis of a spatially regulated phosphotyrosine phosphatase identifies tyrosine phosphorylation as a key regulatory pathway in *Dictyostelium*. *Cell* **71**: 637–647.

278. **Howard, P. K., B. M. Sefton, and R. A. Firtel.** 1993. Tyrosine phosphorylation of actin in *Dictyostelium* associated with cell-shape changes. *Science* **259**: 241–244.

279. **Hurley, T. R., R. Hyman, and B. M. Sefton.** 1993. Differential effects of expression of the CD45 tyrosine protein phosphatase on the tyrosine phosphorylation of the lck, fyn, and c-src tyrosine protein kinases. *Mol. Cell. Biol.* **13**: 1651–1656.

280. **Itoh, M., M. Streuli, N. X. Krueger, and H. Saito.** 1992. Purification and characterization of the catalytic domains of the human receptor-linked protein tyrosine phosphatases HPTPb, leukocyte common antigen (LCA) and leukocyte common antigen-related molecule (LAR). *J. Biol. Chem.* **267**: 12356–12363.

281. **Johnson, P., H. L. Ostergaard, C. Wasden, and I. S. Trowbridge.** 1992. Mutational analysis of CD45. A leukocyte-specific protein tyrosine phosphatase. *J. Biol. Chem.* **267**: 8035–8041.

282. **Justement, L. B., K. S. Campbell, N. C. Chien, and J. C. Cambier.** 1991. Regulation of B-cell antigen receptor signal transduction and phosphorylation by CD45. *Science* **252**: 1839–1842.

283. **Kawauchi, K., A. H. Lazarus, M. J. Rapoport, A. Harwood, J. C. Cambier, and T. L. Delovitch.** 1994. Tyrosine kinase and CD45 tyrosine phosphatase activity mediate p21ras activation in B cells stimulated through the antigen receptor. *J. Immunol.* **152**: 3306–3316.

284. **Kishihara, K., J. Penninger, V. A. Wallace, T. M. Kundig, K. Kawai, A. Wakeham, E. Timms, K. Pfeffer, P. S. Ohashi, M. L. Thomas, C. Furlonger, C. J. Paige, and T. W. Mak.** 1993. Normal B-lymphocyte development but impaired T-cell maturation in CD45-exon 6 protein tyrosine phosphatase-deficient mice. *Cell* **74**: 143–156.

285. **Koretzky, G. A., M. Kohmetscher, and S. Ross.** 1993. CD45–associated kinase activity requires lck but not T cell receptor expression in the Jurkat T cell line. *J. Biol. Chem.* **268**: 8958–8964.

286. **Koretzky, G. A., J. Picus, T. Schultz, and A. Weiss.** 1991. Tyrosine phosphatase CD45 is required for T-cell antigen receptor and CD2-mediated activation of a protein tyrosine kinase and interleukin 2 production. *Proc. Natl Acad. Sci. USA* **88**: 2037–2041.

287*. **Koretzky, G. A., J. Picus, M. L. Thomas, and A. Weiss.** 1990. Tyrosine phosphatase CD45 is essential for coupling T-cell antigen receptor to the phosphatidyl inositol pathway. *Nature* **346**: 66–68.

288*. **Kozlowski, M., I. Mlinaricrascan, G. S. Feng, R. Shen, T. Pawson, and K. A. Siminovitch.** 1993. Expression and catalytic activity of the tyrosine phosphatase PTP1C is severely impaired in motheaten and viable motheaten mice. *J. Exp. Med.* **178**: 2157–2163.

289. **Kuhné, M. R., Z. Z. Zhao, J. Rowles, B. E. Lavan, S. H. Shen, E. H. Fischer, and G. E. Lienhard.** 1994. Dephosphorylation of insulin receptor substrate 1 by the tyrosine phosphatase PTP2C. *J. Biol. Chem.* **269**: 15833–15837.

290*. **Kulas, D. T., W. R. Zhang, B. J. Goldstein, R. W. Furlanetto, and R. A. Mooney.** 1995. Insulin receptor signalling is augmented by antisense inhibition of the protein-tyrosine phosphatase LAR. *J. Biol. Chem.* **270**: 2435–2438.

291. **Kumagai, A. and W. G. Dunphy.** 1991. The cdc25 protein controls tyrosine dephosphorylation of the cdc2 protein in a cell-free system. *Cell* **64**: 903–914.

292. **Kumagai, A. and W. G. Dunphy.** 1992. Regulation of the cdc25 protein during the cell cycle in *Xenopus* extracts. *Cell* **70**: 139–151.

293*. **Lammers, R., B. Bossenmaier, D. E. Cool, N. K. Tonks, J. Schlessinger, E. H. Fischer, and A. Ullrich.** 1993. Differential activities of protein tyrosine phos-phatases in intact cells. *J. Biol. Chem.* **268**: 22456–22462.

294. **Lazaruk, K. D., J. Dissing, and G. F. Sensabaugh.** 1993. Exon structure at the human ACP1 locus supports alternative splicing model for f and s isozyme generation. *Biochem. Biophys. Res. Comm.* **196**: 440–446.

295. **Lee, J. P., H. Cho, W. Bannwarth, E. A. Kitas, and C. T. Walsh.** 1992. NMR analysis of regioselectivity in dephosphorylation of a triphosphotyrosyl dodecapeptide autophosphorylation site of the insulin receptor by a catalytic fragment of LAR phosphotyrosine phosphatase. *Prot. Sci.* **1**: 1353–1362.

296. **Lee, M. S., S. Ogg, M. Xu, L. L. Parker, D. J. Donoghue, J. L. Maller, and H. Piwnica-Worms.** 1992. cdc25+ encodes a protein phosphatase that dephosphorylates p34cdc2. *Mol. Biol. Cell* **3**: 73–84.

297. **Liao, K., R. D. Hoffman, and M. D. Lane.** 1991. Phosphotyrosyl turnover in insulin signaling – characterization of two membrane-bound pp15 protein tyrosine phosphatases from 3T3-L1 adipocytes. *J. Biol. Chem.* **266**: 6544–6553.

298. **Lorenz, H. M., A. S. Lagoo, and K. J. Hardy.** 1994. The cell and molecular basis of leukocyte common antigen (CD45)-triggered, lymphocyte function-associated antigen-1-intercellular adhesion molecule-1-dependent, leukocyte adhesion. *Blood* **83**: 1862–1870.

299. **Maeda, T., A. Y. Tsai, and H. Saito.** 1993. Mutations in a protein tyrosine phosphatase gene (PTP2) and a protein serine/threonine phosphatase gene (PTC1) cause a synthetic growth defect in *Saccharomyces cerevisiae*. *Mol. Cell Biol.* **13**: 5408–5417.

300. **McFarland, E. D., T. R. Hurley, J. T. Pingel, B. M. Sefton, A. Shaw, and M. L. Thomas.** 1993. Correlation between Src family member regulation by the protein-tyrosine-phosphatase CD45 and transmembrane signaling through the T-cell receptor. *Proc. Natl Acad. Sci. USA* **90**: 1402–1406.

301. **Mei, L., C. A. Doherty, and R. L. Huganir.** 1994. RNA splicing regulates the activity of a SH2 domain-containing protein tyrosine phosphatase. *J. Biol. Chem.* **269**: 12254–12262.

302. **Milarski, K. L. and A. R. Saltiel.** 1994. Expression of catalytically inactive Syp phosphatase in 3T3 cells blocks stimulation of mitogen-activated protein kinase by insulin. *J. Biol. Chem.* **269**: 21239–21243.

303*. **Millar, J. B., J. Blevitt, L. Gerace, K. Sadhu, C. Featherstone, and P. Russell.** 1991. p55CDC25 is a nuclear protein required for the initiation of mitosis in human cells. *Proc. Natl Acad. Sci. USA* **88**: 10500–10504.

304. **Millar, J. B., C. H. McGowan, G. Lenaers, R. Jones, and P. Russell.** 1991. p80cdc25 mitotic inducer is the tyrosine phosphatase that activates p34cdc2 kinase in fission yeast. *EMBO J.* **10**: 4301–4309.

305. **Mittler, R. S., G. L. Schieven, P. M. Dubois, K. Klussman, M. P. O'Connell, P. A. Kiener, and V. Herndon.** 1994. CD45-mediated regulation of extracellular calcium influx in a CD4-transfected human T cell line. *J. Immunol.* **153**: 84–96.

306. **Morikawa, K., F. Oseko, and S. Morikawa.** 1991. The role of CD45 in the activation, proliferation and differentiation of human B lymphocytes. *Int. J. Hematol.* **54**: 495–504.

307*. **Mustelin, T. and A. Altman.** 1990. Dephosphorylation and activation of the T cell tyrosine kinase pp56lck by the leukocyte common antigen (CD45). *Oncogene* **5**: 809–813.

308. **Mustelin, T., K. M. Coggeshall, and A. Altman.** 1989. Rapid activation of the T-cell tyrosine protein kinase Pp56Lck by the CD45 phosphotyrosine phosphatase. *Proc. Natl Acad. Sci. USA* **86**: 6302–6306.

309. **Mustelin, T., T. Pessa-Morikawa, M. Autero, M. Gassmann, L. C. Andersson, C. G. Gahmberg, and P. Burn.** 1992. Regulation of the p59fyn protein tyrosine kinase by the CD45 phosphotyrosine phosphatase. *Eur. J. Immunol.* **22**: 1173–1178.

310. **Nel, A. E., J. A. Ledbetter, K. Williams, P. Ho, B. Akerley, K. Franklin, and R. Katz.** 1991. Activation of MAP-2 kinase activity by the CD2 receptor in Jurkat T cells can be reversed by CD45 phosphatase. *Immunology* **73**: 129–133.

311. **Ogimoto, M., T. Katagiri, K. Mashima, K. Hasegawa, K. Mizuno, and H. Yakura.** 1994. Negative regulation of apoptotic death in immature B cells by CD45. *Int. Immunol.* **6**: 647–654.

312. **Ong, C. J., D. Chui, H. S. Teh, and J. D. Marth.** 1994. Thymic CD45 tyrosine phosphatase regulates apoptosis and MHC-restricted negative selection. *J. Immunol.* **152**: 3793–3805.

313. **Oravecz, T., E. Monostori, E. Kurucz, L. Takacs, and I. Ando.** 1991. CD3-induced T-cell proliferation and interleukin-2 secretion is modulated by the CD45 antigen. *Scand. J. Immunol.* **34**: 531–537.

314. **Ostergaard, H. L., D. A. Shackelford, T. R. Hurley, P. Johnson, R. Hyman, B. M. Sefton, and I. S. Trowbridge.** 1989. Expression of CD45 alters phosphorylation of the *lck*-encoded tyrosine protein kinase in murine lymphoma T-cell lines. *Proc. Natl Acad. Sci. USA* **86**: 8959–8963.

315. **Ostergaard, H. L. and I. S. Trowbridge.** 1991. Negative regulation of CD45 protein tyrosine phosphatase activity by ionomycin in T-cells. *Science* **253**: 1423–1425.

316. **Ota, I. M. and A. Varshavsky.** 1992. A gene encoding a putative tyrosine phosphatase suppresses lethality of an N-end rule-dependent mutant. *Proc. Natl Acad. Sci. USA* **89**: 2355–2359.

317. **Pacitti, A., P. Stevis, M. Evans, I. Trowbridge, and T. J. Higgins.** 1994. High level expression and purification of the enzymatically active cytoplasmic region of human CD45 phosphatase from yeast. *Biochim. Biophys. Acta* **1222**: 277–286.

318. **Pallen, C. J., D. S. Lai, H. P. Chia, I. Boulet, and P. H. Tong.** 1991. Purification and characterization of a higher-molecular-mass form of protein phosphotyrosine phosphatase (PTP 1B) from placental membranes. *Biochem. J.* **276**: 315–323.

319. **Pei, D. H., B. G. Neel, and C. T. Walsh.** 1993. Overexpression, purification, and characterization of SHPTP1, a src homology 2-containing protein-tyrosine-phosphatase. *Proc. Natl Acad. Sci. USA* **90**: 1092–1096.

320. **Peyron, J. F., S. Verma, R. de Waal Malefyt, J. Sancho, C. Terhorst, and H. Spits.** 1991. The CD45 protein tyrosine phosphatase is required for the completion of the activation program leading to lymphokine production in the Jurkat human T cell line. *Int. Immunol.* **3**: 1357–1366.

321. **Pingel, J. T., E. D. McFarland, and M. L. Thomas.** 1994. Activation of CD45-deficient T cell clones by lectin mitogens but not anti-Thy-1. *Int. Immunol.* **6**: 169–178.

322*. **Pingel, J. T. and M. L. Thomas.** 1989. Evidence that the leukocyte-common antigen is required for antigen-induced T lymphocyte proliferation. *Cell* **58**: 1055–1065.

323. **Poggi, A., R. Pardi, N. Pella, L. Morelli, S. Sivori, M. Vitale, V. Revello, A. Moretta, and L. Moretta.** 1993. CD45-mediated regulation of LFA1 function in human natural killer cells. Anti-CD45 monoclonal antibodies inhibit the calcium mobilization induced via LFA1 molecules. *Eur. J. Immunol.* **23**: 2454–2463.

324. **Pollack, S., J. A. Ledbetter, R. Katz, K. Williams, B. Akerley, K. Franklin, G. Schieven, and A. E. Nel.** 1991. Evidence for involvement of glycoprotein-CD45 phosphatase in reversing glycoprotein-CD3-induced microtubule-associated protein-2 kinase activity in Jurkat T-cells. *Biochem. J.* **276**: 481–485.

325*. **Pot, D. A. and J. E. Dixon.** 1992. Active site labeling of a receptor-like protein tyrosine phosphatase. *J. Biol. Chem.* **267**: 140–143.

326. **Pot, D. A., T. A. Woodford, E. Remboutsika, R. S. Haun, and J. E. Dixon.** 1991. Cloning, bacterial expression, purification, and characterization of the cytoplasmic domain of rat LAR, a receptor-like protein tyrosine phosphatase. *J. Biol. Chem.* **266**: 19688–19696.

327. **Ramachandran, C., R. Aebersold, N. K. Tonks, and D. A. Pot.** 1992. Sequential dephosphorylation of a multiply phosphorylated insulin receptor peptide by protein tyrosine phosphatases. *Biochemistry* **31**: 4232–4238.

328. **Ramponi, G., G. Manao, G. Camici, G. Cappugi, M. Ruggiero, and D. P. Bottaro.** 1989. The 18 kDa cytosolic acid phosphatase from bovine liver has phosphotyrosine phosphatase activity on the autophosphorylated epidermal growth factor receptor. *FEBS Lett.* **250**: 469–473.

329. **Rankin, B. M., S. A. Yocum, R. S. Mittler, and P. A. Kiener.** 1993. Stimulation of tyrosine phosphorylation and calcium mobilization by Fc γ receptor cross-linking. Regulation by the phosphotyrosine phosphatase CD45. *J. Immunol.* **150**: 605–616.

330. **Reynolds, W. F.** 1993. The tyrosine phosphatase cdc25 selectively inhibits transcription of the *Xenopus* oocyte-type $tRNA^{tyrC}$ gene. *Nucl. Acids Res.* **21**: 4372–4377.

331. **Ruzzene, M., A. Donella-Deana, O. Marin, J. W. Perich, P. Ruzza, G. Borin, A. Calderan, and L. A. Pinna.** 1993. Specificity of T-cell protein tyrosine phosphatase toward phosphorylated synthetic peptides. *Eur. J. Biochem.* **211**: 289–295.

332. **Saeed, A., E. Tremori, G. Manao, G. Camici, G. Cappugi, and G. Ramponi.** 1990. Bovine brain low M_r acid phosphatase: purification and properties. *Physiol. Chem. Phys. Med. NMR* **22**: 81–94.

333. **Sarcevic, B., E. Erikson, and J. L. Maller.** 1993. Purification and characterization of a mitogen- activated protein kinase tyrosine phosphatase from *Xenopus* eggs. *J. Biol. Chem.* **268**: 25075–25083.

334. **Sebastian, B., A. Kakizuka, and T. Hunter.** 1993. Cdc25M2 activation of cyclin-dependent kinases by dephosphorylation of threonine-14 and tyrosine-15. *Proc. Natl Acad. Sci. USA* **90**: 3521–3524.

335. **Sheng, Z. Q. and H. Charbonneau.** 1993. The baculovirus *Autographa californica* encodes a protein tyrosine phosphatase. *J. Biol. Chem.* **268**: 4728–4733.

336. **Shimohama, S., S. Fujimoto, T. Taniguchi, and J. Kimura.** 1994. The endogenous substrate of low molecular weight acid phosphatase in the brain is an epidermal growth factor receptor. *Brain Res.* **662**: 185–188.

337. **Shiroo, M., L. Goff, M. Biffen, E. Shivnan, and D. Alexander.** 1992. CD45 tyrosine phosphatase-activated p59fyn couples the T cell antigen receptor to pathways of diacylglycerol production, protein kinase C activation and calcium influx. *EMBO J.* **11**: 4887–4897.

338. **Sieh, M., J. B. Bolen, and A. Weiss.** 1993. CD45 specifically modulates binding of Lck to a phosphopeptide encompassing the negative regulatory tyrosine of Lck. *EMBO J.* **12**: 315–321.

339. **Southey, M. C., L. J. Gonez, M. S. Sandrin, and B. E. Kemp.** 1990. Active tyrosine phosphatase in immunoprecipitates of multiple isoforms of Ly-5. *Cell Signal.* **2**: 299–304.

340. **Stader, C. and H. W. Hofer.** 1992. A major lienal phosphotyrosine phosphatase is inhibited by phospholipids and inositol trisphosphate. *Biochem. Biophys. Res. Commun.* **189**: 1404–1409.

341. **Stefani, M., A. Caselli, M. Bucciantini, L. Pazzagli, F. Dolfi, G. Camici, G. Manao, and G. Ramponi.** 1993. Dephosphorylation of tyrosine phosphorylated synthetic peptides by rat liver phosphotyrosine protein phosphatase isoenzymes. *FEBS Lett.* **326**: 131–134.

342. **Strausfeld, U., J. C. Labbe, D. Fesquet, J. C. Cavadore, A. Picard, K. Sadhu, P. Russell, and M. Doree.** 1991. Dephosphorylation and activation of a P34Cdc2 cyclin-B complex *in vitro* by human CDC25 protein. *Nature* **351**: 242–245.

343. **Streuli, M., N. X. Krueger, T. Thai, M. Tang, and H. Saito.** 1990. Distinct functional roles of the two intracellular phosphatase like domains of the receptor-linked protein tyrosine phosphatases LCA and LAR. *EMBO J.* **9**: 2399–2407.

344. **Sugimoto, S., R. J. Lechleider, S. E. Shoelson, B. G. Neel, and C. T. Walsh.** 1993. Expression, purification, and characterization of SH2-containing protein tyrosine phosphatase, SH-PTP2. *J. Biol. Chem.* **268**: 22771–22776.

345*. **Sun, H., C. H. Charles, L. F. Lau, and N. K. Tonks.** 1993. MKP-1 (3CH134), an immediate early gene product, is a dual specificity phosphatase that dephosphorylates MAP kinase *in vivo*. *Cell* **75**: 487–493.

346. **Sun, H., N. K. Tonks, and D. Bar-Sagi.** 1994. Inhibition of Ras-induced DNA synthesis by expression of the phosphates MKP-1. *Science* **266**: 285–288.

347. **Taddei, N., P. Chiarugi, P. Cirri, T. Fiaschi, M. Stefani, G. Camici, G. Raugei, and G. Ramponi.** 1994. Aspartic-129 is an essential residue in the catalytic mechanism of the low *M(r)* phosphotyrosine protein phosphatase. *FEBS Lett.* **350**: 328–332.

348. **Tan, X., D. R. Stover, and K. A. Walsh.** 1993. Demonstration of protein tyrosine phosphatase activity in the second of two homologous domains of CD45. *J. Biol. Chem.* **268**: 6835–6838.

349. **Thuillier, L., C. Hivroz, R. Fagard, C. Andreoli, and P. Mangeat.** 1994. Ligation of CD4 surface antigen induces rapid tyrosine phosphorylation of the cytoskeletal protein ezrin. *Cell. Immunol.* **156**: 322–331.

350*. **Tonks, N. K., H. Charbonneau, C. D. Diltz, E. H. Fischer, and K. A. Walsh.** 1988. Demonstration that the leukocyte common antigen CD45 is a protein tyrosine phosphatase. *Biochemistry* **27**: 8695–8701.

351. **Tonks, N. K., M. F. Cicirelli, C. D. Diltz, E. G. Krebs, and E. H. Fischer.** 1990. Effect of microinjection of a low-M_r human placenta protein tyrosine phosphatase on induction of meiotic cell division in *Xenopus* oocytes. *Mol. Cell. Biol.* **10**: 458–463.

352*. **Tonks, N. K., C. D. Diltz, and E. H. Fischer.** 1988. Characterization of the major protein-tyrosine-phosphatases of human placenta. *J. Biol. Chem.* **263**: 6731–6737.

353. **Tonks, N. K., C. D. Diltz, and E. H. Fischer.** 1990. CD45, an integral membrane protein tyrosine phosphatase: characterization of enzyme activity. *J. Biol. Chem.* **265**: 10674–10680.

354. **Torimoto, Y., N. H. Dang, M. Streuli, D. M. Rothstein, H. Saito, S. F. Schlossman, and C. Morimoto.** 1992. Activation of T cells through a T cell-specific epitope of CD45. *Cell. Immunol.* **145**: 111–129.

355. **Townley, R., S. H. Shen, D. Banville, and C. Ramachandran.** 1993. Inhibition of the activity of protein tyrosine phosphatase 1C by its SH2 domains. *Biochemistry* **32**: 13414–13418.

356. **Tsai, A. Y. M., M. Itoh, M. Streuli, T. Thai, and H. Saito.** 1991. Isolation and characterization of temperature-sensitive and thermostable mutants of the human receptor-like protein tyrosine phosphatase LAR. *J. Biol. Chem.* **266**: 10534–10543.

357. **Volarevic, S., B. B. Niklinska, C. M. Burns, H. Yamada, C. H. June, F. J. Dumont, and J. D. Ashwell.** 1992. The CD45 tyrosine phosphatase regulates phosphotyrosine homeostasis and its loss reveals a novel pattern of late T-cell receptor induced Ca^{2+} oscillations. *J. Exp. Med.* **176**: 835–844.

358. **Waheed, A., P. M. Laidler, Y. Y. Wo, and R. L. Van Etten.** 1988. Purification and physicochemical characterization of a human placental acid phosphatase possessing phosphotyrosyl protein phosphatase activity. *Biochemistry* **27**: 4265–4273.

359. **Wang, Y. and C. J. Pallen.** 1991. The receptor-like protein tyrosine phosphatase HPTP-α has two active catalytic domains with distinct substrate specificities. *EMBO J.* **10**: 3231–3237.

360. **Wang, Y. and C. J. Pallen.** 1992. Expression and characterization of wild type, truncated, and mutant forms of the intracellular region of the receptor-like protein tyrosine phosphatase HPTP-β. *J. Biol. Chem.* **267**: 16696–16702.

361*. **Ward, Y., S. Gupta, P. Jensen, M. Wartmann, R. J. Davis, and K. Kelly.** 1994. Control of map kinase activation by the mitogen-induced threonine/tyrosine phosphatase Pac1. *Nature* **367**: 651–654.

362. **Way, B. A. and R. A. Mooney.** 1993. Activation of phosphatidylinositol-3-kinase by platelet-derived growth factor and insulin-like growth factor-1 is inhibited by a transmembrane phosphotyrosine phosphatase. *J. Biol. Chem.* **268**: 26409–26415.

363. **Weaver, C. T., J. T. Pingel, J. O. Nelson, and M. L. Thomas.** 1991. CD8+ T-cell clones deficient in the expression of the CD45 protein tyrosine phosphatase have impaired responses to T-cell receptor stimuli. *Mol. Cell. Biol.* **11**: 4415–4422.

364. **Webb, D. S., Y. Shimizu, G. A. Van Seventer, S. Shaw, and T. L. Gerrard.** 1990. LFA-3, CD44, and CD45: physiologic triggers of human monocyte TNF and IL-1 release. *Science* **249**: 1295–1297.

365. **Wo, Y. Y., M. M. Zhou, P. Stevis, J. P. Davis, Z. Y. Zhang, and R. L. Van Etten.** 1992. Cloning, expression, and catalytic mechanism of the low molecular weight phosphotyrosyl protein phosphatase from bovine heart. *Biochemistry* **31**: 1712–1721.

366. **Wu, J., L. F. Lau, and T. W. Sturgill.** 1994. Rapid deactivation of MAP kinase in PC12 cells occurs independently of induction of phosphatase MKP-1. *FEBS Lett.* **353**: 9–12.

367*. **Xiao, S., D. W. Rose, T. Sasaoka, H. Maegawa, T. R. Burke, P. P. Roller, S. E. Shoelson, and J. M. Olefsky.** 1994. Syp (SH-PTP2) is a positive mediator of growth factor-stimulated mitogenic signal transduction. *J. Biol. Chem.* **269**: 21244–21248.

368. **Zander, N. F., D. E. Cool, C. D. Diltz, L. R. Rohrschneider, E. G. Krebs, and E. H. Fischer.** 1993. Suppression of v-fms-induced transformation by overexpression of a truncated T-cell protein tyrosine phosphatase. *Oncogene* **8**: 1175–1182.

369. **Zander, N. F., J. A. Lorenzen, D. E. Cool, N. K. Tonks, G. Daum, E. G. Krebs, and E. H. Fischer.** 1991. Purification and characterization of a human recombinant T-cell protein-tyrosine-phosphatase from a baculovirus expression system. *Biochemistry* **30**: 6964–6970.

370. **Zhang, Z. T., E. Harms, and R. L. Van Etten.** 1994. Asp-129 of low molecular weight protein tyrosine phosphatase is involved in leaving group protonation. *J. Biol. Chem.* **269**: 25947–25950.

371. **Zhang, Z. Y., J. C. Clemens, H. L. Schubert, J. A. Stuckey, M. W. F. Fischer, D. M. Hume, M. A. Saper, and J. E. Dixon.** 1992. Expression, purification, and physicochemical characterization of a recombinant *Yersinia* protein tyrosine phosphatase. *J. Biol. Chem.* **267**: 23759–23766.

372. **Zhang, Z. Y., J. P. Davis, and R. L. Van Etten.** 1992. Covalent modification and active site-directed inactivation of a low molecular weight phosphotyrosyl protein phosphatase. *Biochemistry* **31**: 1701–1711.

373. **Zhang, Z. Y. and J. E. Dixon.** 1993. Active site labeling of the *Yersinia* protein tyrosine phosphatase – the determination of the pK_a of the active site cysteine and the function of the conserved histidine-402. *Biochemistry* **32**: 9340–9345.

374. **Zhang, Z. Y., D. Maclean, D. J. Mcnamara, T. K. Sawyer, and J. E. Dixon.** 1994. Protein tyrosine phosphatase substrate specificity – size and phosphotyrosine positioning requirements in peptide substrates. *Biochemistry* **33**: 2285–2290.

375. **Zhang, Z. Y., W. P. Malachowski, R. L. Van Etten, and J. E. Dixon.** 1994. Nature of the rate-determining steps of the reaction catalyzed by the *Yersinia* protein-tyrosine phosphatase. *J. Biol. Chem.* **269**: 8140–8145.

376*. **Zhang, Z. Y., A. M. Thiemesefler, D. Maclean, D. J. Mcnamara, E. M. Dobrusin, T. K. Sawyer, and J. E. Dixon.** 1993. Substrate specificity of the protein tyrosine phosphatases. *Proc. Natl Acad. Sci. USA* **90**: 4446–4450.

377. **Zhang, Z. Y. and R. L. Van Etten.** 1990. Purification and characterization of a low-molecular-weight acid phosphatase – a phosphotyrosyl-protein phosphatase from bovine heart. *Arch. Biochem. Biophys.* **282**: 39–49.

378. **Zhang, Z. Y. and R. L. Van Etten.** 1991. Leaving group dependence and proton inventory studies of the phosphorylation of a cytoplasmic phosphotyrosyl protein phosphatase from bovine heart. *Biochemistry* **30**: 8954–8959.

379. **Zhang, Z. Y. and R. L. Van Etten.** 1991. Pre-steady-state and steady-state kinetic analysis of the low molecular weight phosphotyrosyl protein phosphatase from bovine heart. *J. Biol. Chem.* **266**: 1516–1525.

380*. **Zhang, Z. Y., Y. A. Wang, and J. E. Dixon.** 1994. Dissecting the catalytic mechanism of protein-tyrosine phosphatases. *Proc. Natl Acad. Sci. USA* **91**: 1624–1627.

381. **Zhao, Z. Z., P. Bouchard, C. D. Diltz, S. H. Shen, and E. H. Fischer.** 1993. Purification and characterization of a protein tyrosine phosphatase containing SH2 domains. *J. Biol. Chem.* **268**: 2816–2820.

382. **Zhao, Z. Z., S. H. Shen, and E. H. Fischer.** 1993. Stimulation by phospholipids of a protein-tyrosine-phosphatase containing 2 src homology-2 domains. *Proc. Natl Acad. Sci. USA* **90**: 4251–4255.

383. **Zhao, Z. Z., R. Larocque, W. T. Ho, E. H. Fischer, and S. H. Shen.** 1994. Purification and characterization of PTP2C, a widely distributed protein tyrosine phosphatase containing two SH2 domains. *J. Biol. Chem.* **269**: 8780–8785.

384. **Zheng, C. F. and K. L. Guan.** 1993. Dephosphorylation and inactivation of the mitogen-activated protein kinase by a mitogen-induced Thr/Tyr protein phosphatase. *J. Biol. Chem.* **268**: 16116–16119.

385. **Zhou, M. M., J. P. Davis, and R. L. Van Etten.** 1993. Identification and pKa determination of the histidine residues of human low-molecular-weight phosphotyrosyl protein phosphatases: a convenient approach using an MLEV-17 spectral editing scheme. *Biochemistry* **32**: 8479–8486.

Methods for purifying and assaying

386. **Babcook, J., J. Watts, R. Aebersold, and H. J. Ziltener.** 1991. Automated nonisotopic assay for protein-tyrosine kinase and protein-tyrosine phosphatase activities. *Anal. Biochem.* **196**: 245–251.

387. **Brautigan, D. L. and C. L. Shriner.** 1988. Methods to distinguish various types of protein phosphatase activity. *Methods Enzymol.* **159**: 339–346.

388. **Daum, G., F. Solca, C. D. Diltz, Z. Zhao, D. E. Cool, and E. H. Fischer.** 1993. A general peptide substrate for protein tyrosine phosphatases. *Anal. Biochem.* **211**: 50–54.

389. **Dawicki, D. D. and M. Steiner.** 1993. Identification of a protein-tyrosine phosphatase from human platelet membranes by an immobilon-based solid phase assay. *Anal. Biochem.* **213**: 245–255.

390. **Fisher, D. K. and T. J. Higgins.** 1994. A sensitive, high-volume, colorimetric assay for protein phosphatases. *Pharm. Res.* **11**: 759–763.

391. **Hendriks, W., C. Brugman, J. Schepens, and B. Wieringa.** 1994. Rapid assessment of protein-tyrosine phosphatase expression levels by RT-PCR with degenerate primers. *Mol. Biol. Rep.* **19**: 105–108.

392. **Ingebritsen, T. S.** 1991. Resolution and characterization of multiple protein-tyrosine phosphatase activities. *Methods Enzymol.* **201**: 451–465.

393. **King, M. J. and G. J. Sale.** 1988. Assay of phosphotyrosyl protein phosphatase using synthetic peptide 1142–1153 of the insulin receptor. *FEBS Lett.* **237**: 137–140.

394. **Lutz, M. P., D. I. Pinon, and L. J. Miller.** 1994. A nonradioactive fluorescent gel-shift assay for the analysis of protein phosphatase and kinase activities toward protein-specific peptide substrates. *Anal. Biochem.* **220**: 268–274.

395. **Madden, J. A., M. I. Bird, Y. Man, T. Raven, and D. D. Myles.** 1991. Two nonradioactive assays for phosphotyrosine phosphatases with activity toward the insulin receptor. *Anal. Biochem.* **199**: 210–215.

396. **Martin, B., C. J. Pallen, J. H. Wang, and D. J. Graves.** 1985. Use of fluorinated tyrosine phosphates to probe the substrate specificity of the low molecular weight phosphatase activity of calcineurin. *J. Biol. Chem.* **260**: 14932–14937.

397. **Mishra, S. and A. W. Hamburger.** 1993. A microtiter enzyme-linked immunosorbent assay for protein tyrosine phosphatase. *Biochim. Biophys. Acta* **1157**: 93–101.

398. **Nash, K., M. Feldmuller, J. Dejersey, P. Alewood, and S. Hamilton.** 1993. Continuous and discontinuous assays for phosphotyrosyl protein phosphatase activity using phosphotyrosyl peptide substrates. *Anal. Biochem.* **213**: 303–309.

399. **Ottinger, E. A., L. L. Shekels, D. A. Bernlohr, and G. Barany.** 1993. Synthesis of phosphotyrosine-containing peptides and their use as substrates for protein tyrosine phosphatases. *Biochemistry* **32**: 4354–4361.

400. **Pinna, L. A. and A. Donella-Deana.** 1994. Phosphorylated synthetic peptides as tools for studying protein phosphatases. *Bba-Mol. Cell Res.* **1222**: 415–431.

401. **Shacter, E.** 1984. Organic extraction of Pi with isobutanol/toluene. *Anal. Biochem.* **138**: 416–420.

402. **Shriner, C. L. and D. L. Brautigan.** 1987. Metal chelate affinity chromatography of Zn^{2+}-inhibited protein Tyr(P) phosphatases. *J. Biochem. Biophys. Methods.* **14**: 273–278.

403. **Tonks, N. K., C. D. Diltz, and E. H. Fischer.** 1988. Purification of the major protein-tyrosine-phosphatases of human placenta. *J. Biol. Chem.* **263**: 6722–6730.

404*. **Tonks, N. K., C. D. Diltz, and E. H. Fischer.** 1991. Purification of protein-tyrosine phosphatases from human placenta. *Methods Enzymol.* **201**: 427–442.

405*. **Tonks, N. K., C. D. Diltz, and E. H. Fischer.** 1991. Purification and assay of CD45: an integral membrane protein-tyrosine phosphatase. *Methods Enzymol.* **201**: 442–451.

406. **Zhang, Z. Y., D. Maclean, A. M. Thiemesefler, R. W. Roeske, and J. E. Dixon.** 1993. A continuous spectrophotometric and fluorimetric assay for protein tyrosine phosphatase using phosphotyrosine-containing peptides. *Anal. Biochem.* **211**: 7–15.

407. **Zhao, Z. Z., N. F. Zander, D. A. Malencik, S. R. Anderson, and E. H. Fischer.** 1992. Continuous spectrophotometric assay of protein-tyrosine phosphatase using phosphotyrosine. *Anal. Biochem.* **202**: 361–366.

Protein interactions

408. **Altevogt, P., J. Schreck, B. Schraven, S. Meuer, V. Schirrmacher, and A. Mitsch.** 1990. Association of CD2 and T200 (CD45) in mouse T lymphocytes. *Int. Immunol.* **2**: 353–360.

409. **Altin, J. G., E. B. Pagler, B. F. Kinnear, and H. S. Warren.** 1994. Molecular associations involving CD16, CD45 and ζ and γ chains on human natural killer cells. *Immunol. Cell Biol.* **72**: 87–96.

410. **Altin, J. G., E. B. Pagler, and C. R. Parish.** 1994. Evidence for an association of CD45 with 32 000–33 000 MW phosphoproteins on murine T and B lymphocytes. *Immunology* **83**: 420–429.

411. **Bernard, G., D. Zoccola, M. Ticchioni, J. P. Breittmayer, C. Aussel, and A. Bernard.** 1994. Engagement of the CD45 molecule induces homotypic adhesion of human thymocytes through a LFA-1/ICAM-3-dependent pathway. *J. Immunol.* **152**: 5161–5170.

412*. **Brady-Kalnay, S. M., A. J. Flint, and N. K. Tonks.** 1993. Homophilic binding of the receptor-type protein tyrosine phosphatase, PTP-μ, can mediate cell–cell aggregation. *J. Cell Biol.* **122**: 961–972.

413. **Brown, V. K., E. W. Ogle, A. L. Burkhardt, R. B. Rowley, J. B. Bolen, and L. B. Justement.** 1994. Multiple components of the B cell antigen receptor complex associate with the protein tyrosine phosphatase, CD45. *J. Biol. Chem.* **269**: 17238–17244.

414*. **Case, R. D., E. Piccione, G. Wolf, A. M. Benett, R. J. Lechleider, B. G. Neel, and S. E. Shoelson.** 1994. SH-PTP2/Syp SH2 domain binding specificity is defined by direct interactions with platelet-derived growth factor β-receptor, epidermal growth factor receptor, and insulin receptor substrate-1-derived phosphopeptides. *J. Biol. Chem.* **269**: 10467–10474.

415. **Cayla, X., J. Goris, J. Hermann, P. Hendrix, R. Ozon, and W. Merlevede.** 1990. Isolation and characterization of a tyrosyl phosphatase activator from rabbit skeletal muscle and *Xenopus laevis* oocytes. *Biochemistry* **29**: 658–667.

416. **Cayla, X., C. Vanhoof, M. Bosch, E. Waelkens, J. Vandekerckhove, B. Peeters, W. Merlevede, and J. Goris.** 1994. Molecular cloning, expression, and characterization of PTPA, a protein that activates the tyrosyl phosphatase activity of protein phosphatase 2A. *J. Biol. Chem.* **269**: 15668–15675.

417. **Coombe, D. R., S. M. Watt, and C. R. Parish.** 1994. Mac-1 (CD11b/CD18) and CD45 mediate the adhesion of hematopoietic progenitor cells to stromal cell elements via recognition of stromal heparan sulfate. *Blood* **84**: 739–752.

418*. **Desai, D. M., J. Sap, J. Schlessinger, and A. Weiss.** 1993. Ligand-mediated negative regulation of a chimeric transmembrane receptor tyrosine phosphatase. *Cell* **73**: 541–554.

419. **Feng, G. S., R. Shen, H. H. Q. Heng, L. C. Tsui, A. Kazlauskas, and T. Pawson.** 1994. Receptor-binding, tyrosine phosphorylation and chromosome localization of the mouse SH2-containing phosphotyrosine phosphatase Syp. *Oncogene* **9**: 1545–1550.

420. **Furukawa, T., M. Itoh, N. X. Krueger, M. Streuli, and H. Saito.** 1994. Specific interaction of the CD45 protein-tyrosine phosphatase with tyrosine-phosphorylated CD3 ζ chain. *Proc. Natl Acad. Sci. USA* **91**: 10928–10932.

421*. **Gebbink, M. F. B. G., G. C. M. Zondag, R. W. Wubbolts, R. L. Beijersbergen, I. Van Etten, and W. H. Moolenaar.** 1993. Cell–Cell adhesion mediated by a receptor-like protein tyrosine phosphatase. *J. Biol. Chem.* **268**: 16101–16104.

422. **Goldman, S. J., S. Uniyal, L. M. Ferguson, D. E. Golan, S. J. Burakoff, and P. A. Kiener.** 1992. Differential activation of phosphotyrosine protein phosphatase activity in a murine T cell hybridoma by monoclonal antibodies to CD45. *J. Biol. Chem.* **267**: 6197–6204.

423. **Guttinger, M., M. Gassmann, K. E. Amrein, and P. Burn.** 1992. CD45 phosphotyrosine phosphatase and p56lck protein tyrosine kinase: a functional complex crucial in T cell signal transduction. *Int. Immunol.* **4**: 1325–1330.

424. **Habib, T., R. Herrera, and S. J. Decker.** 1994. Activators of protein kinase C stimulate association of Shc and the PEST tyrosine phosphatase. *J. Biol. Chem.* **269**: 25243–25246.

425. **Iida, N., V. B. Lokeshwar, and L. Y. W. Bourguignon.** 1994. Mapping the fodrin binding domain in CD45, a leukocyte membrane-associated tyrosine phosphatase. *J. Biol. Chem.* **269**: 28576–28583.

426. **Jessus, C. and D. Beach.** 1992. Oscillation of MPF is accompanied by periodic association between cdc25 and cdc2-cyclin B. *Cell* **68**: 323–332.

427. **Kazlauskas, A., G. S. Feng, T. Pawson, and M. Valius.** 1993. The 64-kDa protein that associates with the platelet-derived growth factor receptor β-subunit via Tyr-1009 is the SH2-containing phosphotyrosine phosphatase Syp. *Proc. Natl Acad. Sci. USA* **90**: 6939–6942.

428. **Kiener, P. A. and R. S. Mittler.** 1989. CD45-protein tyrosine phosphatase cross-linking inhibits T cell receptor CD3-mediated activation in human T cells. *J. Immunol.* **143**: 23–28.

429. **Kilshaw, P. J. and K. C. Baker.** 1989. A new antigenic determinant on intra-epithelial lymphocytes and its association with CD45. *Immunology* **67**: 160–166.

430*. **Kuhné, M. R., T. Pawson, G. E. Lienhard, and G. S. Feng.** 1993. The insulin receptor substrate-1 associates with the SH2-containing phosphotyrosine phosphatase Syp. *J. Biol. Chem.* **268**: 11479–11481.

431. **Lechleider, R. J., R. M. Freeman, and B. G. Neel.** 1993. Tyrosyl phosphorylation and growth factor receptor association of the human corkscrew homologue, SH-PTP2. *J. Biol. Chem.* **268**: 13434–13438.

432*. **Lechleider, R. J., S. Sugimoto, A. M. Bennett, A. S. Kashishian, J. A. Cooper, S. E. Shoelson, C. T. Walsh, and B. G. Neel.** 1993. Activation of the SH2-containing phosphotyrosine phosphatase SH-PTP2 by its binding site, phosphotyrosine-1009, on the human platelet-derived growth factor receptor β. *J. Biol. Chem.* **268**: 21478–21481.

433. **Ledbetter, J. A., G. L. Schieven, F. M. Uckun, and J. B. Imboden.** 1991. CD45 cross-linking regulates phospholipase-C activation and tyrosine phosphorylation of specific substrates in CD3/Ti-stimulated T-cells. *J. Immunol.* **146**: 1577–1583.

434. **Ledbetter, J. A., N. K. Tonks, E. H. Fischer, and E. A. Clark.** 1988. CD45 regulates signal transduction and lymphocyte activation by specific association with receptor molecules on T or B cells. *Proc. Natl Acad. Sci. USA* **85**: 8628–8632.

435*. **Li, W., R. Nishimura, A. Kashishian, A. G. Batzer, W. J. H. Kim, J. A. Cooper, and J. Schlessinger.** 1994. A new function for a phosphotyrosine phosphatase – linking Grb2-SOS to a receptor tyrosine kinase. *Mol. Cell. Biol.* **14**: 509–517.

436. **Lokeshwar, V. B. and L. Y. W. Bourguignon.** 1992. Tyrosine phosphatase activity of lymphoma CD45 (GP180) is regulated by a direct interaction with the cytoskeleton. *J. Biol. Chem.* **267**: 21551–21557.

437. **Lynes, M. A., D. J. Tibbetts, L. M. Swenson, and C. L. Sidman.** 1993. Dynamic associations of CD45 and Thy-1 on plasma membranes of C3H-gld/gld and C3H-lpr/lpr. I. Potential effects on proliferation and phosphatase activity. *Cell. Immunol.* **151**: 65–79.

438. **Maegawa, H., S. Ugi, O. Ishibashi, R. Tachikawaide, N. Takahara, Y. Tanaka, Y. Takagi, R. Kikkawa, Y. Shigeta, and A. Kashiwagi.** 1993. Src homology-2 domains of protein tyrosine phosphatase are phosphorylated by insulin receptor kinase and bind to the COOH-terminus of insulin receptors *in vitro*. *Biochem. Biophys. Res. Commun.* **194**: 208–214.

439. **Marie-Cardine, A., I. Maridonneau-Parini, and S. Fischer.** 1994. Activation and internalization of $p56^{lck}$ upon CD45 triggering of Jurkat cells. *Eur. J. Immunol.* **24**: 1255–1261.

440*. **Milarski, K. L., G. C. Zhu, C. G. Pearl, D. J. Mcnamara, E. M. Dobrusin, D. Maclean, A. Thiemesefler, Z. Y. Zhang, T. Sawyer, S. J. Decker, J. E. Dixon, and A. R. Saltiel.** 1993. Sequence specificity in recognition of the epidermal growth factor receptor by protein tyrosine phosphatase-1B. *J. Biol. Chem.* **268**: 23634–23639.

441. **Mittler, R. S., B. M. Rankin, and P. A. Kiener.** 1991. Physical associations between CD45 and CD4 or CD8 occur as late activation events in antigen receptor-stimulated human T cells. *J. Immunol.* **147**: 3434–3440.

442. **Oravecz, T., E. Monostori, O. Adrian, E. Kurucz, and I. Ando.** 1994. Novel heterogeneity of the leucocyte common antigen (CD45): disulfide-bound heterodimers between CD45 and an 80 kDa polypeptide. *Immunol. Lett.* **40**: 7–11.

443. **Ostergaard, H. L. and I. S. Trowbridge.** 1990. Coclustering CD45 with CD4 or CD8 alters the phosphorylation and kinase activity of p56lck. *J. Exp. Med.* **172**: 347–350.

444*. **Radha, V., S. Kamatkar, and G. Swarup.** 1993. Binding of a protein-tyrosine phosphatase to DNA through its carboxy-terminal noncatalytic domain. *Biochemistry* **32**: 2194–2201.

445. **Ross, S. E., B. Schraven, F. D. Goldman, J. Crabtree, and G. A. Koretzky.** 1994. The association between CD45 and lck does not require CD4 or CD8 and is independent of T cell receptor stimulation. *Biochem. Biophys. Res. Commun.* **198**: 88–96.

446. **Samelson, L. E., M. C. Fletcher, J. A. Ledbetter, and C. H. June.** 1990. Activation of tyrosine phosphorylation in human T cells via the CD2 pathway: regulation by the CD45 tyrosine phosphatase. *J. Immunol.* **145**: 2448–2454.

447*. **Sap, J., Y. P. Jiang, D. Friedlander, M. Grumet, and J. Schlessinger.** 1994. Receptor tyrosine phosphatase R-PTP-κ mediates homophilic binding. *Mol. Cell. Biol.* **14**: 1–9.

448. **Schraven, B., H. Kirchgessner, B. Gaber, Y. Samstag, and S. Meuer.** 1991. A functional complex is formed in human T lymphocytes between the protein tyrosine phosphatase CD45, the protein tyrosine kinase p56lck and pp32, a possible common substrate. *Eur. J. Immunol.* **21**: 2469–2477.

449. **Schraven, B., Y. Samstag, P. Altevogt, and S. C. Meuer.** 1990. Association of CD2 and CD45 on human T lymphocytes. *Nature* **345**: 71–74.

450. **Schraven, B., A. Schirren, H. Kirchgessner, B. Siebert, and S. C. Meuer.** 1992. Four CD45/P56lck-associated phosphorproteins (pp29-pp32) undergo alterations in human T cell activation. *Eur. J. Immunol.* **22**: 1857–1863.

451. **Schraven, B., D. Schoenhaut, E. Bruyns, G. Koretzky, C. Eckerskorn, R. Wallich, H. Kirchgessner, P. Sakorafas, B. Labkovsky, S. Ratnofsky, and S. Meuer.** 1994. LPAP, a novel 32-kDA phosphoprotein that interacts with CD45 in human lymphocytes. *J. Biol. Chem.* **269**: 29102–29111.

452. **Spertini, F., A. V. Wang, T. Chatila, and R. S. Geha.** 1994. Engagement of the common leukocyte antigen CD45 induces homotypic adhesion of activated human T cells. *J. Immunol.* **153**: 1593–1602.

453*. **Sugimoto, S., T. J. Wandless, S. E. Shoelson, B. G. Neel, and C. T. Walsh.** 1994. Activation of the SH2-containing protein tyrosine phosphatase, SH-PTP2, by phosphotyrosine-containing peptides derived from insulin receptor substrate-1. *J. Biol. Chem.* **269**: 13614–13622.

454. **Takeda, A., A. L. Maizel, K. Kitamura, T. Ohta, and S. Kimura.** 1994. Molecular cloning of the CD45–associated 30-kDa protein. *J. Biol. Chem.* **269**: 2357–2360.

455. **Takeda, A., J. J. Wu, and A. L. Maizel.** 1992. Evidence for monomeric and dimeric forms of CD45 associated with a 30-kDa phosphorylated protein. *J. Biol. Chem.* **267**: 16651–16659.

456. **Tauchi, T., G. S. Feng, M. S. Marshall, R. Shen, C. Mantel, T. Pawson, and H. E. Broxmeyer.** 1994. The ubiquitously expressed Syp phosphatase interacts with c-kit and Grb2 in hematopoietic cells. *J. Biol. Chem.* **269**: 25206–25211.

457. **Tauchi, T., G. S. Feng, R. Shen, H. Y. Song, D. Donner, T. Pawson, and H. E. Broxmeyer.** 1994. SH2-containing phosphotyrosine phosphatase Syp is a target of p210bcr-abl tyrosine kinase. *J. Biol. Chem.* **269**: 15381–15387.

458. **Torimoto, Y., N. H. Dang, E. Vivier, T. Tanaka, S. F. Schlossman, and C. Morimoto.** 1991. Coassociation of CD26 (dipeptidyl peptidase IV) with CD45 on the surface of human T lymphocytes. *J. Immunol.* **147**: 2514–2517.

459. **Ugi, S., H. Maegawa, J. M. Olefsky, Y. Shigeta, and A. Kashiwagi.** 1994. Src homology 2 domains of protein tyrosine phosphatase are associated *in vitro* with both the insulin receptor and insulin receptor substrate-1 via different phosphotyrosine motifs. *FEBS Lett.* **340**: 216–220.

460. **Volarevic, S., C. M. Burns, J. J. Sussman, and J. D. Ashwell.** 1990. Intimate association of Thy-1 and the T-cell antigen receptor with the CD45 tyrosine phosphatase. *Proc. Natl Acad. Sci. USA* **87**: 7085–7089.

461. **Wagner, N., P. Engel, and T. F. Tedder.** 1993. Regulation of the tyrosine kinase-dependent adhesion pathway in human lymphocytes through CD45. *J. Immunol.* **150**: 4887–4899.

462. **Welham, M. J., U. Dechert, K. B. Leslie, F. Jirik, and J. W. Schrader.** 1994. Interleukin (IL)-3 and granulocyte/macrophage colony-stimulating factor, but not IL-4, induce tyrosine phosphorylation, activation, and association of SH-PTP2 with Grb2 and phosphatidylinositol 3′-kinase. *J. Biol. Chem.* **269**: 23764–23768.

463*. **Yi, T. L. and J. N. Ihle.** 1993. Association of hematopoietic cell phosphatase with c-kit after stimulation with c-kit ligand. *Mol. Cell. Biol.* **13**: 3350–3358.

464. **Yi, T. L., A. L. F. Mui, G. Krystal, and J. N. Ihle.** 1993. Hematopoietic cell phosphatase associates with the interleukin-3 (IL-3) receptor-β chain and down-regulates IL-3-induced tyrosine phosphorylation and mitogenesis. *Mol. Cell. Biol.* **13**: 7577–7586.

465. **Zeggari, M., J. P. Esteve, I. Rauly, C. Cambillau, H. Mazarguil, M. Dufresne, L. Pradayrol, J. A. Chayvialle, N. Vaysse, and C. Susini.** 1994. Co-purification of a protein tyrosine phosphatase with activated somatostatin receptors from rat pancreatic acinar membranes. *Biochem. J.* **303**: 441–448.

466. **Zheng, X. F. and J. V. Ruderman.** 1993. Functional analysis of the P box, a domain in cyclin B required for the activation of Cdc25. *Cell* **75**: 155–164.

467. **Zocchi, M. R., A. Poggi, F. Crosti, S. Tongiani, and C. Rugarli.** 1992. Signalling in human tumour infiltrating lymphocytes: the CD28 molecule is functional and is physically associated with the CD45R0 molecule. *Eur. J. Cancer* 28A:749–754.

Sequences

468. **Adachi, M., T. Miyachi, M. Sekiya, Y. Hinoda, A. Yachi, and K. Imai.** 1994. Structure of the human LC-PTP (HePTP) gene: similarity in genomic organization within protein-tyrosine phosphatase genes. *Oncogene* **9**: 3031–3035.

469. **Adachi, M., M. Sekiya, Y. Arimura, M. Takekawa, F. Itoh, Y. Hinoda, K. Imai, and A. Yachi.** 1992. Protein-tyrosine phosphatase expression in pre-B-cell NALM-6. *Cancer Res.* **52**: 737–740.

470. **Adachi, M., M. Sekiya, M. Isobe, Y. Kumura, Z. Ogita, Y. Hinoda, K. Imai, and A. Yachi.** 1992. Molecular cloning and chromosomal mapping of a human protein-tyrosine phosphatase LC-PTP. *Biochem. Biophys. Res. Commun.* **186**: 1607–1615.

471. **Adachi, M., M. Sekiya, T. Miyachi, K. Matsuno, Y. Hinoda, K. Imai, and A. Yachi.** 1992. Molecular cloning of a novel protein-tyrosine phosphatase SH-PTP3 with sequence similarity to the src-homology region-2. *FEBS Lett.* **314**: 335–339.

472*. **Ahmad, S., D. Banville, Z. Z. Zhao, E. H. Fischer, and S. H. Shen.** 1993. A widely expressed human protein-tyrosine phosphatase containing src homology-2 domains. *Proc. Natl Acad. Sci. USA* **90**: 2197–2201.

473*. **Alphey, L., J. Jimenez, H. White-Cooper, I. Dawson, P. Nurse, and D. M. Glover.** 1992. *twine*, a cdc25 homolog that functions in the male and female germline of *Drosophila*. *Cell* **69**: 977–988.

474. **Arimura, Y., Y. Hinoda, F. Itoh, M. Takekawa, M. Tsujisaki, M. Adachi, K. Imai, and A. Yachi.** 1992. cDNA cloning of new protein tyrosine phosphatases in the human colon. *Tumor Biol.* **13**: 180–186.

475*. **Banville, D., S. Ahmad, R. Stocco, and S. H. Shen.** 1994. Novel protein-tyrosine phosphatase with homology to both the cytoskeletal proteins of the band 4.1 family and junction-associated guanylate kinases. *J. Biol. Chem.* **269**: 22320–22327.

476. **Barclay, A. N., D. I. Jackson, A. C. Willis, and A. F. Williams.** 1987. Lymphocyte specific heterogeneity in the rat leucocyte common antigen (T200) is due to differences in polypeptide sequences near the NH_2-terminus. *EMBO J.* **6**: 1259–1264.

477. **Barnea, G., M. Grumet, J. Sap, R. U. Margolis, and J. Schlessinger.** 1994. Close similarity between receptor-linked tyrosine phosphatase and rat brain proteoglycan. *Cell* **76**: 205

478*. **Barnea, G., O. Silvennoinen, B. Shaanan, A. M. Honegger, P. D. Canoll, P. D'Eustachio, B. Morse, J. B. Levy, S. Laforgia, K. Huebner, J. M. Musacchio, J. Sap, and J. Schlessinger.** 1993. Identification of a carbonic anhydrase-like domain in the extracellular region of RPTPγ defines a new subfamily of receptor tyrosine phosphatases. *Mol. Cell. Biol.* **13**: 1497–1506.

479. **Bastien, L., C. Ramachandran, S. Liu, and M. Adam.** 1993. Cloning, expression and mutational analysis of SH-PTP2, human protein-tyrosine phosphatase. *Biochem. Biophys. Res. Commun.* **196**: 124–133.

480. **Beckmann, G. and P. Bork.** 1993. An adhesive domain detected in functionally diverse receptors. *Trends Biochem. Sci.* **18**: 40–41.

481*. **Brown-Shimer, S., K. A. Johnson, J. B. Lawrence, C. Johnson, A. Bruskin, N. R. Green, and D. E. Hill.** 1990. Molecular cloning and chromosome mapping of the human gene encoding protein phosphotyrosyl phosphatase 1B. *Proc. Natl Acad. Sci. USA* **87**: 5148–5152.

482. **Camici, G., G. Manao, G. Cappugi, A. Modesti, M. Stefani, and G. Ramponi.** 1989. The complete amino acid sequence of the low molecular weight cytosolic acid phosphatase. *J. Biol. Chem.* **264**: 2560–2567.

483. **Caselli, A., L. Pazzagli, P. Paoli, G. Manao, G. Camici, G. Cappugi, and G. Ramponi.** 1994. Porcine liver low M_r phosphotyrosine protein phosphatase: the amino acid sequence. *J. Prot. Chem.* **13**: 107–115.

484*. **Charbonneau, H., N. K. Tonks, S. Kumar, C. D. Diltz, M. Harrylock, D. E. Cool, E. G. Krebs, E. H. Fischer, and K. A. Walsh.** 1989. Human placenta protein-tyrosine-phosphatase: amino acid sequence and relationship to a family of receptor-like proteins. *Proc. Natl Acad. Sci. USA* **86**: 5252–5256.

485. **Charbonneau, H., N. K. Tonks, K. A. Walsh, and E. H. Fischer.** 1988. The leukocyte common antigen (CD45): a putative receptor-linked protein tyrosine phosphatase. *Proc. Natl Acad. Sci. USA* **85**: 7182–7186.

486. **Charles, C. H., A. S. Abler, and L. F. Lau.** 1992. cDNA sequence of a growth-factor inducible immediate early gene and characterization of its encoded protein. *Oncogene* **7**: 187–190.

487*. **Chernoff, J., A. R. Schievella, C. A. Jost, R. L. Erikson, and B. G. Neel.** 1990. Cloning of a cDNA for a major human protein-tyrosine-phosphatase. *Proc. Natl Acad. Sci. USA* **87**: 2735–2739.

488*. **Cool, D. E., N. K. Tonks, H. Charbonneau, K. A. Walsh, E. H. Fischer, and E. G. Krebs.** 1989. cDNA isolated from a human T-cell library encodes a member of the protein-tyrosine phosphatase-family. *Proc. Natl Acad. Sci. USA* **86**: 5257–5261.

489. **Courtot, C., C. Fankhauser, V. Simanis, and C. F. Lehner.** 1992. The *Drosophila* cdc25 homolog twine is required for meiosis. *Development* **116**: 405–416.

490. **Del Vecchio, R. L. and N. K. Tonks.** 1994. Characterization of two structurally related *Xenopus laevis* protein tyrosine phosphatases with homology to lipid-binding proteins. *J. Biol. Chem.* **269**: 19639–19645.

491. **Demetrick, D. J. and D. H. Beach.** 1993. Chromosome mapping of human CDC25A and CDC25B phosphatases. *Genomics* **18**: 144–147.

492. **Devries, L., R. Y. Li, A. Ragab, J. M. F. Ragabthomas, and H. Chap.** 1991. Expression of a truncated protein-tyrosine phosphatase messenger RNA in human lung. *FEBS Lett.* **282**: 285–288.

493. **Diamond, R. H., D. E. Cressman, T. M. Laz, C. S. Abrams, and R. Taub.** 1994. PRL-1, a unique nuclear protein tyrosine phosphatase, affects cell growth. *Mol. Cell. Biol.* **14**: 3752–3762.

494. **Dimartino, J. F., P. Hayes, Y. Saga, and J. S. Lee.** 1994. A novel initiator/promoter element within the CD45 upstream region. *Int. Immunol.* **6**: 1279–1283.

495. **Dissing, J. and A. H. Johnsen.** 1992. Human red cell acid phosphatase (ACP1): the primary structure of the two pairs of isozymes encoded by the ACP1*A and ACP1*C alleles. *Biochim. Biophys. Acta* **1121**: 261–268.

496. **Dissing, J., A. H. Johnsen, and G. F. Sensabaugh.** 1991. Human red cell acid phosphatase (ACP1). The amino acid sequence of the two isozymes Bf and Bs encoded by the ACP1*B allele. *J. Biol. Chem.* **266**: 20619–20625.

497. **Doi, K., A. Gartner, G. Ammerer, B. Errede, H. Shinkawa, K. Sugimoto, and K. Matsumoto.** 1994. MSG5, a novel protein phosphatase promotes adaptation to pheromone response in *S. cerevisiae*. *EMBO J.* **13**: 61–70.

498. **Edgar, B. A. and P. H. O'Farrell.** 1989. Genetic control of cell division patterns in the *Drosophila* embryo. *Cell* **57**: 177–187.

499. **Emslie, E. A., T. A. Jones, D. Sheer, and S. M. Keyse.** 1994. The Cl100 gene, which encodes a dual specificity (Tyr/Thr) map kinase phosphatase, is highly conserved and maps to human chromosome 5Q34. *Hum. Genet.* **93**: 513–516.

500. **Fang, K. S., K. Barker, M. Sudol, and H. Hanafusa.** 1994. A transmembrane protein-tyrosine phosphatase contains spectrin-like repeats in its extracellular domain. *J. Biol. Chem.* **269**: 14056–14063.

501*. **Feng, G. S., C. C. Hui, and T. Pawson.** 1993. SH2-containing phosphotyrosine phosphatase as a target of protein-tyrosine kinases. *Science* **259**: 1607–1611.

502*. **Freeman, R. M., J. Plutzky, and B. G. Neel.** 1992. Identification of a human src homology 2-containing protein-tyrosine-phosphatase – a putative homolog of *Drosophila* corkscrew. *Proc. Natl Acad. Sci. USA* **89**: 11239–11243.

503. **Galaktionov, K. and D. Beach.** 1991. Specific activation of cdc25 tyrosine phosphatases by B-type cyclins – evidence for multiple roles of mitotic cyclins. *Cell* **67**: 1181–1194.

504. **Gebbink, M. F. B. G., I. Van Etten, G. Hateboer, R. Suijkerbuijk, R. L. Beijersbergen, A. G. Vankessel, and W. H. Moolenaar.** 1991. Cloning, expression and chromosomal localization of a new putative receptor-like protein tyrosine phosphatase. *FEBS Lett.* **290**: 123–130.

505. **Gerondakis, S., C. Economou, and R. J. Grumont.** 1994. Structure of the gene encoding the murine dual specificity tyrosine-threonine phosphatase PAC1. *Genomics* **24**: 182–184.

506. **Goldner-Sauve, A., C. Szpirer, J. Szpirer, G. Levan, and D. L. Gasser.** 1991. Chromosome assignments of the genes for glucocorticoid receptor, myelin basic protein, leukocyte common antigen, and TRPM2 in the rat. *Biochem. Genet.* **29**: 275–286.

507*. **Gu, M. X., I. Warshawsky, and P. W. Majerus.** 1992. Cloning and expression of a cytosolic megakaryocyte protein-tyrosine phosphatase with sequence homology to retinaldehyde-binding protein and yeast SEC14p. *Proc. Natl Acad. Sci. USA* **89**: 2980–2984.

508. **Gu, M. X., J. D. York, I. Warshawsky, and P. W. Majerus.** 1991. Identification, cloning, and expression of a cytosolic megakaryocyte protein-tyrosine-phosphatase with sequence homology to cytoskeletal protein 4.1. *Proc. Natl Acad. Sci. USA* **88**: 5867–5871.

509. **Guan, K., R. J. Deschenes, H. Qiu, and J. E. Dixon.** 1991. Cloning and expression of a yeast protein tyrosine phosphatase. *J. Biol. Chem.* **266**: 12964–12970.

510. **Guan, K., D. J. Hakes, Y. Wang, H. D. Park, T. G. Cooper, and J. E. Dixon.** 1992. A yeast protein phosphatase related to the Vaccinia virus VH1 phosphatase is induced by nitrogen starvation. *Proc. Natl Acad. Sci. USA* **89**: 12175–12179.

511. **Guan, K. L., S. S. Broyles, and J. E. Dixon.** 1991. A Tyr/Ser protein phosphatase encoded by Vaccinia virus. *Nature* **350**: 359–362.

512. **Guan, K. L., R. J. Deschenes, and J. E. Dixon.** 1992. Isolation and characterization of a 2nd protein tyrosine phosphatase gene, PTP2, from *Saccharomyces cerevisiae*. *J. Biol. Chem.* **267**: 10024–10030.

513*. **Guan, K. L. and J. E. Dixon.** 1990. Protein tyrosine phosphatase activity of an essential virulence determinant in *Yersinia*. *Science* **249**: 553–556.

514. **Guan, K. L., R. S. Haun, S. J. Watson, R. L. Geahlen, and J. E. Dixon.** 1990. Cloning and expression of a protein-tyrosine-phosphatase. *Proc. Natl Acad. Sci. USA* **87**: 1501–1505.

515. **Gyuris, J., E. Golemis, H. Chertkov, and R. Brent.** 1993. Cdi1, a human G1 and S phase protein phosphatase that associates with Cdk2. *Cell* **75**: 791–803.

516. **Hakes, D. J., K. J. Martell, W. G. Zhao, R. F. Massung, J. J. Esposito, and J. E. Dixon.** 1993. A protein phosphatase related to the Vaccinia virus VH1 is encoded in the genomes of several orthopoxviruses and a baculovirus. *Proc. Natl Acad. Sci. USA* **90**: 4017–4021.

517*. **Hall, L. R., M. Streuli, S. F. Schlossman, and H. Saito.** 1988. Complete exon–intron organization of the human leukocyte common antigen (CD45) gene. *J. Immunol.* **141**: 2781–2787.

518. **Hannon, G. J., D. Casso, and D. Beach.** 1994. Kap – a dual specificity phosphatase that interacts with cyclin-dependent kinases. *Proc. Natl Acad. Sci. USA* **91**: 1731–1735.

519. **Harder, K. W., L. L. Anderson, A. M. V. Duncan, and F. R. Jirik.** 1992. The gene for receptor-like protein tyrosine phosphatase (PTPRB) is assigned to chromosome-12q15 → q21. *Cytogenet. Cell Genet.* **61**: 269–270.

520. **Hariharan, I. K., P. T. Chuang, and G. M. Rubin.** 1991. Cloning and characterization of a receptor-class phosphotyrosine phosphatase gene expressed on central nervous system axons in *Drosophila melanogaster*. *Proc. Natl Acad. Sci. USA* **88**: 11266–11270.

521. **Hasegawa, K., T. Ariyama, J. Inazawa, K. Mizuno, M. Ogimoto, T. Katagiri, and H. Yakura.** 1993. Chromosomal assignment of the gene for protein tyrosine phosphatase HPTP δ. *Jpn. J. Cancer Res.* **84**: 1219–1222.

522. **Hiraga, A., H. Munakata, K. Hata, Y. Suzuki, and S. Tsuiki.** 1992. Purification and characterization of a rat liver protein-tyrosine phosphatase with sequence similarity to src-homology region-2. *Eur. J. Biochem.* **209**: 195–206.

523. **Howard, P. K., M. Gamper, T. Hunter, and R. A. Firtel.** 1994. Regulation by protein-tyrosine phosphatase PTP2 is distinct from that by PTP1 during *Dictyostelium* growth and development. *Mol. Cell. Biol.* **14**: 5154–5164.

524. **Ishibashi, T., D. P. Bottaro, A. Chan, T. Miki, and S. A. Aaronson.** 1992. Expression cloning of a human dual-specificity phosphatase. *Proc. Natl Acad. Sci. USA* **89**: 12170–12174.

525. **Ishibashi, T., D. P. Bottaro, P. Michieli, C. A. Kelley, and S. A. Aaronson.** 1994. A novel dual specificity phosphatase induced by serum stimulation and heat shock. *J. Biol. Chem.* **269**: 29897–29902.

526. **Isobe, M., Y. Hinoda, K. Imai, and M. Adachi.** 1994. Chromosomal localization of an SH2 containing tyrosine phosphatase (SH-PTP3) gene to chromosome 12q24.1. *Oncogene* **9**: 1751–1753.

527. **Itoh, F., S. Ikuta, Y. Hinoda, Y. Arimura, M. Ohe, M. Adachi, T. Ariyama, J. Inazawa, K. Imai, and A. Yachi.** 1993. Expression and chromosomal assignment of PTPH1 gene encoding a cytosolic protein tyrosine phosphatase homologous to cytoskeletal-associated proteins. *Int. J. Cancer* **55**: 947–951.

528. **James, P., B. D. Hall, S. Whelen, and E. A. Craig.** 1992. Multiple protein tyrosine phosphatase-encoding genes in the yeast *Saccharomyces cerevisiae*. *Gene* **122**: 101–110.

529*. **Jiang, Y. P., H. Wang, P. D'Eustachio, J. M. Musacchio, J. Schlessinger, and J. Sap.** 1993. Cloning and characterization of R-PTP-κ, a new member of the receptor protein tyrosine phosphatase family with a proteolytically cleaved cellular adhesion molecule-like extracellular region. *Mol. Cell. Biol.* **13**: 2942–2951.

530. **Jimenez, J., L. Alphey, P. Nurse, and D. M. Glover.** 1990. Complementation of fission yeast cdc2ts and cdc25ts mutants identifies two cell cycle genes from *Drosophila*: a cdc2 homologue and *string*. *EMBO J.* **9**: 3565–3571.

531. **Jirik, F. R., L. L. Anderson, and A. M. V. Duncan.** 1992. The human protein-tyrosine phosphatase PTPα/LRP gene is assigned to chromosome-20p13. *Cytogenet. Cell Genet.* **60**: 117–118.

532. **Jirik, F. R., K. W. Harder, I. G. Melhado, L. L. Anderson, and A. M. V. Duncan.** 1992. The gene for leukocyte antigen-related tyrosine phosphatase (LAR) is localized to human chromosome-1p32, a region frequently deleted in tumors of neuroectodermal origin. *Cytogenet. Cell Genet.* **61**: 266–268.

533. **Jirik, F. R., N. M. Janzen, I. G. Melhado, and K. W. Harder.** 1990. Cloning and chromosomal assignment of a widely expressed human receptor-like protein-tyrosine phosphatase. *FEBS Lett.* **273**: 239–242.

534. **Johnson, C. V., D. E. Cool, M. B. Glaccum, N. Green, E. H. Fischer, A. Bruskin, D. E. Hill, and J. B. Lawrence.** 1993. Isolation and mapping of human T-cell protein tyrosine phosphatase sequences – localization of genes and pseudogenes discriminated using fluorescence hybridization with genomic versus cDNA probes. *Genomics* **16**: 619–629.

535. **Kakizuka, A., B. Sebastian, U. Borgmeyer, I. Hermans-Borgmeyer, J. Bolado, T. Hunter, M. F. Hoekstra, and R. M. Evans.** 1992. A mouse cdc25 homolog is differentially and developmentally expressed. *Genes Dev.* **6**: 578–590.

536. **Kaplan, R., B. Morse, K. Huebner, C. Croce, R. Howk, M. Ravera, G. Ricca, M. Jaye, and J. Schlessinger.** 1990. Cloning of three human tyrosine phosphatases reveals a multigene family of receptor-linked protein-tyrosine-phosphatases expressed in brain. *Proc. Natl Acad. Sci. USA* **87**: 7000–7004.

537. **Keyse, S. M. and E. A. Emslie.** 1992. Oxidative stress and heat shock induce a human gene encoding a protein-tyrosine phosphatase. *Nature* **359**: 644–647.

538. **Kim, D. and R. F. Weaver.** 1993. Transcription mapping and functional analysis of the protein tyrosine/serine phosphatase (PTPase) gene of the *Autographa californica* nuclear polyhedrosis virus. *Virology* **195**: 587–595.

539. **Krueger, N. X. and H. Saito.** 1992. A human transmembrane protein-tyrosine-phosphatase, PTPζ, is expressed in brain and has an *N*-terminal receptor domain homologous to carbonic anhydrases. *Proc. Natl Acad. Sci. USA* **89**: 7417–7421.

540*. **Krueger, N. X., M. Streuli, and H. Saito.** 1990. Structural diversity and evolution of human receptor-like protein tyrosine phosphatases. *EMBO J.* **9**: 3241–3252.

541. **Kwak, S. P., D. J. Hakes, K. J. Martell, and J. E. Dixon.** 1994. Isolation and characterization of a human dual specificity protein-tyrosine phosphatase gene. *J. Biol. Chem.* **269**: 3596–3604.

542*. **Laforgia, S., B. Morse, J. Levy, G. Barnea, L. A. Cannizzaro, F. Li, P. C. Nowell, L. Boghosiansell, J. Glick, A. Weston, C. C. Harris, H. Drabkin, D. Patterson, C. M. Croce, J. Schlessinger, and K. Huebner.** 1991. Receptor protein-tyrosine phosphatase-γ is a candidate tumor suppressor gene at human chromosome region 3p21. *Proc. Natl Acad. Sci. USA* **88**: 5036–5040.

543. **Lan, M. S., J. Lu, Y. Goto, and A. L. Notkins.** 1994. Molecular cloning and identification of a receptor-type protein tyrosine phosphatase, IA-2, from human insulinoma. *DNA Cell Biol.* **13**: 505–514.

544. **Latif, F., K. Tory, W. Modi, L. Geil, S. Laforgia, K. Huebner, B. Zbar, and M. I. Lerman.** 1993. A MspI polymorphism and linkage mapping of the human protein-tyrosine phosphatase G (PTPRG) gene. *Hum. Mol. Genet.* **2**: 91.

545. **Lazaruk, K. D., J. Dissing, and G. F. Sensabaugh.** 1993. Exon structure at the human ACP1 locus supports alternative splicing model for f and s isozyme generation. *Biochem. Biophys. Res. Commun.* **196**: 440–446.

546*. **Levy, J. B., P. D. Canoll, O. Silvennoinen, G. Barnea, B. Morse, A. M. Honegger, J. T. Huang, L. A. Cannizzaro, S. H. Park, T. Druck, K. Huebner, J. Sap, M. Ehrlich, J. M. Musacchio, and J. Schlessinger.** 1993. The cloning of a receptor-type protein tyrosine phosphatase expressed in the central nervous system. *J. Biol. Chem.* **268**: 10573–10581.

547. **Lombroso, P. J., G. Murdoch, and M. Lerner.** 1991. Molecular characterization of a protein-tyrosine-phosphatase enriched in striatum. *Proc. Natl Acad. Sci. USA* **88**: 7242–7246.

548. **Lu, J., A. L. Notkins, and M. S. Lan.** 1994. Isolation, sequence and expression of a novel mouse brain cDNA, mIA-2, and its relatedness to members of the protein tyrosine phosphatase family. *Biochem. Biophys. Res. Commun.* **204**: 930–936.

549. **Maekawa, K., N. Imagawa, M. Nagamatsu, and S. Harada.** 1994. Molecular cloning of a novel protein-tyrosine phosphatase containing a membrane-binding domain and Glgf repeats. *FEBS Lett.* **337**: 200–206.

550. **Manao, G., L. Pazzagli, P. Cirri, A. Caselli, G. Camici, G. Cappugi, A. Saeed, and G. Ramponi.** 1992. Rat liver low M_r phosphotyrosine protein phosphatase isoenzymes: purification and amino acid sequences. *J. Prot. Chem.* **11**: 333–345.

551. **Martell, K. J., S. Kwak, D. J. Hakes, J. E. Dixon, and J. M. Trent.** 1994. Chromosomal localization of four human VH1-like protein-tyrosine phosphatases. *Genomics* **22**: 462–464.

552. **Matozaki, T., T. Suzuki, T. Uchida, J. Inazawa, T. Ariyama, K. Matsuda, K. Horita, H. Noguchi, H. Mizuno, C. Sakamoto, and M. Kasuga.** 1994. Molecular cloning of a human transmembrane-type protein tyrosine phosphatase and its expression in gastrointestinal cancers. *J. Biol. Chem.* **269**: 2075–2081.

553*. **Matthews, R. J., D. B. Bowne, E. Flores, and M. L. Thomas.** 1992. Characterization of hematopoietic intracellular protein tyrosine phosphatases: description of a phosphatase containing an SH2 domain and another enriched in proline-, glutamic acid-, serine-, and threonine-rich sequences. *Mol. Cell. Biol.* **12**: 2396–2405.

554*. **Matthews, R. J., E. D. Cahir, and M. L. Thomas.** 1990. Identification of an additional member of the protein-tyrosine-phosphatase family: evidence for alternative splicing in the tyrosine phosphatase domain. *Proc. Natl Acad. Sci. USA* **87**: 4444–4448.

555. **Matthews, R. J., E. Flores, and M. L. Thomas.** 1991. Protein tyrosine phosphatase domains from the protochordate *Styela plicata*. *Immunogenetics* **33**: 33–41.

556*. **Maurel, P., U. Rauch, M. Flad, R. K. Margolis, and R. U. Margolis.** 1994. Phosphacan, a chondroitin sulfate proteoglycan of brain that interacts with neurons and neural cell-adhesion molecules, is an extracellular variant of a receptor-type protein tyrosine phosphatase. *Proc. Natl Acad. Sci. USA* **91**: 2512–2516.

557*. **Mauro, L. J., E. A. Olmsted, B. M. Skrobacz, R. J. Mourey, A. R. Davis, and J. E. Dixon.** 1994. Identification of a hormonally regulated protein tyrosine phosphatase associated with bone and testicular differentiation. *J. Biol. Chem.* **269**: 30659–30667.

558*. **Mclaughlin, S. and J. E. Dixon.** 1993. Alternative splicing gives rise to a nuclear protein tyrosine phosphatase in *Drosophila*. *J. Biol. Chem.* **268**: 6839–6842.

559. **Michiels, T. and G. Cornelis.** 1988. Nucleotide sequence and transcription analysis of yop51 from *Yersinia enterocolitica* W22703. *Microbial Pathogenesis* **5**: 449–459.

560. **Millar, J. B., G. Lenaers, and P. Russell.** 1992. Pyp3 PTPase acts as a mitotic inducer in fission yeast. *EMBO J.* **11**: 4933–4941.

561. **Millar, J. B., P. Russell, J. E. Dixon, and K. L. Guan.** 1992. Negative regulation of mitosis by 2 functionally overlapping PTPases in fission yeast. *EMBO J.* **11**: 4943–4952.

562. **Miyasaka, H. and S. S. L. Li.** 1992. The cDNA cloning, nucleotide sequence and expression of an intracellular protein tyrosine phosphatase from mouse testis. *Biochem. Biophys. Res. Commun.* **185**: 818–825.

563*. **Mizuno, K., K. Hasegawa, T. Katagiri, M. Ogimoto, T. Ichikawa, and H. Yakura.** 1993. MPTP-δ, a putative murine homolog of HPTP-δ, is expressed in specialized regions of the brain and in the B-cell lineage. *Mol. Cell. Biol.* **13**: 5513–5523.

564*. **Moller, N. P. H., K. B. Moller, R. Lammers, A. Kharitonenkov, I. Sures, and A. Ullrich.** 1994. Src kinase associates with a member of a distinct subfamily of protein-tyrosine phosphatases containing an ezrin-like domain. *Proc. Natl Acad. Sci. USA* **91**: 7477–7481.

565. **Moriyama, T., Y. Fujiwara, E. Imai, M. Takenaka, S. Kawanishi, T. Inoue, T. Noguchi, T. Tanaka, T. Kamada, and N. Ueda.** 1992. cDNA cloning of rat LRP, a receptor like protein tyrosine phosphatase, and evidence for its gene regulation in cultured rat mesangial cells. *Biochem. Biophys. Res. Commun.* **188**: 34–39.

566. **Moriyama, T., S. Kawanishi, T. Inoue, E. Imai, T. Kaneko, C. Xia, M. Takenaka, T. Noguchi, T. Kamada, and N. Ueda.** 1994. cDNA cloning of a cytosolic protein tyrosine phosphatase (RKPTP) from rat kidney. *FEBS Lett.* **353**: 305–308.

567. **Mosinger, B., U. Tillmann, H. Westphal, and M. L. Tremblay.** 1992. Cloning and characterization of a mouse cDNA encoding a cytoplasmic protein-tyrosine-phosphatase. *Proc. Natl Acad. Sci. USA* **89**: 499–503.

568. **Nagata, A., M. Igarashi, S. Jinno, K. Suto, and H. Okayama.** 1991. An additional homolog of the fission yeast cdc25+ gene occurs in humans and is highly expressed in some cancer cells. *New Biologist* **3**: 959–968.

569. **Nargi, J. L. and T. A. Woodford-Thomas.** 1994. Cloning and characterization of a cdc25 phosphatase from mouse lymphocytes. *Immunogenetics* **39**: 99–108.

570. **Nishi, M., S. Ohagi, and D. F. Steiner.** 1990. Novel putative protein tyrosine phosphatases identified by the polymerase chain reaction. *FEBS Lett.* **271**: 178–180.

571. **Noguchi, T., R. Metz, L. H. Chen, M. G. Mattei, D. Carrasco, and R. Bravo.** 1993. Structure, mapping, and expression of *erp*, a growth factor-inducible gene encoding a nontransmembrane protein tyrosine phosphatase, and effect of ERP on cell growth. *Mol. Cell. Biol.* **13**: 5195–5205.

572. **O'Connell, M. J., A. H. Osmani, N. R. Morris, and S. A. Osmani.** 1992. An extra copy of nimE/cyclinB elevates pre-MPF levels and partially suppresses mutation of nimT/cdc25 in *Aspergillus nidulans*. *EMBO J.* **11**: 2139–2149.

573*. **O'Grady, P., N. X. Krueger, M. Streuli, and H. Saito.** 1994. Genomic organization of the human LAR protein tyrosine phosphatase gene and alternative splicing in the extracellular fibronectin type-III domains. *J. Biol. Chem.* **269**: 25193–25199.

574. **Ohagi, S., M. Nishi, and D. F. Steiner.** 1990. Sequence of a cDNA encoding human LRP (leukocyte common antigen-related peptide). *Nucl. Acids Res.* **18**: 7159

575. **Oon, S. H., A. Hong, X. H. Yang, and W. Chia.** 1993. Alternative splicing in a novel tyrosine phosphatase gene (DPTP4E) of *Drosophila melanogaster* generates two large receptor-like proteins which differ in their carboxyl termini. *J. Biol. Chem.* **268**: 23964–23971.

576*. **Ostman, A., Q. Yang, and N. K. Tonks.** 1994. Expression of DEP-1, a receptor-like protein-tyrosine-phosphatase, is enhanced with increasing cell density. *Proc. Natl Acad. Sci. USA* **91**: 9680–9684.

577. **Ottilie, S., J. Chernoff, G. Hannig, C. S. Hoffman, and R. L. Erikson.** 1991. A fission-yeast gene encoding a protein with features of protein-tyrosine-phosphatases. *Proc. Natl Acad. Sci. USA* **88**: 3455–3459.

578*. **Ottilie, S., J. Chernoff, G. Hannig, C. S. Hoffman, and R. L. Erikson.** 1992. The fission yeast genes *pyp1+* and *pyp2+* encode protein tyrosine phosphatases that negatively regulate mitosis. *Mol. Cell. Biol.* **12**: 5571–5580.

579. **Pan, M. G., C. Rim, K. P. Lu, T. Florio, and P. J. S. Stork.** 1993. Cloning and expression of 2 structurally distinct receptor-linked protein-tyrosine phosphatases generated by RNA processing from a single gene. *J. Biol. Chem.* **268**: 19284–19291.

580*. **Perkins, L. A., I. Larsen, and N. Perrimon.** 1992. *corkscrew* encodes a putative protein tyrosine phosphatase that functions to transduce the terminal signal from the receptor tyrosine kinase torso. *Cell* **70**: 225–236.

581. **Perrimon, N., L. Engstrom, and A. P. Mahowald.** 1985. Developmental genetics of the 2C-D region of the *Drosophila* X chromosome. *Genetics* **111**: 23–41.

582*. **Plutzky, J., B. G. Neel, and R. D. Rosenberg.** 1992. Isolation of a src homology 2-containing tyrosine phosphatase. *Proc. Natl Acad. Sci. USA* **89**: 1123–1127.

583. **Plutzky, J., B. G. Neel, R. D. Rosenberg, R. L. Eddy, M. G. Byers, S. Janisait, and T. B. Shows.** 1992. Chromosomal localization of an SH2-containing tyrosine phosphatase (PTPN6). *Genomics* **13**: 869–872.

584. **Potts, M., H. Sun, K. Mockaitis, P. J. Kennelly, D. Reed, and N. K. Tonks.** 1993. A protein-tyrosine/serine phosphatase encoded by the genome of the Cyanobacterium Nostoc-Commune UTEX-584. *J. Biol. Chem.* **268**: 7632–7635.

585. **Pulido, R., S. F. Schlossman, H. Saito, and M. Streuli.** 1994. Identification of amino acids at the junction of exons 3 and 7 that are used for the generation of glycosylation-related human Cd45Ro and Cd45Ro-like antigen specificities. *J. Exp. Med.* **179**: 1035–1040.

586. **Rabin, D. U., S. M. Pleasic, J. A. Shapiro, H. Yoowarren, J. Oles, J. M. Hicks, D. E. Goldstein, and P. M. M. Rae.** 1994. Islet cell antigen 512 is a diabetes-specific islet autoantigen related to protein tyrosine phosphatases. *J. Immunol.* **152**: 3183–3188.

587. **Ralph, S. J., M. L. Thomas, C. C. Morton, and I. S. Trowbridge.** 1987. Structural variants of human T200 glycoprotein (leukocyte-common antigen). *EMBO J.* **6**: 1251–1257.

588. **Ramalingam, R., D. R. Shaw, and H. L. Ennis.** 1993. Cloning and functional expression of a *Dictyostelium discoideum* protein tyrosine phosphatase. *J. Biol. Chem.* **268**: 22680–22685.

589*. **Rohan, P. J., P. Davis, C. A. Moskaluk, M. Kearns, H. Krutzsch, U. Siebenlist, and K. Kelly.** 1993. PAC-1 – a mitogen-induced nuclear protein tyrosine phosphatase. *Science* **259**: 1763–1766.

590. **Russell, P., S. Moreno, and S. I. Reed.** 1989. Conservation of mitotic controls in fission and budding yeasts. *Cell* **57**: 295–303.

591. **Russell, P. and P. Nurse.** 1986. cdc25+ functions as an inducer in the mitotic control of fission yeast. *Cell* **45**: 145–153.

592. **Saga, Y., J. S. Tung, F. W. Shen, and E. A. Boyse.** 1986. Sequences of Ly-5 cDNA: isoform-related diversity of Ly-5 mRNA [published erratum appears in *Proc. Natl Acad. Sci. USA* 1987, **84**(7):1991]. *Proc. Natl Acad. Sci. USA* **83**: 6940–6944.

593. **Saga, Y., J. S. Tung, F. W. Shen, T. C. Pancoast, and E. A. Boyse.** 1988. Organization of the Ly-5 gene. *Mol. Cell. Biol.* **8**: 4889–4895.

594. **Sakaguchi, A. Y., V. L. Sylvia, L. Martinez, E. A. Smith, E. S. Han, P. A. Lalley, T. B. Shows, and G. G. Choudhury.** 1992. Assignment of tyrosine-specific T-cell phosphatase to conserved syntenic groups on human chromosome-18 and mouse chromosome-18. *Genomics* **12**: 151–154.

595*. **Sap, J., P. D'Eustachio, D. Givol, and J. Schlessinger.** 1990. Cloning and expression of a widely expressed receptor tyrosine phosphatase. *Proc. Natl Acad. Sci. USA* **87**: 6112–6116.

596. **Saras, J., L. Claessonwelsh, C. H. Heldin, and L. J. Gonez.** 1994. Cloning and characterization of PTPL1, a protein tyrosine phosphatase with similarities to cytoskeletal-associated proteins. *J. Biol. Chem.* **269**: 24082–24089.

597. **Sawada, M., M. Ogata, Y. Fujino, and T. Hamaoka.** 1994. cDNA cloning of a novel protein tyrosine phosphatase with homology to cytoskeletal protein 4.1 and its expression in T-lineage cells. *Biochem. Biophys. Res. Commun.* **203**: 479–484.

598. **Schafer, H., D. Baker, B. Thiele, and R. Burger.** 1990. Structure, cellular distribution, and functional characteristics of the guinea pig leucocyte common antigen. *Cell. Immunol.* **128**: 370–384.

599. **Seki, T., K. Yamashita, H. Nishitani, T. Takagi, P. Russell, and T. Nishimoto.** 1992. Chromosome condensation caused by loss of RCC1 function requires the cdc25C protein that is located in the cytoplasm. *Mol. Biol. Cell* **3**: 1373–1388.

600. **Shekels, L. L., A. J. Smith, R. L. Van Etten, and D. A. Bernlohr.** 1992. Identification of the adipocyte acid phosphatase as a PAO-sensitive tyrosyl phosphatase. *Prot. Science* **1**: 710–721.

601. **Shen, F. W., Y. Saga, G. Litman, G. Freeman, J. S. Tung, H. Cantor, and E. A. Boyse.** 1985. Cloning of Ly-5 cDNA. *Proc. Natl Acad. Sci. USA* **82**: 7360–7363.

602*. **Shen, S. H., L. Bastien, B. I. Posner, and P. Chretien.** 1991. A protein-tyrosine phosphatase with sequence similarity to the SH2 domain of the protein-tyrosine kinases. *Nature* **352**: 736–739.

603*. **Shultz, L. D., P. A. Schweitzer, T. V. Rajan, T. L. Yi, J. N. Ihle, R. J. Matthews, M. L. Thomas, and D. R. Beier.** 1993. Mutations at the murine motheaten locus are within the hematopoietic cell protein-tyrosine phosphatase (Hcph) gene. *Cell* **73**: 1445–1454.

604. **Stoker, A. W.** 1994. Isoforms of a novel cell adhesion molecule-like protein tyrosine phosphatase are implicated in neural development. *Mech. Develop.* **46**: 201–217.

605*. **Streuli, M., N. X. Krueger, L. R. Hall, S. F. Schlossman, and H. Saito.** 1988. A new member of the immunoglobulin superfamily that has a cytoplasmic region homologous to the leukocyte common antigen. *J. Exp. Med.* **168**: 1523–1530.

606*. **Streuli, M., N. X. Krueger, A. Y. Tsai, and H. Saito.** 1989. A family of receptor-linked protein tyrosine phosphatases in humans and *Drosophila*. *Proc. Natl Acad. Sci. USA* **86**: 8698–8702.

607. **Suijkerbuijk, R. F., M. F. B. G. Gebbink, W. H. Moolenaar, and A. G. Vankessel.** 1993. Fine mapping of the human receptor-like protein tyrosine phosphatase gene (PTPRM) to 18p11.2 by fluorescence *in situ* hybridization. *Cytogenet. Cell Genet.* **64**: 245–246.

608. **Swarup, G., S. Kamatkar, V. Radha, and V. Rema.** 1991. Molecular cloning and expression of a protein-tyrosine phosphatase showing homology with transcription factors Fos and Jun. *FEBS Lett.* **280**: 65–69.

609. **Takekawa, M., F. Itoh, Y. Hinoda, M. Adachi, T. Ariyama, J. Inazawa, K. Imai, and A. Yachi.** 1994. Chromosomal localization of the protein tyrosine phosphatase G1 gene and characterization of the aberrant transcripts in human colon cancer cells. *FEBS Lett.* **339**: 222–228.

610. **Takekawa, M., F. Itoh, Y. Hinoda, Y. Arimura, M. Toyota, M. Sekiya, M. Adachi, K. Imai, and A. Yachi.** 1992. Cloning and characterization of a human cDNA encoding a novel putative cytoplasmic protein-tyrosine-phosphatase. *Biochem. Biophys. Res. Commun.* **189**: 1223–1230.

611. **Thomas, M. L., A. N. Barclay, J. Gagnon, and A. F. Williams.** 1985. Evidence from cDNA clones that rat leukocyte common antigen (T200) spans the lipid bilayer and contains a cytoplasmic domain of 80 000 M_r. *Cell* **41**: 83–93.

612. **Thomas, M. L., D. B. Bowne, E. Cahir-McFarland, E. Flores, R. J. Matthews, J. T. Pingel, G. Roy, A. Shaw, and H. Shenoi.** 1993. A diversity of protein tyrosine phosphatases expressed by T lymphocytes. *Prog. Immunol.* **8**: 213–219.

613. **Thomas, P. E., B. L. Wharram, M. Goyal, J. E. Wiggins, L. B. Holzman, and R. C. Wiggins.** 1994. GLEPP1, a renal glomerular epithelial cell (podocyte) membrane protein-tyrosine phosphatase – identification, molecular cloning, and characterization in rabbit. *J. Biol. Chem.* **269**: 19953–19961.

614*. **Tian, S. S., P. Tsoulfas, and K. Zinn.** 1991. Three receptor-linked protein-tyrosine phosphatases are selectively expressed on central nervous system axons in the *Drosophila* embryo. *Cell* **67**: 675–685.

615. **Tung, J. S., Y. Saga, and E. A. Boyse.** 1988. Structural features of Ly-5 glycoproteins of the mouse and counterparts in other mammals. *Immunogenetics* **28**: 271–277.

616*. **Vogel, W., R. Lammers, J. T. Huang, and A. Ullrich.** 1993. Activation of a phosphotyrosine phosphatase by tyrosine phosphorylation. *Science* **259**: 1611–1614.

617. **Wan, J., H. Xu, and M. Grunstein.** 1992. CDC14 of *Saccharomyces cerevisiae*. Cloning, sequence analysis, and transcription during the cell cycle. *J. Biol. Chem.* **267**: 11274–11280.

618. **Wary, K. K., Z. Lou, A. M. Buchberg, L. D. Siracusa, T. Druck, S. Laforgia, and K. Huebner.** 1993. A homozygous deletion within the carbonic anhydrase-like domain of the PTPRG gene in murine L-cells. *Cancer Res.* **53**: 1498–1502.

619. **Wo, Y. Y., A. L. McCormack, J. Shabanowitz, D. F. Hunt, J. P. Davis, G. L. Mitchell, and R. L. Van Etten.** 1992. Sequencing, cloning, and expression of human red cell-type acid phosphatase, a cytoplasmic phosphotyrosyl protein phosphatase. *J. Biol. Chem.* **267**: 10856–10865.

620*. **Wong, E. C. C., J. E. Mullersman, and M. L. Thomas.** 1993. Leukocyte common antigen-related phosphatase (LRP) gene structure – conservation of the genomic organization of transmembrane protein tyrosine phosphatases. *Genomics* **17**: 33–38.

621*. **Yan, H., A. Grossman, H. Wang, P. D'Eustachio, K. Mossie, J. M. Musacchio, O. Silvennoinen, and J. Schlessinger.** 1993. A novel receptor tyrosine phosphatase-s that is highly expressed in the nervous system. *J. Biol. Chem.* **268**: 24880–24886.

622*. **Yang, Q., D. Co, J. Sommercorn, and N. K. Tonks.** 1993. Cloning and expression of PTP-PEST – a novel, human, nontransmembrane protein tyrosine phosphatase. *J. Biol. Chem.* **268**: 6622–6628.

623. **Yang, Q., D. Co, J. Sommercorn, and N. K. Tonks.** 1994. Cloning and expression of PTP-PEST – a novel, human, nontransmembrane protein tyrosine phosphatase [correction]. *J. Biol. Chem.* **268**: 17650

624*. **Yang, Q. and N. K. Tonks.** 1991. Isolation of a cDNA clone encoding a human protein-tyrosine phosphatase with homology to the cytoskeletal-associated proteins band-4.1, ezrin, and talin. *Proc. Natl Acad. Sci. USA* **88**: 5949–5953.

625*. **Yang, X. H., K. T. Seow, S. M. Bahri, S. H. Oon, and W. Chia.** 1991. Two *Drosophila* receptor-like tyrosine phosphatase genes are expressed in a subset of developing axons and pioneer neurons in the embryonic CNS. *Cell* **67**: 661–673.

626. **Yi, T. L., J. L. Cleveland, and J. N. Ihle.** 1991. Identification of novel protein tyrosine phosphatases of hematopoietic cells by polymerase chain reaction amplification. *Blood* **78**: 2222–2228.

627. **Yi, T. L., J. L. Cleveland, and J. N. Ihle.** 1992. Protein tyrosine phosphatase containing SH2 domains – characterization, preferential expression in hematopoietic cells, and localization to human chromosome 12p12-p13. *Mol. Cell. Biol.* **12**: 836–846.

628. **Yi, T. L., D. J. Gilbert, N. A. Jenkins, N. G. Copeland, and J. N. Ihle.** 1992. Assignment of a novel protein tyrosine phosphatase gene (Hcph) to mouse chromosome-6. *Genomics* **14**: 793–795.

629. **Zanke, B., J. Squire, H. Griesser, M. Henry, H. Suzuki, B. Patterson, M. Minden, and T. W. Mak.** 1994. A hematopoietic protein tyrosine phosphatase (HePTP) gene that is amplified and overexpressed in myeloid malignancies maps to chromosome 1q32.1. *Leukemia* **8**: 236–244.

630. **Zanke, B., H. Suzuki, K. Kishihara, L. Mizzen, M. Minden, A. Pawson, and T. W. Mak.** 1992. Cloning and expression of an inducible lymphoid-specific, protein tyrosine phosphatase (HePTPase). *Eur. J. Immunol.* **22**: 235–239.

631*. **Zhang, W. R., N. Hashimoto, F. Ahmad, W. Ding, and B. J. Goldstein.** 1994. Molecular cloning and expression of a unique receptor-like protein-tyrosine phosphatase in the leukocyte-common-antigen-related phosphatase family. *Biochem J.* **302**: 39–47.

Structural studies

632*. **Barford, D., A. J. Flint, and N. K. Tonks.** 1994. Crystal structure of human protein tyrosine phosphatase 1B. *Science* **263**: 1397–1404.

633. **Barford, D., J. C. Keller, A. J. Flint, and N. K. Tonks.** 1994. Purification and crystallization of the catalytic domain of human protein tyrosine phosphatase 1B expressed in *Escherichia coli*. *J. Mol. Biol.* **239**: 726–730.

634. **Hoppe, E., P. F. Berne, D. Stock, J. S. Rasmussen, N. P. H. Moller, A. Ullrich, and R. Huber.** 1994. Expression, purification and crystallization of human phosphotyrosine phosphatase 1B. *Eur. J. Biochem.* **223**: 1069–1077.

635*. **Logan, T. M., M. M. Zhou, D. G. Nettesheim, R. P. Meadows, R. L. Van Etten, and S. W. Fesik.** 1994. Solution structure of a low molecular weight protein tyrosine phosphatase. *Biochemistry* **33**: 11087–11096.

636. **McCall, M. N., D. M. Shotton, and A. N. Barclay.** 1992. Expression of soluble isoforms of rat CD45. Analysis by electron microscopy and use in epitope mapping of anti-CD45R monoclonal antibodies. *Immunology* **76**: 310–317.

637*. **Stuckey, J. A., H. L. Schubert, E. B. Fauman, Z. Y. Zhang, J. E. Dixon, and M. A. Saper.** 1994. Crystal structure of *Yersinia* protein tyrosine phosphatase at 2.5 angstrom and the complex with tungstate. *Nature* **370**: 571–575.

638. **Su, X. D., E. G. Agango, N. Taddei, M. Bucciantini, M. Stefani, G. Ramponi, and P. Nordlund.** 1994. Crystallisation of a low molecular weight phosphotyrosine protein phosphatase from bovine liver. *FEBS Lett.* **343**: 107–108.

639*. **Su, X. D., N. Taddei, M. Stefani, G. Ramponi, and P. Nordlund.** 1994. The crystal structure of a low-molecular-weight phosphotyrosine protein phosphatase. *Nature* **370**: 575–578.

640. **Zhang, M., R. L. Van Etten, C. M. Lawrence, and C. V. Stauffacher.** 1994. Crystallization and preliminary X-ray analysis of the low molecular weight phosphotyrosyl protein phosphatase from bovine heart. *J. Mol. Biol.* **238**: 281–283.

641*. **Zhang, M., R. L. Van Etten, and C. V. Stauffacher.** 1994. Crystal structure of bovine heart phosphotyrosyl phosphatase at 2.2–angstrom resolution. *Biochemistry* **33**: 11097–11105.

642. **Zhou, M. M., T. M. Logan, Y. Theriault, R. L. Van Etten, and S. W. Fesik.** 1994. Backbone H-1, C-13, and N-15 assignments and secondary structure of bovine low molecular weight phosphotyrosyl protein phosphatase. *Biochemistry* **33**: 5221–5229.

Toxins and inhibitors

643. **Andrews, D. F., M. B. Lilly, C. K. Tompkins, and J. W. Singer.** 1992. Sodium vanadate, a tyrosine phosphatase inhibitor, affects expression of hematopoietic growth factors and extracellular matrix RNAs in SV40-transformed human marrow stromal cells. *Exp. Hematol.* **20**: 449–453.

644. **Atkinson, T. P., C. W. Lee, S. G. Rhee, and R. J. Hohman.** 1993. Orthovanadate induces translocation of phospholipase C-γ 1 and -γ 2 in permeabilized mast cells. *J. Immunol.* **151**: 1448–1455.

645. **Barbera, A., J. E. Rodriguezgil, and J. J. Guinovart.** 1994. Insulin-like actions of tungstate in diabetic rats – normalization of hepatic glucose metabolism. *J. Biol. Chem.* **269**: 20047–20053.

646. **Becker, D. J., L. N. Ongemba, and J. C. Henquin.** 1994. Comparison of the effects of various vanadium salts on glucose homeostasis in streptozotocin-diabetic rats. *Eur. J. Pharmacol.* **260**: 169–175.

647. **Begum, N.** 1994. Phenylarsine oxide inhibits insulin-stimulated protein phosphatase 1 activity and GLUT-4 translocation. *Am. J. Physiol.* **267**: E14–E23.

648. **Bennett, P. A., R. J. Dixon, and S. Kellie.** 1993. The phosphotyrosine phosphatase inhibitor vanadyl hydroperoxide induces morphological alterations, cytoskeletal rearrangements and increased adhesiveness in rat neutrophil leucocytes. *J. Cell Sci.* **106**: 891–901.

649. **Bergamini, E., V. Detata, M. Novelli, G. Cavallini, P. Masiello, and Z. Gori.** 1994. Beneficial effects of vanadyl sulfate administration on sugar metabolism in the senescent rat. *Ann. N.Y. Acad. Sci.* **717**: 174–179.

650. **Berggren, M. M., L. A. Burns, R. T. Abraham, and G. Powis.** 1993. Inhibition of protein tyrosine phosphatase by the antitumor agent gallium nitrate. *Cancer Res.* **53**: 1862–1866.

651. **Bhanot, S., M. Bryerash, A. Cheung, and J. H. McNeill.** 1994. Bis(maltolato)oxovanadium(IV) attenuates hyperinsulinemia and hypertension in spontaneously hypertensive rats. *Diabetes* **43**: 857–861.

652. **Bianchini, L., A. Nanda, S. Wasan, and S. Grinstein.** 1994. Activation of multiple pH-regulatory pathways in granulocytes by a phosphotyrosine phosphatase antagonist. *Biochem. J.* **301**: 539–544.

653. **Bosch, F., M. Hatzoglou, E. A. Park, and R. W. Hanson.** 1990. Vanadate inhibits expression of the gene for phosphoenolpyruvate carboxykinase (GTP) in rat hepatoma cells. *J. Biol. Chem.* **265**: 13677–13682.

654. **Brichard, S. M., F. Assimacopoulos-Jeannet, and B. Jeanrenaud.** 1992. Vanadate treatment markedly increases glucose utilization in muscle of insulin-resistant fa/fa rats without modifying glucose transporter expression. *Endocrinology* **131**: 311–317.

655. **Brichard, S. M., B. Desbuquois, and J. Girard.** 1993. Vanadate treatment of diabetic rats reverses the impaired expression of genes involved in hepatic hlucose metabolism – effects on glycolytic and gluconeogenic enzymes, and on glucose transporter GLUT2. *Mol. Cell Endocrinol.* **91**: 91–97.

656. **Brichard, S. M., L. N. Ongemba, J. Girard, and J. C. Henquin.** 1994. Tissue-specific correction of lipogenic enzyme gene expression in diabetic rats given vanadate. *Diabetologia* **37**: 1065–1072.

657. **Bruck, R., H. Prigozin, Z. Krepel, P. Rotenberg, Y. Shechter, and S. Barmeir.** 1991. Vanadate inhibits glucose output from isolated perfused rat liver. *Hepatology* **14**: 540–544.

658. **Burke, T. R., H. K. Kole, and P. P. Roller.** 1994. Potent inhibition of insulin receptor dephosphorylation by a hexamer peptide containing the phosphotyrosyl mimetic F(2)Pmp. *Biochem. Biophys. Res. Commun.* **204**: 129–134.

659*. **Burke, T. R., M. S. Smyth, A. Otaka, M. Nomizu, P. P. Roller, G. Wolf, R. Case, and S. E. Shoelson.** 1994. Nonhydrolyzable phosphotyrosyl mimetics for the preparation of phosphatase-resistant SH2 domain inhibitors. *Biochemistry* **33**: 6490–6494.

660. **Cam, M. C., R. A. Pederson, R. W. Brownsey, and J. H. McNeill.** 1993. Long-term effectiveness of oral vanadyl sulphate in streptozotocin-diabetic rats. *Diabetologia* **36**: 218–224.

661. **Chao, W., H. L. Liu, and M. S. Olson.** 1993. Effect of orthovanadate on tyrosine Phosphorylation of P120 GTPase-activating protein in rat liver macrophages (Kupffer cells). *Biochem. Biophys. Res. Commun.* **191**: 55–60.

662. **Chatterjee, S., B. J. Goldstein, P. Csermely, and S.E. Shoelson.** 1992. Design and synthesis of potent substrates and inhibitors of protein-tyrosine phosphatases. In *Peptides: Chemistry and Biology*. J. E. Rivier and J. A. Smith, eds. Escom Science Publishers, Leiden, Netherlands. 553–555.

663. **Cordera, R., G. Andraghetti, R. A. DeFronzo, and L. Rossetti.** 1990. Effect of *in vivo* vanadate treatment on insulin receptor tyrosine kinase activity in partially pancreatectomized diabetic rats. *Endocrinology* **126**: 2177–2183.

664. **Detata, V., M. Novelli, G. Cavallini, P. Masiello, Z. Gori, and E. Bergamini.** 1993. Beneficial effects of the oral administration of vanadyl sulphate on glucose metabolism in senescent rats. *J. Gerontol.* **48**: B191–B195.

665. **Donofrio, F., M. Q. U. Le, J. L. Chiasson, and A. K. Srivastava.** 1994. Activation of mitogen activated protein (Map) kinases by vanadate is independent of insulin receptor autophosphorylation. *FEBS Lett.* **340**: 269–275.

666. **Elberg, G., J. Li, and Y. Shechter.** 1994. Vanadium activates or inhibits receptor and non-receptor protein tyrosine kinases in cell-free experiments, depending on its oxidation state. *J. Biol. Chem.* **269**: 9521–9527.

667. **Fantus, I. G., F. Ahmad, and G. Deragon.** 1990. Vanadate augments insulin binding and prolongs insulin action in rat adipocytes. *Endocrinology* **127**: 2716–2725.

668. **Fantus, I. G., F. Ahmad, and G. Deragon.** 1994. Vanadate augments insulin-stimulated insulin receptor kinase activity and prolongs insulin action in rat adipocytes – evidence for transduction of amplitude of signaling into duration of response. *Diabetes* **43**: 375–383.

669. **Fantus, I. G., S. Kadota, G. Deragon, B. Foster, and B. I. Posner.** 1989. Pervanadate [peroxide(s) of vanadate] mimics insulin action in rat adipocytes via activation of the insulin receptor tyrosine kinase. *Biochemistry* **28**: 8864–8871.

670. **Ferber, S., J. Meyerovitch, K. M. Kriauciunas, and C. R. Kahn.** 1994. Vanadate normalizes hyperglycemia and phosphoenolpyruvate carboxykinase mRNA levels in ob/ob mice. *Metabolism* **43**: 1346–1354.

671. **Ganguli, S., D. J. Reuland, L. A. Franklin, D. D. Deakins, W. J. Johnston, and A. Pasha.** 1994. Effects of maternal vanadate treatment on fetal development. *Life Sci.* **55**: 1267–1276.

672. **Ghosh, J. and R. A. Miller.** 1993. Suramin, an experimental chemotherapeutic drug, irreversibly blocks T cell CD45-protein tyrosine phosphatase *in vitro*. *Biochem. Biophys. Res. Commun.* **194**: 36–44.

673*. **Gordon, J. A.** 1991. Use of vanadate as protein-phosphotyrosine phosphatase inhibitor. *Methods Enzymol.* **201**: 477–482.

674. **Grinstein, S., W. Furuya, D. J. Lu, and G. B. Mills.** 1990. Vanadate stimulates oxygen consumption and tyrosine phosphorylation in electropermeabilized human neutrophils. *J. Biol. Chem.* **265**: 318–327.

675. **Hadari, Y. R., B. Geiger, O. Nadiv, I. Sabanay, C. T. Roberts, D. LeRoith, and Y. Zick.** 1993. Hepatic tyrosine-phosphorylated proteins identified and localized following *in vivo* inhibition of protein tyrosine phosphatases – effects of H_2O_2 and vanadate administration into rat livers. *Mol. Cell Endocrinol.* **97**: 9–17.

676. **Hamel, P. and Y. Girard.** 1994. Synthesis of dephostatin, a novel protein tyrosine phosphatase inhibitor. *Tetrahedron Lett.* **35**: 8101–8102.

677. **Hecht, D. and Y. Zick.** 1992. Selective inhibition of protein tyrosine phosphatase activities by H_2O_2 and vanadate *in vitro*. *Biochem. Biophys. Res. Commun.* **188**: 773–779.

678. **Heffetz, D., I. Bushkin, R. Dror, and Y. Zick.** 1990. The insulinomimetic agents H_2O_2 and vanadate stimulate protein tyrosine phosphorylation in intact cells. *J. Biol. Chem.* **265**: 2896–2902.

679. **Heffetz, D., W. J. Rutter, and Y. Zick.** 1992. The insulinomimetic agents H_2O_2 and vanadate stimulate tyrosine phosphorylation of potential target proteins for the insulin receptor kinase in intact cells. *Biochem J.* **288**: 631–635.

680. **Henquin, J. C., F. Carton, L. N. Ongemba, and D. J. Becker.** 1994. Improvement of mild hypoinsulinaemic diabetes in the rat by low non-toxic doses of vanadate. *J. Endocrinol.* **142**: 555–561.

681*. **Hiriyanna, K. T., D. Baedke, K. H. Baek, B. A. Forney, G. Kordiyak, and T. S. Ingebritsen.** 1994. Thiophosphorylated substrate analogs are potent active site-directed inhibitors of protein-tyrosine phosphatases. *Anal. Biochem.* **223**: 51–58.

682. **Hutchison, K. A., L. F. Stancato, R. Jove, and W. B. Pratt.** 1992. The protein–protein complex between pp60v-src and hsp90 is stabilized by molybdate, vanadate, tungstate, and an endogenous cytosolic metal. *J. Biol. Chem.* **267**: 13952–13957.

683. **Iivanainen, A. V., C. Lindqvist, T. Mustelin, and L. C. Andersson.** 1991. Phosphotyrosine phosphatases are involved in reversion of T-lymphoblastic proliferation. *Eur. J. Immunol.* **20**: 2509–2512.

684. **Imbert, V., J. F. Peyron, D. Farahi Far, B. Mari, P. Auberger, and B. Rossi.** 1994. Induction of tyrosine phosphorylation and T-cell activation by vanadate peroxide, an inhibitor of protein tyrosine phosphatases. *Biochem. J.* **297**: 163–173.

685. **Imoto, M., H. Kakeya, T. Sawa, C. Hayashi, M. Hamada, T. Takeuchi, and K. Umezawa.** 1993. Dephostatin, a novel protein tyrosine phosphatase inhibitor produced by *Streptomyces*. I. Taxonomy, isolation, and characterization. *J. Antibiotics* **46**: 1342–1346.

686. **Ingebritsen, T. S.** 1989. Phosphotyrosyl-protein phosphatases. II. Identification and characterization of two heat-stable protein inhibitors. *J. Biol. Chem.* **264**: 7754–7759.

687. **Johnson, T. M., M. H. Meisler, M. I. Bennett, and G. R. Willsky.** 1990. Vanadate induction of pancreatic amylase mRNA in diabetic rats. *Diabetes* **39**: 757–759.

688. **Kadota, S., I. G. Fantus, G. Deragon, H. J. Guyda, and B. I. Posner.** 1987. Stimulation of insulin-like growth factor II receptor binding and insulin receptor kinase activity in rat adipocytes. Effects of vanadate and H_2O_2. *J. Biol. Chem.* **262**: 8252–8256.

689. **Kakeya, H., M. Imoto, Y. Takahashi, H. Naganawa, T. Takeuchi, and K. Umezawa.** 1993. Dephostatin, a novel protein tyrosine phosphatase inhibitor produced by *Streptomyces*. II. Structure determination. *J. Antibiotics* **46**: 1716–1719.

690. **Kindberg, G. M., O. Gudmundsen, and T. Berg.** 1990. The effect of vanadate on receptor-mediated endocytosis of asialoorosomucoid in rat liver parenchymal cells. *J. Biol. Chem.* **265**: 8999–9005.

691. **Leighton, B., G. J. Cooper, C. DaCosta, and E. A. Foot.** 1991. Peroxovanadates have full insulin-like effects on glycogen synthesis in normal and insulin-resistant skeletal muscle. *Biochem. J.* **276**: 289–292.

692. **Lerea, K. M., N. K. Tonks, E. G. Krebs, E. H. Fischer, and J. A. Glomset.** 1989. Vanadate and molybdate increase tyrosine phosphorylation in a 50-kilodalton protein and stimulate secretion in electropermeabilized platelets. *Biochemistry* **28**: 9286–9292.

693. **Liotta, A. S., H. K. Kole, H. M. Fales, J. Roth, and M. Bernier.** 1994. A synthetic tris-sulfotyrosyl dodecapeptide analogue of the insulin receptor 1146-kinase domain inhibits tyrosine dephosphorylation of the insulin receptor *in situ*. *J. Biol. Chem.* **269**: 22996–23001.

694. **Lonnroth, P., J. W. Eriksson, B. I. Posner, and U. Smith.** 1993. Peroxovanadate but not vanadate exerts insulin-like effects in human adipocytes. *Diabetologia* **36**: 113–116.

695. **Meyerovitch, J., P. L. Rothenberg, Y. Shechter, S. Bonner-Weir, and C. R. Kahn.** 1991. Vanadate normalizes hyperglycemia in two mouse models of non-insulin dependent diabetes mellitus. *J. Clin. Invest.* **87**: 1286–1294.

696. **Miralpeix, M., J. F. Decaux, A. Kahn, and R. Bartrons.** 1991. Vanadate induction of L-type pyruvate kinase mRNA in adult rat hepatocytes in primary culture. *Diabetes* **40**: 462–464.

697. **Montesano, R., M. S. Pepper, D. Belin, J. D. Vassalli, and L. Orci.** 1988. Induction of angiogenesis *in vitro* by vanadate, an inhibitor of phosphotyrosine phosphatases. *J. Cell Physiol.* **134**: 460–466.

698. **Mountjoy, K. G. and J. S. Flier.** 1990. Vanadate regulates glucose transporter (GLUT-1) expression in NIH3T3 mouse fibroblasts. *Endocrinology* **127**: 2025–2034.

699. **Munoz, P., A. Guma, M. Camps, M. Furriols, X. Testar, M. Palacin, Zorzano, and A.** 1992. Vanadate stimulates system A amino acid transport activity in skeletal muscle. Evidence for the involvement of intracellular pH as a mediator of vanadate action. *J. Biol. Chem.* **267**: 10381–10388.

700. **Oster, M. H., J. M. Llobet, J. L. Domingo, J. B. German, and C. L. Keen.** 1993. Vanadium treatment of diabetic sprague-dawley rats results in tissue vanadium accumulation and pro-oxidant effects. *Toxicology* **83**: 115–130.

701. **Ozcelikay, A. T., C. Pekiner, N. Ari, Y. Ozturk, A. Ozuari, and V. M. Altan.** 1994. The effect of vanadyl treatment on vascular responsiveness of streptozotocin-diabetic rats. *Diabetologia* **37**: 572–578.

702*. **Posner, B. I., R. Faure, J. W. Burgess, A. P. Bevan, D. Lachance, G. Y. Zhangsun, I. G. Fantus, J. B. Ng, D. A. Hall, B. S. Lum, and A. Shaver.** 1994. Peroxovanadium compounds – a new class of potent phosphotyrosine phosphatase inhibitors which are insulin mimetics. *J. Biol. Chem.* **269**: 4596–4604.

703. **Pugazhenthi, S. and R. L. Khandelwal.** 1990. Insulin-like effects of vanadate on hepatic glycogen metabolism in nondiabetic and streptozocin-induced diabetic rats. *Diabetes* **39**: 821–827.

704. **Pugazhenthi, S. and R. L. Khandelwal.** 1993. Does the insulin-mimetic action of vanadate involve insulin receptor kinase. *Mol. Cell. Biochem.* **128**: 211–218.

705. **Pugazhenthi, S., R. L. Khandelwal, and J. F. Angel.** 1991. Insulin-like effect of vanadate on malic enzyme and glucose-6-phosphate dehydrogenase activities in streptozotocin-induced diabetic rat liver. *Biochim. Biophys. Acta* **1083**: 310–312.

706. **Pumiglia, K. M., L. F. Lau, C. K. Huang, S. Burroughs, and M. B. Feinstein.** 1992. Activation of signal transduction in platelets by the tyrosine phosphatase inhibitor pervanadate (vanadyl hydroperoxide). *Biochem. J.* **286**: 441–449.

707. **Rodriguez-Gil, J. E., A. M. Gomez-Foix, C. Fillat, F. Bosch, and J. J. Guinovart.** 1991. Activation by vanadate of glycolysis in hepatocytes from diabetic rats. *Diabetes* **40**: 1355–1359.

708. **Rodriguezgil, J. E., A. M. Gomezfoix, C. Fillat, F. Bosch, and J. J. Guinovart.** 1991. Activation by vanadate of glycolysis in hepatocytes from diabetic rats. *Diabetes* **40**: 1355–1359.

709. **Secrist, J. P., L. A. Burns, L. Karnitz, G. A. Koretzky, and R. T. Abraham.** 1993. Stimulatory effects of the protein tyrosine phosphatase inhibitor, pervanadate, on T-cell activation events. *J. Biol. Chem.* **268**: 5886–5893.

710. **Shisheva, A., O. Ikonomov, and Y. Shechter.** 1994. The protein tyrosine phosphatase inhibitor, pervanadate, is a powerful antidiabetic agent in streptozotocin-treated diabetic rats. *Endocrinology* **134**: 507–510.

711. **Shisheva, A. and Y. Shechter.** 1993. Mechanism of pervanadate stimulation and potentiation of insulin-activated glucose transport in rat adipocytes – dissociation from vanadate effect. *Endocrinology* **133**: 1562–1568.

712. **Shisheva, A. and Y. Shechter.** 1993. Role of cytosolic tyrosine kinase in mediating insulin-like actions of vanadate in rat adipocytes. *J. Biol. Chem.* **268**: 6463–6469.

713. **Strout, H. V., P. P. Vicario, C. Biswas, R. Saperstein, E. J. Brady, P. F. Pilch, and J. Berger.** 1990. Vanadate treatment of streptozotocin diabetic rats restores expression of the insulin-responsive glucose transporter in skeletal muscle. *Endocrinology* **126**: 2728–2732.

714. **Swarup, G., S. Cohen, and D. L. Garbers.** 1982. Inhibition of membrane phosphotyrosyl-protein phosphatase activity by vanadate. *Biochem. Biophys. Res. Commun.* **107**: 1104–1109.

715. **Tchou Wong, K. M. and I. B. Weinstein.** 1992. Altered expression of protein kinase-C, lck, and CD45 in a 12-*O*-tetradecanoylphorbol-13-acetate-dependent leukemic T-cell variant that expresses a high level of interleukin-2 receptor. *Mol. Cell. Biol.* **12**: 394–401.

716. **Thompson, K. H., J. Leichter, and J. H. McNeill.** 1993. Studies of vanadyl sulfate as a glucose-lowering agent in STZ-diabetic rats. *Biochem. Biophys. Res. Commun.* **197**: 1549–1555.

717. **Thompson, W. J., B. H. Tan, and S. J. Strada.** 1991. Activation of rabbit liver high affinity cAMP (type IV) phosphodiesterase by a vanadyl-glutathione complex. Characterization of the role of the sulfhydryl. *J. Biol. Chem.* **266**: 17011–17019.

718. **Totpal, K., S. Agarwal, and B. B. Aggarwal.** 1992. Phosphatase inhibitors modulate the growth-regulatory effects of human tumor necrosis factor on tumor and normal cells. *Cancer Res.* **52**: 2557–2562.

719. **Venkatesan, N., A. Avidan, and M. B. Davidson.** 1991. Antidiabetic action of vanadyl in rats independent of *in vivo* insulin-receptor kinase activity. *Diabetes* **40**: 492–498.

720. **Wang, Q. P., U. Dechert, F. Jirik, and S. G. Withers.** 1994. Suicide inactivation of human prostatic acid phosphatase and a phosphotyrosine phosphatase. *Biochem. Biophys. Res. Commun.* **200**: 577–583.

721. **Weinstock, R. S. and J. L. Messina.** 1992. Vanadate and insulin stimulate gene-33 expression. *Biochem. Biophys. Res. Commun.* **189**: 931–937.

722. **Yuen, V. G., C. Orvig, and J. H. McNeill.** 1993. Glucose-lowering effects of a new organic vanadium complex, bis(maltolato)-oxovanadium(IV). *Can. J. Physiol. Pharmacol.* **71**: 263–269.

723. **Yuen, V. G., C. Orvig, K. H. Thompson, and J. H. McNeill.** 1993. Improvement in cardiac dysfunction in streptozotocin-induced diabetic rats following chronic oral administration of bis(maltolato)oxovanadium(IV). *Can. J. Physiol. Pharmacol.* **71**: 270–276.

724. **Zhang, A. Q., Z. Y. Gao, P. Gilon, M. Nenquin, G. Drews, and J. C. Henquin.** 1991. Vanadate stimulation of insulin release in normal mouse islets. *J. Biol. Chem.* **266**: 21649–21656.

Additional general references

725. **Aposto, I., R. Kuciel, E. Wasylewska, and W. S. Ostrowski.** 1985. Phosphotyrosine as a substrate of acid and alkaline phosphatases. *Acta Biochim. Pol.* **32**: 187–197.

726. **Begum, N., A. L. Graham, K. E. Sussman, and B. Draznin.** 1992. Role of cAMP in mediating effects of fasting on dephosphorylation of insulin receptor. *Am. J. Physiol.* **262**: E142–E149.

727. **Begum, N., K. E. Sussman, and B. Draznin.** 1991. High levels of cytosolic free calcium inhibit dephosphorylation of insulin receptor and glycogen synthase. *Cell Calcium* **12**: 423–430.

728. **Begum, N., K. E. Sussman, and B. Draznin.** 1991. Differential effects of diabetes on adipocyte and liver phosphotyrosine and phosphoserine phosphatase activities. *Diabetes* **40**: 1620–1629.

729. **Bernier, M., A. S. Liotta, H. K. Kole, D. D. Shock, and J. Roth.** 1994. Dynamic regulation of intact and *C*-terminal truncated insulin receptor phosphorylation in permeabilized cells. *Biochemistry* **33**: 4343–4351.

730. **Boehm, T.** 1993. Analysis of multigene families by DNA fingerprinting of conserved domains – directed cloning of tissue-specific protein tyrosine phosphatases. *Oncogene* **8**: 1385–1390.

731. **Bohmer, F. D., S. A. Bohmer, and C. H. Heldin.** 1993. The dephosphorylation characteristics of the receptors for epidermal growth factor and platelet-derived growth factor in Swiss 3T3-cell membranes suggest differential regulation of receptor signalling by endogenous protein-tyrosine phosphatases. *FEBS Lett.* **331**: 276–280.

732. **Boivin, P. and C. Galand.** 1986. The human red cell acid phosphatase is a phosphotyrosine protein phosphatase which dephosphorylates the membrane protein band 3. *Biochem. Biophys. Res. Commun.* **134**: 557–564.

733. **Boivin, P., C. Galand, and O. Bertrand.** 1987. Protein band 3 phosphotyrosyl phosphatase. Purification and characterization. *Int. J. Biochem.* **19**: 613–618.

734. **Bottari, S. P., I. N. King, S. Reichlin, I. Dahlstroem, N. Lydon, and M. de Gasparo.** 1992. The angiotensin AT2 receptor stimulates protein tyrosine phosphatase activity and mediates inhibition of particulate guanylate cyclase. *Biochem. Biophys. Res. Commun.* **183**: 206–211.

735. **Boylan, J. M., D. L. Brautigan, J. Madden, T. Raven, L. Ellis, and P. A. Gruppuso.** 1992. Differential regulation of multiple hepatic protein tyrosine phosphatases in alloxan diabetic rats. *J. Clin. Invest.* **90**: 174–179.

736. **Brautigan, D. L., P. Bornstein, and B. Gallis.** 1981. Phosphotyrosyl-protein phosphatase. Specific inhibition by Zn. *J. Biol. Chem.* **256**: 6519–6522.

737. **Brautigan, D. L. and F. M. Pinault.** 1991. Activation of membrane protein-tyrosine phosphatase involving cAMP-dependent and Ca^{2+}/phospholipid-dependent protein kinases. *Proc. Natl Acad. Sci. USA* **88**: 6696–6700.

738. **Brunati, A. M. and L. A. Pinna.** 1985. Isolation and partial characterization of distinct species of phosphotyrosyl protein phosphatases from rat spleen. *Biochem. Biophys. Res. Commun.* **133**: 929–936.

739. **Buscail, L., N. Delesque, J. P. Esteve, N. Saintlaurent, H. Prats, P. Clerc, P. Robberecht, G. I. Bell, C. Liebow, A. V. Schally, N. Vaysse, and C. Susini.** 1994. Stimulation of tyrosine phosphatase and inhibition of cell proliferation by somatostatin analogues – mediation by human somatostatin receptor subtypes Sstr1 and Sstr2. *Proc. Natl Acad. Sci. USA* **91**: 2315–2319.

740. **Butler, M. T., A. Ziemiecki, B. Groner, and R. R. Friis.** 1989. Characterization of a membrane-associated phosphotyrosyl protein phosphatase from the A431 human epidermoid carcinoma cell line. *Eur. J. Biochem.* **185**: 475–483.

741. **Butler, T. M., A. Ziemiecki, and R. R. Friis.** 1990. Megakaryocytic differentiation of K562 cells is associated with changes in the cytoskeletal organization and the pattern of chromatographically distinct forms of phosphotyrosyl-specific protein phosphatases. *Cancer Res.* **50**: 6323–6329.

742. **Carpenter, G., L. King, Jr., and S. Cohen.** 1979. Rapid enhancement of protein phosphorylation in A-431 cell membrane preparations by epidermal growth factor. *J. Biol. Chem.* **254**: 4884–4891.

743. **Cayla, X., K. Ballmerhofer, W. Merlevede, and J. Goris.** 1993. Phosphatase 2A associated with polyomavirus small-T or middle-T antigen is an okadaic acid-sensitive tyrosyl phosphatase. *Eur. J. Biochem.* **214**: 281–286.

744. **Chackalaparampil, I., S. Bagrodia, and D. Shalloway.** 1994. Tyrosine dephosphorylation of pp60(c-src) is stimulated by a serine/threonine phosphatase inhibitor. *Oncogene* **9**: 1947–1955.

745. **Chan, C. P., B. Gallis, D. K. Blumenthal, C. J. Pallen, J. H. Wang, and E. G. Krebs.** 1986. Characterization of the phosphotyrosyl protein phosphatase activity of calmodulin-dependent protein phosphatase. *J. Biol. Chem.* **261**: 9890–9895.

746. **Chen, J., B. L. Martin, and D. L. Brautigan.** 1992. Regulation of protein serine-threonine phosphatase type-2A by tyrosine phosphorylation. *Science* **257**: 1261–1264.

747. **Cheng, H. F. and M. Tao.** 1989. Purification and characterization of a phosphotyrosyl-protein phosphatase from wheat seedlings. *Biochim. Biophys. Acta* **998**: 271–276.

748. **Chernoff, J. and H. C. Li.** 1983. Multiple forms of phosphotyrosyl- and phosphoseryl-protein phosphatase from cardiac muscle: partial purification and characterization of an EDTA-stimulated phosphotyrosyl-protein phosphatase. *Arch. Biochem. Biophys.* **226**: 517–530.

749. **Chernoff, J. and H. C. Li.** 1985. A major phosphotyrosyl-protein phosphatase from bovine heart is associated with a low-molecular-weight acid phosphatase. *Arch. Biochem. Biophys.* **240**: 135–145.

750. **Chernoff, J., M. A. Sells, and H. C. Li.** 1984. Characterization of phosphotyrosyl-protein phosphatase activity associated with calcineurin. *Biochem. Biophys. Res. Commun.* **121**: 141–148.

751. **Chevalier, S., D. Landry, and A. Chapdelaine.** 1988. Phosphotyrosine phosphatase activity of human and canine acid phosphatases of prostatic origin. *Prostate* **12**: 209–219.

752. **Clari, G., A. M. Brunati, and V. Moret.** 1986. Partial purification and characterization of phosphotyrosyl-protein phosphatase(s) from human erythrocyte cytosol. *Biochem. Biophys. Res. Commun.* **137**: 566–572.

753. **Clari, G., A. M. Brunati, and V. Moret.** 1987. Membrane-bound phosphotyrosyl-protein phosphatase activity in human erythrocytes. Dephosphorylation of membrane band 3 protein. *Biochem. Biophys. Res. Commun.* **142**: 587–594.

754. **Colas, B., C. Cambillau, L. Buscail, M. Zeggari, J. P. Esteve, V. Lautre, F. Thomas, N. Vaysse, and C. Susini.** 1992. Stimulation of a membrane tyrosine phosphatase activity by somatostatin analogues in rat pancreatic acinar cells. *Eur. J. Biochem.* **207**: 1017–1024.

755. **Cool, D. E. and J. J. Blum.** 1993. Protein tyrosine phosphatase activity in *Leishmania donovani*. *Mol. Cell. Biochem.* **128**: 143–149.

756. **Cunningham, B. A., J. J. Hemperly, B. A. Murray, E. A. Prediger, R. Brackenbury, and G. M. Edelman.** 1987. Neural cell adhesion molecule: structure, immunoglobulin-like domains, cell surface modulation, and alternative RNA splicing. *Science* **236**: 799–806.

757. **Damuni, Z., H. S. Xiong, and M. Li.** 1994. Autophosphorylation-activated protein kinase inactivates the protein tyrosine phosphatase activity of protein phosphatase 2A. *FEBS Lett.* **352**: 311–314.

758. **David, M., P. M. Grimley, D. S. Finbloom, and A. C. Larner.** 1993. A nuclear tyrosine phosphatase downregulates interferon-induced gene expression. *Mol. Cell. Biol.* **13**: 7515–7521.

759. **David, M., G. Romero, Z. Y. Zhang, J. E. Dixon, and A. C. Larner.** 1993. *In vitro* activation of the transcription factor ISGF3 by interferon-α involves a membrane-associated tyrosine phosphatase and tyrosine kinase. *J. Biol. Chem.* **268**: 6593–6599.

760. **DeSeau, V., N. Rosen, and J. B. Bolen.** 1987. Analysis of $pp60^{c\text{-}src}$ tyrosine kinase activity and phosphotyrosyl phosphatase activity in human colon carcinoma and normal human colon mucosal cells. *J. Cell. Biochem.* **35**: 113–128.

761. **Dissing, J. and O. Svensmark.** 1990. Human red cell acid phosphatase: purification and properties of the A, B and C isozymes. *Biochim. Biophys. Acta* **1041**: 232–242.

762. **Donella-Deana, A., K. Lopandic, S. Barbaric, and L. A. Pinna.** 1986. Distinct specificities of repressible acid phosphatase from yeast toward phosphoseryl and phosphotyrosyl phosphopeptides. *Biochem. Biophys. Res. Commun.* **139**: 1202–1209.

763. **Edgar, B. A., F. Sprenger, R. J. Duronio, P. Leopold, and P. H. O'Farrell.** 1994. Distinct molecular mechanism regulate cell cycle timing at successive stages of *Drosophila* embryogenesis. *Genes Dev.* **8**: 440–452.

764. **Errasfa, M. and A. Stern.** 1993. Inhibition of protein tyrosine phosphatase activity in HER14 cells by melittin and Ca^{2+} ionophore-A23187. *Eur. J. Pharmacol-Molec. Pharm.* **247**: 73–80.

765. **Errasfa, M. and A. Stern.** 1994. Inhibition of epidermal growth factor-dependent protein tyrosine phosphorylation by phorbol myristate acetate is mediated by protein tyrosine phosphatase activity. *FEBS Lett.* **339**: 7–10.

766. **Faure, R., G. Baquiran, J. J. M. Bergeron, and B. I. Posner.** 1992. The dephosphorylation of insulin and epidermal growth factor receptors. Role of endosome-associated phosphotyrosine phosphatase(s). *J. Biol. Chem.* **267**: 11215–11221.

767. **Fedde, K. N., M. P. Michel, and M. P. Whyte.** 1993. Evidence against a role for alkaline phosphatase in the dephosphorylation of plasma membrane proteins – hypophosphatasia fibroblast study. *J. Cell. Biochem.* **53**: 43–50.

768. **Florio, T., M. G. Pan, B. Newman, R. E. Hershberger, O. Civelli, and P. J. S. Stork.** 1992. Dopaminergic inhibition of DNA synthesis in pituitary tumor cells is associated with phosphotyrosine phosphatase activity. *J. Biol. Chem.* **267**: 24169–24172.

769. **Florio, T., C. Rim, R. E. Hershberger, M. Loda, and P. J. S. Stork.** 1994. The somatostatin receptor SSTR1 is coupled to phosphotyrosine phosphatase activity in CHO-K1 cells. *Mol. Endocrinol.* **8**: 1289–1297.

770. **Formisano, P., G. Condorelli, and F. Beguinot.** 1991. Antiphosphotyrosine immunoprecipitation of an insulin-stimulated receptor phosphatase activity from FRTL5 cells. *Endocrinology* **128**: 2949–2957.

771. **Foulkes, J. G., E. Erikson, and R. L. Erikson.** 1983. Separation of multiple phosphotyrosyl- and phosphoseryl-protein phosphatases from chicken brain. *J. Biol. Chem.* **258**: 431–438.

772. **Foulkes, J. G., R. F. Howard, and A. Ziemiecki.** 1981. Detection of a novel mammalian protein phosphatase with activity for phosphotyrosine. *FEBS Lett.* **130**: 197–200.

773. **Frangioni, J. V. and B. G. Neel.** 1993. Solubilization and purification of enzymatically active glutathione *S*-transferase (pGEX) fusion proteins. *Anal. Biochem.* **210**: 179–187.

774. **Frank, D. A. and A. C. Sartorelli.** 1986. Regulation of protein phosphotyrosine content by changes in tyrosine kinase and protein phosphotyrosine phosphatase activities during induced granulocytic and monocytic differentiation of HL-60 leukemia cells. *Biochem. Biophys. Res. Commun.* **140**: 440–447.

775. **Frank, D. A. and A. C. Sartorelli.** 1988. Biochemical characterization of tyrosine kinase and phosphotyrosine phosphatase activities of HL-60 leukemia cells. *Cancer Res.* **48**: 4299–4306.

776. **Frank, D. A. and A. C. Sartorelli.** 1988. Alterations in tyrosine phosphorylation during the granulocytic maturation of HL-60 leukemia cells. *Cancer Res.* **48**: 52–58.

777. **Gallis, B., P. Bornstein, and D. L. Brautigan.** 1981. Tyrosylprotein kinase and phosphatase activities in membrane vesicles from normal and Rous sarcoma virus-transformed rat cells. *Proc. Natl Acad. Sci. USA* **78**: 6689–6693.

778. **Gellatly, K. S., G. B. G. Moorhead, S. M. G. Duff, D. D. Lefebvre, and W. C. Plaxton.** 1994. Purification and characterization of a potato tuber acid phosphatase having significant phosphotyrosine phosphatase activity. *Plant Physiol.* **106**: 223–232.

779. **Gentleman, S., T. M. Martensen, J. J. Digiovanna, and G. J. Chader.** 1984. Protein tyrosine kinase and protein phosphotyrosine phosphatase in normal and psoriatic skin. *Biochim. Biophys. Acta* **798**: 53–59.

780. **Gimond, C. and M. Aumailley.** 1992. Cellular interactions with the extracellular matrix are coupled to diverse transmembrane signaling pathways. *Exp. Cell Res.* **203**: 365–373.

781. **Goren, H. J. and D. Boland.** 1991. The 180 000 molecular weight plasma membrane insulin receptor substrate is a protein tyrosine phosphatase and is elevated in diabetic plasma membranes. *Biochem. Biophys. Res. Commun.* **180**: 463–469.

782. **Griswold-Prenner, I., C. R. Carlin, and M. R. Rosner.** 1993. Mitogen-activated protein kinase regulates the epidermal growth factor receptor through activation of a tyrosine phosphatase. *J. Biol. Chem.* **268**: 13050–13054.

783. **Gruppuso, P. A., J. M. Boylan, P. A. Carter, J. A. Madden, and T. Raven.** 1992. Hepatic insulin and EGF receptor phosphorylation and dephosphorylation in Fetal Rats. *Am. J. Physiol.* **262**: E7–E13.

784. **Gruppuso, P. A., J. M. Boylan, B. A. Levine, and L. Ellis.** 1992. Insulin receptor tyrosine kinase domain auto-dephosphorylation. *Biochem. Biophys. Res. Commun.* **189**: 1457–1463.

785. **Gruppuso, P. A., J. M. Boylan, B. I. Posner, R. Faure, and D. L. Brautigan.** 1990. Hepatic protein phosphotyrosine phosphatase. Dephosphorylation of insulin and epidermal growth factor receptors in normal and alloxan diabetic rats. *J. Clin. Invest.* **85**: 1754–1760.

786. **Gruppuso, P. A., J. M. Boylan, B. L. Smiley, R. J. Fallon, and D. L. Brautigan.** 1991. Hepatic protein tyrosine phosphatases in the rat. *Biochem. J.* **274**: 361–367.

787. **Halaban, R., B. L. Fan, J. Ahn, Y. Funasaka, H. Gitaygoren, and G. Neufeld.** 1992. Growth factors, receptor kinases, and protein tyrosine phosphatases in normal and malignant melanocytes. *J. Immunother.* **12**: 154–161.

788. **Hampe, C. S. and I. Pecht.** 1994. Protein tyrosine phosphatase activity enhancement is induced upon Fc ε receptor activation of mast cells. *FEBS Lett.* **346**: 194–198.

789. **Haring, H. U., M. Kasuga, M. F. White, M. Crettaz, and C. R. Kahn.** 1984. Phosphorylation and dephosphorylation of the insulin receptor: evidence against an intrinsic phosphatase activity. *Biochemistry* **23**: 3298–3306.

790. **Hashimoto, N., W. R. Zhang, and B. J. Goldstein.** 1992. Insulin receptor and epidermal growth factor receptor dephosphorylation by three major rat liver protein-tyrosine phosphatases expressed in a recombinant bacterial system. *Biochem. J.* **284**: 569–576.

791. **Hauguel deMouzon, S., P. Peraldi, F. Alengrin, and E. Van Obberghen.** 1993. Alteration of phosphotyrosine phosphatase activity in tissues from diabetic and pregnant rats. *Endocrinology* **132**: 67–74.

792. **Hernandez-Sotomayor, S. M. T., C. L. Arteaga, C. Soler, and G. Carpenter.** 1993. Epidermal growth factor stimulates substrate-selective protein-tyrosine-phosphatase activity. *Proc. Natl Acad. Sci. USA* **90**: 7691–7695.

793. **Hierowski, M. T., C. Liebow, K. du Sapin, and A. V. Schally.** 1985. Stimulation by somatostatin of dephosphorylation of membrane proteins in pancreatic cancer MIA PaCa–2 cell line. *FEBS Lett.* **179**: 252–256.

794. **Hiraga, A., K. Hata, Y. Suzuki, and S. Tsuiki.** 1993. Identification of a rat liver protein-tyrosine phosphatase similar to human placental PTPase-1B using quantitatively phosphorylated protein substrates. *J. Biochem. Tokyo* **113**: 180–188.

795. **Honda, H., M. Shibuya, S. Chiba, Y. Yazaki, and H. Hirai.** 1993. Identification of novel protein-tyrosine phosphatases in a human leukemia cell line, F-36P. *Leukemia* **7**: 742–746.

796. **Horlein, D., B. Gallis, D. L. Brautigan, and P. Bornstein.** 1982. Partial purification and characterization of phosphotyrosyl-protein phosphatase from Ehrlich ascites tumor cells. *Biochemistry* **21**: 5577–5584.

797. **Iivanainen, A. V., C. Lindqvist, T. Mustelin, and L. C. Andersson.** 1991. Phosphotyrosine phosphatases are involved in reversion of T-lymphoblastic proliferation. *Eur. J. Immunol.* **20**: 2509–2512.

798. **Imes, S. S., N. O. Kaplan, and A. F. Knowles.** 1987. Plasma membrane-associated phosphatase activities hydrolyzing [^{32}P]phosphotyrosyl histones and [^{32}P]phosphatidylinositol phosphate. *Anal. Biochem.* **161**: 316–322.

799. **Janckila, A. J., T. A. Woodford, K. W. Lam, C. Y. Li, and L. T. Yam.** 1992. Protein-tyrosine phosphatase activity of hairy cell tartrate-resistant acid phosphatase. *Leukemia* **6**: 199–203.

800. **Jena, B. P., P. J. Padfield, T. S. Ingebritsen, and J. D. Jamieson.** 1991. Protein tyrosine phosphatase stimulates Ca^{2+}-dependent amylase secretion from pancreatic acini. *J. Biol. Chem.* **266**: 17744–17746.

801. **Jones, S. W., R. L. Erikson, V. M. Ingebritsen, and T. S. Ingebritsen.** 1989. Phosphotyrosyl-protein phosphatases. I. Separation of multiple forms from bovine brain and purification of the major form to near homogeneity. *J. Biol. Chem.* **264**: 7747–7753.

802. **Kambayashi, Y., S. Bardhan, K. Takahashi, S. Tsuzuki, H. Inui, T. Hamakubo, and T. Inagami.** 1993. Molecular cloning of a novel angiotensin-II receptor isoform involved in phosphotyrosine phosphatase inhibition. *J. Biol. Chem.* **268**: 24543–24546.

803. **Kansha, M., K. Takeshige, and S. Minakami.** 1993. Decrease in the phosphotyrosine phosphatase activity in the plasma membrane of human neutrophils on stimulation by phorbol 12-myristate 13-acetate. *Biochim. Biophys. Acta* **1179**: 189–196.

804. **Kidd, K. R., B. J. Kerns, R. K. Dodge, and J. R. Wiener.** 1992. Histochemical staining of protein-tyrosine phosphatase activity in primary human mammary carcinoma: relationship with established prognostic indicators. *J. Histochem. Cytochem.* **40**: 729–735.

805. **Kincaid, R. L., T. M. Martensen, and M. Vaughan.** 1986. Modulation of calcineurin phosphotyrosyl protein phosphatase activity by calmodulin and protease treatment. *Biochem. Biophys. Res. Commun.* **140**: 320–328.

806. **King, M. J. and G. J. Sale.** 1988. Insulin-receptor phosphotyrosyl-protein phosphatases. *Biochem. J.* **256**: 893–902.

807. **King, M. J. and G. J. Sale.** 1990. Dephosphorylation of insulin-receptor autophosphorylation sites by particulate and soluble phosphotyrosyl-protein phosphatases. *Biochem. J.* **266**: 251–259.

808. **King, M. J., R. P. Sharma, and G. J. Sale.** 1991. Site-specific dephosphorylation and deactivation of the human insulin receptor tyrosine kinase by particulate and soluble phosphotyrosyl protein phosphatases. *Biochem. J.* **275**: 413–418.

809. **Kolb, J. P. and A. Abadie.** 1993. Inhibitors of protein tyrosine kinases and protein tyrosine phosphatases suppress IL-4-induced CD23 expression and release by human B lymphocytes. *Eur. Cytokine Netw.* **4**: 429–438.

810. **Kornblihtt, A. R., K. Umezawa, K. Vibe Pedersen, and F. E. Baralle.** 1985. Primary structure of human fibronectin: differential splicing may generate at least 10 polypeptides from a single gene. *EMBO J.* **4**: 1755–1759.

811. **Kowalski, A., H. Gazzano, M. Fehlmann, and E. Van Obberghen.** 1983. Dephosphorylation of the hepatic insulin receptor: absence of intrinsic phosphatase activity in purified receptors. *Biochem. Biophys. Res. Commun.* **117**: 885–893.

812. **Kraft, A. S. and R. L. Berkow.** 1987. Tyrosine kinase and phosphotyrosine phosphatase activity in human promyelocytic leukemia cells and human polymorphonuclear leukocytes. *Blood* **70**: 356–362.

813. **Kusari, J., K. A. Kenner, K. I. Suh, D. E. Hill, and R. R. Henry.** 1994. Skeletal muscle protein tyrosine phosphatase activity and tyrosine phosphatase 1B protein content are associated with insulin action and resistance. *J. Clin. Invest.* **93**: 1156–1162.

814. **Lander, H. M., D. M. Levine, and A. Novogrodsky.** 1993. Haemin enhancement of glucose transport in human lymphocytes – stimulation of protein tyrosine phosphatase and activation of p56(lck) tyrosine kinase. *Biochem J.* **291**: 281–287.

815. **Lau, K. H., T. K. Freeman, and D. J. Baylink.** 1987. Purification and characterization of an acid phosphatase that displays phosphotyrosyl-protein phosphatase activity from bovine cortical bone matrix. *J. Biol. Chem.* **262**: 1389–1397.

816. **Lazarovits, A. I., S. Poppema, M. J. White, and J. Karsh.** 1992. Inhibition of alloreactivity *in vitro* by monoclonal antibodies directed against restricted isoforms of the leukocyte-common antigen (CD45). *Transplantation* **54**: 724–729.

817. **Lee, M. T., C. Liebow, A. R. Kamer, and A. V. Schally.** 1991. Effects of epidermal growth factor and analogues of luteinizing hormone-releasing hormone and somatostatin on phosphorylation and dephosphorylation of tyrosine residues of specific protein substrates in various tumors. *Proc. Natl Acad. Sci. USA* **88**: 1656–1660.

818. **Li, H. C., J. Chernoff, L. B. Chen, and A. Kirschonbaum.** 1984. A phosphotyrosyl-protein phosphatase activity associated with acid phosphatase from human prostate gland. *Eur. J. Biochem.* **138**: 45–51.

819. **Liebow, C., M. T. Lee, and A. Schally.** 1990. Antitumor effects of somatostatin mediated by the stimulation of tyrosine phosphatase. *Metabolism* **39**: 163–166.

820. **Liebow, C., C. Reilly, M. Serrano, and A. V. Schally.** 1989. Somatostatin analogues inhibit growth of pancreatic cancer by stimulating tyrosine phosphatase. *Proc. Natl Acad. Sci. USA* **86**: 2003–2007.

821. **Lin, M. F. and G. M. Clinton.** 1986. Human prostatic acid phosphatase has phosphotyrosyl protein phosphatase activity. *Biochem. J.* **235**: 351–357.

822. **Lin, M. F. and G. M. Clinton.** 1988. The epidermal growth factor receptor from prostate cells is dephosphorylated by a prostate-specific phosphotyrosyl phosphatase. *Mol. Cell Biol.* **8**: 5477–5485.

823. **Lu, X., T. B. Chou, N. G. Williams, T. Roberts, and N. Perrimon.** 1993. Control of cell fate determination by p21ras/Ras1, an essential component of torso signalling in *Drosophila*. *Genes Dev.* **7**: 621–632.

824. **Machicao, F., T. Urumow, and O. H. Wieland.** 1982. Phosphorylation-dephosphorylation of purified insulin receptor from human placenta. Effect of insulin. *FEBS Lett.* **149**: 96–100.

825. **Maher, P. A.** 1993. Activation of phosphotyrosine phosphatase activity by reduction of cell–substrate adhesion. *Proc. Natl Acad. Sci. USA* **90**: 11177–11181.

826. **Marvel, J., G. Rimon, P. Tatham, and S. Cockcroft.** 1991. Evidence that the CD45 phosphatase regulates the activity of the phospholipase-C in mouse lymphocytes-T. *Eur. J. Immunol.* **21**: 195–201.

827. **Mashima, K., Y. Okajima, J. Usui, T. Shimizu, and K. Kimura.** 1994. Alteration in protein-tyrosine phosphatase of rat epithelial cells by RSV-transformation: application of phospho-tyrosyl glutamine synthetase to the study of protein-tyrosine phosphatase. *J. Biochem. Tokyo* **115**: 333–337.

828. **Masliah, E., M. Mallory, L. Hansen, M. Alford, T. Albright, R. Terry, P. Shapiro, M. Sundsmo, and T. Saitoh.** 1991. Immunoreactivity of CD45, a protein phosphotyrosine phosphatase, in Alzheimer's disease. *Acta Neuropathol. Berl.* **83**: 12–20.

829. **McGuire, M. C., R. M. Fields, B. L. Nyomba, I. Raz, C. Bogardus, N. K. Tonks, and J. Sommercorn.** 1991. Abnormal regulation of protein tyrosine phosphatase activities in skeletal muscle of insulin-resistant humans. *Diabetes* **40**: 939–942.

830. **Mei, L. and R. L. Huganir.** 1991. Purification and characterization of a protein tyrosine phosphatase which dephosphorylates the nicotinic acetylcholine receptor. *J. Biol. Chem.* **266**: 16063–16072.

831. **Meikrantz, W., D. M. Smith, M. M. Sladicka, and R. A. Schlegel.** 1991. Nuclear localization of an *O*-glycosylated protein phosphotyrosine phosphatase from human cells. *J. Cell Sci.* **98**: 303–307.

832. **Meyerovitch, J., J. M. Backer, P. Csermely, S. E. Shoelson, and C. R. Kahn.** 1992. Insulin differentially regulates protein phosphotyrosine phosphatase activity in rat hepatoma cells. *Biochemistry* **31**: 10338–10344.

833. **Meyerovitch, J., J. M. Backer, and C. R. Kahn.** 1989. Hepatic phosphotyrosine phosphatase activity and its alteration in diabetic rats. *J. Clin. Invest.* **84**: 976–983.

834. **Mimura, T., P. Fernsten, W. Jarjour, and J. B. Winfield.** 1990. Autoantibodies specific for different isoforms of CD45 in systemic lupus erythematosus. *J. Exp. Med.* **172**: 653–656.

835. **Mire-Sluis, A. R. and R. Thorpe.** 1991. Interleukin-4 proliferative signal transduction involves the activation of a tyrosine-specific phosphatase and the dephosphorylation of an 80-kDa protein. *J. Biol. Chem.* **266**: 18113–18118.

836. **Mishra, S. and A. W. Hamburger.** 1993. Role of intracellular Ca^{2+} in the epidermal growth factor induced inhibition of protein tyrosine phosphatase activity in a breast cancer cell line. *Biochem. Biophys. Res. Commun.* **191**: 1066–1072.

837. **Mishra, S. and A. W. Hamburger.** 1993. *O*-Phospho-L-tyrosine inhibits cellular growth by activating protein tyrosine phosphatases. *Cancer Res.* **53**: 557–563.

838. **Monteiro, H. P., Y. Ivaschenko, R. Fischer, and A. Stern.** 1991. Inhibition of protein tyrosine phosphatase activity by diamide is reversed by epidermal growth factor in fibroblasts. *FEBS Lett.* **295**: 146–148.

839. **Monteiro, H. P., Y. Ivaschenko, R. Fischer, and A. Stern.** 1993. Ascorbic acid inhibits protein tyrosine phosphatases in NIH 3T3 cells expressing human epidermal growth factor receptors. *Int. J. Biochem.* **25**: 1859–1864.

840*. **Mooney, R. A. and D. L. Anderson.** 1989. Phosphorylation of the insulin receptor in permeabilized adipocytes is coupled to a rapid dephosphorylation reaction. *J. Biol. Chem.* **264**: 6850–6857.

841. **Mooney, R. A. and K. L. Bordwell.** 1992. Differential dephosphorylation of the insulin receptor and its 160-kDa substrate (pp160) in rat adipocytes. *J. Biol. Chem.* **267**: 14054–14060.

842. **Motoyama, N., K. Takimoto, M. Okada, and H. Nakagawa.** 1987. Phosphotyrosine phosphatase: a novel phosphatase specific for phosphotyrosine, 2′-AMP and *p*-nitrophenylphosphate in rat brain. *J. Biochem. Tokyo* **101**: 939–947.

843. **Nadiv, O., M. Shinitzky, H. Manu, D. Hecht, C. T. Roberts, Jr., D. LeRoith, and Y. Zick.** 1994. Elevated protein tyrosine phosphatase activity and increased membrane viscosity are associated with impaired activation of the insulin receptor kinase in old rats. *Biochem. J.* **298**: 443–450.

844. **Nelson, R. L. and P. E. Branton.** 1984. Identification, purification, and characterization of phosphotyrosine-specific protein phosphatases from cultured chicken embryo fibroblasts. *Mol. Cell. Biol.* **4**: 1003–1012.

845. **Nishimura, H., S. Hattori, M. Abe, S. Hirose, and T. Shirai.** 1992. Differential expression of a CD45R epitope(6B2) on murine CD5+ B cells: possible difference in the post-translational modification of CD45 molecules. *Cell. Immunol.* **140**: 432–443.

846. **Okada, M., K. Owada, and H. Nakagawa.** 1986. [Phosphotyrosine]protein phosphatase in rat brain. A major [phosphotyrosine]protein phosphatase is a 23 kDa protein distinct from acid phosphatase. *Biochem. J.* **239**: 155–162.

847. **Olichon-Berthe, C., S. Hauguel deMouzon, P. Peraldi, E. Van Obberghen, and Y. Le Marchand-Brustel.** 1994. Insulin receptor dephosphorylation by phosphotyrosine phosphatases obtained from insulin-resistant obese mice. *Diabetologia* **37**: 56–60.

848. **Paganelli, R., E. Scala, E. Scarselli, C. Ortolani, A. Cossarizza, D. Carmini, F. Aiuti, and M. Fiorilli.** 1992. Selective deficiency of CD4+/CD45RA+ lymphocytes in patients with ataxia-telangiectasia. *J. Clin. Immunol.* **12**: 84–91.

849. **Pallen, C. J., G. N. Panayotou, L. Sahlin, and M. D. Waterfield.** 1988. Purification of a phosphotyrosine phosphatase that dephosphorylates the epidermal growth factor receptor autophosphorylation sites. *Ann. N. Y. Acad. Sci.* **551**: 299–308.

850. **Pallen, C. J., L. Sahlin, G. Panayotou, and M. D. Waterfield.** 1988. Purification and characterization of placental membrane phosphotyrosine phosphatases. *Cold Spring Harb. Symp. Quant. Biol.* **53**: 447–454.

851*. **Pallen, C. J. and P. H. Tong.** 1991. Elevation of membrane tyrosine phosphatase activity in density-dependent growth-arrested fibroblasts. *Proc. Natl Acad. Sci. USA* **88**: 6996–7000.

852. **Pan, M. G., T. Florio, and P. J. S. Stork.** 1992. G protein activation of a hormone-stimulated phosphatase in human tumor cells. *Science* **256**: 1215–1217.

853. **Partanen, S. and F. Pekonen.** 1989. Histochemically demonstrable phosphotyrosyl-protein phosphatase in normal human breast, in benign breast diseases and in breast cancer. *Anticancer Res.* **9**: 667–671.

854. **Patterson, W. P., C. W. Caldwell, and Y. W. Yesus.** 1993. *In vivo* upregulation of CD45RA in neutrophils of acutely infected patients. *Clin. Immunol. Immunopathol.* **68**: 35–40.

855. **Patthy, L.** 1990. Homology of a domain of the growth hormone/prolactin receptor family with type III modules of fibronectin. *Cell* **61**: 13–14.

856. **Peraldi, P., S. Hauguel deMouzon, F. Alengrin, and E. Van Obberghen.** 1992. Dephosphorylation of human insulin-like growth factor-I (IGF-I) receptors by membrane-associated tyrosine phosphatases. *Biochem J.* **285**: 71–78.

857. **Pernelle, J. J., C. Creuzet, J. Loeb, and G. Gacon.** 1991. Phosphorylation of the lymphoid cell kinase p56lck is stimulated by micromolar concentrations of Zn^{2+}. *FEBS Lett.* **281**: 278–282.

858. **Pulido, R., V. Alvarez, F. Mollinedo, and F. Sanchezmadrid.** 1992. Biochemical and functional characterization of the leucocyte tyrosine phosphatase CD45 (CD45RO, 180 kD) from human neutrophils – *in vivo* upregulation of CD45RO plasma membrane expression on patients undergoing haemodialysis. *Clin. Exp. Immunol.* **87**: 329–335.

859. **Purushotham, K. R., G. A. Paul, P. L. Wang, and M. G. Humphreysbeher.** 1994. Characterization of an Sh2 containing protein tyrosine phosphatase in rat parotid gland acinar cells. *Life Sci.* **54**: 1185–1194.

860. **Puzas, J. E. and J. S. Brand.** 1985. Bone cell phosphotyrosine phosphatase: characterization and regulation by calcitropic hormones. *Endocrinology* **116**: 2463–2468.

861. **Raid, A., B. Oliver, A. Abdelrahman, R. I. Shaafi, and J. J. Hajjar.** 1993. Role of tyrosine kinase and phosphotyrosine phosphatase in growth of the intestinal crypt cell (IEC-6) line. *Proc. Soc. Exp. Biol. Med.* **202**: 435–439.

862. **Rijksen, G., M. C. W. Voller, and E. J. J. Vanzoelen.** 1993. The role of protein tyrosine phosphatases in density-dependent growth control of normal rat kidney cells. *FEBS Lett.* **322**: 83–87.

863. **Rivard, N., D. Lebel, J. Laine, and J. Morisset.** 1994. Regulation of pancreatic tyrosine kinase and phosphatase activities by cholecystokinin and somatostatin. *Am. J. Physiol.* **266**: G1130–G1138.

864. **Roome, J., T. O'Hare, P. F. Pilch, and D. L. Brautigan.** 1988. Protein phosphotyrosine phosphatase purified from the particulate fraction of human placenta dephosphorylates insulin and growth-factor receptors. *Biochem. J.* **256**: 493–500.

865. **Rotenberg, S. A. and D. L. Brautigan.** 1987. Membrane protein phosphotyrosine phosphatase in rabbit kidney. Proteolysis activates the enzyme and generates soluble catalytic fragments. *Biochem. J.* **243**: 747–754.

866. **Rotin, D., B. Margolis, M. Mohammadi, R. J. Daly, G. Daum, N. Li, E. H. Fischer, W. H. Burgess, A. Ullrich, and J. Schlessinger.** 1992. SH2 domains prevent tyrosine dephosphorylation of the EGF receptor: identification of Tyr-992 as the high-affinity binding site for SH2 domains of phospholipase C γ. *EMBO J.* **11**: 559–567.

867. **Sako, F., N. Taniguchi, and A. Makita.** 1985. Phosphotyrosine phosphatase activity in human placenta. *Jpn. J. Exp. Med.* **55**: 21–27.

868. **Shibata, K., M. Noda, Y. Sawa, and T. Watanabe.** 1994. Acid phosphatase purified from *Mycoplasma fermentans* has protein tyrosine phosphatase-like activity. *Infect. Immun.* **62**: 313–315.

869. **Shriner, C. L. and D. L. Brautigan.** 1984. Cytosolic protein phosphotyrosine phosphatases from rabbit kidney. *J. Biol. Chem.* **259**: 11383–11390.

870. **Singh, T. J.** 1990. Characterization of a bovine brain magnesium-dependent phosphotyrosine protein phosphatase that is inhibited by micromolar concentrations of calcium. *Biochem. Biophys. Res. Commun.* **167**: 621–627.

871. **Smilowitz, H. M., L. Aramli, D. Xu, and P. M. Epstein.** 1991. Phosphotyrosine phosphatase activity in human platelets. *Life Sci.* **49**: 29–37.

872. **Sparks, J. W. and D. L. Brautigan.** 1985. Specificity of protein phosphotyrosine phosphatases: comparison with mammalian alkaline phosphatase using polypeptide substrates. *J. Biol. Chem.* **260**: 2042–2045.

873. **Spivak, J. L., J. Fisher, M. A. Isaacs, and W. D. Hankins.** 1992. Protein kinases and phosphatases are involved in erythropoietin-mediated signal transduction. *Exp. Hematol.* **20**: 500–504.

874. **Sredy, J., D. R. Sawicki, B. R. Flam and D. Sullivan.** 1995. Insulin resistance is associated with abnormal dephosphorylation of a synthetic phosphopeptide corresponding to the major autophosphorylation sites of the insulin receptor. *Metabolism* **44**: 1074–1081.

875. **Starling, G. C. and D. N. Hart.** 1990. CD45 molecule cross-linking inhibits natural killer cell-mediated lysis independently of lytic triggering. *Immunology* **71**: 190–195.

876. **Stoscheck, C. M., D. B. Friedman, and L. E. King.** 1989. Identification of a phosphotyrosyl-protein phosphatase in mouse epidermis. *J. Invest. Dermatol.* **92**: 379–384.

877. **Strout, H. V., P. P. Vicario, R. Saperstein, and E. E. Slater.** 1988. A protein phosphotyrosine phosphatase distinct from alkaline phosphatase with activity against the insulin receptor. *Biochem. Biophys. Res. Commun.* **151**: 633–640.

878. **Sullivan, S. G., D. T. Y. Chiu, M. Errasfa, J. M. Wang, J. S. Qi, and A. Stern.** 1994. Effects of H_2O_2 on protein tyrosine phosphatase activity in Her14 cells. *Free Radical Biol. Med.* **16**: 399–403.

879. **Sun, X. J., D. L. Crimmins, M. G. Myers, M. Miralpeix, and M. F. White.** 1993. Pleiotropic insulin signals are engaged by multisite phosphorylation of IRS-1. *Mol. Cell. Biol.* **13**: 7418–7428.

880. **Swarup, G., K. V. Speeg, Jr., S. Cohen, and D. L. Garbers.** 1982. Phosphotyrosyl-protein phosphatase of TCRC-2 cells. *J. Biol. Chem.* **257**: 7298–7301.

881. **Swarup, G. and G. Subrahmanyam.** 1989. Purification and characterization of a protein-phosphotyrosine phosphatase from rat spleen which dephosphorylates and inactivates a tyrosine-specific protein kinase. *J. Biol. Chem.* **264**: 7801–7808.

882. **Tahirijouti, N., C. Cambillau, N. Viguerie, C. Vidal, L. Buscail, N. Saintlaurent, N. Vaysse, and C. Susini.** 1992. Characterization of a membrane tyrosine phosphatase in AR42J cells – regulation by somatostatin. *Am. J. Physiol.* **262**: G1007–G1014.

883. **Takahasi, K., S. Bardhan, Y. Kambayashi, H. Shirai, and T. Inagami.** 1994. Protein tyrosine phosphatase inhibition by angiotensin II in rat pheochromocytoma cells through type 2 receptor, at(2). *Biochem. Biophys. Res. Commun.* **198**: 60–66.

884. **Tamura, S., Y. Suzuki, K. Kikuchi, and S. Tsuiki.** 1986. Identification and characterization of Mg^{2+}-dependent phosphotyrosyl protein phosphatase from rat liver cytosol. *Biochem. Biophys. Res. Commun.* **140**: 212–218.

885. **Tappia, P. S., P. G. P. Atkinson, R. P. Sharma, and G. J. Sale.** 1993. Regulation of an hepatic low-M_r membrane-associated protein-tyrosine phosphatase. *Biochem J.* **292**: 1–5.

886. **Tappia, P. S., R. P. Sharma, and G. J. Sale.** 1991. Dephosphorylation of autophosphorylated insulin and epidermal-growth-factor receptors by two major subtypes of protein-tyrosine-phosphatase from human placenta. *Biochem. J.* **278**: 69–74.

887. **Tchou Wong, K. M. and I. B. Weinstein.** 1992. Altered expression of protein kinase-C, lck, and CD45 in a 12-*O*-tetradecanoylphorbol-13-acetate-dependent leukemic T-cell variant that expresses a high level of interleukin-2 receptor. *Mol. Cell. Biol.* **12**: 394–401.

888. **Titus, L., L. G. Marzilli, J. Rubin, M. S. Nanes, and B. D. Catherwood.** 1993. Rat osteoblasts and Ros 17/2.8 cells contain a similar protein tyrosine phosphatase. *Bone Miner.* **23**: 267–284.

889. **Tomaska, L. and R. J. Resnick.** 1993. Involvement of a phosphotyrosine protein phosphatase in the suppression of platelet-derived growth factor receptor autophosphorylation in ras-transformed cells. *Biochem J.* **293**: 215–221.

890. **Tung, H. Y. and L. J. Reed.** 1987. Identification and purification of a cytosolic phosphotyrosyl protein phosphatase from bovine spleen. *Anal. Biochem.* **161**: 412–419.

891. **Umezawa, K., A. R. Kornblihtt, and F. E. Baralle.** 1985. Isolation and characterization of cDNA clones for human liver fibronectin. *FEBS Lett.* **186**: 31–34.

892. **Vergnes, H., A. Brisson Lougarre, P. Limouzy, C. Dray, and J. Grozdea.** 1992. Phosphotyrosine phosphatase activity and haematologic changes in Down's syndrome patients. *Br. J. Haematol.* **80**: 157–159.

893. **Vicendo, P., J. Fauvel, J. M. F. Ragabthomas, and H. Chap.** 1991. Identification, characterization and purification to near-homogeneity of a novel 67 kDa phosphotyrosyl protein phosphatase associated with pig lung annexin extract. *Biochem. J.* **278**: 435–440.

894. **Villa-Moruzzi, E., S. Lapi, M. Prat, G. Gaudino, and P. M. Comoglio.** 1993. A protein tyrosine phosphatase activity associated with the hepatocyte growth factor/scatter factor receptor. *J. Biol. Chem.* **268**: 18176–18180.

895. **Wilson, G. F. and L. K. Kaczmarek.** 1993. Mode-switching of a voltage-gated cation channel is mediated by a protein kinase-A regulated tyrosine phosphatase. *Nature* **366**: 433–438.

896. **Yamada, K. M.** 1991. Adhesive recognition sequences. *J. Biol. Chem.* **266**: 12809–12812.

897. **Yamauchi, K.** 1994. SH-PTP2 regulates insulin-dependent transcriptional activation. *Diabetes* **43**: 2A(Abstr.)

898. **Yoshioka, T., O. Tanaka, H. Otani, and H. Shinohara.** 1991. Histochemically demonstrable phosphotyrosine protein phosphatase in the rat hippocampal formation. *Brain Res.* **555**: 177–179.

899. **Zafriri, D., M. Argaman, E. Canaani, and A. Kimchi.** 1993. Induction of protein-tyrosine-phosphatase activity by interleukin-6 in M1 myeloblastic cells and analysis of possible counteractions by the BCR-ABL oncogene. *Proc. Natl Acad. Sci. USA* **90**: 477–481.

References for second edition

Reviews

900. **Bliska, J. B.** 1995. Crystal structure of the *Yersinia* tyrosine phosphatase. *Trends Microbiol.* **3**: 125–127.

901*. **Brady-Kalnay, S. M., and N. K. Tonks.** 1995. Protein tyrosine phosphatases as adhesion receptors. *Curr. Opin. Cell Biol.* **7**: 650–657.

902. **Cobb, M. H. and E. J. Goldsmith.** 1995. How MAP kinases are regulated. *J. Biol. Chem.* **270**: 14843–14846.

903. **Coghlan, V. M., Z. E. Hausken, and J. D. Scott.** 1995. Subcellular targeting of kinases and phosphatases by association with bifunctional anchoring proteins. *Biochem. Soc. Trans.* **23**: 592–596.

904. **Defranco, A. L. and D. A. Law.** 1995. Tyrosine phosphatases and the antibody response. *Science* **268**: 263–264.

905. **Edgar, B. A., D. A. Lehman, and P. H. O'Farrell.** 1994. Transcriptional regulation of string (cdc25): a link between developmental programming and the cell cycle. *Development* **120**: 3131–3143.

906. **Hoffmann, I. and E. Karsenti.** 1994. The role of cdc25 in checkpoints and feedback controls in the eukaryotic cell cycle. *J. Cell Sci.* **18**: 75–79.

907*. **Hunter, T.** 1995. Protein kinases and phosphatases: the yin and yang of protein phosphorylation and signaling. *Cell* **80**: 225–236.

908. **Imboden, J. B. and G. A. Koretzky.** 1995. Intracellular signalling – switching off signals. *Current Biol.* **5**: 727–729.

909*. **Keyse, S. M.** 1995. An emerging family of dual specificity MAP kinase phosphatases. *Biochim. Biophys. Acta–Mol. Cell Res.* **1265**: 152–160.

910. **McCulloch, J. and K. A. Siminovitch.** 1994. Involvement of the protein tyrosine phosphatase PTP1C in cellular physiology, autoimmunity and oncogenesis. *Adv. Exp. Med. Biol.* **365**: 245–254.

911. **Okumura, M. and M. L. Thomas.** 1995. Regulation of immune function by protein tyrosine phosphatases. *Curr. Opin. Immunol.* **7**: 312–319.

912. **Penninger, J. M. and T. W. Mak** 1994. Signal transduction, mitotic catastrophes, and death in T-cell development. *Immunol. Rev.* **142**: 231–272.

913. **Southey, M. C., D. M. Findlay, and B. E. Kemp.** 1995. Regulation of membrane-associated tyrosine phosphatases in UMR 106.06 osteoblast-like cells. *Biochem. J.* **305**: 485–490.

914. **Stork, P. J. S., A. Misrapress, and M. G. Pan.** 1995. Receptor-activated tyrosine phosphatases: activity assays and molecular cloning. *Receptor Mol. Biol.* **25**: 242–260.

915. **Yakura, H.** 1994. The role of protein tyrosine phosphatases in lymphocyte activation and differentiation. *Crit. Rev. Immunol.* **14**: 311–336.

Derivatization

916. **Gerlach, M. S., A. Kharitonenkov, W. Vogel, S. Ali, and A. Ullrich.** 1995. Protein-tyrosine phosphatase 1D modulates its own state of tyrosine phosphorylation. *J. Biol. Chem.* **270**: 24635–24637.

917. **Giordanengo, V., M. Limouse, J. F. Peyron, and J. C. Lefebvre.** 1995. Lymphocytic CD43 and CD45 bear sulfate residues potentially implicated in cell to cell interactions. *Eur. J. Immunol.* **25**: 274–278.

918. **Hoffmann, I., G. Draetta, and E. Karsenti.** 1994. Activation of the phosphatase activity of human cdc25A by a cdk2-cyclin E dependent phosphorylation at the G1/S transition. *EMBO J.* **13**: 4302–4310.

919. **Izumi, T. and J. L. Maller.** 1995. Phosphorylation and activation of the *Xenopus* cdc25 phosphatase in the absence of cdc2 and cdk2 kinase activity. *Mol. Biol. Cell* **6**: 215–226.

920. **Li, R. Y., F. Gaits, A. Ragab, J. M. F. Ragabthomas, and H. Chap.** 1995. Tyrosine phosphorylation of an SH2-containing protein tyrosine phosphatase is coupled to platelet

thrombin receptor via a pertussis toxin-sensitive heterotrimeric G-protein. *EMBO J.* **14**: 2519–2526.

921. **Meyerputtlitz, B., P. Milev, E. Junker, I. Zimmer, R. U. Margolis, and R. K. Margolis.** 1995. Chondroitin sulfate and chondroitin/keratan sulfate proteoglycans of nervous tissue – developmental changes of neurocan and phosphacan. *J. Neurochem.* **65**: 2327–2337.

922. **Ogg, S., B. Gabrielli, and H. Piwnica-Worms.** 1994. Purification of a serine kinase that associates with and phosphorylates human Cdc25C on serine 216. *J. Biol. Chem.* **269**: 30461–30469.

923. **Swieter, M., E. H. Berenstein, W. D. Swaim, and R. P. Siraganian.** 1995. Aggregation of IgE receptors in rat basiphilic leukemia 2H3 cells induces tyrosine phosphorylation of the cytosolic protein-tyrosine phosphatase HePTP. *J. Biol. Chem.* **270**: 21902–21906.

924. **Tracy, S., P. Vandergeer, and T. Hunter.** 1995. The receptor-like protein-tyrosine phosphatase, RPTP alpha, is phosphorylated by protein kinase C on two serines close to the inner face of the plasma membrane. *J. Biol. Chem.* **270**: 10587–10594.

925. **Vambutas, V., D. R. Kaplan, M. A. Sells, and J. Chernoff.** 1995. Nerve growth factor stimulates tyrosine phosphorylation and activation of Src homology-containing protein-tyrosine phosphatase 1 in PC12 cells. *J. Biol. Chem.* **270**: 25629–25633.

Expression

926. **Adachi, M., T. Torigoe, M. Sekiya, Y. Minami, T. Taniguchi, Y. Hinoda, A. Yachi, J. C. Reed, and K. Imai.** 1995. IL-2-induced gene expression of protein tyrosine phosphatase LC-PTP requires acidic and serine-rich regions within IL-2 receptor beta chain. *FEBS Lett.* **372**: 113–118.

927. **Ahmad, F., R. V. Considine, and B. J. Goldstein.** 1995. Increased abundance of the receptor-type protein-tyrosine phosphatase LAR accounts for the elevated insulin receptor dephosphorylating activity in adipose tissue of obese human subjects. *J. Clin. Invest.* **95**: 2806–2812.

928. **Ahmad, F. and B. J. Goldstein.** 1995. Purification, identification and subcellular distribution of three predominant protein-tyrosine phosphatase enzymes in skeletal muscle tissue. *Biochim. Biophys. Acta–Protein Struct. Mol. Enzym.* **1248**: 57–69.

929. **Bieber, T., M. Jurgens, A. Wollenberg, E. Sander, D. Hanau, and H. Delasalle.** 1995. Characterization of the protein tyrosine phosphatase CD45 on human epidermal Langerhans cells. *Eur. J. Immunol.* **25**: 317–321.

930. **Bissen, S. T.** 1995. Expression of the cell cycle control gene, *cdc25*, is constitutive in the segmental founder cells but is cell-cycle-regulated in the micromeres of leech embryos. *Development* **121**: 3035–3043.

931. **Bonini, J. A., J. R. Colca, C. Dailey, M. White, and C. Hofmann.** 1995. Compensatory alterations for insulin signal transduction and glucose transport in insulin-resistant diabetes. *Am. J. Physiol. – Endocrinol. Metab.* **32**: E759–E765.

932. **Celler, J. W., X. M. Luo, L. J. Gonez, and F. D. Bohmer.** 1995. mRNA expression of two transmembrane protein tyrosine phosphatases is modulated by growth factors and growth arrest in 3T3 fibroblasts. *Biochem. Biophys. Res. Commun.* **209**: 614–621.

933. **Chida, D., T. Kume, Y. Mukouyama, S. Tabata, N. Nomura, M. L. Thomas, T. Watanabe, and M. Oishi.** 1995. Characterization of a protein tyrosine phosphatase (RIP) expressed at a very early stage of differentiation in both mouse erythroleukemia and embryonal carcinoma cells. *FEBS Lett.* **358**: 233–239.

934. **Cornelius, G. and M. Engel.** 1995. Stress causes induction of MAP kinase-specific phosphatase and rapid repression of MAP kinase activity in *Drosophila*. *Cell. Signal.* **7**: 611–615.

935. **Cyster, J. G., D. Fowell, and A. N. Barclay.** 1994. Antigenic determinants encoded by alternatively spliced exons of CD45 are determined by the polypeptide but influenced by glycosylation. *Int. Immunol.* **6**: 1875–1881.

936. **Ezumi, Y., H. Takayama, and M. Okuma.** 1995. Differential regulation of protein-tyrosine phosphatases by integrin alpha(IIb)beta(3) through cytoskeletal reorganization and tyrosine phosphorylation in human platelets. *J. Biol. Chem.* **270**: 11927–11934.

937. **Festin, R., A. Bjorkland, and T. H. Totterman.** 1994. Multicolor flow cytometric analysis of the CD45 antigen provides improved lymphoid cell discrimination in bone marrow and tissue biopsies. *J. Immunol. Methods* **177**: 215–224.

938. **Fitzpatrick, K. A., S. M. Gorski, Z. Ursuliak, and J. V. Price.** 1995. Expression of protein tyrosine phosphatase genes during oogenesis in *Drosophila melanogaster*. *Mech. Dev.* **53**: 171–183.

939. **Fricker, R. A., E. M. Torres, P. J. Lombroso, and S. B. Dunnett.** 1994. The localization of an antibody to STEP in embryonic striatal tissue grafts. *Neuroreport* **5**: 2638–2640.

940. **Gaits, F., R. Y. Li, A. Ragab, J. M. F. Ragabthomas, and H. Chap.** 1995. Increase in receptor-like protein tyrosine phosphatase activity and expression level on density-dependent growth arrest of endothelial cells. *Biochem. J.* **311**: 97–103.

941*. **Gebbink, M. F., G. C. M. Zondag, G. M. Koningstein, E. Feiken, R. W. Wubbolts, and W. H. Moolenaar.** 1995. Cell surface expression of receptor protein tyrosine phosphatase RPTP-mu is regulated by cell–cell contact. *J. Cell Biol.* **131**: 251–260.

942. **Hansson, J., M. Dohlsten, H. O. Sjogren, and G. Hedlund.** 1995. Distinct splicing of CD45 mRNA in activated rat gamma delta cytotoxic T lymphocytes. *Eur. J. Immunol.* **25**: 75–79.

943. **Heimerl, P., C. Stader, R. Willmann, and H. W. Hofer.** 1995. Phorbol ester-dependent regulation of nuclear protein-tyrosine phosphatase *in situ*. *Cell. Signal.* **7**: 341–350.

944. **Ikuta, S., F. Itoh, Y. Hinoda, M. Toyota, Y. Makiguchi, K. Imai, and A. Yachi.** 1994. Expression of cytoskeletal-associated protein tyrosine phosphatase PTPH1 mRNA in human hepatocellular carcinoma. *J. Gastroenterol.* **29**: 727–732.

945. **Johannisson, A. and R. Festin.** 1995. Phenotype transition of CD4(+) T cells from CD45RA to CD45rO is accompanied by cell activation and proliferation. *Cytometry* **19**: 343–352.

946. **Kaneko, T., T. Moriyama, E. Imai, Y. Akagi, M. Arai, T. Inoue, C. Xia, T. Noguchi, T. Kamada, and N. Ueda.** 1995. Expression of transmembrane-type protein tyrosine phosphatase mRNA along rat nephron segments. *Am. J. Physiol.–Renal. Fl. Elect.* **37**: F1102–F1108.

947. **Katsura, H., M. C. Williams, J. S. Brody, and Q. Yu.** 1995. Two closely related receptor-type tyrosine phosphatases are differentially expressed during rat lung development. *Dev. Dynamics* **204**: 89–97.

948. **Kawano, S., E. Tatsumi, N. Yoneda, A. Tani, and F. Nakamura.** 1995. Expression pattern of CD45 RA/RO isoformic antigens in T-lineage neoplasms. *Am. J. Hematol.* **49**: 6–14.

949. **Kitamura, T., K. Nakamura, Y. Mizuno, and K. Kikuchi.** 1995. Gene expressions of protein tyrosine phosphatases in regenerating rat liver and rat ascites hepatoma cells. *Japanese J. Cancer Res.* **86**: 811–818.

950*. **Kwak, S. P. and J. E. Dixon.** 1995. Multiple dual specificity protein tyrosine phosphatases are expressed and regulated differentially in liver cell lines. *J. Biol. Chem.* **270**: 1156–1160.

951*. **Lorenzen, J. A., C. Y. Dadabay, and E. H. Fischer.** 1995. COOH-terminal sequence motifs target the T cell protein tyrosine phosphatase to the ER and nucleus. *J. Cell Biol.* **131**: 631–643.

952. **Lucibello, F. C., M. Truss, J. Zwicker, F. Ehlert, M. Beato, and R. Muller.** 1995. Periodic *cdc25c* transcription is mediated by a novel cell cycle-regulated repressor element (CDE). *EMBO J.* **14**: 132–142.

953. **Maegawa, H., R. Ide, M. Hasegawa, S. Ugi, K. Egawa, M. Iwanishi, R. Kikkawa, Y. Shigeta, and A. Kashiwagi.** 1995. Thiazolidine derivatives ameliorate high glucose-induced insulin resistance via the normalization of protein-tyrosine phosphatase activities. *J. Biol. Chem.* **270**: 7724–7730.

954. **Mizuno, K., K. Hasegawa, M. Ogimoto, T. Katagiri, and H. Yakura**. 1994. Developmental regulation of gene expression for the MPTP delta isoforms in the central nervous system and the immune system. *FEBS Lett.* **355**: 223–228.

955. **Mok, S. C., T. T. Kwok, R. S. Berkowitz, A. J. Barrett, and F. W. Tsui.** 1995. Overexpression of the protein tyrosine phosphatase, nonreceptor type 6 (PTPN6), in human epithelial ovarian cancer. *Gynecol. Oncol.* **57**: 299–303.

956. **Nambirajan, S., R. S. Reddy, and G. Swarup.** 1995. Enhanced expression of a chromatin associated protein tyrosine phosphatase during G(0) to S transition. *J. Biosci.* **20**: 461–471.

957. **Ottenhoffkalff, A. E., B. A. Vanoirschot, A. Hennipman, R. A. Deweger, G. E. J. Staal, and G. Rijksen.** 1995. Protein tyrosine phosphatase activity as a diagnostic parameter in breast cancer. *Beast Cancer Res. Treat.* **33**: 245–256.

958. **Payton, M. A., C. J. Hawkes, and M. R. Christie**. 1995. Relationship of the 37 000- and 40 000-*M*(r) tryptic fragments of islet antigens in insulin-dependent diabetes to the protein-tyrosine phosphatase-like molecule IA-2 (ICA512). *J. Clin. Invest.* **96**: 1506–1511.

959. **Pulido, R., N. X. Krueger, C. Serra-Pages, H. Saito, and M. Streuli.** 1995. Molecular characterization of the human transmembrane protein-tyrosine phosphatase delta – evidence for tissue-specific expression of alternative human transmembrane protein-tyrosine phosphatase delta isoforms. *J. Biol. Chem.* **270**: 6722–6728.

960. **Raschke, W. C., M. Hendricks, and C. M. Chen.** 1995. Genetic basis of antigenic differences between three alleles of *ly5* (*CD45*) in mice. *Immunogenetics* **41**: 144–147.

961. **Rivard, N., F. R. Mckenzie, J. M. Brondello, and J. Pouyssegur.** 1995. The phosphotyrosine phosphatase PTP1d, but not PTP1c, is an essential mediator of fibroblast proliferation induced by tyrosine kinase and G protein-coupled receptors. *J. Biol. Chem.* **270**: 11017–11024.

962. **Sahin, M., J. J. Dowling, and S. Hockfield.** 1995. Seven protein tyrosine phosphatases are differentially expressed in the developing rat brain. *J. Comp. Neurol.* **351**: 617–631.

963. **Sahin, M., S. A. Slaugenhaupt, J. F. Gusella, and S. Hockfield.** 1995. Expression of PTPH1, a rat protein tyrosine phosphatase, is restricted to the derivatives of a specific diencephalic segment. *Proc. Natl Acad. Sci. USA* **92**: 7859–7863.

964. **Sakai, T., T. Agui, and K. Matsumoto.** 1995. Abnormal CD45RC expression and elevated CD45 protein tyrosine phosphatase activity in LEC rat peripheral CD4(+) T cells. *Eur. J. Immunol.* **25**: 1399–1404.

965. **Shock, L, P., D. J. Bare, S. G. Klinz, and P. F. Maness.** 1995. Protein tyrosine phosphatases expressed in developing brain and retinal Müller glia. *Mol. Brain Res.* **28**: 110–116.

966. **Solca, F. F., D. I. Lurie, C. D. Diltz, R. S. Johnson, S. Kumar, E. W. Rubel, and E. H. Fischer.** 1994. Identification and purification of a chicken brain neuroglia-associated protein. *J. Biol. Chem.* **269**: 27559–27565.

967. **Stoker, A. W., B. Gehrig, F. Haj, and B. H. Bay.** 1995. Axonal localisation of the CAM-like tyrosine phosphatase CRYP alpha: a signalling molecule of embryonic growth cones. *Development* **121**: 1833–1844.

968. **Suzuki, T., T. Matozaki, A. Mizoguchi, and M. Kasuga.** 1995. Localization and subcellular distribution of SH-PTP2, a protein-tyrosine phosphatase with src homology-2 domains, in rat brain. *Biochem. Biophys. Res. Commun.* **211**: 950–959.

969. **Tabiti, K., D. R. Smith, H. S. Goh, and C. J. Pallen.** 1995. Increased mRNA expression of the receptor-like protein tyrosine phosphatase alpha in late stage colon carcinomas. *Cancer Lett.* **93**: 239–248.

970. **Tagawa, M., T. Shirasawa, S. Fujimura, and S. Sakiyama.** 1994. Identification of a rat protein tyrosine phosphatase gene preferentially expressed in the embryonal brain. *Cell Mol. Biol. Res.* **40**: 627–631.

971. **Tagawa, M., T. Shirasawa, S. Fujimura, and S. Sakiyama.** 1994. Expression of protein tyrosine phosphatase genes in the developing brain of mouse and rat. *Biochem. Mol. Biol. Int.* **33**: 1221–1227.

972. **Takano, S., H. Fukuyama, M. Fukumoto, K. Hirashimizu, T. Higuchi, J. Takenawa, H. Nakayama, J. Kimura, and J. Fujita.** 1995. Induction of CL100 protein tyrosine phosphatase following transient forebrain ischemia in the rat brain. *J. Cereb. Blood Flow Metab.* **15**: 33–41.

973. **Thude, H., J. Hundrieser, K. Wonigeit, and R. Schwinzer.** 1995. A point mutation in the human *CD45* gene associated with defective splicing of exon A, *Eur. J. Immunol.* **25**: 2101–2106.

974. **Vantus, T., P. Csermely, I. Teplan, and G. Keri.** 1995. The tumor-selective somatostatin analog, TT2-32 induces a biphasic activity in human colon tumor cell line, SW620. *Tumour Biol.* **16**: 261–267.

975. **Wang, H., H. Yan, P. D. Canoll, O. Silvennoinen, J. Schlessinger, and J. M. Musacchio.** 1995. Expression of receptor protein tyrosine phosphatase-sigma (RPTP-sigma) in the nervous system of the developing and adult rat. *J. Neurosci. Res.* 41: 297–310.

976. **Wickramasinghe, D., S. Becker, M. K. Emst, J. L. Resnick, J. M. Centanni, L. Tessarollo, L. B. Grabel, and P. J. Donovan.** 1995. Two CDC25 homologues are differentially expressed during mouse development. *Development* **121**: 2047–2056.

977. **Wiessner, C.** 1995. The dual specificity phosphatase PAC-1 is transcriptionally induced in the rat brain following transient forebrain ischemia. *Mol. Brain Res.* **28**: 353–356.

978. **Wiessner, C., T. Neumannhaefelin, P. Vogel, T. Back, and K. A. Hossmann.** 1995. Transient forebrain ischemia induces an immediate-early gene encoding the mitogen-activated protein kinase phosphatase 3CH134 in the adult rat brain. *Neuroscience* **64**: 959–966.

979. **Wu, S. and D. J. Wolgemuth.** 1995. The distinct and developmentally regulated patterns of expression of members of the mouse *cdc25* gene family suggest differential firnctions during gametogenesis. *Dev. Biol.* **170**: 195–206.

980*. **Zhang, J. S. and F. M. Longo.** 1995. LAR tyrosine phosphatase receptor: alternative splicing is preferential to the nervous system, coordinated with cell growth and generates novel isoforms containing extensive CAG repeats. *J. Cell Biol.* **128**: 415–431.

Function

981. **Adamczewski, M., R. P. Numerof, G. A. Koretzky, and J. P. Kinet.** 1995. Regulation by CD45 of the tyrosine phosphorylation of high affinity IgE receptor beta- and gamma-chains. *J. Immunol.* **154**: 3047–3055.

982*. **Ahmad, F., P. M. Li, J. Meyerovitch, and B. J. Goldstein.** 1995. Osmotic loading of neutralizing antibodies defines a role for protein-tyrosine phosphatase 1B in negative regulation of the insulin action pathway. *J. Biol. Chem.* **270**: 20503–20508.

982a. **Alessi, D. R., C. Smythe, and S. M. Keyse.** 1993. The human CL100 gene encodes a Tyr/Thr-protein phosphatase which potently and specifically inactivates MAP kinase and suppresses its activation by oncogenic ras in *Xenopus* oocyte extracts. *Oncogene* **8**: 2015–2020.

983. **Aroca, P., D. P. Bottaro, T. Ishibashi, S. A. Aaronson, and E. Santos.** 1995. Human dual specificity phosphatase VHR activates maturation promotion factor and triggers meiotic maturation in *Xenopus* oocytes. *J. Biol. Chem.* **270**: 14229–14234.

984. **Bevan, A. P., J. W. Burgess, J. F. Yale, P. G. Drake, D. Lachance, G. Baquiran, A. Shaver, and B. I. Posner.** 1995. *In vivo* insulin mimetic effects of PV compounds: role for tissue targeting in determining potency. *Am. J. Physiol–Endocrinol. Metab.* **31**: E60–E66.

985. **Bliska, J. B. and D. S. Black.** 1995. Inhibition of the Fc receptor-mediated oxidative burst in macrophages by the *Yersinia pseudotuberculosis* tyrosine phosphatase. *Infect. Immun.* **63**: 681–685.

986*. **Brady-Kalnay, S. M. and N. K. Tonks.** 1994. Identification of the homophilic binding site of the receptor protein tyrosine phosphatase PTP-mu. *J. Biol. Chem.* **269**: 28472–28477.

987. **Brondello, J. M., F. R. Mckenzie, H. Sun, N. K. Tonks, and J. Pouyssegur.** 1995. Constitutive MAP kinase phosphatase (MKP-I) expression blocks G1 specific gene transcription and S-phase entry in fibroblasts. *Oncogene* **10**: 1895–1904.

988. **Chiarugi, P., P. Cirri, G. Raugei, G. Camici, F. Dolfi, A. Berti, and G. Ramponi.** 1995. PDGF receptor as a specific *in vivo* target for low *M*(r) phosphotyrosine protein phosphatase. *FEBS Lett.* **372**: 49–53.

989. **Cirri, P., A. Caselli, G. Manao, G. Camici, R. Polidori, G. Cappugi, and G. Ramponi.** 1995. Kinetic studies on rat liver low *M*(r) phosphotyrosine protein phosphatases. The activation mechanism of the isoenzyme acp2 by cGMP. *Biochim. Biophys. Acta–Gen. Subjects* **1243**: 129–135.

990. **Cyster, J. G. and C. C. Goodnow.** 1995. Protein tyrosine phosphatase 1C negatively regulates antigen receptor signaling in B lymphocytes and determines thresholds for negative selection. *Immunity* **2**: 13–24.

991. **Dechert, U., M. Affolter, K. W. Harder, J. Matthews, P. Owen, I. Clark-Lewis, M. L. Thomas, R. Aebersold, and F. R. Jirik.** 1995. Comparison of the specificity of bacterially expressed cytoplasmic protein-tyrosine phosphatases SHP and SH-PTP2 towards synthetic phosphopeptide substrates. *Eur. J. Biochem.* **231**: 673–681.

992. **den Hertog, J., J. Sap, C. E. G. M. Pals, J. Schlessinger, and W. Kruijer.** 1995. Stimulation of receptor protein-tyrosine phosphatase alpha activity and phosphorylation by phorbol ester. *Cell Growth Differ.* **6**: 303–307.

993*. **Denu, J. M. and J. E. Dixon.** 1995. A catalytic mechanism for the dual-specific phosphatases. *Proc. Natl Acad. Sci. USA* **92**: 5910–5914.

994. **Denu, J. M., G. C. Zhou, Y. P. Guo, and J. E. Dixon.** 1995. The catalytic role of aspartic acid-92 in a human dual-specific protein-tyrosine-phosphatase. *Biochemistry* **34**: 3396–3403.

995. **Denu, J. M., G. C. Zhou, L. Wu, R. Zhao, J. D. Yuvaniyama, M. A. Saper, and J. E. Dixon.** 1995. The purification and characterization of a human dual-specific protein tyrosine phosphatase. *J. Biol. Chem.* **270**: 3796–3803.

996. **Denu, J. M., G. C. Zhou, L. Wu, R. Zhao, J. Yuvaniyama, M. A. Saper, and J. E. Dixon.** 1995. The purification and characterization of a human dual-specific protein tyrosine phosphatase (correction). *J. Biol. Chem.* **270**: 10358

997. **Doody, G. M., L. B. Justement, C. C. Delibrias, R.J. Matthews, J. J. Lin, M. L. Thomas, and D. T. Fearon.** 1995. A role in B cell activation for CD22 and the protein tyrosine phosphatase SHP. *Science* **269**: 242–244.

998. **Duff, J. L., B. P. Monia, and B. C. Berk.** 1995. Mitogen-activated protein (MAP) kinase is regulated by the MAP kinase phosphatase (MKP-1) in vascular smooth muscle cells. *J. Biol. Chem.* **270**: 7161–7166.

999. **Fallman, M., K. Andersson, S. Hakansson, K. E. Magnusson, O. Stendahl, and H. Wolfwatz.** 1995. *Yersinia pseudotuberculosis* inhibits Fc receptor-mediated phagocytosis on J774 cells. *Infect. Immun.* **63**: 3117–3124.

1000. **Galaktionov, K., A. K. Lee, J. Eckstein, G. Draetta, J. Meckler, M. Loda, and D. Beach.** 1995. CDC25 phosphatases as potental human oncogenes. *Science* **269**: 1575–1577.

1001. **Green, S. P., E. L Hartland, R. M. Robinsbrowne, and W. A. Phillips.** 1995. Role of yopH in the suppression of tyrosine phosphorylation and respiratory burst activity in murine macrophages infected with *Yersinia enterocolitica*. *J. Leukocyte Biol.* **57**: 972–977.

1002. **Hamilton, B. A., A. Ho, and K. Zinn.** 1995. Targeted mutagenesis and genetic analysis of a *Drosophila* receptor-linked protein tyrosine phosphatase gene. *Roux. Arch. Dev. Biol.* **204**: 187–192.

1003. **Hanaoka, K., N. Fujita, S. H. Lee, H. Seimiya, M. Naito, and T. Tsuruo.** 1995. Involvement of CD45 in adhesion and suppression of apoptosis of mouse malignant T-lymphoma cells. *Cancer Res.* **55**: 2186–2190.

1004. **Hausdorff, S. F., A. M. Bennett, B. G. Neel, and M. J. Birnbaum.** 1995. Different signaling roles of SHPTP2 in insulin-induced GLUT1 expression and GLUT4 translocation. *J. Biol. Chem.* **270**: 12965–12968.

1005. **Hoffmeyer, F., K. Witte, U. Gebhardt, and R. E. Schmidt.** 1995. The low affinity Fc-gamma-RIIA and Fc-gamma-RIIIB on polymorphonuclear neutrophils are differentially regulated by CD45 phosphatase. *J. Immunol.* **155**: 4016–4023.

1006. **Katagiri, T., M. Ogimoto, K. Hasegawa, K. Mizuno, and H. Yakura.** 1995. Selective regulation of Lyn tyrosine kinase by CD45 in immature B cells. *J. Biol. Chem.* **270**: 27987–27990.

1007. **Kawai, K., K. Kishihara, T. J. Molina, V. A. Wallace, T. W. Mak, and P. S. Ohashi.** 1995. Impaired development of v gamma 3 dendritic epidermal T cells in p56(lck) protein tyrosine kinase-deficient and CD45 protein tyrosine phosphatase-deficient mice. *J. Exp. Med.* **181**: 345–349.

1008. **Kuhne, M. R., Z. H. Zhao, and G. E. Lienhard.** 1995. Evidence against dephosphorylation of insulin-elicited phosphotyrosine proteins *in vivo* by the phosphatase PTP2C. *Biochem. Biophys. Res. Commun.* **211**: 190–197.

1009*. **Kulas, D. T., B. J. Goldstein, and R. A. Mooney.** 1996. The transmembrane protein-tyrosine phosphatase LAR modulates signaling by multiple receptor tyrosine kinases. *J. Biol. Chem.* **271**: 748–754.

1010. **Lamb, P., J. Haslam, L. Kessler, H. M. Seidel, R. B. Stein, and J. Rosen.** 1994. Rapid activation of the interferon-gamma signal transduction pathway by inhibitors of tyrosine phosphatases. *J. Interferon Res.* **14**: 365–373.

1011. **Leitenberg, D., S. Constant, D. D. Lu, B. R. Smith, and K. Bottomly.** 1995. CD4 and CD45 regulate qualitatively distinct patterns of calcium mobilization in individual CD4(+) T cells. *Eur. J. Immunol.* **25**: 2445–2451.

1012. **Liao, K. and M. D. Lane.** 1995. The blockade of preadipocyte differentiation by protein tyrosine phosphatase HA2 is reversed by vanadate. *J. Biol. Chem.* **270**: 12123–12132.

1013. **Lindstrom, E., A. Berglund, K. H. Mild, P. Lindstrom, and E. Lundgren.** 1995. CD45 phosphatase in Jurkat cells is necessary for response to applied ELF magnetic fields. *FEBS Lett.* **370**: 118–122.

1014. **Liu, Y. S., M. Gorospe, C. L. Yang, and N. J. Holbrook.** 1995. Role of mitogen-activated protein kinase phosphatase during the cellular response to genotoxic stress – inhibition of c-jun N-terminal kinase activity and AP-1-dependent gene activation. *J. Biol. Chem.* **270**: 8377–8380.

1015. **Maroun, C. R. and M. Julius.** 1994. Distinct involvement of CD45 in antigen receptor signalling in CD4+ and CD8+ primary T cells. *Eur. J. Immunol.* **24**: 967–973.

1016. **Mckenney, D. W., H. Onodera, L. Gorman, T. Mimura, and D. M. Rothstein.** 1995. Distinct isoforms of the CD45 protein-tyrosine phosphatase differentially regulate interleukin 2 secretion and activation of signal pathways involving vav in T cells. *J. Biol. Chem.* **270**: 24949–24954.

1017. **Menon, S. D., G. R. Guy, and Y. H. Tan.** 1995. Involvement of a puative protein tyrosine phosphatase and I-kappa-B-alpha serine phosphorylation in nuclear factor kappa-B activation by tumor necrosis factor. *J. Biol. Chem.* **270**: 18881–18887.

1018. **Millar, J. B. A., V. Buck, and M. G. Wilkinson.** 1995. Pyp1 and pyp2 PTPases dephosphorylate an osmosensing MAP kinase controlling cell size at division in fission yeast. *Genes Dev.* **9**: 2117–2130.

1019. **Moller, N. P. H., K. B. Moller, R. Lammers, A. Kharitonenkov, E. Hoppe, F. C. Wiberg, I. Sures, and A. Ullrich.** 1995. Selective down-regulation of the insulin receptor signal by protein-tyrosine phosphatases alpha and epsilon. *J. Biol. Chem.* **270**: 23126–23131.

1020. **Mustelin, T., S. Williams, P. Tailor, C. Couture, G. Zenner, P. Burn, J. D. Ashwell, and A. Altman.** 1995. Regulation of the p70zap tyrosine protein kinase in T cells by the CD45 phosphotyrosine phosphatase. *Eur. J. Immunol.* **25**: 942–946.

1021. **Ng, D. H. W., A. Maiti, and P. Johnson.** 1995. Point mutation in the second phosphatase domain of CD45 abrogates tyrosine phosphatase activity. *Biochem. Biophys. Res. Commun.* **206**: 302–309.

1022. **Niklinska, B. B., D. Hou, C. June, A. M. Weissman, and J. D. Ashwell.** 1994. CD45 tyrosine phosphatase activity and membrane anchoring are required for T-cell antigen receptor signaling. *Mol. Cell. Biol.* **14**: 8078–8084.

1023. **Novak, T. J., D. Farber, D. Leitenberg, S. C. Hong, P. Johnson, and K. Bottomly.** 1994. Isoforms of the transmembrane tyrosine phosphatase CD45 differentially affect T cell recognition. *Immunity* **1**: 109–119.

1024. **O'Connor, T. J., J. D. Bjorge, H. C. Cheng, J. H. Wang, and D. J. Fujita.** 1995. Mechanism of c-SRC activation in human melanocytes: elevated level of protein tyrosine phosphatase activity directed against the carboxy-terminal regulatory tyrosine. *Cell Growth Differ.* **6**: 123–130.

1025. **Ogimoto, M., T. Katagiri, K. Mashima, K. Hasegawa, K. Mizuno, and H. Yakura.** 1995. Antigen receptor-initiated growth inhibition is blocked in CD45-loss variants of a mature B lymphoma, with limited effects on apoptosis. *Eur. J. Immunol.* **25**: 2265–2271.

1026. **Passini, N., J. D. Larigan, S. Genovese, E. Appella, F. Sinigaglia, and L. Rogge.** 1995. The 37/40-kilodalton autoantigen in insulin-dependent diabetes mellitus is the putative tyrosine phosphatase IA-2. *Proc. Natl Acad. Sci. USA* **92**: 9412–9416.

1027. **Pei, D. H., U. Lorenz, U. Klingmuller, B. G. Neel, and C. T. Walsh.** 1994. Intramolecular regulation of protein tyrosine phosphatase SH-PTP1: a new function for src homology 2 domains. *Biochemistry* **33**: 15483–15493.

1028. **Poon, R. Y. C. and T. Hunter.** 1995. Dephosphorylation of cdk2 thr(160) by the cyclin-dependent kinase-interacting phosphatase KAP in the absense of cyclin. *Science* **270**: 90–93.

1029. **Reeves, S. A., B. Sinha, I. Baur, D. Reinhold, and G. Harsh.** 1995. An alternative role for the Src-homology-domain-containing phosphotyrosine phosphatase (SH-PTP2) in regulating epidermal-growth-factor-dependent cell growth. *Eur. J. Biochem.* **233**: 55–61.

1030. **Rime, H., D. Huchon, V. De Smedt, C. Thibier, K. Galaktionov, C. Jessus, and R. Ozon.** 1994. Microinjection of Cdc25 protein phosphatase into *Xenopus* prophase oocyte activates MPF and arrests meiosis at metaphase I. *Biol. Cell* **82**: 11–22.

1031. **Sapru, M. K., G. C. Zhou, and D. Goldman.** 1994. Protein-tyrosine phosphatases specifically regulate muscle adult-type nicotinic acetylcholine receptor gene expression. *J. Biol. Chem.* **269**: 27811–27814.

1032. **Sawada, T., K. L. Milarski, and A. R. Saltiel.** 1995. Expression of a catalytically inert Syp blocks activation MAP kinase pathway downstream of *p21(Ras)*. *Biochem. Biophys. Res. Comm.* **214**: 737–743.

1033. **Shen, F., X. L. Xu, L. H. Graf, and A. S. F. Chong.** 1995. CD45-cross-linking stimulates IFN-gamma production in NK cells. *J. Immunol.* **154**: 644–652.

1034. **Sigrist, S., G. Ried, and C. F. Lehner.** 1995. DMcdc2 kinase is required for both meiotic divisions during *Drosophila* spermatogenesis and is activated by the twine/cdc25 phosphatase. *Mech. Dev.* **53**: 247–260.

1035. **Sorio, C., J. Mendrola, Z. W. Lou, S. Laforgia, C. M. Croce, and K. Huebner.** 1995. Characterization of the receptor protein-tyrosine phosphatase gene product PTP-gamma – binding and activation by triphosphorylated nucleosides. *Cancer Res.* **55**: 4855–4864.

1036. **Stein Gerlach, M., A. Kharitonenkov, W. Vogel, S. Ali, and A. Ullrich.** 1995. Protein-tyrosine phosphatase 1D modulates its own state of tyrosine phosphorylation. *J. Biol. Chem.* **270**: 24635–24637.

1037. **Su, X., T. Zhou, Z. Wang, P. Yang, R. S. Jope, and J. D. Mountz.** 1995. Defective expression of hematopoietic cell protein tyrosine phosphatase (HCP) in lymphoid cells blocks Fas-mediated apoptosis. *Immunity* **2**: 353–362.

1038*. **Tang, T. L., R. M. Freeman, A. M. O'Reilly, B. G. Neel, and S. Y. Sokol.** 1995. The SH2-containing protein-tyrosine phosphatase SH-PTP2 is required upstream of MAP kinase for early *Xenopus* development. *Cell* **80**: 473–483.

1039. **Vanhoof, C., X. Cayla, M. Bosch, W. Merlevede, and J. Goris.** 1994. The phosphotyrosyl phosphatase activator of protein phosphatase 2a – a novel purification method, immunological and enzymic characterization. *Eur. J. Biochem.* **226**: 899–907.

1040. **Weigt, C., H. Korte, R. Pogge, R. P. Vonstrandmann, W. Hengstenberg, and H. E. Meyer.** 1995. Identification of phosphocysteine by electrospray mass spectroscopy combined with Edman degradation. *J. Chromatogr.* **712**: 141–147.

1041. **Wilson, L. K., B. M. Benton, S. Zhou, J. Thorner, and G. S. Martin.** 1995. The yeast immunophilin FPR3 is a physiological substrate of the tyrosine-specific phosphoprotein phosphatase PTP1. *J. Biol. Chem.* **270**: 25185–25193.

1042*. **Yamauchi, K., K. L. Miharski, A. R. Saltiel, and J. E. Pessin.** 1995. Protein-tyrosine phosphatase SHPTP2 is a required positive effector for insulin downstream signaling. *Proc. Natl Acad. Sci. USA* **92**: 664–668.

1043. **Zhang, S. H., W. R. Eckberg, Q. Yang, A. A. Samatar, and N. K. Tonks.** 1995. Biochemical characterization of a human band 4.1-related protein-tyrosine phosphatase, PTPH1. *J. Biol. Chem.* **270**: 20067–20072.

1044. **Zhang, Z. Y.** 1995. Kinetic and mechanistic characterization of a mammalian protein-tyrosine phosphatase, PTP1. *J. Biol. Chem.* **270**: 11199–11204.

1045. **Zhang, Z. Y.** 1995. Are protein-tyrosine phosphatases specific for phosphotyrosine? *J. Biol. Chem.* **270**: 16052–16055.

1046*. **Zhang, Z. Y., Y. Wang, L. Wu, E. B. Fauman, J. A. Stuckey, H. L. Schubert, M. A. Saper, and J. E. Dixon.** 1994. The cys(x)(5)arg catalytic motif in phosphoester hydrolysis. *Biochemistry* **33**: 15266–15270.

1047. **Zhao, Z. Z., Z. Tan, J. H. Wright, C. D. Diltz, S. H. Shen, E. G. Krebs, and E. H. Fischer.** 1995. Altered expression of protein-tyrosine phosphatase 2c in 293 cells affects protein tyrosine phosphorylation and mitogen activated protein kinase activation. *J. Biol. Chem.* **270**: 11765–11769.

1048. **Zhou, G. C., J. M. Denu, L. Wu, and J. E. Dixon.** 1994. The catalytic role of Cys(124) in the dual specificity phosphatase VHR. *J. Biol. Chem.* **269**: 28084–28090.

Methods for purifyng and assaying

1049. **Altin, J. G. and E. B. Pagler.** 1995. A one-step procedure for biotinylation and chemical cross-linking of lymphocyte surface and intracellular membrane-associated molecules. *Anal. Biochem.* **224**: 382–389.

1050. **Bohmer, F. D., A. Bohmer, A. Obermeier, and A. Ullrich.** 1995. Use of selective tyrosine kinase blockers to monitor growth factor receptor dephosphorylation in intact cells. *Anal. Biochem.* **228**: 267–273.

1051. **Bult, A., J. R. Naegele, E. Sharma, and P. J. Lombroso.** 1995. IdentifLing and characterizing protein-tyrosine phosphatases in the brain. *Neuroprotocols* **6**: 91–104.

1052. **Cheng, Q., Z. X. Wang, and S. D. Killilea.** 1995. A continuous spectrophotometric assay for protein phosphatases. *Anal. Biochem.* **226**: 68–73.

1053. **Ng, D. H. W., K. W. Harder, I. Clark-Lewis, F. Jirik, and P. Johnson.** 1995. Non radioactive method to measure CD45 protein tyrosine phosphatase activity isolated directly from cells. *J. Immunol. Methods* **179**: 177–185.

1054. **Paoli, P., G. Camici, G. Manao, and G. Ramponi.** 1995. 2-methoxybenzoyl phosphate: a new substrate for continuous fluorimetric and spectrophotometric acyl phosphatase assays. *Experientia* **51**: 57–62.

1055. **Skarnes, W. C., J. Moss, S. M. Hurtley, and R. S. Beddington.** 1995. Capturing genes encoding membrane and secreted proteins important for mouse development. *Proc. Natl Acad. Sci. USA* **92**: 6592–6596.

Protein interactions

1056. **Arendt, C. W. and H. L. Ostergaard.** 1995. CD45 protein-tyrosine phosphatase is specifically associated with a 116-kDa tyrosine-phosphorylated glycoprotein. *J. Biol. Chem.* **270**: 2313–2319.

1057*. **Brady-Kalnay, S. M., D. L. Rimm, and N. K. Tonks.** 1995. Receptor protein tyrosine phosphatase PTP-mu associates wih cadherins and catenins *in vivo*. *J. Cell Biol.* **130**: 977–986.

1058. **Campbell, M. A. and N. R. Klinman.** 1995. Phosphotyrosine-dependent association between CD22 and protein tyrosine phosphatase 1C. *Eur. J. Immunol.* **25**: 1573–1579.

1059. **Carmo, A. M. and M. D. Wright.** 1995. Association of the transmembrane 4 superfamily molecule CD53 with a tyrosine phosphatase activity. *Eur. J. Immunol.* **25**: 2090–2095.

1060. **Conklin, D. S., K. Galaktionov, and D. Beach.** 1995. 14-3-3 proteins associate with cdc25 phosphatases. *Proc. Natl Acad. Sci. USA* **92**: 7892–7896.

1061. **Corvaia, N., I. G. Reischl, E. Kroemer, and G. C. Mudde.** 1995. Modulation of Fc gamma receptor-mediated early events by the tyrosine phosphatase CD45 in primary human monocytes. Consequences for interleukin-6 production. *Eur. J. Immunol.* **25**: 738–744.

1062. **D'Ambrosio, D., K. L. Hippen, S. A. Minskoff, I. Mellman, G. Pani, K. A. Siminovitch, and J. C. Cambier.** 1995. Recruitment and activation of PTP1c in negative regulation of antigen receptor signaling by Fc gamma RIIB 1. *Science* 268: 293–297.

1063. **Fuhrer, D. K., G. S. Feng, and Y. C. Yang.** 1995. Syp associates with gp130 and Janus kinase 2 in response to interleukin-11 in 3T3-L1 mouse preadipocytes. *J. Biol. Chem.* **270**: 24826–24830.

1064. **Galaktionov, K., C. Jessus, and D. Beach.** 1995. Raf1 interaction with cdc25 phosphatase ties mitogenic signal transduction to cell cycle activation. *Gene Dev.* **9**: 1046–1058.

1065. **Gervais, F. G. and A. Veillette.** 1995. The unique amino-terminal domain of p56(lck) regulates interactions with tyrosine protein phosphatases in T lymphocytes. *Mol. Cell. Biol.* **15**: 2393–2401.

1066. **Huyer, G., Z. M. Li, M. Adam, W. R. Huckle, and C. Ramachandran.** 1995. Direct determination of the sequence recognition requirements of the SH2 domains of SH-PTP2. *Biochemistry* **34**: 1040–1049.

1067. **Kitamura, K., A. Maiti, D. H. W. Ng, P. Johnson, A. L. Maizel, and A. Takeda.** 1995. Characterization of the interaction between CD45 and CD45-AP. *J. Biol. Chem.* **270**: 21151–21157.

1068. **Klinghoffer, R. A. and A. Kazlauskas.** 1995. Identification of a putative Syp substrate, the PDGF-beta receptor. *J. Biol. Chem.* **270**: 22208–22217.

1069*. **Klingmuller, U., U. Lorenz, L. C. Cantley, B. G. Neel, and H. F. Lodish.** 1995. Specific recruitment of SH-PTP1 to the erythropoietin receptor causes inactivation of JAK2 and termination of proliferative signals. *Cell* **80**: 729–738.

1070. **Lankester, A. C., G. M. W. Vanschijndel, and R. A. W. Vanlier.** 1995. Hematopoietic cell phosphatase is recruited to CD22 following B cell antigen receptor ligation. *J. Biol. Chem.* **270**: 20305–20308.

1071. **Maclean, D., A. M. Sefler, G. C. Zhu, S. J. Decker, A. R. Saltiel, J. Singh, D. Mcnamara, E. M. Dobrusin, and T. K. Sawyer.** 1995. Differentiation of peptide molecular recognition by phospholipase C gamma-1 src homology-2 domain and a mutant tyrosine phosphatase PTP1b(c215s). *Protein Sci.* **4**: 13–20.

1072*. **Milev, P., D. R. Friedlander, T. Sakurai, L. Karthikeyan, M. Flad, R. K. Margolis, M. Grumet, and R. U. Margolis.** 1994. Interactions of the chondroitin sulfate proteoglycan phosphacan, the extracellular domain of a receptor-type protein tyrosine phosphatase, with neurons, glia, and neural cell adhesion molecules. *J. Cell Biol.* **127**: 1703–1715.

1073. **Milev, P., B. Meyerputtlitz, R. K. Margolis, and R. U. Margolis.** 1995. Complex-type asparagine-linked oligosaccharides on phosphacan and protein-tyrosine phosphatase-zeta/beta mediate their binding to neural cell adhesion molecules and tenascin. *J. Biol. Chem.* **270**: 24650–24653.

1074. **Pani, G., M. Kozlowski, J. C. Cambier, G. B. Mills, and K. A. Siminovitch.** 1995. Identification of the tyrosine phosphatase PTP1c as a B cell antigen receptor-associated protein involved in the regulation of B cell signaling. *J. Exp. Med.* **181**: 2077–2084.

1075*. **Peles, E., M. Nativ, P. L. Campbell, T. Sakurai, R. Martinez, S. Lev, D. O. Clary, J. Schilling, G. Barnea, G. D. Plowman, M. Grumet, and J. Schlessinger.** 1995. The carbonic anhydrase domain of receptor tyrosine phosphatase beta is a functional ligand for the axonal cell recognition molecule contactin. *Cell* **82**: 251–260.

1076*. **Pluskey, S., T. J. Wandless, C. T. Walsh, and S. E. Shoelson.** 1995. Potent stimulation of SH-PTP2 phosphatase activity by simultaneous occupancy of both SH2 domains. *J. Biol. Chem.* **270**: 2897–2900.

1077. **Seely, B. L., D. R. Reichart, P. A. Staubs, B. H. Jhun, D. Hsu, H. Maegawa, K. L. Milarski, A. R. Saltiel, and J. M. Olefsky.** 1995. Localization of the insulin-like growth factor I receptor binding sites for the SH2 domain proteins p85, Syp, and GTPase activating protein. *J. Biol. Chem.* **270**: 19151–19157.

1078*. **Serra-Pages, C., N. L. Kedersha, L. Fazikas, Q. Medley, A. Debant, and M. Streuli.** 1995. The LAR transmembrane protein tyrosine phosphatase and a coiled-coil LAR-interacting protein co-localize at focal adhesions. *EMBO J.* **14**: 2827–2838.

1079. **Sgroi, D., G. A. Koretzky, and I. Stamenkovic.** 1995. Regulation of CD45 engagement by the B-cell receptor CD22. *Proc. Natl Acad. Sci. USA* **92**: 4026–4030.

1079a. **Staubs, P. A., D. R. Reichart, A. R. Saltiel, K. L. Milarski, H. Maegawa, P. Berhanu, J. M. Olefsky, and B. L. Seely.** 1994. Localization of the insulin receptor binding sites for the SH2 domain proteins p85, Syp, and GAP. *J. Biol. Chem.* **269**: 27186–27192.

1080*. **Tauchi, T., G. S. Feng, R. Shen, M. Hoatlin, G. C. Bagby, D. Kabat, L. Lu, and H. E. Broxmeyer.** 1995. Involvement of SH2-containing phosphotyrosine phosphatase syp in erythropoietin receptor signal transduction pathways. *J. Biol. Chem.* **270**: 5631–5635.

1081. **Tian, S. S. and K. Zinn.** 1994. An adhesion molecule-like protein that interacts with and is a substrate for a *Drosophila* receptor-linked protein tyrosine phosphatase. *J. Biol. Chem.* **269**: 28478–28486.

1082. **Tomic, S., U. Greiser, R. Lammers, A. Kharitonenkov, E. Imyanitov, A. Ullrich, and F. D. Bohmer.** 1995. Association of SH2 domain protein tyrosine phosphatases with the epidermal growth factor receptor in human tumor cells – phosphatidic acid activates receptor dephosphorylation by PTP1C. *J. Biol. Chem.* **270**: 21277–21284.

1083. **Wu, Y. J., G. Pani, K. A. Siminovitch, and N. Hozumi.** 1995. Antigen receptor-triggered apoptosis in immature B cell lines is associated with the binding of a 44-kDa phosphoprotein to the PTP1C tyrosine phosphatase. *Eur. J. Immunol.* **25**: 2279–2284.

1084. **Yamauchi, K. and J. E. Pessin.** 1995. Epidermal growth factor-induced association of the SHPTP2 protein tyrosine phosphatase with a 115-kDa phosphotyrosine protein. *J. Biol. Chem.* **270**: 14871–14874.

1085. **Yamauchi, K., V. Ribon, A. R. Saltiel, and J. E. Pessin.** 1995. Identification of the major SH-PTP2-binding protein that is tyrosine phosphorylated in response to insulin. *J. Biol. Chem.* **270**: 17716–17722.

1086. **Yetter, A., S. Uddin, J. J. Krolewski, H. Y. Jiao, T. L. Yi, and L. C. Platanias.** 1995. Association of the interferon-dependent tyrosine kinase TYK-2 with the hematopoietic cell phosphatase. *J. Biol. Chem.* **270**: 18179–18182.

1087. **Yi, T. L., J. L. Zhang, O. Miura, and J. N. Ihle.** 1995. Hematopoietic cell phosphatase associates with erythropoietin (epo) receptor after epo-induced receptor tyrosine phosphorylation: identification of potential binding sites. *Blood* **85**: 87–95.

1088. **Zapata, J. M., M. R. Campanero, M. Marazuela, F. Sanchezmadrid, and M. O. Delandazuri.** 1995. B-cell homotypic adhesion through exon-A restricted epitopes of CD45 involves LPA-1/ICAM-1, ICAM-3 interactions, and induces co-clustering of CD45 and LFA-1. *Blood* **86**: 1861–1872.

1089. **Zondag, G. C. M., G. M. Koningstein, Y. P. Jiang, J. Sap, W. H. Moolenaar, and M. F. Gebbink.** 1995. Homophilic interactions mediated by receptor tyrosine phosphatases mu and kappa – a critical role for the novel extracellular MAM domain. *J. Biol. Chem.* **270**: 14247–14250.

1090. **Zondag, G. C. M., G. M. Koningstein, Y. P. Jiang, J. Sap, W. H. Moolenaar, and M. F. Gebbink.** 1995. Homophilic interactions mediated by receptor tyrosine phosphatases mu and kappa – a critical role for the novel extracellular MAM domain (correction). *J. Biol. Chem.* **270**: 24621.

Sequences

1091. **Ariyama, T., K. Hasegawa, J. Inazawa, K. Mizuno, M. Ogimoto, T. Katagiri, and H. Yakura.** 1995. Assignment of the human protein tyrosine phosphatase, receptor-type, zeta (*PTPRZ*) gene to chromosome band 7q31.3. *Cytogenet. Cell Genet.* **70**: 52–54.

1092. **Banville, D., S. Ahmad, R. Stocco, and S. H. Shen.** 1995. A novel protein-tyrosine phosphatase with homology to both the cytoskeletal proteins of the band 4.1 family and junction-associated guanylate kinases (correction). *J. Biol. Chem.* **270**: 10359.

1093. **Banville, D., R. Stocco, and S. H. Shen.** 1995. Human protein tyrosine phosphatase 1c (*PTPN6*) gene structure: alternate promoter usage and exon skipping generate multiple transcripts. *Genomics* **27**: 165–173.

1094. **Boldog, F. L., B. Waggoner, T. W. Glover, I. Chumakov, D. Lepaslier, D. Cohen, R. M. Gemmill, and H. A. Drabkin.** 1994. Integrated YAC contig containing the 3p14.2 hereditary renal carcinoma 3;8 translocation breakpoint and the fragile site FRA3B. *Gene Chrom. Cancer* **11**: 216–221.

1095. **Boulanger, L. M., P. J. Lombroso, A. Raghunathan, M. J. During, P. Wahle, and J. R. Naegele.** 1995. Cellular and molecular characterization of a brain-enriched protein tyrosine phosphatase. *J. Neurosci.* **15**: 1532–1544.

1096. **Charest, A., J. Wagner, E. S. Muise, H. H. Q. Heng, and M. L. Tremblay.** 1995. Structure of the murine MPTP-PEST gene – genomic organization and chromosomal mapping. *Genomics* **28**: 501–507.

1097. **Charest, A., J. Wagner, S. H. Shen, and M. L. Tremblay.** 1995. Murine protein tyrosine phosphatase-PEST, a stable cytosolic protein tyrosine phosphatase. *Biochem. J.* **308**: 425–432.

1098. **Cui, H., W. X. Xu, D. Powell, and B. Newman.** 1995. Cloning and sequencing of a pig cdc25 tyrosine phosphatase cDNA. *J. Anim. Sci.* **73**: 630.

1099. **Desai, C. J., E. Popova, and K. Zinn.** 1994. A *Drosophila* receptor tyrosine phosphatase expressed in the embryonic CNS and larval optic lobes is a member of the set of proteins bearing the 'HRP' carbohydrate epitope. *J. Neurosci.* **14**: 7272–7283.

1100. **Guan, K. L. and E. Butch.** 1995. Isolation and characterization of a novel dual specific phosphatase, HVH2, which selectively dephosphorylates the mitogen-activated protein kinase. *J. Biol. Chem.* **270**: 7197–7203.

1101. **Harder, K. W., J. Saw, N. Miki, and F. Jirik.** 1995. Coexisting amplifications of the chromosome 1p32 genes (*PTPRF* and *MYCL1*) encoding protein tyrosine phosphatase LAR and l-myc in a small cell lung cancer line. *Genomics* **27**: 552–553.

1102. **Haring, M. A., M. Siderius, C. Jonak, H. Hirt, K. M. Walton, and A. Musgrave.** 1995. Tyrosine phosphatase signalling in a lower plant: cell-cycle and oxidative stress-regulated expression of the *Chlamydomonas eugametos VH-PTP13* gene. *Plant J.* **7**: 981–988.

1103. **Hendriks, W., J. Schepens, C. Brugman, P. Zeeuwen, and B. Wieringa.** 1995. A novel receptor-type protein tyrosine phosphatase with a single catalytic domain is specifically expressed in mouse brain. *Biochem. J.* **305**: 499–504.

1104. **Higashituji, H., S. Arii, M. Furutani, M. Imamura, Y. Kaneko, J. Takenawa, H. Nakayama, and J. Fujita.** 1995. Enhanced expression of multiple protein tyrosine phosphatases in the regenerating mouse liver: isolation of PTP-RL10, a novel cytoplasmic-type phosphatase with sequence homology to cytoskeletal protein 4.1. *Oncogene* **10**: 407–414.

1105. **Honda, H., J. Inazawa, J. Nishida, Y. Yazaki, and H. Hirai.** 1994. Molecular cloning, characterization, and chromosomal localization of a novel protein-tyrosine phosphatase, HPTP eta. *Blood* **84**: 4186–4194.

1106. **Kamb, A., P. A. Futreal, J. Rosenthal, C. Cochran, K. D. Harshman, Q. Liu, R. S. Phelps, S. V. Tavtigian, T. Tran, C. Hussey, *et al.*** 1994. Localization of the VHR phosphatase gene and its analysis as a candidate for BRCA1. *Genomics* **23**: 163–167.

1107. **Labbe, D., D. Banville, Y. Tong, R. Stocco, S. Masson, S. Y. Ma, G. Fantus, and S. H. Shen.** 1994. Identification of a novel protein tyrosine phosphatase with sequence homology to the cytoskeletal proteins of the band 4.1 family. *FEBS Lett.* **356**: 351–356.

1108. **Lewis, T., L. A. Groom, A. A. Sneddon, C. Smythe, and S. M. Keyse.** 1995. XCL100, an inducible nuclear MAP kinase phosphatase from *Xenopus laevis* – its role in MAP kinase inactivation during early development. *J. Cell Sci.* **108**: 2885–2896.

1109. **Li, X., J. Luna, P. J. Lombroso, and U. Francke.** 1995. Molecular cloning of the human homolog of a striatum-enriched phosphatase (STEP) and chromosomal mapping of the human and murine loci. *Genomics* **28**: 442–449.

1110. **Martell, K. J., A. F. Seasholtz, S. P. Kwak, K. K. Clemens, and J. E. Dixon.** 1995. HVH-5 – a protein tyrosine phosphatase abundant in brain that inactivates mitogen-activated protein kinase. *J. Neurochem.* **65**: 1823–1833.

1111. **Misra-Press, A., C. S. Rim, H. Yao, M. S. Roberson, and P. J. S. Stork.** 1995. A novel mitogen-activated protein kinase phosphatase – structure, expression, and regulation. *J. Biol. Chem.* **270**: 14587–14596.

1112. **Mondesert, O., S. Moreno, and P. Russell.** 1994. Low molecular weight protein-tyrosine phosphatases are highly conserved between fission yeast and man. *J. Biol. Chem.* **269**: 27996–27999.

1113. **Mukouyama, Y., T. Watanabe, T. Kume, and M. Oishi.** 1995. Genetic mapping of *PTPRM* on mouse chromosome 17. *Mamm. Genome* **6**: 757–758.

1114. **Nehls, M., M. Schorpp, and T. Boehm.** 1995. An intragenic deletion in the human *PTPN6* gene affects transcriptional activity. *Hum. Genet.* **95**: 713–715.

1115. **Ogata, M., M. Sawada, Y. Fujino, and T. Hamaoka.** 1995. cDNA cloning and characterization of a novel receptor-type protein tyrosine phosphatase expressed predominantly in the brain. *J. Biol. Chem.* **270**: 2337–2343.

1116. **Ostanin, K., C. Pokalsky, S. Wang, and R. L. Van Etten.** 1995. Cloning and characterization of a *Saccharomyces cerevisiae* gene encoding the low molecular weight protein-tyrosine phosphatase. *J. Biol. Chem.* **270**: 18491–18499.

1117*. **Pulido, R., C. Serra-Pages, M. Tang, and M. Streuli.** 1995. The LAR/PTP-DELTA/PTP-SIGMA subfamily of transmembrane protein-tyrosine phosphatases – multiple human LAR, PTP-DELTA, and PTP-SIGMA isoforms are expressed in a tissue specific manner and associate with the LAR-interacting protein LIP.1. *Proc. Natl Acad. Sci. USA* **92**: 11686–11690.

1118*. **Sato, T., S. Irie, S. Kitada, and J. C. Reed.** 1995. FAP-1: a protein tyrosine phosphatase that associates with fas. *Science* **268**: 411–415.

1119. **Schaapveld, R. Q. J., A. M. J. M. Vandenmaagdenberg, J. T. G. Schepens, D. O. Weghuis, A. G. Vankessel, B. Wieringa, and W. J. A. J. Hendriks.** 1995. The mouse gene PTPRF encoding the leukocyte common antigen-related molecule LAR: cloning, characterization, and chromosomal localization. *Genomics* **27**: 124–130.

1120. **Seimiya, H., T. Sawabe, J. Inazawa, and T. Tsuruo.** 1995. Cloning, expression and chromosomal localization of a novel gene for protein tyrosine phosphatase (PTP-u2) induced by various differentiation-inducing agents. *Oncogene* **10**: 1731–1738.

1121. **Sharma, E. and P. J. Lombroso.** 1995. A neuronal protein tyrosine phosphatase induced by nerve growth factor. *J. Biol. Chem.* **270**: 49–53.

1122. **Sharma, E. and P. J. Lombroso.** 1995. A neuronal protein tyrosine phosphatase induced by nerve growth factor (correction). *J. Biol. Chem.* **270**: 23234.

1123. **Sharma, E., F. S. Zhao, A. Bult, and P. J. Lombroso.** 1995. Identification of two alternatively spliced transcripts of STEP – a subfamily of brain-enriched protein tyrosine phosphatases. *Mol. Brain Res.* **32**: 87–93.

1124. **Shiozuka, K., Y. Watanabe, T. Ikeda, S. Hashimoto, and H. Kawashima.** 1995. Cloning and expression of PCPTP1-encoding protein-tyrosine phosphatase. *Gene* **162**: 279–284.

1125. **Smith, A. L, P. J. Mitchell, J. Shipley, B. A. Gusterson, M. V. Rogers, and M. R. Crompton.** 1995. Pez: a novel human cDNA encoding protein tyrosine phosphatase- and ezrin-like domains. *Biochem. Biophys. Res. Commun.* **209**: 959–965.

1126. **Vandenmaagdenberg, A. M. J. M., H. H. V. Vandenhurk, D. O. Weghuis, B. Wieringa, A. G. Vankessel, and W. J. A. J. Hendriks.** 1995. Assignment of the human protein tyrosine phosphatase epsilon (*PTPRE*) gene to chromosome 10q26 by fluorescence *in situ* hybridization. *Genomics* **30**: 128–129.

1127. **Vanhoof, C., M. S. Aly, A. Garcia, X. Cayla, J. J. Cassiman, W. Merlevede, and J. Goris.** 1995. Structure and chromosomal localization of the human gene of the phosphotyrosyl phosphatase activator (PTPA) of protein phosphatase 2A. *Genomics* **28**: 261–272.

1128. **Wagner, J., D. Boerboom, and M. L. Tremblay.** 1994. Molecular cloning and tissue-specific RNA processing of a murine receptor-type protein tyrosine phosphatase. *Eur. J. Biochem.* **226**: 773–782.

1129. **Watanabe, T., Y. Mukouyama, M. Rhodes, M. Thomas, T. Kume, and M. Oishi.** 1995. Chromo-somal location of murine protein tyrosine phosphatase (*PTPRJ* and *PTPRE*) genes. *Genomics* **29**: 793–795.

1130. **Wiggins, R. C., J. E. Wiggins, M. Goyal, B. L. Wharram, and P. E. Thomas.** 1995. Molecular cloning of cDNAs encoding human GLEPP1, a membrane protein tyrosine phosphatase: characterization of the GLEPP1 protein distribution in human kidney and assignment of the GLEPP1 gene to human chromosome 12p12–p13. *Genomics* **27**: 174–181.

1131. **Yi, H. F., C. C. Morton, S. Weremowicz, O. W. Mcbride, and K. Kelly.** 1995. Genomic organization and chromosomal localization of the *DUSP2* gene, encoding a MAP kinase phosphatase, to human 2p11.2-q11. *Genomics* **28**: 92–96.

1132. **Zhang, Z. Y., G. C. Zhou, J. M. Denu, L. Wu, X. J. Tang, O. Mondesert, P. Russell, E. Butch, and K. L. Guan.** 1995. Purification and characterization of the low molecular weight protein tyrosine phosphatase, STP1, from the fission yeast *Schizosaccharomyces pombe*. *Biochemistry* **34**: 10560–10568.

Structural studies

1133*. **Jia, Z. C., D. Barford, A. J. Flint, and N. K. Tonks.** 1995. Structural basis for phosphotyrosine peptide recognition by protein tyrosine phosphatase 1B. *Science* **268**: 1754–1758.

1134. **Lee, C. H., D. Kominos, S. Jacques, B. Margolis, J. Schlessinger, S. E. Shoelson, and J. Kuriyan.** 1994. Crystal structures of peptide complexes of the amino-terminal SH2 domain of the Syp tyrosine phosphatase. *Structure* **2**: 423–438.

1135. **Schubert, H. L., E. B. Fauman, J. A. Stuckey, J. E. Dixon, and M. A. Saper.** 1995. A ligand-induced conformational change in the *Yersinia* protein tyrosine phosphatase. *Prot. Sci.* **4**: 1904–1913.

Toxins and inhibitors

1136. **Barnes, D. M., D. B. Sykes, Y. Shechter, and D. S. Miller.** 1995. Multiple sites of vanadate and peroxovanadate action in *Xenopus oocytes*. *J. Cell. Physiol.* **162**: 154–161.

1137. **Cam, M. C., J. Faun, and J. H. McNeill.** 1995. Concentration-dependent glucose-lowering effects of oral vanadyl are maintained following treatment withdrawal in streptozotocin-diabetic rats. *Metabolism* **44**: 332–339.

1138. **Cohen, N., M. Halberstam, P. Shlimovich, C. J. Chang, H. Shamoon, and L. Rossetti.** 1995. Oral vanadyl sulfate improves hepatic and peripheral insulin sensitivity in patients with non-insulin-dependent diabetes mellitus. *J. Clin. Invest.* **95**: 2501–2509.

1139. **Fukui, K., H. Ohyanishiguchi, M. Nakai, H. Sakurai, and H. Kamada.** 1995. Detection of vanadyl–nitrogen interaction in organs of the vanadyl-treated rat – electron spin echo envelope modulation study. *FEBS Lett.* **368**: 31–35.

1140. **Gokita, T., Y. Miyauchi, and M. K. Uchida.** 1994. Effects of tyrosine kinase inhibitor, genistein, and phosphotyrosine-phosphatase inhibitor, orthovanadate, on Ca^{2+}-free contraction of uterine smooth muscle of the rat. *Gen. Pharmacol.* **25**: 1673–1677.

1141. **Hamaguchi, T., T. Sudo, and H. Osada.** 1995. RK-682, a potent inhibitor of tyrosine phosphatase, arrested the mammalian cell cycle progression at G(1) phase. *FEBS Lett.* **372**: 54–58.

1142. **Haque, S. J., V. Flati, A. Deb, and B. R. G. Williams.** 1995. Roles of protein-tyrosine phosphatases in Stat1-alpha-mediated cell signalling. *J. Biol. Chem.* **270**: 25709–25714.

1143. **Hiriyanna, K. T., W. R. Buck, S. S. Shen, and T. S. Ingebritsen.** 1995. Thiophosphorylated RCM-lysozyme, an active site-directed protein tyrosine phosphatase inhibitor, inhibits G2/M transition during mitotic cell cycle and uncouples MPF activation from G2 M transition. *Exp. Cell Res.* **216**: 21–29.

1144. **Horiguchi, T., K. Nishi, S. Hakoda, S. Tanida, A. Nagata, and H. Okayama.** 1994. Dnacin A1 and dnacin B1 are antitumor antibiotics that inhibit cdc25b phosphatase activity. *Biochem. Pharmacol.* **48**: 2139–2141.

1145. **Kole, H. K., M. Akamatsu, B. Ye, X. J. Yan, D. Barford, P. P. Roller, and T. R. Burke.** 1995. Protein-tyrosine phosphatase inhibition by a peptide containing the phosphotyrosyl mimetic, L-*O*-malonyltyrosine. *Biochem. Biophys. Res. Commun.* **209**: 817–822.

1146. **Madsen, K. L., D. Ariano, and R. N. Fedorak.** 1995. Vanadate treatment rapidly improves glucose transport and activates 6-phosphofructo-1-kinase in diabetic rat intestine. *Diabelologia* **38**: 403–412.

1147. **Miski, M., X. Shen, R. Cooper, A. M. Gillum, D. K. Fisher, R. W. Miller, and T. J. Higgins.** 1995. Apomorphine alkaloids, CD45 protein tyrosine phosphatase inhibitors, from *Rollinia ulei*. *Bioorg. Med. Chem. Lett.* **5**: 1519–1522.

1148. **Rogers, M. V., C. Buensuceso, F. Montague, and L. Mahadevan.** 1994. Vanadate stimulates differentiation and neurite outgrowth in rat pheochromocytoma PC12 cells and neurite extension in human neuroblastoma SH-SY5Y cells. *Neuroscience* **60**: 479–494.

1149. **Rozsnyay, Z., G. Sarmay, and J. Gergely.** 1995. Rapid desensitization of B-cell receptor by a dithiol-reactive protein tyrosine phosphatase inhibitor: uncoupling of membrane IgM from syk inhibits signals leading to Ca^{2+} mobilization. *Immunol. Lett.* **44**: 149–156.

1150. **Sandirasegarane, L. and V. Gopalakrishnan.** 1995. Vanadate increases cytosolic free calcium in rat aortic smooth muscle cells. *Life Sci.* **56**: PL169–PL174.

1151. **Schieven, G. L., A. F. Wahl, S. Myrdal, L. Grosmaire, and J. A. Ledbetter.** 1995. Lineage-specific induction of B-cell apoptosis and altered signal transduction by the phosphotyrosine phosphatase inhibitor, *bis*(maltolato)oxovanadium(IV). *J. Biol. Chem.* **270**: 20824–20831.

1152. **Shimizu, H., H. Takayama, J. D. Lee, K. Satake, T. Taniguchi, H. Yamamura, and T. Nakamura.** 1994. Effects of vanadate on prostacyclin and endothelin-1 production and protein-tyrosine phosphorylation in human endothelial cells. *Thromb. Haemost.* **72**: 973–978.

1153. **Singh, S. and B. B. Aggarwal.** 1995. Protein-tyrosine phosphatase inhibitors block tumor necrosis factor-dependent activation of the nuclear transcription factor NF-kappa B. *J. Biol. Chem.* **270**: 10631–10639.

1154. **Teshima, R., H. Ikebuchi, M. Nakanishi, and J. Sawada.** 1994. Stimulatory effect of pervanadate on calcium signals and histamine secretion of RBL-2H3 cells. *Biochem. J.* **302**: 867–874.

1155. **Wenzel, U. O., B. Fouqueray, P. Biswas, G. Grandaliano, G. G. Choudhury, and H. E. Abboud.** 1995. Activation of mesangial cells by the phosphatase inhibitor vanadate – potential implications for diabetic nephropathy. *J. Clin. Invest.* **95**: 1244–1252.

1156. **Wilden, P. A. and D. Broadway.** 1995. Combination of insulinomimetic agents H_2O_2 and vanadate enhances insulin receptor mediated tyrosine phosphorylation of IRS-1 leading to IRS-1 association with the phosphatidylinositol 3-kinase. *J. Cell. Biochem.* **58**: 279–291.

Additional general references

1157. **Blanquet, P. R. and F. Croquet.** 1995. Activation of phosphotyrosine phosphatase activity is associated with decreased differentiation in adult bovine lens. *J. Cell. Physiol.* **165**: 358–366.

1158. **Guo, Y. L. and S. J. Roux.** 1995. Partial purification and characterization of an enzyme from pea nuclei with protein tyrosine phosphatase activity. *Plant Physiol.* **107**: 167–175.

1159. **Kambayashi, Y., K. Takahashi, S. Bardhan, and T. Inagami.** 1995. Cloning and expression of protein tyrosine phosphatase-like protein derived from a rat pheochromocytoma cell line. *Biochem. J.* **306**: 331–335.

1160. **Mei, L., K. Liao, and R. L. Huganir.** 1994. Characterization of substrate specificity of the protein tyrosine phosphatase purified from the electric organ of *Torpedo californica*. *Neurosci. Lett.* **182**: 21–24.

1161. **Nahmias, C., S. M. Cazaubon, M. M. Briendsutren, D. Lazard, P. Villageois, and A. D. Strosberg.** 1995. Angiotensin II AT2 receptors are functionally coupled to protein tyrosine dephosphorylation in n1e-115 neuroblastoma cells. *Biochem. J.* **306**: 87–92.

1162. **Salamero, J., M. Fougereau, and P. Seckinger**. 1995. Internalization of B cell and pre-B cell receptors is regulated by tyrosine kinase and phosphatase activities. *Eur. J. Immunol.* **25**: 2757–2764.

1163. **Suzuki, K., T. Okamoto, Y. Yoshimura, Y. Deyama, Y. Hisada, and A. Matsumoto.** 1995. Phosphotyrosyl protein phosphatase-like activity of a clonal osteoblastic cell line (MC3T3-E1 cell). *Arch. Oral Biol.* **40**: 825–830.

Alignment of PTPase sequences

The sequences for alignment were extracted from available PIR (release 43) or SwissProt (release 29) sequences, or if necessary, translated from the cDNA sequence in the GenBank/EMBL database (GenBank release 87). In order to provide an informative as well as comprehensive set of PTPase sequences, 17 separate alignments are presented on the fold-out pages. These alignments (designated A–Q) correspond to subgroups of PTPase homologues that have been associated by overall structural homology and the presence of specialized functional domains as indicated in Table 1. The individual sequences are also coded to Table 2, where detailed information, including database accession numbers, are provided.

Sequences were aligned using PILEUP from the GCG program package (version 8.0), and displayed using the program PRETTY. In order to identify only the most highly conserved residues among the compared sequences, the parameters for PRETTY were set to display as the 'Consensus' only those residues that appeared in each of the aligned sequences, except for alignment O, where a plurality of 5 was assigned.

Non-transmembrane, single domain PTPases are represented in alignments A–J, with description and enzyme prototype as follows: A, C-terminal functional segment (PTP1B); B, small dual-specificity (VH1); C, small, dual-specificity (cdc25); D, SH2-domains (SH-PTP1); E, cytoskeletal protein motif (MEG1); F, retinaldehyde binding motif (MEG2); G, 'PEST' domain (PEST); H, small, unique homology (low M_r PTP); I, PTPases from yeasts and *Dictyostelium*; J, *Yersinia* PTPases and as-yet unclassified intracellular PTPases.

Transmembrane PTPases are shown in alignments K–Q. Enzymes with a single cytoplasmic PTPase homology domain are depicted in alignments K and L: K, only Fn-III domains in extracellular segment (RPTP-β); L, neuroendocrine cell PTPases (ICA512). Enzymes with tandem intracellular PTPase domains are shown in alignments M–Q: M, short extracellular domain (RPTP-α); N, large extracellular domain with Fn-III and Ig-like repeats (LAR); O, in D2, catalytic Cys replaced by Asp (RPTP-γ); P, single Fn-III domain and spectrin repeats in extracellular domain (CD45); Q, MAM-type adhesive domain (RPTP-μ).

In each alignment, the catalytic (CXXXXXR) signature motif is highlighted in the shaded box in the single domain enzymes and where this motif occurs for the tandem domain PTPases. Other highly conserved residues can be identified by their appearance in the consensus presented for each alignment subgroup.

```
                            1                                                                   51
PTP1B_HUMAN       A1  .......... ....MEMEKE FEQIDKSGSW AAIYQDIRHE AS......DF PCRVAKLPKN KNRNRYRDVS PFDHSRIKLH QEDNDYINAS
PTP1B_MOUSE       A2  .......... ....MEMEKE FEEIDKAGNW AAIYQDIRHE AS......3F PCKVAKLPKN KNRNRYRYVS PFDHSRIKLH QEDNDYINAS
PTP1B_RAT         A3  .......... ....MEMEKE FEQIDKAGNW AAIYQDIRHE AS......DF PCRIAKLPKN KNRNRYRDVS PFDHSRIKLH QEDNDYINAS
TCPTP_HUMAN       A4  .......... ..MPTTIERE FEELDTQRRW QPLYLEIRNE SH......DY PHRVAKFPEN RNRNRYRDVS PYDHSRVKLQ NAENDYINAS
TCPTP_MOUSE       A5  .......... ..MSATIERE FEELDAQCRW QPLYLEIRNE SH......DY PHRVAKFPEN RNRNRYRDVS PYDHSRVKLQ STENDYINAS
TCPTP_RAT         A6  .......... ..MSATIERE FEELDAQCRW QPLYLEIRNE SH......DY PHRVAKFPEN RNRNRYRDVS PYDHSRVKLQ SAENDYINAS
DPTP61F 'm'      A7  MSEQKTSGSG SAAAARLQIE AEYKDKGPQW HRFYKEICET CDREAKEKQF STSESERHTN RGLNRYRDVN PYDHSRIVLK RGSVDYINAN
HA2               A8  .......... ....MEMEKE FEEIDKAGNW AAIYQDIRHE AS......DF PCKVAKLPKN KNRNRYRDVS PFDHSRIKLH QEDNDYINAS
Consensus             ---------- ---------E -E--D----W ---Y--I--- ---------- ---------N ---NRYR-V- P-DHSR--L- ----DYINA-

CL100_HUMAN       B1  .....MVMEV GTLDAGGLRA LLG....... .......... ..ERAAQCLL LDCRSFFAFN AGHIAGSVNV RFSTIVRRRA KG..AMGLEH
MKP-1_MOUSE       B2  .....MVMEV GILDAGGLRA LLR....... .......... ..EGAAQCLL LDCRSFFAFN AGHIAGSVNV RFSTIVRRRA KG..AMGLEH
PAC1_HUMAN        B3  ....MGLEAA RELECAALGT LLR....... .......... DPREAERTLL LDCRPFLAFC RRHVRAARPV PWNALLRRRA RGPPAAVLAC
PAC1_MOUSE        B4  MPIAMGLETA CELECAALGA LLR....... .......... EPREAERTLL LDCRPFLAFC RSHVRAARPV PWNALLRRRA PGTPAAALAC
H23_HUMAN         B5  .......MKV TSLDGGHVRK MLRK...... .......... ..EAAARCVV LDCRPYLAFA ASNVRGSLNV NLNSVVLRRA RGG.AVSARY
HVB-P             B6  ...MVTMEEL REMDCSVLKR LMNRDENGGG AGGTGSHGTL GLPSGGKCLL LDCRPFLAHS AGYILGSVNV RCNTIVRRRA KG..SVSLEQ
HVH-3             B7  .......MKV TSLDGRQLRK MLRK...... .......... ..EAAARCVV LDCRPYLAFA ASNVRGSLNV NLNSVVLRRA RGG.AVSARY
MKP-2_RAT         B8  ...MVTMEEL REMDCSVLKR LMNRDENGGG AGGSGSHGTL GLPSGGKCLL LDCRPFLAHS AGYILGSVNV RCNTIVRRRA KG..SVSLEQ
Consensus                                                                 LDCR---A-- ---------V -------RRA -G--------

VBR               B9  .......... .......... .......... .......... .......... .......... .......... .......... ..........
VH1_VACCC         B10 .......... .......... .......... .......... .......... .......... .......... .......... ..........
VH1_VACCV         B11 .......... .......... .......... .......... .......... .......... .......... .......... ..........
VH1_VARIOLA       B12 .......... .......... .......... .......... .......... .......... .......... .......... ..........
VH1_RACCOONPOX    B13 .......... .......... .......... .......... .......... .......... .......... .......... ..........
VH1_YEAST         B14 .......... .......... .......... .......... .......... .......... .......... .......... ..........
BVH1              B15 .......... .......... .......... .......... .......... .......... .......... .......... ..........
MSG5              B16 MQFHSDKQHL DSKTDIDFKP NSPRSLQNRN TKNLSLDIAA LHPLMEFSSP SQDVPGSVKF PSPTPLNLFM KPKPIVLEKC PPKVSPRPTP
Consensus             ---------- ---------- ---------- ---------- ---------- ---------- ---------- ---------- ----------

cdc25Dm (TWINE)   C1  .......... .......... .......... .......... .......... .......... .......... .......... ..........
cdc25Dm1 (STRING) C2  .......... .......... .......... .......... .......... .......... .......... .......... ..........
cdc25B_HUMAN      C3  .......... .......... .......... .......... .......MEV PQPEPAPGSA LSPAGVCGGA QRPGHLPGLL LGSHGLLGSP
cdc25M2_MOUSE     C4  .......... .......... .......... .......... .......MEV PLQKSAPGSA LSPARVLGGI QRPRHLSVFE FESDGFLGSP
cdc25A_HUMAN      C5  .......... .......... .......... .......... .......MEL G.PSPAPRRL LFACSPPPAS QPVV...... ...KALFGAS
cdc25A_XENLA      C6  .......... .......... .......... .......... .........M AESHIMSSEA PPKTNTGLNF RTNCRMVLNL LREKDCSVTF
cdc25B_XENLA      C7  MYILLTLLTK STMESSCDNF EGLEHYYYTD AEPSSKCNKF QASGGSGVVM AESHIMSSEA PPKSNPGLNI RTNCRMILNL LREKDCSVTF
cdc25C_XENLA      C8  .......... .......... .......... MIKERVTPPF QASGGSGVVM AESHIISSEA PPKSNPGLNI RTNCRMILNL LREKDCSVTF
cdc25M1_MOUSE     C9  .......... .......... .......... .......... .........M STGPIPPASE EGSFVSAPSF RSKQRKILHL LLERNTSFTI
cdc25c_HAMST      C10 .......... .......... .......... .......... .........M STGPFPSSRR EESSVSAPSF RFSQRKMLNL LLERNTSFT.
cdc25C_HUMAN      C11 .......... .......... .......... .......... .........M STELFSSTRE EGSSGSGPSF RSNQRKMLNL LLERDTSFTV
cdc25 (nimT)      C12 .......... .......... .......... .......... .......MEH SSPLAAMQPP SVMLG..HCF RSDAPTSYHG FSPLPGLGPG
cdc25+            C13 .......... .......... .......... .......... .........M DSPLSSLSFT NTLSGKRNVL RPAARELKLM SDRNANQELD
cdc25 (MIH1)      C14 .......... .......... .......... .......... .......... .......... .......... .......... ..........
Consensus             ---------- ---------- ---------- ---------- ---------- ---------- ---------- ---------- ----------

SH-PTP2_HUMAN       D1 MTSRRWFHPN ITGVEAENLL LTRGVDGSFL ARPSKSNPGD FTLSVRRNGA VTHIKIQNTG DYYDLYGGEK FATLAELVQY YMEHHGQLKE
SH-PTP2_RAT ADIPOSE D2 MTSRRWFHPN ITGVEAENLL LTRGVDGSFL ARPSKSNPGD FTLSVRRNGA VTHIKIQNTG DYYDLYGGEK FATLPELVQY YMEHHGQLKE
SH-PTP2_RAT BRAIN   D3 MTSRRWFHPN ITGVEAENLL LTRGVDGSFL ARPSKSNPGD FTLSVRRNGA VTHIKIQNTG DYYDLYGGEK FATLAELVQY YMEHHGQLKE
SH-PTP1_HUMAN       D4 ..MVRWFHRD LSGLDAETLL KGRGVHGSFL ARPSRKNQGD FSLSVRVGDQ VTHIRIQNSG DFYDLYGGEK FATLTELVEY YTQQQGVLQD
SH-PTP1_MOUSE       D5 ..MVRWFHRD LSGPDAETLL KGRGVPGSFL ARPSRKNQGD FSLSVRVDDQ VTHIRIQNSG DFYDLYGGEK FATLTELVEY YTQQQGILQD
CSW                 D6 MSSRRWFHPT ISGIEAEKLL QEQGFDGSFL ARLSSSNPGA FTLSVRRGNE VTHIKIQNNG DFFDLYGGEK FATLPELVQY YME.NGELKE
Consensus              ----RWFH-- --G--AE-LL ---G--GSFL AR-S--N-G- F-LSVR---- VTHI-IQN-G D--DLYGGEK FATL-ELV-Y Y----G-L--

PTP36_MOUSE       E1  .......... .......... .......... .......... .......... .......... .......... .......... ..........
PTPD1_HUMAN       E2  .......... .......... .......... .......... .......... .......... .......... .......... ..........
MEG1_HUMAN        E3  .......... .......... .......... .......... .......... .......... .......... .......... ..........
PTPH1_HUMAN       E4  .......... .......... .......... .......... .......... .......... .......... .......... ..........
PTP-BAS_HUMAN     E5  MHVSLAEALE VRGGPLQEEE IWAVLNQSAE SLQELFRKVS LADPAALGFI ISPWSLLLLP SGSVSFTDEN ISNQDLRAFT APEVLQNQSL
PEZ_HUMAN         E6  .......... .......... .......... .......... .......... .......... .......... .......... ..........
PTPRL10_MOUSE     E7  .......... .......... .......... .......... .......... .......... .......... .......... ..........
PTP2E_RAT         E8  .......... .......... .......... .......... .......... .......... .......... .......... ..........
RIP_MOUSE         E9  MHVSLAEALE VRGGPLQEEE IWAVLNQSAE SLQEVFRRVS IADPAALGFI ISPWSLLLLP SGSVSFTDEN VSNQDLRAFT GPEVLQSHSL
Consensus             ---------- ---------- ---------- ---------- ---------- ---------- ---------- ---------- ----------

PTPX1             F1  .......... MAENQLTAQE ERAIDEFLAE LKTREQPQLV SVPP.NTALK FLMARKFDVS RAIDLFQAYR NTRLKEGIYN INPDEEPLRA
PTPX10            F2  .......... MAENQLTAPE ERAIEEFLAE LKNREQPQLV SEPP.DTALK FLMARKFDVS RAIDLFQAYR NTRLKEGIYN INPDEELLRT
MEG-2             F3  MEPATAPRPD MAP.ELTPEE EQATKQFLEE INKWTVQYNV SPLSWNVAVK FLMARKFDVL RAIELFHSYR ETRRKEGIVK LKPHEEPLRS
Consensus             ---------- MA---LT--E E-A---FL-E ---------V S------A-K FLMARKFDV- RAI-LF--YR -TR-KEGI-- --P-EE-LR-

PTP-PEST_MOUSE    G1  MEQVEILRRF IQRVQAMKSP DHNGEDNFAR DFMRLRRLST KYRTEKIYPT ATGEKEENVK KNRYKDILPF DHSRVKLTLK TPSQDSDYIN
PTP-PEST_HUMAN    G2  MEQVEILRKF IQRVQAMKSP DHNGEDNFAR DFMRLRRLST KYRTEKIYPT ATGEREENVK KNRYKDILPF DHSRVKLTLK TPSQDSDYIN
PEP_MOUSE         G3  MDQREILQQL LKEAQKKK.. ..LNSEEFAS EFLKLKRQST KYKADKIYPT TVAQRPKNIK KNRYKDILPY DHSLVELSLL TSDEDSSYIN
Consensus             N-Q-EIL--- ----Q--K-- -------FA- -F--L-R-ST KY---KIYPT -------N-K KNRYKDILP- DHS-V-L-L- T---DS-YIN

Low Mr PTP_BOVINE H1  ..AEQVTKSV LFV[CLGNICR] SPIAEAVFRK LVTDQNISDN W.VIDSGAVS DWNVGRSPDP RAVSCLRNHG INTAHKARQV TKEDFVTFDY
Low Hr PTP_HUMAN1 H2  ..AEQATKSV LFV[CLGNICR] SPIAEAVFRK LVWDQNISEN W.VIDSGAVS DWNVGRSPDP RAVSCLRNHG IHTAHKARQI TKEDFATFDY
Low Mr PTP_HUMAN2 H3  ..AEQATKSV LFV[CLGNICR] SPIAEAVFRK LVWDQNISEN W.RVDSAATS GYEIGNPPDY RGQSCMKRHG IPMSHVARQI TKEDFATFDY
LTP1_YEAST        H4  MTIEKPKISV AFI[CLGNFCR] SPMAEAIFKH EVEKANLENR FNKIDSFGTS NYHVGESPDH RTVSICKQHG VKINHKGKQI KTKHFDEYDY
STP1_YEAST        H5  ...MTKNIQV LFV[CLGNICR] SPMAEAVFRN EVEKAGLEAR FDTIDSCGTG AWHVGNRPDP RTLEVLKKNG IHTKHLARKL STSDFKNFDY
Consensus             --------V -F-[CLGN-CR] SP-AEA-F-- -V-------- ----DS---- ----G--PD- R-------G ----H----- ----F---DY

pyp2+             I1  MLHLLSKDEF NSTLKSFEEQ TESVSWIIDL RLHSKYAVSH IKNAINVSLP TALLRRPSFD IGKVFACIKC NVKVSLDEIN AIFLYDSSMA
pyp3+             I2  .......... .......... .......... .......... .......... .......... .......... .......... ..........
pyp1+             I3  .......... .......... .......... .......... .......... .......... .......... .......... ..........
YEAST PTP1        I4  .......... .......... .......... .......... .......... .......... .......... .......... ..........
DdPTP1            I5  .......... .......... .......... .......... .......... .......... .......... .......... ..........
DdPTP2            I6  .......... .......... .......... .......... .......... .......... .......... .......... ..........
YEAST PTP2        I7  .......... .......... .......... .......... .......... ........MD RIAQQYRNGK RDNNGNRMAS SAISEKGHIQ
Candida PTP       I8  .......... .......... .......... .......... .......... .......... .......... .......... ..........
Consensus             ---------- ---------- ---------- ---------- ---------- ---------- ---------- ---------- ----------
```

```
           101                                                       151
LIKMEEAQRS YILTQGPLPN TCGHFWEMVW EQKSRGVVML NRVMEKGSLK CAQYWPQK.. EEKEMIFEDT NLKLTLISED IKSYYTVRQL ELENLTTQET REILHFHYTT
LIKMEEAQRS YILTQGPLPN TCGHFWEMVW EQKPRGVVML NRIMEKGSLK CAQYWPQQ.. EEKEMVFDDT GLKLTLISED VKSYYTVRQL ELENLTTKET REILHFHYTT
LIKMEEAQRS YILTQGPLPN TCGHFWEMVW EQKSRGVVML NRIMEKGSLK CAQYWPQK.. EEKEMVFDDT NLKLTLISED VKSYYTVRQL ELENLATQEA REILHFHYTT
LVDIEEAQRS YILTQGPLPN TCCHFWLMVW QQKTKAVVML NRIVEKESVK CAQYWP.T.. DDQEMLFKET GFSVKLLSED VKSYYTVHLL QLENINSGET RTISHFHYTT
LVDIEEAQRS YILTQGPLPN TCCHFWLMVW QQKTKAVVML NRTVEKESVK CAQYWP.T.. DDREMVFKET GFSVKLLSED VKSYYTVHLL QLENINTGET RTISHFHYTT
LVDIEEAQRS YILTQGPLPN TCCHFWLMVW QQKTRAVVML NRTVEKESVK CAQYWP.T.. DDREMVFKET GFSVKLLSED VKSYYTVHLL QLENINSGET RTISHFHYTT
LVQLERAERQ YILTQGPLVD TVGHFWLMVW EQKSRAVLML NKLMEKKQIK CHLYWPNEMG ADKALKLPHV KLTVELVRLE TYQNFVRRWF KLTDLETQQS REVMQFHYTT
LIKMEEAQRS YILTQGPLPN TCGHFWEMVW EQKSRGVVML NRIMEKGSLK CAQYWPQQ.. EEKEMVFDDT GLKLTLISED VKSYYTVRQL ELENLTTKET REILPFHYTT
L---E-A-R- YILTQGPL-- T--HFW-MVW -QK---V-ML N---EK---K C--YWP---- ---------- -----L---- ---------- -L-------- R----FHYTT

IVP.NAELRG RLL...AGAY HAVVLLDERS AALDGAKRDG TLALAAGALC RE..ARAAQV FFLKGGYEAF SASCPELCSK QSTPMGLSLP LSTS.VPDSA ESGCSSCSTP
IVP.NAELRG RLL...AGAY HAVVLLDERS ASLDGAKRDG TLALAAGALC RE..ARSTQV FFLQGGYEAF SASCPELCSK QSTPTGLSLP LSTS.VPDSA ESGCSSCSTP
LLPDRA.LRT RLV...RGEL ARAVVLDEGS ASVAELRPDS PAHVLLAALL HETRAGPTAV YFLRGGFDGF QGCCPDLCSE APAPALPPTG DKTS.RSDS. .......RAP
LLPDRA.LRA RLG...RGEL ARAVVLDESS ASVTELPPDG PAHLLLAALQ HEMRGGPTTV CFLRGGFKSF QTYCPDLCSE APAQALPPAG AENS.NSDP. .......RVP
VLPDEAR.RA RLLQEGGGGV AAVVVLDQGS RHWQKLREES .AFVVLTSLL ACLPAGP.RV YFLKGGYETF YSEYPECCVD VKPISQEKIE SERALISQCG KPVVNVSYRP
ILPAEEEVRA RLR...SGLY SAVIVYDERS PRAESLREDS TVSLVVQALR RN..AERTDI CLLKGGYERF SSEYPEFCSK TKALAAIPPP VPPS.ATEPL DLGCSSCGTP
VLPDEAA.RA RLLQEGGGGV AAVVVLDQGS RHWQKLREES AARVVLTSLL ACLPAGP.RV YFLKGGYETF YSEYPECCVD VKPISQEKIE SERALISQCG KPVVNVSYRP
ILPAEEEVRA RLR...SGLY SAVIVYDEGS PRAESLREDS TVSLVVQALR RN..AERTDI CLLKGGYERF SSEYPEFCSK TKALAAIPPP VPPS.ATEPL DLGCSSCGTP
--P-----R- RL-----G-- ------D--S ---------- --------L- ---------- --L-GG---F ----P--C-- ---------- ---------- ---------P

.......... .......... .......... .......... .......... .......... .....MSGSF ELSVQDLNDL LSDGSGCY.. ..SLPSQP.. CNEVTPRIYV
.......... .......... .......... .......... .......... .......... .......... .MDKKSLYKY LLLRSTGDMH KAKSPTIM.. TRVTNNVYLG
.......... .......... .......... .......... .......... .......... .......... .MDKKSLYKY LLLRSTGDMH KAKSPTIM.. TRVTNNVYLG
.......... .......... .......... .......... .......... .......... .......... .MDKKSLYKY LLLRSTGDMR RAKSPTIM.. TRVTNNVYLG
.......... .......... .......... .......... .......... .......... .......... .MDKKSLYKY LLLRSTGDIH RAKSPTMM.. TRVTNNVYLG
.......... .......... .......... .......... .......... .......... .......... .......... ..MAGNAN.. ..SVDEEV.. TRILGGIYLG
.......... .......... .......... .......... .......... .......... .......... .......... ........MF PARWHNYLQC GQVIKDSNLI
PSLSMRRSEA SIYTLPTSLK NRTVSPSVYT KSSTVSSISK LSSSSPLSSF SEKPHLNRVH SLSVKTKDLK LKGIRGRSQT ISGLETSTPI SSTREGTLDS TDVNRFSNQK
---------- ---------- ---------- ---------- ---------- ---------- ---------- ---------- ---------- ---------- ----------

.......... .......... .......... .......... .......... .......... .......... .......... .......... .......... ..........
.......... .......... .......... .......... .......... .......... .......... .......... .......... .......... ...MLWETIV
VRAAASSPVT TLTQTMHDLA GLGS...... ......RSRL THLSLSRRAS ESSLSSESSE SSDAGLCMDS PSPMDPHMAE QTFEQAIQAA SRIIRNEQFA IRRFQSMPVR
EPTASSSPVT TLTQTMHNLA GLGSEPPKAQ VGSLSFQNRL ADLSLSRRTS ECSLSSESSE SSDAGLCMDS PSPVDPQMAE RTFEQAIQAA SRVIQNEQFT IKRFRSLPVR
A.AGGLSPVT NLTVTMDQLQ GLGSD..... .......... YEQPLEVKNN SNLQRMGSSE STDSGFCLDS PGPLDSK... .......... ....ENLENP MRRIHSLPQK
SPEQPLTPVT DLAVGFSNLS TFSGETPKRC .......... ..LDLSNL.. GDETAPLPTE SPD....RIS SGKVESPKAQ FVQFDGLFTP DLGWKAKKCP RGNMNSVLPR
SPEQPLTPVT DLAVGFSNLS TFSGETPKRC .......... ..LDLSNL.. GDETAPLPTE SPD....RMS SGKLESPKTQ FVQFGGLFTP DLAWKAKKCP KRNMNSVLPH
SPEQPLTPVT DLAVGFSNLS TFSGETPKRC .......... ..LDLSNL.. GDETAPLPTE SPD....RMS SGKLESPKTQ FVQFDGLFTP DLAWKAKKCP KRNMNSVLPH
RSDFPESPKD KLH.DSANLS ILSGGTPKCC .......... ..LDLSNLSS GEMSASPLTT SAD....LED NGSLDSSGPL DRQLTG.... ......KDFH QDLMKGIPVQ
.QDFPRSPGD KLL.DSTNLS ILSGGTPKRC .......... ..LDLSNLSN GEMSASPLIT SAD....FDD TGSLDSSGPQ DVQLTE.... ......KNHH QDPMKGIPVQ
CPDVPRTPVG KFLGDSANLS ILSGGTPKCC .......... ..LDLSNLSS GEITATQLTT SAD....LDE TGHLDSSGLQ EVHLAG.... ......MNHD QHLMKCSPAQ
GFNFKDLSMK RSNGDYFGTK VVRGSSPTAS LAA....... ...DLSQNFH IDQSPQVATP RRSLFSACLL GNGNRRGVDD AMTTPPLPSS SPAPAM.... .........D
FFFPKSKHIA STLVDPFG.K TCSTASPASS LAA....... ...DMSMNMH IDESPALPTP RRTLFRS..L SCTVETPLAN KTIVSPLPES PSNDALTESY FFRQPASKYS
.......... .......... .......... .......... .......... .......... .......... .......... .......... .......... ..........
---------- ---------- ---------- ---------- ---------- ---------- ---------- ---------- ---------- ---------- ----------

KNGDVIELKY PLNCADPTSE RWFHGHLSGK EAEKLLTEKG KHGSFLVRES QSHPGDFVLS VRTGDDKGES NDGKSKVTHV MIRCQELKYD VGGGERFDSL TDLVEHYKKN
KNGDVIELKY PLNCADPTSE RWFHGHLSGK EAEKLLTEKG KHGSFLVRES QSHPGDFVLS VRTGDDKGES NDSKSKVTHV MIRCQELKYD VGGGERFDSL TDLVEHYKKN
KNGDVIELKY PLNCADPTSE RWFHGHLSGK EAEKLLTEKG KHGSFLVRES QSHPGDFVLS VRTGDDKGES NDSKSKVTHV MIRCQELKYD VGGGERFDSL TDLVEHYKKN
RDGTIIHLKY PLNCSDPTSE RWYHGHMSGG QAETLLQAKG EPWTFLVRES LSQPGDFVLS VLSDQPKAGP .GSPLRVTHI KVMCEGGRYT VGGLETFDSL TDLVEHFKKT
RDGTIIHLKY PLNCSDPTSE RWYHGHISGG QAESLLQAKG EPWTFLVRES LSQPGDFVLS VLNDQPKAGP .GSPLRVTHI KVMCEGGRYT VGGSETFDSL TDLVEHFKKT
KNGQAIELKQ PLICAEPTTE RWFHGNLSGK EAEKLILERG KNGSFLVRES QSKPGDFVLS VRTDD..... .....KVTHV MIRWQDKKYD VGGGESFGTL SELIDHYKRN
--G--I-LK- PL-C--PT-E RW-HG--SG- -AE-L----G ----FLVRES -S-PGDFVLS V--------- ------VTH- --------Y- VGG-E-F--L --L--H-K--

.......... .......... .......... .......... .......... .......... .......... .......... .......... .......... ..........
.......... .......... .......... .......... .......... .......... .......... .......... .......... .......... ..........
.......... .......... .......... .......... .......... .......... .......... .......... .......... .......... ..........
.......... .......... .......... .......... .......... .......... .......... .......... .......... .......... ..........
TSLSDVEKIH IYSLGMTLYW GADYEVPQSQ PIKLGDHLNS ILLGMCEDVI YARVSVRTVL DACSAHIRNS NCAPSFSYVK HLVKLVLGNL SGTDQLSCNS EQKPDRSQAI
.......... .......... .......... .......... .......... .......... .......... .......... .......... .......... ..........
.......... .......... .......... .......... .......... .......... .......... .......... .......... .......... ..........
.......... IYSLGMTLYW GADHEVPQSQ PIKLGDHLNS ILLGMCEDVI YARVSVRTVL .......... .......... .......... .......... ..........
TSLADVEKIH ---------- ---------- ---------- ---------- ---------- DACSA.TSGT ASRAFVSYVK QLVKLVLGNI SGTDPLSRSS EQKPDRSQAI
----------                                                       ---------- ---------- ---------- ---------- ----------

ELLSGKFTVL KQPADNPSIP PLPQKSRPLP PGMEL.SIHM PEDGGMMVLG LVQHVKRMKK RGIFQEYEEI RKDPPVGSFD ISKKSHNQAK NRYSDVLCLD QSRVKLGLVG
ELLSGKFTVF .......... ....KSRPLP PGMEL.SLHM PEDGGMMVLD LVQHVKRMKK RGIFQEYEEI RKDPPVGSFD ISKKSHNQVK NRYSDVLCLD QSRVKLGLVG
EILSGKFTIL .......... .......SLP PALDWDSVHV PGPHAMTIQE LVDYVNARQK QGIYEEYEDI RRENPVGTFH CSMSPGNLEK NRYGDVPCLD QTRVKLTKRS
E-LSGKFT-- ---------- --------LP P-----S-H- P----M---- LV--V----K -GI--EYE-I R---PVG-F- -S----N--K NRY-DV-CLD Q-RVKL----

ANFIKGVYGP KAYVATQGPF RNTVIDFWRM IWEYNVVMIV MACREFEMGR KKCERYWPLY GEDPITFAPF KISCENEQAR TDYFIRTLLL EFQNESRRLY QFHYVNWPDH
ANFIKGVYGP KAYVATQGPL ANTVIDFWRM VWEYNVVIIV MACREFEMGR KKCERYWPLY GEDPITFAPF KISCEDEQAR TDYFIRTLLL EFQNESRRLY QFHYVNWPDH
ASFIKGVYGP KAYIATQGPL STTLLDFWRM IWEYRILVIV MACMEFEHGR KKCERYWAEP GETQLQFGPF SISCEAEKKK SDYKIRTLKA KFNNETRIIY QFHYKNWPDH
A-FIKGVYGP KAY-ATQGP- --T--DFWRM -WEY----IV MAC-EFEMG- KKCERYW--- GE----F-PF -ISCE-E--- -DY-IRTL-- -F-NE-R--Y QFHY-NWPDH

ILCMDESNLR DLNRKSNQVK NCRAKIELLG SY...DPQKQ LIIEDPYYGN DADFETVYQQ CVRCCRAFLE KVR H1  Low Mr PTP_BOVINE
ILCMDESNLR DLNRKSNQVK TCKAKIELLG SY...DPQKQ LIIEDPYYGN DSDFETVYQQ CVRCCRAFLE KAH H2  Low Hr PTP_HUMAN1
ILCMDESNLR DLNRKSNQVK TCKAKIELLG SY...DPQKQ LIIEDPYYGN DSDFETVYQQ CVRCCRAFLE KAH H3  Low Mr PTP_HUMAN2
IIGMDESNIN NLKKI..QPE GSKAKVCLFG DWNTNDGTVQ TIIEDPWYGD IQDFEYNFKQ ITYFSKQFLK KEL H4  LTP1_YEAST
IFAMDSSNLR NINRV..KPQ GSRAKVMLFG EYAS..PGVS KIVDDPYYGG SDGFGDCYIQ LVDFSQNFLK SIA H5  STP1_YEAST
I--MD-SN-- ---------- ---AK--L-G ---------- -I--DP-YG- ---F-----Q -------FL- --- Consensus

GMNRIYDLVQ KFRRGGYSKK IYLLSNGFEA FASSHPDAIV STEMVKESVP YKIDINENCK LDILHLSDPS AVSTPISPDY SFPLRVPINI PPPLCTPSVV SDTFSEFASH
.......... .......... .......... .......... .......... .......... .......... .......... .......... .......... ..........
.......... .......... .......... .......... .......... .......... ...MNFSNGS KSSTFTIAPS GSCIALPPQR GVATSKYAVH ASCLQEYLDK
.......... .......... .......... .......... .......... .......... .......... .......... .......... .......... ..........
.......... .......... .......... .......... .......... .......... .......... .......... .......... .......... ..........
.......... .......... .......... .......... .......... .......... .......... .......... .......... .......... ..........
VNQTRTPGQM PVYRGETINL SNLPQNQIKP CKDLDDVNIR RNNSNRHSKI LLLDLCAGPN TNSFLGNTNA KDITVLSLPL PSTLVKRSNY PFENLLKNYL GSDEKYIEFT
.......... .......... .......... .......... .......... .......... .......... .......... .......... .......... ..........
---------- ---------- ---------- ---------- ---------- ---------- ---------- ---------- ---------- ---------- ----------
```

```
                            201                                                           251
         PTP1B_HUMAN  A1 WPDFGVPESP ASFLNFLFKV RESGSLSPEH GPVVVHCSAG IGRSGTFCLA DTCLLLMDKR KDPSSVDIKK VLLEMRKFRM GLIQTADQLR
         PTP1B_MOUSE  A2 WPDFGVPESP ASFLNFLFKV RESGSLSLEH GPIVVHCSAG IGRSGTFCLA DTCLLLMDKR KDPSSVDIKK VLLEMRRFRM GLIQTADQLR
           PTP1B_RAT  A3 WPDFGVPESP ASFLNFLFKV RESGSLSPEH GPIVVHCSAG IGRSGTFCLA DTCLLLMDKR KDPSSVDIKK VLLEMRRFRM GLIQTADQLR
         TCPTP_HUMAN  A4 WPDFGVPESP ASFLNFLFKV RESGSLNPDH GPAVIHCSAG IGRSGTFSLV DTCLVLMEKG DD...INIKQ VLLNMRKYRM GLIQTPDQLR
         TCPTP_MOUSE  A5 WPDFGVPESP ASFLNFLFKV RESGCLTPDH GPAVIHCSAG IGREGTFSLV DTCLVLMEKG ED...VNVKQ LLLNMRKYRM GLIQTPDQLR
           TCPTP_RAT  A6 WPDFGVPESP ASFLNFLFKV RESGSLNPDH GPAVIHCSAG IGRSGTFSLV DTCLVLMEKG ED...VNVKQ ILLSMRKYRM GLIQTPDQLR
        DPTP61F 'm'  A7 WPDFGIPSSP NAFLKFLQQV RDSGCLSRDV GPAVVHCSAG IGRSGTFCLV DCCLVLIDKY GE...CNVSK VLCELRTYRM GLIQTADQLD
                 HA2  A8 WPDFGVPESP ASFLNFLFKV RESGSLSLEH GPIVVHCSAG IGRSGTFCLA DTCLLLMDKR KDPSSVDIKK VLLEMRRFRM GLIQTADHVR
           Consensus     WPDFG-P-SP --FL-FL--V R-SG-L---- GP-V-HCSAG IGRSGTF-L- D-CL-L--K- ---------- -L---R--RM GLIQT-D---

         CL100_HUMAN  B1 LYDQGGPVEI LPFLYLGSAY HASRKDMLDA LGITALINVS ANCPNHFEGH YQYKSIPVED NHKADISSWF NEAIDFIDSI KNAGGRVFVH
         MKP-1_MOUSE  B2 LYDQGGPVEI LSFLYLGSAY HASRKDMLDA LGITALINVS ANCPNHFEGH YQYKSIPVED NHKADISSWF NEAIDFIDSI KDAGGRVFVH
          PAC1_HUMAN  B3 VYDQGGPVEI LPYLFLGSCS HSSDLQGLQA CGITAVLNVS ASCPNHFEGL FRYKSIPVED NQMVEISAWF QEAIGFIDWV KNSGGRVLVH
          PAC1_MOUSE  B4 IYDQGGPVEI LPYLYLGSCN HSSDLQGLQA CGITAVLNVS ASCPNHFEGL FHYKSIPVED NQMVEISAWF QEAISFIDSV KNSGGRVLVH
           H23_HUMAN  B5 AYDQGGPVEI LPFLYLGSAY HASKCEFLAN LHITALLNVS RRTSEACMTH LHYKWIPVED SHTADISSHF QEAIDFIDCV REKGGKVLVH.
               HVB-P  B6 LHDQGGPVEI LPFLYLGSAY HAARRDMLDA LGITALLNVS SDCPNHFEGH YQYKCIPVED HHKADISSWF MEAIEYIDAV KDCRGRVLVH
               HVH-3  B7 AYDQGGPVEI LPFLYLGSAY HASKCEFLAN LHITALLNVS RRTSEACMTH LHYKWIPVED SHTADISSHF QEAIDFIDCV REKGGKVLVH
           MKP-2_RAT  B8 LHDQGGPVEI LPFLYLGSAY HAARRDMLDA LGITALLNVS SDCPNHFEGH YQYKCIPVED NHKADISSWF MEAIEYIDAV KDCRGRVLVH
           Consensus     --DQGGPVEI L--L-LGS-- H------L-- --ITA--NVS ---------- --YK-IPVED -----IS--F -EAI--ID-- ----G-V-VH

                 VBR  B9 GNASVAQDIP KLQKLGITHV LNAAEGRSFM HVNTNANFYK DSGITYLGIK AND....TQE FNLSAYFERA ADFIDQAL.. ..........
           VH1_VACCC B10 NYKNAMDAPS SEVKFKY... ........VL NLTMDKYTLP NSNINIIHIP LVD....DTT TDISKYFDDV TAFLSKC... ..........
           VH1_VACCV B11 NYKNAMDAPS SEVKFKY... ........VL NLTMDKYTLP NSNINIIHIP LVD....DTT TDISKYFDDV TAFLSKC... ..........
         VH1_VARIOLA B12 NYKNAMNAPS SEVKFKY... ........VL NLTMDKYTLP NSNINIIHIP LVD....DTT TDISKYFDDV TAFLSKC... ..........
     VH1_RACCOONPOX B13 NYKNAMEAPS SEVKFKY... ........IL NLTMDKYSFT NSNINIIHVP MVD....DTS TDISIYFDDI TAFLSKC... ..........
           VH1_YEAST B14 GIRPIIDHRP LGAEFNITHI L......SVI KFQVIPEYLI RKGYTLKNIP IDD....DDV TDVLQYFDET NRFIDQCLFP NEVEYSPRLV
                BVH1 B15 CFKTPL.... .......... RPELF...AY .VTSEEDVWT AEQIVKQNPS IGAIIDL... TNTSKYYDGV HFLRAGLLYK KIQVPGQTLP
                MSG5 B16 NMQTTLIFPE EDSDLNIDMV HAEIYQRTVY .LDGPLLVLP PNLYLYSEPK LEDILSFDLV INVAKEIPNL EFLIPPEMAH KIKYYHIEWT
           Consensus     ---------- ---------- ---------- ---------- ---------- ---------- ---------- ---------- ----------

    cdc25Dm (TWINE)  C1 ......MEY. ..QAKRRKSA VQETPMQWML KRHIPA.... .......STT VLSPITELSQ NMNGARLDGT PKSTQRIPAN RTLNNFNSLS
  cdc25Dm1 (STRING)  C2 EENNCSMDCN ISNNTSSSSS INKMSGSRRA RRSLELMSMD QEELSFYDDD VVPQDQQRSA SPELMGLLSP EGSPQRFQIV RQPKILPAMG
        cdc25B_HUMAN  C3 LLGHSPVL.R NITNSQAPDG RRKSEAG.SG AASSSGEDKE NDGFVF.... ..KMPWKPTH PSSTHALAEW ASRREAFAQR PSSAP.DLMC
      cdc25M2_MOUSE  C4 LLEHSPVL.Q SITNSRALDS WRKTEAGYRA AANSPGEDKE NDGYIF.... ..KMPQELPH SSSAQALAEW VSRRQAFTQR PSSAP.DLMC
        cdc25A_HUMAN  C5 LLGCSPALKR SHSDSLDHDI FQLID..... ....PDENKE NEAFEF.... ..KKPVRPVS RGCLHSHG.L QEGKDLFTQR QNSAQLGMLS
        cdc25A_XENLA  C6 LLCSTP.... .........S FKKTSGG... ..QRSVSNKE NEGELF.... .........K SPNCKPVALL LPQEVVDSQF SPTPENKVDI
        cdc25B_XENLA  C7 LLCSTP.... .........S FKKASGG... ..QRSLSNKE NEGELF.... .........K NPNCKPVALL LPQEVVDSQL SPTPENKVDI
        cdc25C_XENLA  C8 LLCSTP.... .........S FKKASGG... ..QRSLSNKE NEGELF.... .........K NPNCKPVALL LPQEVVDSQL SPTPENKVDI
      cdc25M1_MOUSE  C9 LLCSTP.... ....NAMNHG HRKKIAK... ..RSTSAHKE NAAFLLLSLQ LIDAGPSCAD KHQFKGIGMG GTRTPRFRKM PGGPLTSPLC
       cdc25c_HAMST C10 LLCSTP.... ....NALDHS HRKKDAV... ..RGLSANKE NI........ .......... NTNLKTLQWE SPRIPRFQNT PGDPLASPLP
       cdc25C_HUMAN C11 LLCSTP.... ....NGLDRG HRKRDAM... ..CSSSANKE N......... .......... .......... .......... ..........
       cdc25 (nimT) C12 IMDMSP.... .....LPHKP PF.ISTPEIE LDSPTLESSP MDTTMM.... .......... .......... ........ST DGLVPDSPTV
              cdc25+ C13 ITQDSPRVSS TIAYSFKPKA SIALNTTKSE ATRSSLSSSS FDSYLRPNVS RSRSSGNAPP FLRSRSSSSY SINKKKGTSG GQATRHLTYA
       cdc25 (MIH1) C14 .......... .......... ........MN NIFHGTEDEC ANEDVLSFQK ISLKSPFGKK KNIFRNVQTF FKSKSKHSNV DDDLINK...
           Consensus     ---------- ---------- ---------- ---------- ---------- ---------- ---------- ---------- ----------

       SH-PTP2_HUMAN  D1 PMVETLGTVL QLKQPLNTTR INAAEIESRV RELSKLAETT DKVKQGFWEE FETLQQQECK LLYSRKEGQR QENKNKNRYK NILPFDHTRV
SH-PTP2_RAT ADIPOSE  D2 PMVETLGTVL QLKQPLNTTR INAAEIESRV RELSKLAETT DKVKQGFWEE FETLQQQECK LLYSRKEGQR QENKNKNRYK NILPFDHTRV
  SH-PTP2_RAT BRAIN  D3 PMVETLGTVL QLKQPLNTTR INAAEIESRV RELSKLAETT DKVKQGFWEE FETLQQQECK LLYSRKEGQR QENKNKNRYK NILPFDHTRV
       SH-PTP1_HUMAN  D4 GIEEASGAFV YLRQPYYATR VNAADIENRV LELNKKQESE DTAKAGFWEE FESLQKQEVK NLHQRLEGQR PENKGKNRYK NILPFDHSRV
       SH-PTP1_MOUSE  D5 GIEEASGAFV YLRQPYYATR VNAADIENRV LELNKKQESE DTRKAGFWEE FESLQKQEVK NLHQRLEGQR PENKSKNRYK NILPFDHSRV
                 CSW  D6 PMVETCGTVV HLRQPFNATR ITAAGINARV EQL....... ..VKGGFWEE FESLQ.QDSR DTFSRNEGYK QENRLKNRYR NILPYDHTRV
           Consensus     ---E--G--- -L-QP---TR --AA-I--RV --L------- ---K-GFWEE FE-LQ-Q--- ----R-EG-- -EN--KNRY- NILP-DH-RV

         PTP36_MOUSE  E1 .......... .......... .......... .......... .......... .......... .......... .......... ..........
         PTPD1_HUMAN  E2 .......... .......... .......... .......... .......... .......... .......... .......... ..........
          MEG1_HUMAN  E3 .......... .......... .......... .......... .......... .......... .......... .......... ..........
         PTPH1_HUMAN  E4 .......... .......... .......... .......... .......... .......... .......... .......... ..........
      PTP-BAS_HUMAN  E5 RDRLRGKGLP TGRSSTSDVL DIQKPPLSHQ TFLNKGLSKS MGFLSIKDTQ D.ENYFKDIL SDNSGR.EDS ENTFSPYQFK TSGPEKKPIP
           PEZ_HUMAN  E6 .......... .......... .......... .......... .......... .......... .......... .......... ..........
       PTPRL10_MOUSE  E7 .......... .......... .......... .......... .......... .......... .......... .......... ..........
           PTP2E_RAT  E8 .......... .......... .......... .......... .......... .......... .......... .......... ..........
           RIP_MOUSE  E9 RDRLRGKGLP TGRSSTSDAL DTHEAPLSQQ TFLNKGLSKS MGFLSIRDTR DEEDYLKDTP SDNNSRHEDS ETFSSPYQFK TSTPQ.....
           Consensus     ---------- ---------- ---------- ---------- ---------- ---------- ---------- ---------- ----------

               PTPX1  F1 TDETTDYTNA SFMDGYKRKN AYIATQGPLP KTFDDFWRMV WEQKVLIIVM TTRVIERGRI KCGQYWPLEA GRSEDTGHFI IRNIHIDLFQ
              PTPX10  F2 TDETTDYINA SFMDGYKRKN AYIATQGPLP KTFDDFWRMV WEQKVLIIVM TTRVIERGRI KCGQYWPLEA GRSEDTGHFI IRNIHIDLFQ
               MEG-2  F3 GHTQTDYINA SFMDGYKQKN AYIGTQGPLE NTYRDFWLMV WEQKVLVIVM TTRFEEGGRR KCGQYWPLEK DSRIRFGFLT VTNLGVENMN
           Consensus     ----TDY-NA SFMDGYK-KN AYI-TQGPL- -T--DFW-MV WEQKVL-IVM TTR--E-GR- KCGQYWPLE- ------G--- --N-------

    PTP-PEST_MOUSE  G1 DVPSSFDSIL DHISLPRKYQ EHEDVPICIH CSAGCGRTGA ICAIDYTWNL LKAGKIPEEF NVFNLIQEMR TQRHSAVQTK EQYELVHRAI
    PTP-PEST_HUMAN  G2 DVPSSFDSIL DMISLMRKYQ EHEDVPICIH CSAGCGRTGA ICAIDYTWNL LKAGKIPEEF NVFNLIQEMR TQRHSAVQTK EQYELVHRAI
           PEP_MOUSE  G3 DVPSSIDPIL QLIWDMRCYQ EDDCVPICIH CSAGCGRTGV ICAVDYTWML LKDGIIPKNF SVFNLIQEMR TQRPSLVQTQ EQYELVYSAV
           Consensus     DVPSS-D-IL --I--MR-YQ E---VPICIH CSAGCGRTG- ICA-DYTW-L LK-G-IP--F -VFNLIQEMR TQR-S-VQT- EQYELV--A-


                pyp2+  I1 AEYPGFSGLT PFSIHSPTAS SVRSCQSIYG SPLSPPNSAF QAEMPYFPIS PAISCASSCP STPDEQKNFF IVGNAPQQTP ARPSLRSVPS
                pyp3+  I2 .......... .......... .......... .......... .......... .......... .......... .......... ..........
                pyp1+  I3 EAWKDDTLII DL........ .....RPVSE FSKSRIKGSV NLSLPATLIK RPAFSVARII SNLHDVDDKR DFQNWQEFSS ILVCVPAWIA
          YEAST PTP1  I4 .......... .......... .......... .......... .......... .......... .......... .......... ..........
               DdPTP1  I5 .......... .......... .......... .......... .......... .......... .......... .......... ..........
               DdPTP2  I6 .......... .......... .......... .......... .......... .......... .......... .......... ..........
          YEAST PTP2  I7 KIIKDYDIFI FSDSFSRISS CLKTTFCLIE KFKKFICHFF PSPYLKFFLL EGSLNDSKAP SLGKNKKNCI LPKLDLNLNV NLTSRSTLNL
          Candida PTP  I8 .......... .......... .......... .......... .......... .......... .......... .......... MTTPLSSYST
            Consensus     ---------- ---------- ---------- ---------- ---------- ---------- ---------- ---------- ----------
```

```
           301                                                   351
FSYLAVIEGA KFIMGDSSVQ DQWKELSHED LEPPPEHIPP PPRPPKRIL. .......... .......... .......... .......... ......EPHN GKCREFFPNH
FSYLAVIEGA KFIMGDSSVQ DQWKELSRED LDLPPEHVPP PPRPPKRTL. .......... .......... .......... .......... ......EPHN GKCKELFSSH
FSYQAIIEGI KFIMGDSSVQ DQWKELSHED LEPPPEHVPP PPRPPKRTL. .......... .......... .......... .......... ......EPHN GKCKELFSNH
FSYMAIIEGA KCIKGDSSIQ KRWKELSKED LSPAFDH... ..SPNKIMT. .......... .......... .......... .......... ......EKYN GN........
FSYMAIIEGA KYTKGDSNIQ KRWKELSKED LSPICDH... ..SQNRVMV. .......... .......... .......... .......... ......EKYN GK........
FSYMAIIEGA KYTKGDSNI. .......... .......... ...QNRTMT. .......... .......... .......... .......... ......EKYN GK........
FSYQAIIEGI KKLHDPTFLD AEEPLISNDT ETHTLDELPP PLPPRVQSLN LPLAPNSGGI LSLNMRAAQA NGAESIGKEL SKDALNNFIN QHDMFHDAEV ADSRPLPPLP
FSYLAVIEGA KFIMGDSSVQ DQWKELSRED LDLPPEHVPP PPRPPKRTL. .......... .......... .......... .......... ......EPHN GKCKELFSSH
FSY-A-IEG- K--------- ---------- ---------- ---------- ---------- ---------- ---------- ---------- ---------- ----------

CQAGISRSAT ICLAYLMRTN RVKLDEAFEF VKQRRSIISP NFSFMGQLLQ FESQVL.... ...APHCSAE AGSPAMAVLD RGTSTTT... ..VFNFPVSI PV....HSTN
CQAGISRSAT ICLAYLMRTN RVKLDEAFEF VKQRRSIISP NFSFMGQLLQ FESQVL.... ...APHCSAE AGSPAMAVLD RGTSTTT... ..VFNFPVSI PV....HPTN
CQAGISRSAT ICLAYLMQSR RVRLDEAFDF VKQRRGVISP NFSFMGQLLQ FETQVL.... ---CH----- ---------- ---------- ---------- ----------
CQAGISRSAT ICLAYLIQSH RVRLDEAFDF VKQRRGVISP NFSFMGQLLQ LETQVL.... ---CH----- ---------- ---------- ---------- ----------
CEAGISRSPT ICMAYLMKTK QFRLKEAFDY IKQRRSMVSP NFGFMGQLLQ YESEILPSTP NPQPPSCQGE AAGSSLIGHL QTLSPDMQGA YCTFPASVLA RCLPTQQSQS
CQAGISRSAT ICLAYLMMKK RVRLEEAFEF VKQRRSIISP NFSFMGQLLQ FESQVL.... ...ATSCAAE AASPSGPLRE RGKTPATPTS QFVFSFPVSV GV....HSAP
CEAGISRSPT ICMAYLMKTK QFRLKEAFDY IKQRRSMVSP NFGFMGQLLQ YESEILPSTP NPQPPSCQGE AAGSSLIGHL QTLSPDMQGA YCTFPASVLA PV....PTHS
CQAGISRSAT ICLAYLMMKK RVRLEEAFEF VKQRRSIISP NFSFMGQLLQ FESQVL.... ...ATSCAAE AASPSGPLRE RGKTPATPTS QFVFSFPVSV GV....HSAP
C-AGISRS-T IC-AYL---- ---L-EAF-- -KQRR---SP NF-FMGQLLQ -E---L---- ---------- ---------- ---------- ---------- ----------

....AQKN.. .......... ...GRVLVHC REGYSRSPTL VIAYLMMRQK MDVKS..... AL.SIVRQNR EIGPNDGFLA QL........ ..CQLNDRLA KEGKLKP...
....DQRN.. .......... ...EPVLVHC AAGVNRSGAM ILAYLMSKNK ESSPMLYFLY VYHSMRDLRG AFVENPSFKR QIIEKYVIDK N......... ..........
....DQRN.. .......... ...EPVLVHC AAGVNRSGAM ILAYLMSKNK ESLPMLYFLY VYHSMRDLRG AFVENPSFKR QIIEKYVIDK N......... ..........
....DQRN.. .......... ...EPVLVHC VAGVNRSGAM ILAYLMSKNK ESSPMLYFLY VYHSMRDLRG AFVENPSFKR QIIEKYVIDK N......... ..........
....DQRN.. .......... ...EPVLVHC AAGVNRSGAM ILAYLMSKNK ESSPMLYFLY VYHSMRDLRG AFVENPSFKR QIIEKYVIDK N......... ..........
DFKKKPQR.. .......... ...GAVFAHC QAGLSRSVTF IVAYLMYRYG LSLSM..... AMHAVKRKKP SVEPNENFME QLHLFEKMGG DFVDFDNPAY KQWKLKQSIK
PES.IVQEFI DTVKEF.TEK CPGMLVGVHC THGINRTGYM VCRYLMHTLG IAPQE..... AIDRFEKARG HKIERQNYVQ DLLI...... .......... ..........
HTSKIVKDLS RLTRIIHTAH SQGKKILVHC QCGVSRSASL IVAYIMRYYG LSLND..... AYDELKGVAK DISPNMGLIF QLMEWGTMLS KNSPGEEGET VHMPEEDDIG
---------- ---------- --------HC --G--R---- ---Y-M---- ---------- ---------- ---------- ---------- ---------- ----------

.......SRT .LGSFSSSCS SYESGNSLDD E......... YMDMFEMESA ENH.N....L ELPDDLEVLL SGQLKSESNL ...EEMSNKK GSLRRCLSMY PS........
VSSDHTPARS .FRIFNSLSS TCSMESSMDD E......... YMELFEMESQ SQQTA....L GFPSGLNSLI SGQIKEQPAA KSPAGLSMRR PSVRRCLSMT ESNTNSTTTP
LSPDRKMEVE ELSPLALGRF SLTPAEGDTE ED......DG FVDILESD.L KDDD...... AVPPGMESLI SAPLVKTLEK EEEKDLVMYS KCQRLFRS.. ..........
LTTEWKMEVE ELSPVAQSS. SLTPVERASE ED......DG FVDILESD.L KDDE...... KVPAGMENLI SAPLVKKLDK EEEQDLIMFS KCQRLFRS.. ..........
SNERDSSEPG NFIPLFTPQS PVTAT..LSD ED......DG FVDLLDGENL KNEE...... ETPSCMASLW TAPLVMRTTN LDNRC..... ...KLFDS.. ..........
SLDE....DC EMNILGSPIS ADPPCLDGAH DDIKMQNLDG FADFFSVDEE EMENPPGAVG NLSSSMAILL SGPLLNQDIE VSNVNNISLN RS.RLYRS.. ..........
SLEE....DC EMNILGSPIS ADPPCLDGAH DDIKMQNLDG FADFFSVDEE EMENPPGAVG NLSCSMAILL SGPLLNQDVE ISNVNNISLN RS.RLYRS.. ..........
SLDE....DC EMNILGSPIS ADPPCLDGAH DDIKMQNLDG FADFFSVDEE EMENPPGAVG NLSCSMAILL SGPLLNQDVE ISNVNNISLN RS.RLYRS.. ..........
EL........ EMKHLGSPIT TVPK...... .......... LSQNVKLEDQ ..ERISEDPM ECS....... ...LGDQDA. .......... .......... ..........
LLGNGVSMDT EVRSLGSPIT AVPK...... .......... LSKNLNLEDQ ..EEISEEPM EFS....... ...LEDHDT. .......... .......... ..........
..DNGNLVDS EMKYLGSPIT TVPK...... .......... LDKNPNLGED QAEEISDELM EFS....... ...LKDQEAK VS........ RS.GLYRS.. ..........
LPKDGKQERR RPTFLRPSLA .........R SKAQSFQVGM TRPAPESQGP PFKFQTNGIN KTSSGVAA.. ..SLEDMFGE SPQRERPMM. RINS...... .........T
LSRTCSQSSN TTSLLESCLT DDTDDFELMS DHEDTFTMGK VADLPES.SV ELVEDAASIQ RPNSDFGACN DNSLDDLFQA SPIKPIDMLP KINK...... .........D
.......... ..ENLAFDKS PLLTNHRSKE IDGPSPNIKQ LGHRDELDEN ENEND..... ......DIVL SMHFASQTLQ SPTRNSSRRS LTNNRDNDLL SRIKYPGSPQ
---------- ---------- ---------- ---------- ---------- ---------- ---------- ---------- ---------- ---------- ----------

VLHDGDPNEP VSDYINANII M......... .......... .......... .......... .......... .......... .......... .......... ..........
VLHDGDPNEP VSDYINANII M......... .......... .......... .......... .......... .......... .......... .......... ..........
VLHDGDPNEP VSDYINANII M......... .......... .......... .......... .......... .......... .......... .......... ..........
ILQGRDSNIP GSDYINANYI K......... .......... .......... .......... .......... .......... .......... .......... ..........
ILQGRDSNIP GSDYINANYV K......... .......... .......... .......... .......... .......... .......... .......... ..........
KLLDVEHSVA GAEYINANYI RLPTDGDLYN MSSSSESLNS SVPSCPACTA AQTQRNCSNC QLQNKTCVQC AVKSAILPYS NCATCSRKSD SLSKHKRSES SASSSPSSGS
-L-------- ---YINAN-- ---------- ---------- ---------- ---------- ---------- ---------- ---------- ---------- ----------

.......... .......... .......... .......... .......... .......... .......... .......... .......... .......... ..........
.......... .......... .......... .......... .......... .......... .......... .......... .......... .......... ..........
.......... .......... .......... .......... .......... .......... .......... .......... .......... .......... ..........
.......... .......... .......... .......... .......... .......... .......... .......... .......... .......... ..........
GIDVLSKKKI WASSMDLLCT ADRDFSSGET ATYRRCHPEA VTVRTSTTPR KKEARYSDGS IALDIFGPQK MDPIYHTREL PTSSAISSAL DRIRERQKKL QVLREAMNVE
.......... .......... .......... .......... .......... .......... .......... .......... .......... .......... ..........
.......... .......... .......... .......... .......... .......... .......... .......... .......... .......... ..........
.......... .......... .......... .......... .......... .......... .......... .......... .......... .......... ..........
.MDALSKKKT WASSMDLLCA AIRDI.SGET GRYQRCDPET VTGRTSITPR KKEGRYSDGS IALDIFGPQK VEPVIHTREL PTSTAVSSAL DRIRERQKKL QVLREAMNVE
---------- ---------- ---------- ---------- ---------- ---------- ---------- ---------- ---------- ---------- ----------

DFKLTHLEVY NKQTDESQSV AHYQYMSWPD FGVPKSASAM LDFRTQVKQH QAVAVQALGM AWTGHPAGPP VVVHCSAGIG RTGTFCTLDI CLSRLEDIGT VDVLQTVKRM
DFKLTHFEVY NKQTDESRSV AHYQYMSWPD FGVPKSASAM LDFRSQVKQH QAVAVQNLGM EWTGHPAGPP IVVHCSAGIG RTGTFCTLDI CLSRLEDIGT VDVLQTVKRM
HYKKTTLEIH NTEERQKRQV THFQFLSWPD YGVPSSAASL IDFLRVVRNQ QSLAVSNMGA RSKGQCPEPP IVVHCSAGIG RTGTFCSLDI CLAQLEELGT LNVFQTVSRM
--K-T--E-- N--------V -H-Q--SWPD -GVP-SA--- -DF---V--- Q--AV---G- ---G----PP -VVHCSAGIG RTGTFC-LDI CL--LE--GT --V-QTV-RM

AQLFENQLQL YEIHGAQKIR DG.NEITTGT MVSSIDSEDE TS.PPKPPRT RSCLVE..GD AKEEILQPPE PHPVPPILTP SPPSAFPTVT TVVQDSDRYH PKPVLHMA..
AQLFEKQLQL YEIHGAQKIA DGVNEINTEN MVSSIEPEKQ DSPPPKPPRT RSCLVE..GD AKEEILQPPE PHPVPPILTP SPPSAFPTVT TVWQDNDRYH PKPVLHMV..
LELFKRHMDV .......... .....ISDNH LGREIQAQCS IPEQSLTVEA DSCPLDLPKN AMRDVKTTNQ HSKQGAEAES TGGSSLGLRT STMNAEEEL. ...VLHSAKS
--LF------ ---------- -----I---- ----I----- ---------- -SC------- A--------- ---------- ---S-----T ---------- ---VLH----

YPSSNNQRRP SASRVRSFSN YVKSSNVVNP SLSQASLEII PRKSMKRDSN AQNDGTSTMT SKLKPSVGLS NTRDAPKPGG LRRANKPCFN KETKGSIFSK ENKGPFTCNP
.......... .......... .......... .......... .......... .......... .......... .......... .......... .......... ..........
NYVTNAEVIG EKFRKESYSG DFGILDLDYS KVSGKYPSVI DNSPVKSKLG ALPSARPRLS YSAAQTAPIS LSSEGSDYFS RPPPTPNVAG LSLNNFFCPL PENKDNKSSP
.......... .......... .......... .......... .......... .......... .......... .......... .......... .......... ..........
.......... .......... .......... .......... ..MGSVESSN QMNGSIENKT NKI.DISVLR PLPT.RSNSS ISLSSSSHSS FSRMGSLGSL PTNSGSSSPY
.......... .......... .........M DESFDAFEIK PNSSIIEHFN NHPTCIDKEF NFILQKTELR ALETDKFRSA LEAKNKLKNR YSNVLPFEET RVKINIDD..
RINIPPPNDS NKIFLQSLKK DLIHYSPNSL QKFFQFNMPA DLAPNDTILP NWLKFCSVKE NEKVILKKLF NNFETLENFE MQRLEKCLKF KKKPLHQKQL SQKQRGPQST
TVTNHHPTFS FESLNSISSN NSTRNNQSNS VNSLLYFNSS GSSMVSSSSD AAPTSISTTT TSTTSMTDAS ANADNQQVYT ITEEDSINDI NRKEQNSFSI QPNQTPTMLP
---------- ---------- ---------- ---------- ---------- ---------- ---------- ---------- ---------- ---------- ----------
```

```
                             401                                                                                          451
        PTP1B_HUMAN A1  QWVKEETQED KDCPIKEEKG SPLN...... .......... ..........  .......... .......... .....AAPYG IESMSQDTEV
        PTP1B_MOUSE A2  QWVSEETCGD EDSLAREE.G RAQS...... .......... ..........  .......... .......... .....SAMHS VSSMSPDTEV
          PTP1B_RAT A3  QWVSEESCED EDILAREE.S RAPS...... .......... ..........  .......... .......... .....IAVHS MSSMSQDTEV
        TCPTP_HUMAN A4  ......RIGL EEEKLTGDRC TGLS...... .......... ..........  .......... .......... .....SKMQD TMEENSESAL
        TCPTP_MOUSE A5  ......RIGS EDEKL..... TGLP...... .......... ..........  .......... .......... .....SKVQD TVEESSESIL
          TCPTP_RAT A6  ......RIGS EDEKL..... TGLS...... .......... ..........  .......... .......... .....SKVPD TVEESSESIL
       DPTP61F 'm'  A7  VRAFNDSDSD EDYLLDDDDE DDTDEDEEYE TINEHDADPV NGHVPATTQP  HADDVNANNE KPAVPVDEQH KANGIDPIPG QLPASPENEL
                HA2 A8  QWVSEETCGD EDSLAREE.G RAQS...... .......... ..........  .......... .......... .....SAMHS VSSMSPDTEV
          Consensus     ---------- ---------- ---------- ---------- ----------  ---------- ---------- ---------- ----------

         CL100_HUMAN B1 SALSYLQSPI TTSPSC.... ...... B1 CL100_HUMAN
         MKP-1_MOUSE B2 SALNYLKSPI TTSPSC.... ...... B2 MKP-1_MOUSE
          PAC1_HUMAN B3 .......... .......... ...... B3 PAC1_HUMAN
          PAC1_MOUSE B4 .......... .......... ...... B4 PAC1_MOUSE
           H23_HUMAN B5 SAEALWQRPN PAKTGMEESA QPQEQL B5 H23_HUMAN
               HVB-P B6 SSLPYLHSPI TTSPSC.... ...... B6 HVB-P
               HVH-3 B7 TVSELSRSPV ATATSC.... ...... B7 HVH-3
           MKP-2_RAT B8 SSLPYLHSPI TTSPSC.... ...... B8 MKP-2_RAT
           Consensus    ---------- ---------- ------ Consensus

                 VBR B9  .......... .......... .......... .......... ..........  .......... .......... .......... ..........
           VH1_VACCC B10 .......... .......... .......... .......... ..........  .......... .......... .......... ..........
           VH1_VACCV B11 .......... .......... .......... .......... ..........  .......... .......... .......... ..........
         VH1_VARIOLA B12 .......... .......... .......... .......... ..........  .......... .......... .......... ..........
     VH1_RACCOONPOX B13  .......... .......... .......... .......... ..........  .......... .......... .......... ..........
           VH1_YEAST B14 LDPSGSELVS NSGMFKDSES SQDLDKLTEA EKSKVTAVRC KKCRTKLALS  TSFIAHDPPS KESSEGHFIK RAANSHRIID IQESQANCSH
                BVH1 B15 .......... .......... .......... .......... ..........  .......... .......... .......... ..........
                MSG5 B16 NNEVSSTTKS YSSASFRSFP MVTNLSSSPN DSSVNSSEVT PRTPATLTGA  RTALATERGE DDEHCKSLSQ PADSLEASVD NESISTAPEQ
           Consensus     ---------- ---------- ---------- ---------- ----------  ---------- ---------- ---------- ----------

    cdc25Dm (TWINE) C1  .....EQPEE AVQEPDQETN MPMKKMQ... .......... ..........  ........RK  TLSMNDAEIM RALG.....D EPELIGDLSK
 cdc25Dm1 (STRING) C2  PPKTPETARD CFKRPEPPAS ANCSPIQSKR HRCATVEKEN CPAPSPLSQV  TISHPPPLRK  CMSLNDAEIM SALARSENRN EPELIGDFSK
       cdc25B_HUMAN C3  PSMPCSVIRP ILKRLERPQD RDTPVQNKRR RSVTPPEEQQ EAEEPKAR..  VL.....RSK  SLC.....H. DEIENLLDSD HRELIGDYSK
      cdc25M2_MOUSE C4  PSMPCSVIRP ILKRLERPQD RDVPVQSKRR KSVTPLEEQQ .LEEPKAR..  VF.....RSK  SLC.....H. .EIENILDSD HRGLIGDYSK
       cdc25A_HUMAN C5  PSLCSSSTRS VLKRPERSQE ESPPGSTKRR KSMSGASPKE STNPEKAH..  ET.....LHQ  SLSLASSPK. GTIENILDND PRDLIGDFSK
       cdc25A_XENLA C6  PSMPEKLDRP MLKRPVRPLD SETPVRVKRR RSTSSSLQPQ EENFQPQRRG  TS.....LKK  TLSLCDVDI. STVLDE.DCG HRQLIGDFTK
       cdc25B_XENLA C7  PSMPEKLDRP MLKRPVRPLD SETPVRVKRR RSTSSPLQPE EENCQPQRRG  TS.....LKK  TLSLCDVDI. STVLDE.DCG HRQLIGDFSK
       cdc25C_XENLA C8  PSMPEKLDRP MLKRPVRPLN SETPVRVKRR RSTSSPLQPE EENCQPQRRG  TS.....LKK  TLSLCDVDI. SSVLDE.DCG HRQLIGDFSK
      cdc25M1_MOUSE C9  .......... .......... .......... .......... ........KG  LS.....LRK  MVPLCDMNA. IQMEEE.ESG SELLIGDFSK
      cdc25c_HAMST C10  .......... .......... .......... .......... ........KE  ..........  .......... .......... ..........
      cdc25C_HUMAN C11  PSMPENLNRP RLKQVEKFKD NTIPDKVKKK YFSGQG.... .....KLRKG  LC.....LKK  TVSLCDITI. TQMLEE.DSN QGHLIGDFSK
       cdc25 (nimT) C12  SGLNS.RLRP PLGSGSHVRG NGSPSAASVR KSAHPNMRPR KQCRRSLSMY  EHPEDVIADS  EVSYTSNAPL QSI.....SD FE.....ETQ
            cdc25+ C13  IAFPSLKVRS P......... ..SPMAFAMQ EDAEYDEQDT PVVRRTQSMF  LNSTRLGLFK  SQDLVCVTPK QST.....KE SERFISSHVE
       cdc25 (MIH1) C14  RSSSFSRSRS LSRKPSMNSS SNSSRRVQR. ........QD GKIPRSSRKS  SQKFSNITQN  TLNFTSASSS PLAPNSVG.. ....VKCFES
          Consensus     ---------- ---------- ---------- ---------- ----------  ----------  ---------- ---------- ----------

       SH-PTP2_HUMAN D1  .......... .......... .......... ...PEFETKC NNSKPK....  ..........  ...KSYIATQ GCL....QNT VNDFWRMVFQ
 SH-PTP2_RAT ADIPOSE D2  .......... .......... .......... ...PEFETKC NNSKPK....  ..........  ...KSYIATQ GCL....QNT VNDFWRMVFQ
   SH-PTP2_RAT BRAIN D3  .......... .......... .......... ...PEFETKC NNSKPK....  ..........  ...KSYIATQ GCL....QNT VNDFWRMVFQ
       SH-PTP1_HUMAN D4  .......... .......... .......... ...NQLLGPD ENA.......  ..........  ...KTYIASQ GCL....EAT VNDFWQMAWQ
       SH-PTP1_MOUSE D5  .......... .......... .......... ...NQLLGPD ENS.......  ..........  ...KTYIASQ GCL....DAT VNDFWQMAWQ
                 CSW D6  GSGPGSSGTS GVSSVNGPGT PTNLTSGTAG CLVGLLKRHS NDSSGAVSIS  MAERERERER  EMFKTYIATQ GCLLTQQVNT VTDFWNMVWQ
           Consensus     ---------- ---------- ---------- ---------- ----------  ----------  ---K-YIA-Q GCL------T V-DFW-M--Q

         PTP36_MOUSE E1  .......... .......... .......... .......... ..........  ..........  .......... .......... ..........
         PTPD1_HUMAN E2  .......... .......... .......... .......... ..........  ..........  .......... .......... ..........
          MEG1_HUMAN E3  .......... .......... .......... .......... ..........  ..........  .......... .......... ..........
         PTPH1_HUMAN E4  .......... .......... .......... .......... ..........  ..........  .......... .......... ..........
       PTP-BAS_HUMAN E5  EPVRRYKTYH GDVFSTSSES PSIISSESDF RQVRRSEASK RFESSSGLPG  VDETLSQGQS  QRPSRQYETP FEGNLINQEI MLKRQEEELM
           PEZ_HUMAN E6  .......... .......... .......... .......... ..........  ..........  .......... .......... ..........
       PTPRL10_MOUSE E7  .......... .......... .......... .......... ..........  ..........  .......... .......... ..........
           PTP2E_RAT E8  .......... .......... .......... .......... ..........  ..........  .......... .......... ..........
           RIP_MOUSE E9  EPVRRYKTYH SDIFSISSES PSVISSESDF RQVRKSEASK RFESSSGLPG  VDET...GQT  .RPSRQYETS LEGNLINQDI MLRRQEEEMM
           Consensus     ---------- ---------- ---------- ---------- ----------  ----------  ---------- ---------- ----------

               PTPX1 F1  RTQRAFSIQT WDQYYFCYMA IIEYAQRKGI IAPVEWSDTD FETDSE F1 PTPX1
              PTPX10 F2  RTQASFSIQT WDQYYFCYMA IIEYAQRKGI IAPVEWSDTD YETDSE F2 PTPX10
               MEG-2 F3  RTQRAFSIQT PEQYYFCYKA ILEFAEKEGM ...VSSGQNL LAVESQ F3 MEG-2
           Consensus     RTQ--FSIQT --QYYFCY-A I-E-A---G- ---V------ ----S- Consensus

     PTP-PEST_MOUSE G1  SPEQHPADLN RSY.DKSADQ WGKSESAIEH IDKKLERNLS FEIKKVPLQE  GPKSFDGNTL  ....LNRGHA IKIKSASSSV VDRTSKPQEL
     PTP-PEST_HUMAN G2  SSEQHSADLN RNY.SRSTEL PGKNESTIEQ IDKKLERNLS FEIKKVPLQE  GPKSFDGNTL  ....LNRGHA IKIKSASPCI ADKISKPQEL
           PEP_MOUSE G3  SPSFNCLELN CGCNNKAVIT RNGQARASPV VGEPLQKYQS LDFGSMLFGS  CPSALPINTA  DRYHNSKGPV KRTKSTPFEL IQQRKTNDLA
           Consensus     S-------LN -----K---- ---------- ----L----S ----------  -P-----NT-  -------G-- ---KS----- ----------


              pyp2+ I1  WGAKKVSPPP CEVLADLNTA .SI...FYKF KRLEEMEMTR SLAFNDSKSD WCCLASSRST  SISRK....N RYTDIVPYDK TRVRL.AVP.
              pyp3+ I2  .......... .MSFKEVSTE NGV...LTPL ITIKEKAYMI IEGLNEEEIE ...LLNTRLP  KLSKKALARN RYSNIVPYEN TRVRLDPMW.
              pyp1+ I3  FGSATVQTPC LHSVPDAFTN PDVATLYQKF LRLQSLEHQR LVSCSDRNSQ WSTV.....D  SLSNTSYKKN RYTDIVPYNC TRVHLKRTS.
         YEAST PTP1 I4  .......... .MAAAPWYIR QRDTDLLGKF KFIQNQEDGR LREATNGTVN SR...WSLGV  SIEPRNDARN RYVNIMPYER NRVHLKTLS.
             DdPTP1 I5  YNNSSFD... ..LVDEERIK SSIYNLKNHI KCIHKIKEEF RLLEESVGPS ETSE......  GDKKHNTSKN RYTNILPVNH TRVQLKKIQD
             DdPTP2 I6  .DDDDED... ..DNEDDIIV SNNNNNNNN. .......NEK RIKRNSIGSS GQSDVMSNSS  DEEDHGGSGD EGTTL..... ..........
         YEAST PTP2 I7  DDSKLYSLTS LQRQYKSSLK SNIQKNQKLK LIIPKNNTSS SPSPLSSDDT IMSPINDYEL  TEGIQSFTKN RYSNILPYEH SRVKLPHSPK
       Candida PTP I8  TSSYTLQRPP GLHEYTSSIS SISSTSSNST SAPVSPALIN YSPKHSRKPN SLNLNRNMKN  LSLNLHDSTN GYTSPLPKST NSNQPRGNFI
          Consensus     ---------- ---------- ---------- ---------- ---------- ----------  ---------- ---------- ----------
```

```
                  501                                                                 551
RSRVVGGSLR GAQAASPAKG EPSLPEKDE. DHALSYWKPF LVNMCVATVL TA.GAYLCYR FLFNSNT. A1 PTP1B_HUMAN
RRRMVGGGLQ SAQASVPTEE ELSSTEEEHK AHWPSKWKPF LVNVCMATLL AT.GAYLCYR VCFH.... A2 PTP1B_MOUSE
RKRMVGGGLQ SAQASVPTEE ELSPTEEEQK AHRPVHWKPF LVNVCMATAL AT.GAYLCYR VCFH.... A3 PTP1B_RAT
RKR.IREDRK ATTAQKVQQM KQRLNENERK RKRWLYWQPI LTKMGFMSVI LV.GAFVGWR LFFQQNAL A4 TCPTP_HUMAN
RKR.IREDRK ATTAQKVQQM KQRLNETERK RKR.....PR LTDT...... .......... ........ A5 TCPTP_MOUSE
RKR.IREDRK ATTAQKVQQM RQRLNETERK RKR.....PR LTDT...... .......... ........ A6 TCPTP_RAT
KRRK.RNEYQ ASLEQKVNDM KRKQRENEDK QLAAKKRRSL LTYIAAGVVV GVICAYAYTK LG...... A7 DPTP61F 'm'
RRRMVGGGLQ SAQASVPTEE ELSSTEEEHK AHWPSHWKPF LVNVCMATLL AT.GAYLCYR VCFH.... A8 HA2
--R------- ---------- -----E---- ---------- L--------- ---------- -------- Consensus
```

```
.......... .......... .......... .......... .......... .......... .......... .......... .. B9 VBR
.......... .......... .......... .......... .......... .......... .......... .......... .. B10 VH1_VACCC
.......... .......... .......... .......... .......... .......... .......... .......... .. B11 VH1_VACCV
.......... .......... .......... .......... .......... .......... .......... .......... .. B12 VH1_VARIOLA
.......... .......... .......... .......... .......... .......... .......... .......... .. B13 VH1_RACCOONPOX
FFIEPLKWMQ PELQGKQELE GKFSCPGCSS KVGGYNWKGS RCSCGKWVIP AIHLQTSKVD QFPLQSTALP NMVNFESEKV NR B14 VH1_YEAST
.......... .......... .......... .......... .......... .......... .......... .......... .. B15 BVH1
MMFLP..... .......... .......... .......... .......... .......... .......... .......... .. B16 MSG5
---------- ---------- ---------- ---------- ---------- ---------- ---------- ---------- -- Consensus
```

```
PCTLPCLATG IRHRDLKTIS SDTLARLIQG EFDEQLGSQ. .GGYEIIDCR YPYEFLGGHI RGAKNLYTRG QIQEAFPTL. .........T SNQENRRIYV FHCEFSSERG
AYALPLMEG. .RHRDLKSIS SETVARLLKG EFSDKVAS.. ...YRIIDCR YPYEFEGGHI EGAKNLYTTE QILDEFLTVQ QTELQQQQNA ESGHKRNIII FHCEFSSERG
AFLLQTVDG. .KHQDLKYIS PETMVALLTG KFSNIVDK.. ...FVIVDCR YPYEYEGGHI KTAVNLPLER DAESFL.... ...LKSPIAP CSLDKRVILI FHCEFSSERG
AFLLQTVDG. .KHQDLKYIS PETMVALLTG KFSNIVEK.. ...FVIVDCR YPYEYEGGHI KNAVNLPLER DAETFL.... ...LQRPIMP CSLDKRIILI FHCEFSSERG
GYLFHTVAG. .KHQDLKYIS PEIMASVLNG KFANLIKE.. ...FVIIDCR YPYEYEGGHI KGAVNLHMEE EVEDFL.... ...LKKPIVP TD.GKRVIVV FHCEFSSERG
VYALPTVTG. .RHQDLRYIT GETLAALIHG DFSSLVEK.. ...IFIIDCR YPYEYDGGHI KGALNLHRQE EVTDYF.... ...LKQPLTP TMAQKRLIII FHCEFSSERG
VYALPTVTG. .RHQDLRYIT GETLAALMHG DFNSLVEK.. ...FFIIDCR YPYEYDGGHI KSAFNLHRQE EVTDYF.... ...LQQPLTP LMVQKRLIII FHCEFSSERG
VYALPTVTG. .RHQDLRYIT GETLAALMHG DFNSLVEK.. ...FFIIDCR YPYEYDGGHI KSAFNLHRQD EVTDYF.... ...LQQPLTP LMAQKRLIII FHCEFSSERG
VCVLPTVPG. .KHPDLKYIS PDTVAALLSG KFQSVIER.. ...FYIIDCR YPYEYLGGHI LGALNLHSQK ELHEFF.... ...LRKPVVP LDIQKRVIIV FLCEFSSERG
.CVLPTVSG. .KHQDLKYIT PDTVAALLSG KFQGLIEK.. ...FYIIDCR YPYEYLGGHI LGAINLCSQK ELHEFF.... ...LKKPIVP LDIQKRVIIV FLCEFSSERG
VCALPTVSG. .KHQDLKYVN PETVAALLSG KFQGLIEK.. ...FYVIDCR YPYEYLGGHI QGALNLYSQE ELFNFF.... ...LKKPIVP LDTQKRIIIV FHCEFSSERG
ALQLPHFIPE EQADNLGRID KATLVDIKEG KYDNMFDN.. ...IMIIDCR FEYEYDGGHI VGAVNYNDKE NLAAEL.... ...FADP.KP .....RTAIV FHCEYSVHRA
DLSLPCFAVK E..DSLKRIT QETLLGLLDG KFKDIFDK.. ...CIIIDCR FEYEYLGGHI STAVNLNTKQ AIVDAF.... ...LSKPLTH .....RVALV FHCEHSAHRA
CLAKTQIPYY YDDRNSMTFS LEFLQKRLKN ILQNNMCESF YNSCRIIDCR FEYEYTGGHI INSVNIHSRD ELEYEFI... .HKVLHSDTS NNNTLPTLLI IHCEFSSHRG
---------- ---------- ---------- ---------- -------DCR --YE--GGHI ----N----- ---------- ---------- ---------- --CE-S--R-
```

```
ENSRVIVMTT KEVERGKSKC VKYWPDEYAL KEYGVMRVRN VKESAAHDYT LRELKLSKVG ....QGNTER TVWQYHFRTW PDHGVPSDPG GVLDFLEEVH HKQESIMDA.
ENSRVIVMTT KEVERGKSKC VKYWPDECAL KEYGVMRVRN VRESAAHDYT LRELKLSKVG ....QGNTER TVWQYHFRTW PDHGVPSDPG GVLDFLEEVH HKQESIVDA.
ENSRVIVMTT KEVERGKSKC VKYWPDECAL KEYGVMRVRN VRESAAHDYT LRELKLSKVG QALLQGNTER TVWQYHFRTW PDHGVPSDPG GVLDFLEEVH HKQESIVDA.
ENSRVIVMTT REVEKGRNKC VPYWPEVGMQ RAYGPYSVTN CGEHDTTEYK LRTLQVSPLD ....NGDLIR EIWHYQYLSW PDHGVPSEPG GVLSFLDQIN QRQESLPHA.
ENTRVIVMTT REVEKGRNKC VPYWPEVGTQ RVYGLYSVTN SREHDTAEYK LRTLQISPLD ....NGDLVR EIWHYQYLSW PDHGVPSEPG GVLSFLDQIN QRQESLPHA.
ENTRVIVMTT KEYERGKEKC ARYWPDEGRS EQFGHARIQC VSENSTSDYT LREFLVS... ...WRDQPAR RIFHYHFQVW PDHGVPADPG CVLNFLQDVN TRQSHLAQAG
EN-RVIVMTT -E-E-G--KC --YWP----- ---G------ --E-----Y- LR----S--- ---------R ----Y----W PDHGVP--PG -VL-FL---- --Q-----A-
```

```
.......... .......... .......... .......MPF GLKLRRTRRY .......... .......... NVLS...... ..KNCFVARI RLLDSNVIEC TLSVESTGQE
.......... .......... .......... .....MPLPF GLKLKRTRRY .......... .......... TVSS...... ..KSCLVARI QLLNNEFVEF TLSVESTGQE
.......... .......... .......... .......MTS RFRLPAGRTY .......... .......... NVRASELARD RQHTEVVCNI LLLDNTVQAF KVNKHDQGQV
.......... .......... .......... .......MTS RLRALGGRIN .......... .......... NIRTSELPKE KTRSEVICSI HFLDGVVQTF KVTKQDTGQV
QLQAKMALRQ SRLSLYPGDT IKASMLDITR DPLREIALET AMTQRKLRNF FGPEFVKMTI EPFISLDLPR SIL..TKKGK NEDNRRKVNI MLLNGQRLEL TCDTKTICKD
.......... .......... .......... .......MPF GLKLRRTRRY .......... .......... NVLS...... ..KNCFVTRI RLLDSNVIEC TLSVESTGQE
.......... .......... .......... .....MPLPF GLKLKRTRRY .......... .......... TVSS...... ..KSCLVARI QLLNNEFVEF TLSVESTGQE
.......... .......... .......... .....MPLPF GLKLKRTRRY .......... .......... TVSS...... ..KSCLVARI QLLNNEFVEF TLSVESTGQE
QLQARMALRQ SRLSLYPGDT VKASMLDISR DPLREMALET AMTQRKLRNF FGPENRKMTV EPFVSLDLPR SILSQTKKGK SEDQRRKVNI RLLSGQRLEL TCDTKTICKD
---------- ---------- ---------- ---------- -------R-- ---------- ---------- ---------- ---------I --L------- ----------
```

```
SAGALKVDDV SQNSCADCSA AHSHRAAESS EES.....QS NSHTPPRPDC LPLDKKGHVT WSL.HGPENA TPVPDSPDGK .SPDNHSQTL KTVSSTPNST AEEEAHDLTE
SSD.LNVGDT SQNSCVDCSV TQSNKVSVTP PEE.....SQ NSDTPPRPDR LPLDEKGHVT WSF.HGPENA IPIPDLSEGN .SSDINYQTR KTVSLTPSPT TQVETPDLVD
VGDGFSCLES QLHEHYSLRE LQVQRVAHVS SEELNYSLPG ACDASCVPRH SPGALRVHLY TSLAEDPYFS SSPPNSADSK MSFDLPEKQD GATSPGALLP ASSTTSFFYS
---------- -E-------- -------P-- -P-----H-- ---------- ---------- -S----P--- ---P------ -S-D------ ---S------ ----------
```

```
KGCS...... .......... .......... ...DYINASH IDV.GNKK.. ....YIACQA PKPGTLLDFW EMVWHNSG.T NGVIVMLTNL YEAGSEKCSQ YWPDNKDHA.
KEAC...... .......... .......... ...DYINASI VKIPSGKT.. ....FIATQG PTSNSIDVFW KMVWQSVP.K SGIIVMLTKL RERHRLKCDI YWPVELFET.
PSEL...... .......... .......... ...DYINASF IKTETSN... ....YIACQG SISRSISDFW HMVWDNVE.N IGTIVMLGSL FEAGREMCTA YWPSNGIGD.
GN........ .......... .......... ...DYINASY VKVNVPGQSI EPGYYIATQG PTRKTWDQFW QMCYHNCPLD NIVIVMVTPL VEYNREKCYQ YWPRGGVDDT
KEGS...... .......... .......... ...DYINANY IDGAYPKQ.. ....FICTQG PLPNTIADFW RMVWEN...R CRIIVMLSRE SENCRIKCDR YWPEQIGGEQ
...S...... .......... .......... ...DYINASF INNG...T.. ....YICTQG PLLNTIVDFW KMIWEQ...N SNIIVMLTRE EENFKTKCDK YWPD......
PPAVSEASTT ETKTDKSYPM CPVDAKNHSC KPNDYINANY LKLTQINPDF K...YIATQA PLPSTMDDFW KVITLN...K VKVIISLNSD DELNLRKWDI YWNNLSYSNH
MDSPSKKSTP VNRIGNN... .......... NGNDYINATL LQTPSITQTP TMPPPLSLAQ GPPSSVG..S ESVYKFPLIS NACLNYSAGD SDSEVESISM KQAAKNTIIP
---------- ---------- ---------- ---DYINA-- ---------- ---------- ---------- ---------- ---------- ---------- ----------
```

```
                         601                                                651

     cdc25Dm (TWINE) C1  PKLLRYLRSN DRSQHTHNYP ALDYPELYIL HNGYKEFFGL YSQLCQPSQY VPMLAPAH.. ..NDEFRYFR AKTKSWQCGE GGDSGIGGGG
 cdc25Dm1 (STRING) C2   PKMSRFLRNL DRERNTNAYP ALHYPEIYLL HNGYKEFFES HVELCEPHAY RTMLDPAY.. ..NEAYRHFR AKSKSW.... .NGDGLGGAT
     cdc25B_HUMAN C3    PRMCRFIRER DRA..VNDYP SLYYPEMYIL KGGYKEFFPQ HPNFCEPQDY RPMNHEAF.. ..KDELKTFR LKTRSW.... ........AG
     cdc25M2_MOUSE C4   PRMCRFIRER DRA..ANDYP SLYYPEMYIL KGGYKEFFPQ HPNFCEPQDY RPMNHEAF.. ..RDELRNFR LKTRSW.... ........AG
     cdc25A_HUMAN C5    PRMCRYVRER DRL..GNEYP KLHYPELYVL KGGYKEFFMK CQSYCEPPSY RPMHHEDF.. ..KEDLKKFR TKSRTW.... ........AG
     cdc25A_XENLA C6    PKMCRFLREE DRA..RNEYP SLYYPELYLL KGGYKDFFPE YKELCEPQSY CPMHHQDF.. ..REELLKFR TKCKTS.... ........VG
     cdc25B_XENLA C7    PKMCRFLREE DRA..SNDYP SLYYPELYLL KGGYKDFFPE YKELCEPQSY CPMHHQDF.. ..REDLLKFR TKCKTS.... ........VG
     cdc25C_XENLA C8    PKMCRSLREE DRA..SNDYP SLYYPELYLL KGGYKDFFPE YKELCEPQSY CPMHHQDF.. ..REDLLKFR TKCKTS.... ........VG
     cdc25M1_MOUSE C9   PRMCRSLREK DRA..LNQYP ALYYPELYIL KGGYRDFFPE YMELCDPQSY CPMLHQDH.. ..QAELLSWR SQSKAQ.... ........EG
     cdc25c_HAMST C10   PRMCRSLRRK DRA..LNQYP ALYYPELYIL KGGYRDFFPE YTELCEPQGY CPMHHQDH.. ..QAELLMWR NQSKAQ.... ........EG
     cdc25C_HUMAN C11   PRMCRCLREE DRS..LNQYP ALYYPELYIL KGGYRDFFPE YMELCEPQSY CPMHHQDH.. ..KTELLRCR SQSKVQ.... ........EG
     cdc25 (nimT) C12   PLMAKYIRHR DRAYNVDHYP QLSYPDMYIL EGGYSGFFAE HRSLCYPQNY VEMSAKEHEF ACERGLGKVK QRSKLSRAQT FAFGQQSPEM
           cdc25+ C13   PHLALHFRNT DRRMNSHRYP FLYYPEVYIL HGGYKSFYEN HKNRCDPINY VPMNDRSHVM TCTKAMNNFK RNATFMRTKS YTFWPKCVSF
     cdc25 (MIH1) C14   PSLASHLRNC DRIINQDHYP KLFYPDILIL DGGYKAVLT. FPELCYPRQY VGMNSQENLL NCEQEMDKFR RESKRFATKN NSFRKLASPS
        Consensus       P------R-- DR------YP -L-YP----L --GY------ ----C-P--Y --M------- ---------- ---------- ----------

     SH-PTP2_HUMAN D1   ...GPVVVHC SAGIGRTGTF IVIDILIDII REKGVDCDID VPKTIQMVRS QRSGMVQTEA QYRFIYMAVQ HYIETLQRRI EEEQKSKRKG
SH-PTP2_RAT ADIPOSE D2  ...GPVVVHC SAGIGRTGTF IVIDILIDII REKGVDCDID VPKTIQMVRS QRSGMVQTEA QYRFIYMAVQ HYIETLQRRI EEEQKSKRKG
 SH-PTP2_RAT BRAIN D3   ...GPVVVHC SAGIGRTGTF IVIDILIDII REKGVDCDID VPKTIQMVRS QRSGMVQTEA QYRFIYMAVQ HYIETLQRRI EEEQKSKRKG
     SH-PTP1_HUMAN D4   ...GPIIVHC SAGIGRTGTI IVIDMLMENI STKGLDCDID IQKTIQMVRA QRSGMVQTEA QYKFIYVAIA QFIETTKKKL EVLQSQKGQE
     SH-PTP1_MOUSE D5   ...GPIIVHC SAGIGRTGTI IVIDMLMESI STKGLDCDID IQKTIQMVRA QRSGMVQTEA QYKFIYVAIA QFIETTKKKL EIIQSQKGQE
               CSW D6   EKPGPICVHC SAGIGRTGTF IVIDMILDQI VRNGLDTEID IQRTIQMVRS QRSGLVQTEA QYKFVYYAVQ HYIQTLIARK RAEEQSLQVG
         Consensus      ---GP--VHC SAGIGRTGT- IVID-----I ---G-D--ID ---TIQMVR- QRSG-VQTEA QY-F-Y-A-- --I-T----- ----------

       PTP36_MOUSE E1   CLEAVAQRLE LRETHYFGL. ......WFLS KS.QQARWVE LE..KPLKKH LDKFANEPLL FFGVMFYVPN VSRLQQEATR YQYYLQVKKD
       PTPD1_HUMAN E2   SLEAVAQRLE LREVTYFSL. ......WYYN KQ.NQRRWVD LE..KPLKKQ LDKYALEPTV YFGVVFYVPS VSQLQQEITR YQYYLQLKKD
        MEG1_HUMAN E3   LLDVVFKHLD LTEQDYFGL. ......QLAD DSTDNPRWLD PN..KPIRKQ L.KRGSPYSL NFRVKFFVSD PNKLQEEYTR YQYFLQIKQD
       PTPH1_HUMAN E4   LLDMVHNHLG VTEKEYFGL. ......QHDD DSVDSPRWLE AS..KPIRKQ L.KGGFPCTL HFRVRFFIPD PNTLQQEQTR HLYFLQLKMD
     PTP-BAS_HUMAN E5   VFDMVVAHIG LVEHHLFALA TLKDNEYFFV DPDLKLTKVA PEGWKEEPKK KTKATVNFTL FFRIKFFMDD VSLIQHTLTC HQYYLQLRKD
         PEZ_HUMAN E6   CLEAVAQRLE LRETHYFGL. ......WFLS KS.QQARWVE LE..KPLKKH LDKFANEPLL FFGVMFYVPN VSWLQQEATR YQYYLQVKKD
     PTPRL10_MOUSE E7   SLEAVAQRLE LREVTYFSL, ......WYYN KQ.NQRRWVD LE..KPLKKQ LDKHALEPTV YFGVLFYVPS VSQLQQEITR YQYYLQLKKD
         PTP2E_RAT E8   SLEAVAQRLE LREITYFSL. ......WYYN KQ.NQRRWVD LE..KPLKKQ LDKHALEPTV YFGVVFYVPS VSQLQQEITR YQYYLQLKKD
         RIP_MOUSE E9   VFDMVVAHIG LVEHHLFALA TRKENEYFFV DPDLKLTKVA PEGWKEEPKR KGKAAVDFTL FFRIKFFMDD VSLIQHDLTC HQYYLQLRKD
         Consensus      ----V----- --E---F-L- ---------- ---------- ----K---K- --K------- -F---F---- ----Q---T- --Y-LQ---D


  PTP-PEST_MOUSE G1   HHNSSPLLKA PLSFTNPLHS DDW.HSDGGS SDGAVTRNKT SISTASATVS PASSAESACH RRVLPMSI.. .ARQEVAGTP H.........
  PTP-PEST_HUMAN G2   HDNTSPLFRT PLSFTNPLHS DDS.DSDERN SDGAVTQNKT NISTASATVS AATSTESIST RKVLPMSI.. .ARHNIAGTT H.........
       PEP_MOUSE G3   NPHDSLVMNT LTSFSPPLNQ ETAVEAPSRR TDDEIPPPLP ERTPESFIVV EEAGEPSPRV TESLPLVVTF GASPECSGTS EMKSHDSVGF
       Consensus       ----S----- --SF--PL-- ---------- -D-------- -----S--V- ------S--- ---LP----- -A-----GT- ----------


          pyp2+ I1    .......... ........LC LEGGLRISVQ KYETFEDLK. VNTHLFR.LD KPNGPP.KYI HH....FWVH TWFDKTHPD. IESITGIIR
          pyp3+ I2    .......... ........LN I.GDLSVILV KVYTLTSLNE VQVREFE.LN K.DGVK.KKI LH....FYYN GWPDFGAPH. TFSLLSLTR
          pyp1+ I3    .......... ........KQ VYGDYCVKQI SEENVDNSRF I.LRKFE.IQ NANFPSVKKV HH....YQYP NWSDCNSPEN VKSMVEFLK
     YEAST PTP1 I4    VRIASK..WE SPGGANDMTQ FPSDLKIEFV NVHKVKDYYT VTDIKLT.PT DPLVGPVKTV HH....FYFD LWKDMNKPEE VVPIMELCA
         DdPTP1 I5    FSIYGN..GN .........E VFGTYSVELV EVIQDPE... .REIITRNIR LTFEGETRDI TQ....YQYE GWPDHNIPDH TQPFRQLLH
         DdPTP2 I6    .....K..DE .........E RYGNFIVKFD NNITIPDILI RREFTLENLK ...DNKTRKI YH....FQYT TWPDHGTPVS TTGFLKFV.
     YEAST PTP2 I7    TIKLQNTWEN ICNINGCVLR VFQVKKTAPQ NDNISQDCDL PHNGDLTSIT MAVSEPF.IV YQ....LQYK NWLD.SCGVD MNDIIKLHK
    Candida PTP I8    PMAPPFALQS KSSPLSTPPR LHSPLGVDRG LPISMSPIQS SLNQKFNNIT LQTPLNSSFL INNDEATNFN NKNNKNNNNN STATTTITN
      Consensus       ---------- ---------- ---------- ---------- ---------- ---------- ---------- ---------- ---------
```

```
701                                                         751
```

```
S.....RGLR KSRSRLLYAE .......... .......... .......... ........ C1  cdc25Dm (TWINE)
G.....R.LK KSRSRLML.. .......... .......... .......... ........ C2  cdc25Dm1 (STRING)
E.....RSRR ELCSRLQDQ. .......... .......... .......... ........ C3  cdc25B_HUMAN
E.....RSRR ELCSRLQDQ. .......... .......... .......... ........ C4  cdc25M2_MOUSE
E.....KSKR EMYSRLKKL. .......... .......... .......... ........ C5  cdc25A_HUMAN
D.....RKRR EQIARIMKL. .......... .......... .......... ........ C6  cdc25A_XENLA
D.....RKRR EQVARLMKL. .......... .......... .......... ........ C7  cdc25B_XENLA
D.....RKRR EPEFRLTGQR LG........ .......... .......... ........ C8  cdc25C_XENLA
E.....RQLQ GQIALLVKGA SPQ....... .......... .......... ........ C9  cdc25M1_MOUSE
E.....RQLS EQIALLMKKG VSLP...... .......... .......... ........ C10 cdc25c_HAMST
E.....RQLR EQIALLVKDM SP........ .......... .......... ........ C11 cdc25C_HUMAN
E.....DSPT GRCRNNPGDR KLLASPFNDS PGSRFPGRRM LSY....... ........ C12 cdc25 (nimT)
P.....RR.. .......... .......... .......... .......... ........ C13 cdc25+
NPNFFYRDSH QSSTTMASSA LSFRFEPPPK LSLNHRRVSS GSSLNSSEST GDENFFPS C14 cdc25 (MIH1)
---------- ---------- ---------- ---------- ---------- -------- Consensus
```

```
HEYTNIKYSL ADQTSGDQSP LPPCTP.... .......TPP CAE......M REDSARVYEN VGLMQQQKSF R......... .......... .......... ..........
HEYTNIKYSL VDQTSGDQSP LPPCTP.... .......TPP CAE......M REDSARVYEN VGLMQQQRSF R......... .......... .......... ..........
HEYTNIKSSL VDQTSGDQSP LPPCTP.... .......TPP CAE......M REDSARVYEN VGLMQQQRSF R......... .......... .......... ..........
SEYGNITYPP AMKNAHAKAS RTSSKH.... .......KED VYENLHTKNK REEKVKKQRS ADKEKSKGSL KRK....... .......... .......... ..........
SEYGNITYPP AVRSAHAKAS RTSSKH.... .......KEE VYENVHSKSQ KEEKVKKQRS ADKEKNKGSL KRK....... .......... .......... ..........
REYTNIKYTG EIGNDSQRSP LPPAISSISL VPSKTPLTPT SADLGTGMGL SMGVGMGVGN KHASKQQPPL PVVNCNNNNN GIGNSGCSNG GGSSTTSSSN GSSNGNINAL
-EY-NI---- ---------- ---------- ---------- ---------- ---------- ---------- ---------- ---------- ---------- ----------
```

```
VLEGRLRCSL EQVIRLAGLA VQADFGDYNQ FDSQ.EFLRE YVLFPMDLAM EEAALEELTQ KVAQEHKAHS GILPAEAELM YINEVERLDG FGQEIFPVKD SHGN..SVHL
ILEGSIPCTL EQAIQLAGLA VQADFGDFDQ YESQ.DFLQK FALFPVGWLQ DEKVLEEATQ KVALLHQKYR GLTAPDAEML YMQEVERMDG YGEESYPAKD SQGS..DITI
ILTGRLPCPS NTAALLASFA VQSELGDYDQ SENLSGYLSD YSFIP..... ..NQPQDFEK EIAKLHQQHI GLSPAEAEFN YLNTARTLEL YGVEFHYARD QSNN..EIMI
ICEGRLTCPL NSAVVLASYA VQSHFGDYNS SIHHPGYLSD SHFIP..... ..DQNEDFLT KVESLHEQHS GLKQSEAESC YINIARTLDF YGVELHSGRD LHNL..DLMI
ILEERMHCDD ETSLLLASLA LQAEYGDYQP EVHGVSYFRM EHYLP.ARVM EKLDLSYIKE ELPKLHNTYV GASEKETELE FLKVCQRLTE YGVHFHRVHP EKKSQTGILL
VLEGRLRCTL DQVIRLAGLA VQADFGDYNQ FDSQ.DFLRE YVLFPMDLAL EEAVLEELTQ KVAQEHKAHS GILPAEAELM YINEVERLDG FGQEIFPVKD NHGN..CVHL
ILEGNLPCTL EHAIQLAGLA VQADFGDFDQ YESQ.DFLQK FALLPVAWLQ DEKVLEEAAQ KVALLHQKYR GLTAPEAELL YMQEVERMDG YGEESYPAKD SQGS..DISI
VLEGNLPCTL EQAIQLAGLA VQADFGDFDQ YESQ.DFLQK FALLPVGWLQ DEKLLEEAAQ KVALLHQKYR GLTAPEAEML YMQEVERMDG YGEESYPAKD SQGS..DISI
LLDERVHCDD EAALLLASLA LQAEYGDYQP EVHGVSYFRL EHYLP.ARVM EKLDVSYIKE ELPKLHNTYA GASEKETELE FLKVCQRLTE YGVHFHRVHP EKKSQTGILL
-------C-- -----LA--A -Q---GD--- ---------- ----P----- ---------- -----H---- G------E-- ---------- -G-------- ----------
```

```
........SG AEKDADVSEE SPPPLPERTP ESFVLA.... ..DMPVRPEW HELPNQEWSE QRESEGLTTS GN....EK.. HDAGGIHTEA SADSPPAFSD KKDQITKSPA
........SG AEKDVDVSED SPPPLPERTP ESFVLASE.. .HNTPVRSEW SELQSQERSE QKKSEGLITS EN....EKCD HPAGGIHYEM CIECPPTFSD KREQISENPT
TPSKNVKLRS PRSDRHQDGS PPPPLPERTL ESFFLADEDC IQAQAVQTSS TSYPETTENS TSSKQTLRTP GKSFTRSKSL KIFRNMKKSV CNSSSPSKPT ERVQ.PKNSS
---------- ---D------ -PPPLPERT- ESF-LA---- -----V---- ---------- ------L-T- -------K-- ---------- -----P---- ---Q------
```

```
IDKV...... .......... ........P. .......NDG PMFVHCSAGV GRTGTFIAVD QILQ...... .......... .......... .......... ..........
IKSL...... .......... ........SY S...PDFETA PIIVHCSAGC GRTGTFMALF EILS...... .......... .......... .......... ..........
VNN....... .......... .......... .....SHGSG NTIVHCSAGV GRTGTFIVLD TILR...... .......... .......... .......... ..........
SHSL...... .......... .......... .....NSRGN PIIVHCSAGV GRTGTFIALD HLMHDTLDF. .......... .......... .......... ..........
ITNR...... .......... ........QN QIIPSSDRNV PIIVHCSAGV GRTGTFCTAV IMMKKLDHYF KQLDATPIDQ VVDPFTHLPI TEYQSDNLDL KGLGYHFKSS
.......... .......... .......... SFVDHEKRSG PIVVHCSAGI GRSGTFVAIH SIVAKFAKHY .......... .......... .......... ..........
KNSLLFNPQS FITSLEKDVC KPDLIDDNNS ELHLDTANSS PLLVHCSAGC GRTGVFVTLD FLLSIL.... .......... .......... .......... ..........
ILSTPQNVRY NSKKFHPPEE LQELTSINAY PNGPKNVLNN LIYLYSDPAQ GKIDINKFDL VINVAKECDN MSLQYMNQVP NQREYVYIPW SHNSNISKDL FQITNKIDKF
---------- ---------- ---------- ---------- ---------- G--------- ---------- ---------- ---------- ---------- ----------
```

```
                              801                                              851

           SH-PTP2_HUMAN D1 .......... .......... .......... .......... .......... .......... ........ D1 SH-PTP2_HUMAN
    SH-PTP2_RAT ADIPOSE D2 .......... .......... .......... .......... .......... .......... ........ D2 SH-PTP2_RAT ADIPOSE
      SH-PTP2_RAT BRAIN D3 .......... .......... .......... .......... .......... .......... ........ D3 SH-PTP1_HUMAN
           SH-PTP1_HUMAN D4 .......... .......... .......... .......... .......... .......... ........ D4 SH-PTP1_HUMAN
           SH-PTP1_MOUSE D5 .......... .......... .......... .......... .......... .......... ........ D5 SH-PTP1_MOUSE
                     CSW D6 LGGIGLGLGG NMRKSNFYSD SLKQQQQREE QAPAGAAKFK NIPKDMIGLR PPSHAPALPP PPTPPRKT D6 CSW
               Consensus    ---------- ---------- ---------- ---------- ---------- ---------- -------- Consensus

      PTP36_MOUSE E1 GIFFMGIFVR ..NRVGRQAV I.YRWNDIGS VTHSKAAILL EL.....IDK EETALFHTDD IENAKYISRL FTTRHKF... ..........
      PTPD1_HUMAN E2 GACLEGIFVK ..HKNGRHPV V.FRWHDIAN MSHNKSFFAL EL.....ANK EETIQFQTED METAKYIWRL CVARHKF... ..........
       MEG1_HUMAN E3 GVMSGGILIY ..KNRVRMN. T.FPWLKIVK ISFKCKQFFI QLRKELHESR ETLLGFNMVN YRACKNLWKA CVEHHTF... ..........
      PTPH1_HUMAN E4 GIASAGVAVY ..RKYICTS. F.YPWVNILK ISFKRKKFFI HQRQKQAESR EHIVAFNMLN YRSCKNLWKS CVEHHTF... ..........
  PTP-BAS_HUMAN E5 GVCSKGVLVF EVHNGVRTLV LRFPWRETKK ISFSKKKITL Q.....NTSD GIKHGFQTDN SKICQYLLHL CSYQHKFQLQ MRARQSNQDA
        PEZ_HUMAN E6 GIFFMGIFVR ..NRIGRQAV I.YRWNDMGN ITHNKSTILV EL.....INK EETALFHTDD IENAKYISRL FATRHKF... ..........
   PTPRL10_MOUSE E7 GACLDGIFVK ..HKNGRPPV V.FRWHDIAN MSHNKSFFAL EL.....ANK EETIQFQTED METAKYVWRL CVARHKF... ..........
       PTP2E_RAT E8 GACLDGIFVK ..HKNGRPPV V.FRWHDIAN MSHNKSFFAL EL.....ANK EETIQFQTED METAKYVWRL CVARHKF... ..........
       RIP_MOUSE E9 GVCSKGVLVF EVHNGVRALV LRFPWRETKK ISFLKKKITL Q.....NTSD GIKHAFQTDS SKACQYLLHL CSSQHKFQLQ MRARQSNQDA
       Consensus    G----G---- ---------- ----W----- ---------- ---------- -----F---- ---------- ----H-F--- ----------

 PTP-PEST_MOUSE G1 EVTDIGFGNR CGKPKGPREP PSEWT.  G1 PTP-PEST_MOUSE
 PTP-PEST_HUMAN G2 EATDIGFGNR CGKPRGPRDP PSEWT.  G2 PTP-PEST_HUMAN
      PEP_MOUSE G3 SFLNFGFGNR FSKPKGPRNP PSAWNM  G3 PEP_MOUSE
      Consensus    -----GFGNR --KPKGPR-P PS-W--  Consensus

        pyp2+ I1 .......... .......... ......VPKN ILPK.TTNLE DSKD...... .......... ...FIFNCVN SLRSQRMKMV QNFEQFKFLY
        pyp3+ I2 .......... .......... ......QTDD S.TS.TSKFE ..VD...... .......... ...NIANIVS SLRSQRMQSV QSVDQLVFLY
        pyp1+ I3 .......... .......... ......FPES KLSGFNPSVA DSSD...... .......... ...VVFQLVD HIRKQRMKMV QTFTQFKYVY
   YEAST PTP1 I4 .......... .......... ....KNITER SRHSDRATEE YTRD...... .......... ...LIEQIVL QLRSQRMKMV QTKDQFLFIY
       DdPTP1 I5 IYNSNGINNN NNNNLNNNNN INNNSNGSNN TPQTEPNNEE DDDDAAESTK YAIMDKYNSR IDFNLFSIVL KLREQRPGMV QQLEQYLFCY
       DdPTP2 I6 .......... .......... .......... .......... DEKKQAPS.. .......... ..INLPKLVV EMRNERPGMV QTRDQYRFCY
   YEAST PTP2 I7 .......... .......... .......... SPTTNHSNKI DVWNMTQD.. .......... ...LIFIIVN ELRKQRISMV QNLTQYIACY
 Candida PTP I8 FTNGRKILIH [CQCGVSR]SAC VVVAFYMKKF QLGVNEAYEL LKNGDQKYID ACDRICPNMN LIFELMEFGD KLNNNEISTQ QLLMNSPPTI
    Consensus    ---------- ---------- ---------- ---------- ---------- ---------- ---------- ---------- Q---------
```

```
901                                                     951

.......... .......... .......... ....YKQNKI CTEQSNSPPP IRRQPTWSRS  SLPRQQPYIL .PPMHVQCSE HYSETHT.SQ DSI....FPG NEEALYCR..
.......... .......... .......... ....YRLNQC NLQTQTVTVN PIRRRSSSRM  SLPKPQPYVM PPPPQLHYNG HYTEPYASSQ DNL....FVP NQNGYYCH..
.......... .......... .......... ....FRLDR. .......... ..........  .......... .......... .......... .......PLP PQKNFFAH..
.......... .......... .......... ....FQAKK. .......... ..........  .......... .......... ................LLP QEKNVLSQ..
QDIERASFRS LNLQAESVRG FNMGRAISTG SLASSTLNKL AVRPLSVQAE ILKRLSCSEL  SLYQPLQNSS KEKNDKASWE EKPREMSKSY HDLSQASLYP HRKNVIVNME
.......... .......... .......... ....YKQNKI CTEQSNSPPP IRRQPTWSRS  SLPRQQPYIL .PPVHVQCGE HYSETHT.SQ DSI....FHG NEEALYCN..
.......... .......... .......... ....YRLNQC SLQTQAATLN SVRRDSSSRM  SLPKPQPYAM PPPPQLHYNG HYTEPFASSQ DNI....FVP NKNGFYCH..
.......... .......... .......... ....YRLNQC NLQTQAATLN SVRRGSSSRM  SLPKPQPYAM PPPPQLHYNG HYTEPFASSQ DNV....FVP NKNGFYCH..
QDIERASFRS LNLQAESVRG FNMGRAISTG SLASSTINKL AVRPLSVQAE ILKRLSSSEW  SLYQPLQNSS KEKTDKASWE EKPRGMSKSY HDLSQASLCP HRKQVI.NME
---------- ---------- ---------- ---------- ---------- ----------  ---------- ---------- ---------- ---------- ----------

DVVDYLNSGV NQASKPLMT. .......... .......... .......... ....... I1 pyp2+
TVSQELLQGK EFLLPQL... .......... .......... .......... ....... I2 pyp3+
DLIDSLQKSQ VHF..PVLT. .......... .......... .......... ....... I3 pyp1+
HAAKYLNSLS VNQ....... .......... .......... .......... ....... I4 YEAST PTP1
KTILDEIYHR LNCKLGFSLP HVNNINNYNN YSNTTTTTTS SLASTTIIHP STNSKLN I5 DdPTP1
.......... ....LAISEA MNTVLKKEQK KRKGLSYSYS SIPLTGPEHD ....... I6 DdPTP2
EALLNYFALQ KQIKNALPC. .......... .......... .......... ....... I7 YEAST PTP2
NL........ .......... .......... .......... .......... ....... I8 Candida PTP
---------- ---------- ---------- ---------- ---------- ------- Consensus
```

```
                  1001                                                              1051
   PTP36_MOUSE E1 .......... SHNSLDLN.. YLNGTVTNGS VCSVHSVNSL SCSQSFIQAS PVSSNLSIPG SDIMRADYIP SHRHSTIIVP ..........
   PTPD1_HUMAN E2 .......... SQTSLDRAQI DFNGRIRNGS VYSAHSTNSL NNPQPYLQPS PMSSNPSITG SDVMRPDYLP SHRHSAVIPP ..........
    MEG1_HUMAN E3 .......... YFT....... .LGSKFRYCG RTEVQSVQYG KEKANKDRVF ARSPSKPLAR KLMDWE.... .......... ..........
   PTPH1_HUMAN E4 .......... YWT....... .MGS...... .......... ..RNTKKSVN NQYCKKVIGG ..MVWN.... .......... ..........
 PTP-BAS_HUMAN E5 PPPQTVAELV GKPSHQMSRS DAESLAGVTK LNNSKSVASL NRSPERRKHE SDSSSIEDPG QAYVLGMTMH SSGNSSSQVP LKENDVLHKR
     PEZ_HUMAN E6 .......... SHNSLDLN.. YLNGTVTNGS VCSVDSVNSL NCSQSFIQAS PVSSNLSIPG SDIMRADYIP SHRHSAIIVP ..........
 PTPRL10_MOUSE E7 .......... SQTSLDRTQI DLSGRIRNGS VYSAHSTNSL NTLQPYLQPS PMSSNPSITG SDVMRPDSLP SHRHSALIPP ..........
     PTP2E_RAT E8 .......... SQTSLDRTQI DLSGRIRNGS VYSAHSTNSL NTPQPYLQPS PMSSNPSIPG SDVMRPDYIP SHRHSALIPP ..........
     RIP_MOUSE E9 ALPQAFAELV GKPLYPMARS DTESLAGLPK LDNSKSVASL NRSPERRNHE SDSST.EDPG QAYVVGMSLP SSGKSSSQVP FKDNDTLHKR
     Consensus    ---------- ---------- ---------- ---------- ---------- ---------- ---------- ---------- ----------

                  1201                                                              1251
   PTP36_MOUSE E1 .......... .......... ........ER HPYTVPYAHQ GCYGHKLVSP SDQMNPQNCA MPIKPGASSI SHTVSTPELA NMQLQGAQHY
   PTPD1_HUMAN E2 .......... .......... ........EH AQLPSPAAAH CPFSLSYSFH SP......SP YPYPAERRPV VGAVSVPELT NAQLQ.AQDY
    MEG1_HUMAN E3 .......... .......... ........QE G......... .......... .......... .......... .......... ..........
   PTPH1_HUMAN E4 .......... .......... ........HE I......... .......... .......... .......... .......... ..........
 PTP-BAS_HUMAN E5 HLTNEMKNYM KKSSYMQDSA IDSSSKDHHW SRGTLRHISE NSFGPSGGLR EGSLSSQDSR TESASLSQSQ VNGFFASHLG DQTWQESQHG
     PEZ_HUMAN E6 .......... .......... ........ER HPYTVPYGPQ GVYSNKLVSP SDQRNPKNNV VPSKPGASAI SHTVSTPELA NMQLQGSHNY
 PTPRL10_MOUSE E7 .......... .......... ........EH PHLTSPQSAH YPFNLNYSFH SQ......SP YPYPAERRPV VGAVSVPELT NVQLQ.AQDY
     PTP2E_RAT E8 .......... .......... ........EH PHLASPQSAH YPFNLNYSFH SQ......AP YPYPVERRPV VGAVSVPELT NVQLQ.AQDY
     RIP_MOUSE E9 HFANGMKSYT KKPAYMQDSA MD.PSEDQPW PRGTLRHIPE SPFGLXGGLR EGSLSSQDSR TESASLSQSQ VNGFFASHLG DRGWQEPQHS
     Consensus    ---------- ---------- ---------- ---------- ---------- ---------- ---------- ---------- ----------

                  1401                                                              1451
   PTP36_MOUSE E1 KHHGGGGGTV NKRHSLEVMN SMVRGMEAMT LKSLNIP... ..MARRNTLR EQG....... .......... .....PS... .EET.GGHEV
   PTPD1_HUMAN E2 RH.....AQL HKRNSIEV.A GLSHGLEGLR LKERTLSASA AEVAPRAVSV GSQ....... .......... .....PSVFT ERTQREGPEE
    MEG1_HUMAN E3 ..LPPKQSKK NS.WNQIHYS HSQQDLESHI NETF...... .......... .......... .......... .......... D...IPSSPE
   PTPH1_HUMAN E4 SYLTQKSSSS VSPSSNAPGS CSPDGVDQQL LDDF...... .......... .......... .......... .......... HRVTKGGSTE
 PTP-BAS_HUMAN E5 RHGGIYVKAV IPQGAAESDG RIHKGDRVLA VNGVSLEGAT HKQAVETLRN TGQVVHLLLE KGQSPTSKEH VPVTPQCTLS DQNAQGQGPE
     PEZ_HUMAN E6 KHH....GTV NKRHSLEVMN SMVRGMEAMT LKSLHLP... ..MARRNTLR EQG....... .......... .....PP... .EEGSGSHEV
 PTPRL10_MOUSE E7 RH.....AHL QKRNSIEI.A GLTHGFEGLR LKERTVSASA ADVAPRTFSA GSQ....... .......... .....SSVFS DKMKQEGTEE
     PTP2E_RAT E8 RH.....AHL QKRNSIEI.A GLTHGFEGLR LKEETMSASA ADVAPRTFSA GSQ....... .......... .....SSVFS DKVKQEGTEE
     RIP_MOUSE E9 RHGGIYVKAI IPKGAAESDG RIHKGDRVLA VNGVSLEGAT HKQAVETLRN TGQWHLLLE KGQVPTSREQ DPAGPQSPPP DQDAQRQAPE
     Consensus    ---------- ---------- ---------- ---------- ---------- ---------- ---------- ---------- ----------

                  1601                                                              1651
   PTP36_MOUSE E1 .......... .......... .......... .......... .......... .......... .....HSSES .EEE...... ..........
   PTPD1_HUMAN E2 .......... .......... .......... .......... .......... .......... .....HSSEE EEDE...... ..........
    MEG1_HUMAN E3 .......... .......... .......... .......... .......... .......... .....IRMKP DENG...... ..........
   PTPH1_HUMAN E4 .......... .......... .......... .......... .......... .......... .....IRITP DEDG...... ..........
 PTP-BAS_HUMAN E5 PGVLPEIDTA LLTPLQSPAQ VLPNSSKDSS QP.SCVEQST SSDENEMSDK SKKQCKSPSR RDSYSDSSGS GEDDLVTAPA NISNSTWSSA
     PEZ_HUMAN E6 .......... .......... .......... .......... .......... .......... .....HSSES EEEE...... ..........
 PTPRL10_MOUSE E7 .......... .......... .......... .......... .......... .......... .....DSS.. EEDE...... ..........
     PTP2E_RAT E8 .......... .......... .......... .......... .......... .......... .....HSS.. EEDE...... ..........
     RIP_MOUSE E9 PGVLPEIDTT FLNPLYSPAN SFLNSSKETS QPSSSVEQGA SSHDNGVSGK TKNHCRAPSR RESYSDHSES GEDDSVRAPA KMPNVTRVAA
     Consensus    ---------- ---------- ---------- ---------- ---------- ---------- ---------- -E-------- ----------

                  1801                                                              1851
   PTP36_MOUSE E1 .......... ...ALARIPN RPPPEYPGPR KSVSNGALRQ DQGTPLPAMA RCRVLRHGPS KALSVSRAEQ LAVNGAS... ..........
   PTPD1_HUMAN E2 .......... ...GLA...Q DP....PGCP RVLLAGPLHI LEPKAHVPDA EKRMMDSSP. ..VRTTAEAQ RPWRDGL... ..........
    MEG1_HUMAN E3 .......... ...IVSRVAP GTPADLCVP. ........RL NEGDQVVLIN GRDIAEHTHD QVVLFIKASC ERHSGEL... ..........
   PTPH1_HUMAN E4 .......... ...VVSRINP ESPADTCIP. ........KL NEGDQIVLIN GRDISEHTHD QVVMFIKASR ESHSREL... ..........
 PTP-BAS_HUMAN E5 ITLIKSEKGS LGFTVTKGNQ RIGCYVHDVI QDPAKSDGRL KPGDRLIKVN DTDVTNMTHT DAVNLLRAAS KTVRLVIGRV LELPRIPMLP
     PEZ_HUMAN E6 .......... ...ALARIPN KPPPEYPGPR KSVSNGALRQ DQASLPPAMA RARVLRHGPA KAISMSRTDP PAVNGAS... ..........
 PTPRL10_MOUSE E7 .......... ...AFS...Q E..LNYPCAS ATPITGPLHI FEPKPHVTEP EKRAKDISP. ..VHLVVETH RPRRDGL... ..........
     PTP2E_RAT E8 .......... ...AFS...Q EQQLNYPCAS VTPVTGPLHI FEPKSHVTEP EKRAKDISP. ..VHLVMETH QPRRHGL... ..........
     RIP_MOUSE E9 ITLVKSEKGS LGFTVTKGSQ SIGCYVHDVI QDPAKGDGRL KAGDRLIKVN DTDVTNMTHT DAVNLLRAAP KTVRLVLGRI LELPRMPVFP
     Consensus    ---------- ---------- ---------- ---------- ---------- ---------- ---------- ---------- ----------

                  2001                                                              2051
   PTP36_MOUSE E1 VSEMFSLEDS IIEREMMIRN LEK....... .......... .......... .......... .......... ...QKMTCPG AQK.RPLMLA
   PTPD1_HUMAN E2 VSDLLSGKKN IVE...GLPP LGG....... .......... .......... .......... .......... ...MKKTRVD AKKIGPLKLA
    MEG1_HUMAN E3 VYDVVEEKLE NEPDFQYIPE KAP....... .......... .......... .......... .......... ...LDSVHQD DH........
   PTPH1_HUMAN E4 VRSFADFKSE DELN.QLFPE .AI....... .......... .......... .......... .......... ...FPMCPEG GD........
 PTP-BAS_HUMAN E5 SYSVGSCSQP ALTPNDSFST VAGEEINEIS YPKGKCSTYQ IKGSPNLTLP KESYIQEDDI YDDSQEAEVI QSLLDVVDEE AQNLLNENNA
     PEZ_HUMAN E6 VSEMFSLEDS IIEREMMIRN LEK....... .......... .......... .......... .......... ...QKMAGLE AQK.RPLMLA
 PTPRL10_MOUSE E7 VSDLLSGKKS AVE...GLPP LGG....... .......... .......... .......... .......... ...MKKTRAD AKKIGPLKLA
     PTP2E_RAT E8 VSDLLSGKKN TVE...GLPP LGG....... .......... .......... .......... .......... ...MKKTRAD AKKIGPLKLA
     RIP_MOUSE E9 LTSMEPSGQP ALMPKNSFSK VNGEGVHEAV CPAGEGSSSQ MKESAGLTET KESNSRDDDI YDDPQEAEVI QSLLDVVDEE AQNLLNQRHA
     Consensus    ---------- ---------- ---------- ---------- ---------- ---------- ---------- ---------- ----------

                  2201                                                              2251
   PTP36_MOUSE E1 ........ID ERLRALKKKL EDGMVFTEYE QIPNKK.ANG VFSTATLPEN AERSRIREVV PYEENRVELI PTKENNTGYI NASHIKVVVG
   PTPD1_HUMAN E2 ........ND ERCKILEQRL EQGMVFTEYE RILKKRLVDG ECSTARLPEN AERNRFQDVL PYDDARVELV PTKENNTGYI NASHIKVSVS
    MEG1_HUMAN E3 ........LR ESMIQLAEGL ITGTVLTQFD QLYRKK.PGM TMSCAKLPQN ISKNRYRDIS PYDATRVIL. ...KGNEDYI NANYINMEIP
   PTPH1_HUMAN E4 ........LE GSMAQLKKGL ESGTVLIQFE QLYRKK.PGL AITFAKLPQN LDKNRYKDVL PYDTTRVLL. ...QGNEDYI NASYVNMEIP
 PTP-BAS_HUMAN E5 SGKYTGANLK SVIRVLRGLL DQGIPSKELE NLQELKPLD. QCLIGQTKEN RRKNRYKNIL PYDATRVPL. ...GDEGGYI NASFIKIPVG
     PEZ_HUMAN E6 ........MD ERFRTLKKKL EEGMVFTEYE QIPKKK.ANG IFSTAALPEN AERSRIREVV PYEENRVELI PTKENNTGYI NASHIKVVVG
 PTPRL10_MOUSE E7 ........ND ERCKVLEQRL EQGMVFTEYE RILKKRLVDG ECSTARLPEN AERNRFQDVL PYDDARVELV PTKENNTGYI NASHIKVSVS
     PTP2E_RAT E8 ........ND ERCKVLEQRL EQGTVFTEYE RILKKRLVDG ECSTARLPEN AERNRFQDVL PYDDARVELV PTKENNTGYI NASHIKVSVS
     RIP_MOUSE E9 SGKYTGTQLQ ATIRTLQGLL DQGIPSKELE NLQELKPLD. QCLIGQTKEN RRKNRYKNIL PYDTTRVPL. ...GDEGGYI NATFIRIPVG
     Consensus    ---------- -----L---L --G------- ---------- ---------N ----R----- PY---RV-L- --------YI NA--------

                  2401                                                              2451
   PTP36_MOUSE E1 CPEDVQGFLS YLEEIQSVRR HTNSVLEGIR TRHPPIVVHC SAGVGRTGVV ILSELMIYCL EHNEKVEVPT MLRFLREQRM FMIQTIAQYK
   PTPD1_HUMAN E2 CPEDLKGFLS YLEEIQSVRR HTNSTSDP.Q SPNPPLLVHC SAGVGRTGVV ILSEIMIACL EHNEVLDIPR VLDMLRQQRM MLVQTLCQYT
    MEG1_HUMAN E3 VPDDSSDFLD FVCHVRNKR. .........A GKEEPVVVHC SAGIGRTGVL ITMETAMCLI ECNQPVYPLD IVRTMRDQRA MMIQTPSQYR
   PTPH1_HUMAN E4 IPDDSSDFLE FVNYVRSLR. .........V D.SEPVLVHC SAGIGRTGVL VTMETAMCLT ERNLPIYPLD IVRKMRDQRA MMVQTSSQYK
 PTP-BAS_HUMAN E5 TPSQPDDLLT FISYMRHIHR .......... ..SGPIITHC SAGIGRSGTL ICIDVVLGLI SQDLDFDISD LVRCMRLQRH GMVQTEDQYI
     PEZ_HUMAN E6 CPEDVQGFLS YLEEIQSVRR HTNSMLEGTK NRHPPIVVYC SAGVGRTGVL ILSELMIYCL EHNEKVEVPM MLRLLREQRM FMIQTIAQYK
 PTPRL10_MOUSE E7 CPEDLKGFLS YLEEIQSVRR HTNSTSEP.K SHNPPLLVHC SAGVGRTGVV ILSEIMVACL EHNEVLDIPR VLDMLRQQRM MLVQTLGQYT
     PTP2E_RAT E8 CPEDLKGFLS YLEEIQSVRR HTNSTSEP.R SPNPPLLVHC SAGVGWTGVV ILSEIMVACL EHNEVLDIPR VLELLRQQRM MLVQTLSQYT
     RIP_MOUSE E9 TPSQPDDLLT FISYMRHIRR .......... ..SGPVITHC SAGIGRIGTL ICIDVVLGLI SQDLEFDISD LVRCMRLQRH GMVQTEGQYV
     Consensus    -P------L- ---------- ---------- ----P---HC SAG-GR-G-- ---------- ---------- -----R-QR- ---QT--QY-
```

```
           1101                                                                 1151
.SYRPTPDYE TVMRQMKRGL MHA....... .......... .......... .......DSQ SRSLRNLNII NTHAYNQPEE .......... ..LVYSQPEM R.........
.SYRPTPDYE TVMKQLNRGL VHA....... .......... .......... .......ERQ SHSLRNLNIG SSYAYSRPAA .......... ..LVYSQPEI R.........
.......... .......... .VV....... .......... .......... SRN        SISDDRLETQ SLPSRSPPGT .......... ..PNHRNSTF T.........
.......... .......... .PA....... .......... .......... .......MRR SLSVEHLETK SLPSRSPPIT .......... ..PNWRSPRL R.........
WSIVSSPERE ITLVNLKKDA KYGLGFQIIG GEKMGRLDLG IFISSVAPGG PADLDGCLKP GDRLISVNSV SLEGVSHHAA IEILQNAPED VTLVISQPKE KISKVPSTPV
.SYRPTPDYE TVMRQMKRGI LHT....... .......... .......... .......DSQ SQSLRNLNII NTHAYNQPED .......... ..LVYSQPEM R.........
.SYRPTPDYE TVMKQLNRGM VHA....... .......... .......... .......DRH SHSLRNLNIG SSYAYSRPDA .......... ..LVYSQPEI R.........
.SYRPTPDYE SVMKRLNRGM VHA....... .......... .......... .......DRH SHSLRNLNIG SSYAYSRPDA .......... ..LVYSQPEI R.........
WSIVSSPERE ITLVNLKKDP KHGLGFQIIG GEKMGRLDLG VFISAVTPGG PADLDGCLKP GDRLISVNSV SLEGVSHHAA VDILQNAPED VTLVISQPKE KPSKVPSTPV
---------- ---------- ---------- ---------- ---------- ---------- ---------- ---------- ---------- ---------- ----------

           1301                                                                 1351
STAHMLKNYL FRPPPPYPRP RPATSTPDLA S......... ...HRHKYVS GSSPDLVTRK VQLSVKTFQE DSSPVVHQSL QEVSEPL... .......... .......TAT
PSPNIMRTQV YRPPPPYPPP RPANSTPDLS .......... ....RHLYIS SSNPDLITRR VHHSVQTFQE DSLPVAH.SL QEVSEPL... .......... .......TAA
.......... .........T RLRPSSVGHL V......... ...DHMVH.T SPSEVFVNQR SPSSTQ..AN SIVLESSPSQ ETPGDGK... .......... .......PPA
.......... .........R KPRHSSADNL A......... ...NEMTYIT ETEDVFYTYK GSLAPQDSDS EVSQNRSPHQ ESLSENN... .......... .......PAQ
SPSPSVISKA TEKETFTDSN QSKTKKPGIS DVTDYSDRGD SDMDEATYSS SQDHQTPKQE SSSSVNTSNK MNFKTFSSSP PKPGDIFEVE LAKNDNSLGI SVTGGVNTSV
STAHMLKNYL FRPPPPYPRP RPATSTPDLA S......... ...HRHKYVS GSSPDLVTRK VQLSVKTFQE DSSPVVHQSL QEVSEPL... .......... .......TAT
PAPNIMRTQV YRPPPPYPYP RPANSTPDLS .......... ....RHLYIS SSNPDLITRR VHHSVQTFQE DSLPVAH.SL QEVSEPL... .......... .......TAA
PAPNIMRTQV YRPPPPYPYP RPANSTPDLS .......... ....RHLYIS SSNPDLITRR VHHSVQTFQE DSLPVAH.SL QEVSEPL... .......... .......TAA
SPSPSVTTKV NEK.TFSDSN RSKAKRRGIS DLIEHLDCAD SDKDDSTYTS SQDHQTSKQE PSSSLSTSNK TSFPTSSASP PKPGDTFEVE LAKTDGXLGI SVTGGVNTSV
---------- ---------- ---------- ---------- ---------- ---------- ---------- ---------- ---------- ---------- ----------

           1501                                                                 1551
HGLPQYHHKK TFS....... .......... .......... .......... .......... .......... .......... .......... .......... DATMLI....
AEGLRYGHKK SLS....... .......... .......... .......... .......... .......... .......... .......... .......... DATMLI....
KPTPNGGIPH DN........ .......... .......... .......... .......... .......... .......... .......... .......... ...LVL....
DASQYYCDKN DNG....... .......... .......... .......... .......... .......... .......... .......... .......... DSYLVL....
KV.KKTTQVK DYSFVTEENT FEVKLFKNSS GLGFSFSRED NLIPEQINAS IVRVKKLFPG QPAAESGKID VGDVILKVNG ASLKGLSQQE VISALRGTAP EVFLLLCRPP
PQLPQYHHKK TFS....... .......... .......... .......... .......... .......... .......... .......... .......... DATMLI....
QEGGRYSHKK SLS....... .......... .......... .......... .......... .......... .......... .......... .......... DATMLI....
QGSGGYSHKK SLS....... .......... .......... .......... .......... .......... .......... .......... .......... DATMLI....
KVAKQTPHVK DYSFVTEDNT FEVKLFKNSS GLGFSFSRED NLIPEQINGS IVRVKKLFPG QPAAESGKID VGDVILKVNG APLKGLSQQD VISALRGTAP EVSLLLCRPA
---------- ---------- ---------- ---------- ---------- ---------- ---------- ---------- ---------- ---------- ----------

           1701                                                                 1751
.......... .......... .......... .......... .......... .......... .......... .......EET LEAAPQVPVL REKVEYSAQL QA........
.......... .......... .......... .......... .......... .......... .......... .......DFE EESGARAPPA RAR...EPR. .P........
.......... .......... .......... .......... .......... .......... .......... .......RFG .........F NVKGGYDQKM PV........
.......... .......... .......... .......... .......... .......... .......... .......KFG .........F NLKGGVDQKM PL........
LHQTLSNMVS QAQSHHEAPK SQEDTICTMF YYPQKIPNKP EFEDSNPSPL PPDMAPGQSY QPQSESASSS SMDKYHIHHI SEPTRQENWT PLKNDLENHL EDFELEVELL
.......... .......... .......... .......... .......... .......... .......... .......EEA PESVPQIPML REKMEYSAQL QA........
.......... .......... .......... .......... .......... .......... .......... .......DLE EDSSREQAIS AVS...EPRL TA........
.......... .......... .......... .......... .......... .......... .......... .......DLE DDSSREH... AVS...EPRL TA........
F......... ....PHEAPR SQEESICAMF YLPRKIPGKL ESESSHPPPL ..DVSPGQTC QPPAECAPSD ATGKHFTHLA SQLSKEENIT TLKNDLGNHL EDSELEVELL
---------- ---------- ---------- ---------- ---------- ---------- ---------- ---------- ---------- ---------- ----------

           1901                                                                 1951
.......... .......... .......... .......... .......... ....LGPSIS EPDL.TSVKE RVKKEPV... .......... .......... ......KERP
.......... .......... .......... .......... .......... ....LMPSMS ESDLTTSGRY RARRDSL... .......... .......... ......KKRP
.......... .......... .......... .......... .......... ....MLL... .......... ......V... .......... .......... ......RPNA
.......... .......... .......... .......... .......... ....ALV... .......... ......I... .......... .......... ......RRRA
HLLPDITLTC NKEELGFSLC GGHDSLYQVV YISDINPRSV AAIEGNLQLL DVIHYVNGVS TQGMTLEEVN RALDMSLPSL VLKATRNDLP VVPSSKRSAV SAPKSTKGNG
.......... .......... .......... .......... .......... ....LGPSIS EPDL.TSVKE RVKKEPV... .......... .......... ......KERP
.......... .......... .......... .......... .......... ....LTPSMS ESDLTTSGRY RARRDSV... .......... .......... ......KKRP
.......... .......... .......... .......... .......... ....LTPSMS ESDLTTSGRY RARRDSL... .......... .......... ......KKRP
HLLPDITVTC HGEELGFSLS GGQGSPHGVV YISDINPRSA AAVDGSLQLL DIIHYVNGVS TQGMTLEDAN RALDLSLPSV VLKVTRDGCP WPTT.RAAI SAPRFTKANG
---------- ---------- ---------- ---------- ---------- ---------- ---------- ---------- ---------- ---------- ----------

           2101                                                                 2151
ALNGLSVARV SGREDGRHDA TRVP...... .......... .......... .......... .......... .......... .......... .......... ..........
ALNGLSLSRV PLPDEGKEVA TRAT...... .......... .......... .......... .......... .......... .......... .......... ..........
.......... .......... ...S...... .......... .......... .......... .......... .......... .......... .......... ..........
.......... .......... ...T...... .......... .......... .......... .......... .......... .......... .......... ..........
AGYSCGPGTL KMNGKLSEER TEDTDCDGSP LPEYFTEATK MNGCEEYCEE KVKSESLIQK PQEKKTDDDE ITWGNDELPI ERTNHEDSDK DHSFLTNDEL AVLPVVKVLP
ALNGLSVARV SGREENRVDA TRVP...... .......... .......... .......... .......... .......... .......... .......... ..........
ALNGLSLSRL PLPDEGKEVS TRAT...... .......... .......... .......... .......... .......... .......... .......... ..........
ALNGLSLSRL PLPDEGKEVS TRAT...... .......... .......... .......... .......... .......... .......... .......... ..........
TRRACSPDPL RTNGEAPEE. .GDTDYNGSP LPEDVPESVS SG......EG KVDLASLTAA SQEEKPIEED ATQESRNSTT ETTDGEDSSK DPPFLTNEEL AALPVVRVPP
---------- ---------- ---------- ---------- ---------- ---------- ---------- ---------- ---------- ---------- ----------

           2301                                                                 2351
GSEW..HYIA TQGPLPHTCH DFWQMVWEQG VNVIAMVTAE EEGGRTKSHR YWPKLGSKHS SATYGKFKVT TKFRTDSGCY ATTGLKVKHL LSGQERTVWH LQYTDWPHHG
GIEW..DYIA TQGPLQNTCQ DFWQMVWEQG IAIIAMVTAE EEGGREKSFR YWPRLGSRHN TVTYGRFKIT TRFRTDSGCY ATTGLKMKHL LTGQERTVWH LQYTDWPEHG
SSSIINQYIA CQGPLPHTCT DFWQMTWEQG SSMVVMLTTQ VERGRVKCHQ YWP...EPTG SSSYGCYQVT CHSEEGNTAY IFRKMTLFNQ EKNESRPLTQ IQYIAWPDHG
AANLVNKYIA TQGPLPHTCA QFWQVVWDQK LSLIVMLTTL TERGRTKCHQ YWP...DPPD VMNHGGFHIQ CQSEDCTIAY VSREMLVTNT QTGEEHTVTH LQYVAWPDHG
KEEFV..YIA CQGPLPTTVG DFWQMIWEQK STVIAMMTQE VEGEKIKCQR YWPNILGK.T TMVSNRLRLA LVRMQQLKGF VVRAMTLEDI QTREVRHISH LNFTAWPDHD
GAEW..HYIA TQGPLPHTCH DFWQMVWEQG VNVIAMVTAE EEGGRTKSHR YWPKLGSKHS SATYGKFKVT TKFRTDSVCY ATTGLKVKHL LSGQERTVWH LQYTDWPDHG
GIEW..DYIA TQGPLQNTCQ DFWQMVWEQG VAIIAMVTAE EEGGREKSFR YWPRLGSRHN TVTYGRFKIT TRFRTDSGCY ATTGLKMKHL LTGQERTVWH LQYTDWPEHG
GIEW..DYIA TQGPLQNTCQ DFWQMVWEQG VAIIAMVTAE EEGGREKSFR YWPRLGSRHN TVTYGRFKIT TRFRTDSGCY ATTGLKMKHL LTGQERTVWH LQYTDWPEHG
TQEFV..YIA CQGPLPTTVG DFWQMVWEQN STVIAMMTQE VEGEKIKCQR YWPSILGT.T TMANERLRLA LLRMQQLKGF IVRVMALEDI QTGEVRHISH LNFTAWPDHD
-------YIA -QGPL--T-- -FWQ--W-Q- -----M-T-- -E----K--- YWP------- ---------- ---------- ---------- ---------- ----------

           2501
FVYQVLVQFL QNSRLI.... ....... E1 PTP36_MOUSE
FVYRVLIQFL KSSRLI.... ....... E2 PTPD1_HUMAN
FVCEAILKVY EEGFVKPLTT STNK... E3 MEG1_HUMAN
FVCEAILRVY EEGLVQMLDP S...... E4 PTPH1_HUMAN
FCYQVILYVL TRLQAEEEQK QQPQLLK E5 PTP-BAS_HUMAN
FVYQVLIQFL QNSRLI.... ....... E6 PEZ_HUMAN
FVYRVLIQFL KSSRLI.... ....... E7 PTPRL10_MOUSE
FVYRVLIQFL KSSRLI.... ....... E8 PTP2E_RAT
FCYQVILYVL THLQA.EEQK AQPGLPQ E9 RIP_MOUSE
F--------- ---------- ------- Consensus
```

```
                     1                                                                   51
          YOP51 J1   MNLSLSDLHR QVSRLVQQES GDCTGKLRGN VAANKETTFQ GLTIASGARE SEKVFAQTVL SHVANIVLTQ EDTAKLLQST VKHNLNNYEL
          YOP2B J2   MNLSLSDLHR QVSRLVQQES GDCTGKLRGN VAANKETTFQ GLTIASGARE SEKVFAQTVL SHVANVVLTQ EDTAKLLQST VKHNLNNYDL
           STEP J3   .......... .......... .......... .......... .......... .......... .......... ........ME EKVEDDFLDL
          HePTP J4   .......... .......... .......... .......... .......... .......... .......... .......MVQ AHGGRSRAQP
          RKPTP J5   .......... .......... .......... .......... .......... .......... .......... .......... ..........
          Cdi-1 J6   .......... .......... .......... .......... .......... .......... .......... .......... ..........
            KAP J7   .......... .......... .......... .......... .......... .......... .......... .......... ..........
          PRL-1 J8   .......... .......... .......... .......... .......... .......... .......... .......... ..........
           IphP J9   .......... .......... .......... .......... .......... .......... .......... .......... ..........
       PC12-PTP J10  .......... .......... .......... .......... ...MIIYRLK ERLQLSLRQD KEKNQEIHLS PIALQQAQSE AKAAHSMVQP
      Consensus      ---------- ---------- ---------- ---------- ---------- ---------- ---------- ---------- ----------

        DPTP10D K1   .......... .......... .......... .......... .......... .......... .......... .......... ..........
         DPTP4E K2   .......... .......... .......... .......... .......... .......... .......... .......... ..........
          DEP-1 K3   .......... .......... .......... .......... .......... .......... .......... .......... ..........
          SAP-1 K4   .......... .......... .......... .......... .......... .......... .......... .......... ..........
         RPTP-β K5   MLSHGAGLAL WITLSLLQTG LAEPERCNFT LAESKASSHS VSIQWRILGS PCNFSLIYSS DTLGAALCPT FRIDNTTYGC NLQDLQAGTI
        GLEPP-1 K6   .......... .......... .......... .......... .......... .......... .......... .......... ..........
         RPTP-η K7   .......... .......... .......... .......... .......... .......... .......... .......... ..........
        RPTP-U2 K8   .......... .......... .......... .......... .......... .......... .......... .......... ..........
       RPTP-BR7 K9   .......... .......... .......... .......... .......... .......... .......... .......... ..........
      Consensus      ---------- ---------- ---------- ---------- ---------- ---------- ---------- ---------- ----------

         ICA512 L1   .......... .......... .......... .......... .......... .......... .......... .......... ..........
           IA-2 L2   MRRPRRPGGL GGSGGLRLLL CLLLLSSRPG GCSAVSAHGC LFDRRLCSHL EVCIQDGLFG QCQVGVGQAR PLLQVTSPVL QRLQGVLRQL
      Consensus      ---------- ---------- ---------- ---------- ---------- ---------- ---------- ---------- ----------

   RPTP-α_HUMAN M1   MDSWFILVLL GSGLICVSAN NATTVAPSVG ITRLINSSTA EPVKEEAKTS NPTSSLTSLS VAPTFSPNIT LGPTYLTTVN SSDSDNGTTR
   RPTP-α_MOUSE M2   MDSWFILVLF GSGLIHVSAN NATTVSPSLG TTRLIKTSTT ELAKEENKTS NSTSSVISLS VAPTFSPNLT LEPTYVTTVN SSHSDNGTRR
   RPTP-ε_HUMAN M3   MEPLCPLLLV G......... .......... .......... .......... ......FSLP LARALRGNET .......... ..........
      Consensus      M-----L-L- G--------- ---------- ---------- ---------- -------SL- -A-----N-T ---------- ----------

   LAR-PTP2_RAT N1   ...MAPTWRP SVVSVVGPVG LFLVLLARGC LAEEPPRFIR EPKDQIGVSG GVASFVCQAT GDPKPRVTWN KKGKKVNSQR FETIDFDESS
   RPTP-σ_MOUSE N2   ...MAPTWSP SVVSVVGPVG LFLVLLARGC LAEEPPRFIR EPKDQIGVSG GVASFVCQAT GDPKPRVTWN KKGKKVNSQR FETIDFDESS
          CRYPα N3   ...MRILPSP GMPALLSLVS LLSVLL.MGC VAESPPVFIK KPVDQIGVSG GVASFVCQAT GDPKPRVTWN KKGKKVNSQR FETIEFDESA
   RPTP-δ_MOUSE N4   .......... .......... .......... .......... .......... .......... .......... .......... ..........
      LAR_HUMAN N5   .......... ...MVPLVPA LVMLGLVAGA HGDSKPVFIK VPEDQTGLSG GVASFVCQAT GEPKPRITWM KKGKKVSSQR FEVIEFDDGA
           DLAR N6   MGLQMTAARP IAALSLLVLS LLTWTHPTIV DAAHPPEIIR KPQNQGVRVG GVASFYCAAR GDPPPSIVWR KNGKKVSGTQ SRYTVLEQPG
      Consensus      ---------- ---------- ---------- ---------- ---------- ---------- ---------- ---------- ----------

   RPTP-γ_HUMAN O1   MRRLLEPCWW ILFLKITSSV LHYVVCFPAL TEGYVGALHE NRHGSAVQIR R..RKASGDP YWAYSGAYGP EHWVTSSVSC GSRHQSPIDI
   RPTP-γ_MOUSE O2   MRRLLEPCWW ILFLKITSSV LHYVVCFPAL TEGYVGTLQE SRQDSSVQIR R..RKASGDP YWAYSGAYGP EHWVTSSVSC GGSHQSPIDI
RPTP-ζ_RAT (full) O3   .......... ...MRILQSF L......... ..ACVQLLCV CRLDWAYGYY RQQRKLVEEI GWSYTGALNQ KNWGKKYPIC NSPKQSPINI
RPTP-ζ_RAT (short) O4  .......... ...MRILQSF L......... ..ACVQLLCV CRLDWAYGYY RQQRKLVEEI GWSYTGALNQ KNWGKKYPIC NSPKQSPINI
   RPTP-ζ_HUMAN O5   .......... ...MRILKRF L......... ..ACIQLLCV CRLDWANGYY RQQRKLVEEI GWSYTGALNQ KNWGKKYPTC NSPKQSPINI
       DPTP99A O6   .......... .......... .......... .......... .......... .......... .......... .......... ..........
      Consensus      ---------- ---M-I---- L--------- ---CV--L-- -R-------- R--RK---E- -W-Y-GA--- --W------C ----QSPI-I

     CD45_MOUSE P1   .......... .......... .......... .......... .......... .......... .......... .......... ..........
       CD45_RAT P2   .......... .......... .......... .......... .......... .......... .......... .......... ..........
     CD45_HUMAN P3   .......... .......... .......... .......... .......... .......... .......... .......... .....MYLWL
         ChPTPλ P4   .......... .......... .......... .......... .......... .......... .......... .......... ..........
           DPTP P5   MALLYRRMSM LLNIILAYIF LCAICVQGSV KQEWAEIGKN VSLECASENE AVAWKLGNQT INKNHTRYKI RTEPLKSNDD GSENNDSQDF
      Consensus      ---------- ---------- ---------- ---------- ---------- ---------- ---------- ---------- ----------

   RPTP-μ_HUMAN Q1   MRTLGTCLAT LAGLLL.... ....TAAGET FSGGCLFDEP YSTCGYSQSE GDDFNWEQVN TLTKPTSDPW MPSGSLMLVN ASGRPEGQRA
   RPTP-μ_MOUSE Q2   MRTLGTCLVT LAGLLL.... ....TAAGET FSGGCLFDEP YSTCGYSQAD EDDFNWEQVN TLTKPTSDPW MPSGSFMLVN TSGKPEGQRA
   RPTP-κ_MOUSE Q3   MDVAAAALPA FVALWLLYPW PLLGSALGQF SAGGCTFDDG PGACDYHQDL YDDFEWVHVS AQEPHYLPPE MPQGSYMVVD SSNHDPGEKA
      Consensus      M-----L-- ---L-L---- -----A-G-- --GGC-FD-- ---C-Y-Q-- -DDF-W--V- --------P- MP-GS-M-V- -S----G--A
```

```
                     101                                                                151
RSVGNGNSVL VSLRSDQMTL QDAKVLLEAA LRQESGARGH VSSHSHSVLH APGTPVREGL RSHLDPRTPP LPPRERPHTS GHHGAGEARA TAPSTVSPYG PEARAELSSR
RSVGNGNSVL VSLRSDQMTL QDAKVLLEAA LRQESGARGH VSSHSHSALH APGTPVREGL RSHLDPRTPP LPPRERPHTS GHHGAGEARA TAPSTVSPYG PEARAELSSR
DAVPETPVFD CVMDIKPEAD .......... ..PTSLTVKS MGLQERRGSN VSLTLDMCTP GCNEEGFGYL VSPRE..... ..ESAHEYLL SASRVLRAEE LHEKALDPFL
LTLSLGAAMT QPPPEKTPAK .......... ........KH VRLQERRGSN VALMLDVRSL G.AVEPICSV NTPRE..... ...VTLHFLR TAGHPLTRWA LQRQPPSPKQ
.......... .......... .......... .......... .......... ...MEQVEIL RRFIQRVQAM KSPDH..... ..NGEDNF.. .ARDFMRLRR LSTKYRTEKI
.......... .......... .......... .......... .......... .......... .......... .......... .......... .......... ..........
.......... .......... .......... .......... .......... .......... .......... .......... .......... .......... ..........
.......... .......... .......... .......... .......... .......... .......... .......... .......... .......... ..........
.......... .......... .......... .......... .......... .......... .......... .......... .......... .......... ..........
DQAPK..VLN VVVDPQGQCT PEIRNTASTS VCPSPFRMKP IGLQERRGSN VSLTLDMSSL G.NVEPFVAV STPRE..... ..KVAMEYLQ SASRVLTRPQ LRDVVASSHL
---------- ---------- ---------- ---------- ---------- ---------- ---------- ---------- ---------- ---------- ----------

.......... .......... .......... .......... .......... .......... .......... .......... .......... .......... ..........
.......... .......... .......... .......... .......... .......... .......... .......... .......... .......... ..........
.......... .......... .......... .......... .......... .......... .......... .......... .......... .......... ..........
.......... .......... .......... .......... .......... .......... .......... .......... .......... .......... ..........
YNFKIISLDE ERTVVLQTDP LPPARFGVSK EKTTSTGLHV WWTPSSGKVT SYEVQLFDEN NQKIQGVQIQ ESTSWNEYTF FNLTAGSKYN IAITAVSGGK RSFSVYTNGS
.......... .......... .......... .......... .......... .......... .......... .......... .......... .......... ..........
.......... .......... .......... .......... .......... .......... .......... .......... .......... .......... ..........
.......... .......... .......... .......... .......... .......... .......... .......... .......... .......... ..........
.......... .......... .......... .......... .......... .......... .......... .......... .......... .......... ..........
---------- ---------- ---------- ---------- ---------- ---------- ---------- ---------- ---------- ---------- ----------

.......... .......... .......... .......... .......... .......... .......... .......... .......... .......... ..........
MSQGLSWHDD LTQYVISQEM ERIPRLRPPE PRPRDRSGLA PKRPGPAGEL LLQDIPTGSA PAAQHRLPQP PVGKGGAGAS SSLSPLQAEL LPPLLEHLLL PPQPPHPSLS
---------- ---------- ---------- ---------- ---------- ---------- ---------- ---------- ---------- ---------- ----------

TASTNSIGIT ISPNGTWLPD NQFTDARTEP WEGNSSTAAT TPETFPPSDE TPIIAVMVAL SSLLVIVFII IVLYMLRFKK YKQAGSHSNS KQAGSHSNSF RLSNGRTEDV
AASTESGGTT ISPNGSWLIE NQFTDAITEP WEGNSSTAAT TPETFPPADE TPIIAVMVAL SSLLVIVFII IVLYMLRFKK Y......... KQAGSHSNSF RLSNGRTEDV
.......... .......... .......... .TADSNETTT TSGPPDPGAS QPLLAWLL.L PLLLLLLVLL LAAYFFRFRK ........QR KAVVSTSDK. KMPNGILEEQ
---------- ---------- ---------- ----S----T T-----P--- -P--A----L --LL------ ---Y--RF-K ---------- K---S-S--- ---NG--E--

G.AVLRIQPL RTPRDENVYE CVAQNSVGE. ITVHAKLTVL REDQLPPGFP NIDMGPQLKV VERTRTATML CAASGNPDPE ITWFKDFLPV DPSASNGRIK QLRSGALQIE
G.AVLRIQPL RTPRDENVYE CVAQNSVGE. ITIHAKLTVL REDQLPPGFP NIDMGPQLKV VERTRTATML CAASGNPDPE ITWFKDFLPV DPSASNGRIK QLRSGALQIE
G.AVLRIQPL RTPRDENIYE CVAQNPHGE. VTVHAKLTVL REDQLPPGFP NIDMGPQLKV VERTRTATML CAASGNPDPE ITWFKDFLPV DPSTSNGRIK QLRSGGLQIE
.......... .......... .......... .......... .......... .......... .......... .......... .......... .......... ..........
G.SVLRIQPL RVQRDEAIYE CTATNSLGE. INTSAKLSVL EEEQLPPGFP SIDMGPQLKV VEKARTATML CAAGGNPDPE ISWFKDFLPV DPATSNGRIK QLRSGALQIE
GISILRIEPV RAGRDDAPYE CVAENGVGDA VSADATLTIY EGDKTPAGFP VITQGPGTRV IEVGHTVLMT CKAIGNPTPN IYWIKNQTKV DMSNPRYSLK D...GFLQIE
---------- ---------- ---------- ---------- ---------- ---------- ---------- ---------- ---------- ---------- ----------

LDQYARVGEE YQELQLDGFD NESSNKTWMK NTGKTVAILL KDDYFVSGAG LPGRFKAEKV EFHWGHSN.G SAGSEHSING RRFPVEMQIF FYNPDDFDSF QTAISENRII
LDHHARVGDE YQELQLDGFD NESSNKTWMK NTGKTVAILL KDDYFVSGAG LPGRFKAEKV EFHWGHSN.G SAGSEHSVNG RRFPVEMQIF FYNPDDFDSF QTAISENRII
DEDLTQVNVN LKKLKFQGWE KPSLENTFIH NTGKTVEINL TNDYYLSGGL SEKVFKASKM TFHWGKCNVS SEGSEHSLEG QKFPLEMQIY CFDADRFSSF EETVKGKGRL
DEDLTQVNVN LKKLKFQGWE KPSLENTFIH NTGKTVEINL TNDYYLSGGL SEKVFKASKM TFHWGKCNVS SEGSEHSLEG QKFPLEMQIY CFDADRFSSF EETVKGKGRL
DEDLTQVNVN LKKLKFQGWD KTSLENTFIH NTGKTVEINL TNDYRVSGGV SEMVFKASKI TFHWGKCNMS SDGSEHSLEG QKFPLEMQIY CFDADRFSSF EEAVKGKGKL
.......... .......... .......... .......... .......... .......... .......... .......... .......... .......... ..........
-E----V--- ---L-F-GWD --S---TF-- NTGKTV-I-L --DY--SG-- ----FKA-K- -FHWG--N-- S-GSEHS--G --FP-EMQIY -F--D-F-SF ---V------

.......... .......... .......... .......... .......... .......... .......... .......... .......... .......... ..........
.......... ...VFVTGQG TTDDGLDTTE IVLLP...QT DPLPARTTEF TPPSISERGN GSSETT.... .......... .........Y LPGFSSTLMP HLT.......
KLLAFGFAFL DTEVFVTGQS PTPSPTGLTT AKMPSVPLSS DPLPTHTTAF SPASTFEREN DFSETTTSLS PDNTSTQVSP DSLDNASAFN TTGVSSVQTP HLP.......
.......... .......... .......... .......... .......... .......... .......... .......... .......... ..MFLCLKLL AFG.......
MKYKNVLTLL DVNINDSGNY TCTAQTGQNH STEFQVKPYL PSKVLQSTPD RIKRKIKQDV MLYCLIEMYP QNETTNRNLK WLKDGSQFEF LDTFSSISKL NDTHLNFTLE
---------- ---------- ---------- ---------- ---------- ---------- ---------- ---------- ---------- ---------- ----------

HLLLPQLKEN DTHCIDFHYF VSSKSNSPPG LLNVYVKVNN GPLGNPIWNI SGDPTRTWNR AELAISTFWP NFYQVIFEV. ITSGHQGYLA IDEVKVLGHP CTRTPHFLRI
HLLLPQLKEN DTHCIDFHYF VSSKSNAAPG LLNVYVKVNN GPLGNPIWNI SGDPTRTWHR AELAISTFWP NFYQVIFEV. VTSGHQGYLA IDEVKVLGHP CTRTPHFLRI
RLQLPTMKEN DTHCIDFSYL LYSQKGLNPG TLNILVRVNK GPLANPIWNV TGFTGRDWLR AELAVSTFWP NEYQVIFEAE VSGGRSGYIA IDDIQVLSYP CDKSPHFLRL
-L-LP--KEN DTHCIDF-Y- --S-----PG -LN--V-VN- GPL-NPIWN- -G---R-W-R AELA-STFWP N-YQVIFE-- ---G--GY-A ID---VL--P C---PHFLR-
```

```
                       201                                                                  251
          YOP51 J1     LTTLRNTLAP ATNDPRYLQA CGGEKLNRFR DIQCCRQTAV RAD....... ....LNANYI QVGNTRT... IACQYPLQSQ LESHFRMLAE
          YOP2B J2     LTTLRNTLAP RTNDPRYLQA CGGEKLNRFR DIQCCRQTAV RAD....... ....LNANYI QVGNTRT... IACQYPLQSQ LESHFRMLAE
           STEP J3     LQAEFFEIPM NNRDPKEYDI PGLVRKNRYK TILPNPHSRV RLTSPDPF3P LSSYINANYI RGYNGEEKVY IATQGPIVST VVDFWRMVWQ
          HePTP J4     LEEEFLKIPS NFVSPEDLDI PGHASKDRYK TILPNPQSRV CLGRAQSQED .GDYINANYI RGYDGKEKVY IATQGPMPNT VSDFWEMVWQ
          RKPTP J5     YPTATGEKEE N......... ...VKKNRYK DILPFDHSRV KLTLKTPSQD .SDYINANFI KGVYG.PRAY VATQGPLANT VIDFWRMIWE
          Cdi-1 J6     ......MKPP SSIQTSEFDS SDEEPIEDEQ T..PIHISWL SLSRVNC... .SQFLGLCAL PGCKFKDVR. ........RN VQKDTEELKS
            KAP J7     ......MKPP SSIQTSEFDS SDEEPIEDEQ T..PIHISWL SLSRVNC... .SQFLGLCAL PGCKFKDVR. ........RN VQKDTEELKS
          PRL-1 J8     .......... .......... .......... .......... .MARMNR... .P......AP VEVTYKNMRF LITHNPTNAT LNKFIEELKK
           IphP J9     MKTHHANLAL ALMLGLSSSA TAVAADAPQA VATKAAAPNV KPVAADAHGV IPDGAPGMCA RSPACRATAI PADAFVRTAD LGRLTDADRD
       PC12-PTP J10    LQSEFMEIPM NFVDPKEIDI PRHGTKNRYK TILPNPLSRV CLRPKNITDP LSTYINANYI RGYSGKEKAF IATQGPMINT VNDFWQMVWQ
      Consensus        ---------- ---------- ---------- ---------- ---------- ---------- ---------- ---------- ----------

        DPTP1OD K1     .......... .......... .......... .......... .......... .......... .......... .......... ..........
         DPTP4E K2     .......... .......... .......... .......... .......... .......... .......... .......... ..........
          DEP-1 K3     .......... .......... .......... .......... .......... .......... .......... .......... ..........
          SAP-1 K4     .......... .......... .......... .......... .......... .......... .......... .......... ..........
         RPTP-β K5     TVPSPVKDIG ISTKANSLLI SWSHGSGNVE RYRLMLMDKG ILVHGGVVDK HATSYAFHGL SPGYLYNLTV MTEAAGLQNY RWKLVRTAPM
        GLEPP-1 K6     .......... .......... .......... .......... .......... .......... .......... .......... ..........
         RPTP-η K7     .......... .......... .......... .......... .......... .......... .......... .......... ..........
        RPTP-U2 K8     .......... .......... .......... .......... .......... .......... .......... .......... ..........
       RPTP-BR7 K9     .......... .......... .......... .......... .......... .......... .......... .......... ..........
      Consensus        ---------- ---------- ---------- ---------- ---------- ---------- ---------- ---------- ----------

         ICA512 L1     .......... .......... .......... .......... .......... .......... .......... .......... ..........
           IA-2 L2     YEPALLQPYL FHQFGSRDGS RVSEGSPGMV SVGPLPKAEA PALFSRTASK GIFGDHPGHS YGDLPGPSPA QLFQDSGLLY LAQELPAPSR
      Consensus        ---------- ---------- ---------- ---------- ---------- ---------- ---------- ---------- ----------

   RPTP-α_HUMAN M1     EPQSVPLLAR SPSTNRKYPP LPVDKLEEEI NRRMADDNKL FREEFNALPA CPIQATCEAA SKEENKEKNR YVNILPY... ..........
   RPTP-α_MOUSE M2     EPQSVPLLAR SPSTNRKYPP LPVDKLEEEI NRRMADDNKI FREEFNALPA CPIQATCEAA SKEENKEKNR YVNILPFLSL AVSKDAVKAL
   RPTP-ε_HUMAN M3     EQQRVMLLSR SPSGPKKYFP IPVEHLEEEI RIRSADDCKQ FREEFNSLPS GHIQGTFELA NKEENREKNR YPNILP.... ..........
      Consensus        E-Q-V-LL-R SPS---KY-P -PV--LEEEI --R-ADD-K- FREEFN-LP- --IQ-T-E-A -KEEN-EKNR Y-NILP---- ----------

   LAR-PTP2_RAT N1     SSEETDQGKY ECVATNSAGV RYS...SPAN LYVRVRRVAP RFSILP.MSH EIMPGGNVNI TCVAVGSPMP YVKWMQGAED LTPEDDMPVG
   RPTP-σ_MOUSE N2     SSEETDQGKY ECVATNSAGV RYS...SPAN LYVRVRRVAP RFSILP.MSH EIMPGGNVNI TCVAVGSPMP YVKWMQGAED LTPEDDMPVG
          CRYPα N3     SSEETDQGKY ECVASNSAGV RYS...SPAN LYVRVRRVAP RFSILP.VSH EIMPGGNVNI TCVAVGSPMP YVKWMQGAED LTPEDDMPVG
   RPTP-δ_MOUSE N4     .......... ..MCLTSCFI LASHMLSCDL VFVPVRRVPP RFSIPP.TNH EIMPGGSVNI TCVAVGSPMP YVKWMLGAED LTPEDDMPIG
      LAR_HUMAN N5     SSEESDQGKY ECVATNSAGT RYS...APAN LYVRVRRVAP RFSIPP.SSQ EVMPGGSVNL TCVAVGAPMP YVKWMMGAEE LTKEDEMPVG
           DLAR N6     NSREEDQGKY ECVAENSMGT EHS...KATN LYVKVRRVPP TFSRPPETIS EVMLGSNLNL SCIAVGSPMP HVKWMKGSED LTPENEMPIG
      Consensus        ---------- ------S--- --S------- --V-VRRV-P -FS--P---- E-M-G---N- -C-AVG-PMP -VKWM-G-E- LT-E--MP-G

   RPTP-γ_HUMAN 01     GAMAIFFQVS PRDNSALDPI IHGLKGVVHH EKETFLDPFV LRDLLPASLG SYYRYTGSLT TPPCSEIVEW IVFRRPVPIS YHQLEAFYSI
   RPTP-γ_MOUSE 02     GAMAIFFQVS PRDNSALDPI IHGLKGVVHH EKETFLDPFI LRDLLPASLG SYYRYTGSLT TPPCSEIVEW IVFRRPVPIS YHQLEAFYSI
RPTP-ζ_RAT (full) 03  RALSILFEIG VEENLDYKAI IDGTESVSRF GKQAALDPFI LQNLLPNSTD KYYIYNGSLT SPPCTDTVEW IVFKDTVSIS ESQLAVFCEV
RPTP-ζ_RAT (short) 04 RALSILFEIG VEENLDYKAI IDGTESVSRF GKQAALDPFI LQNLLPNSTD KYYIYNGSLT SPPCTDTVEW IVFKDTVSIS ESQLAVFCEV
   RPTP-ζ_HUMAN 05     RALSILFEVG TEENLDFKAI IDGVESVSRF GKQAALDPFI LLNLLPNSTD KYYIYNGSLT SPPCTDTVDW IVFKDTVSIS ESQLAVFCEV
        DPTP99A 06     .......... .......... .......... .......... .......... .......... .......... .......... ..........
      Consensus        -AL-ILF-V- --EN-----I I-G---V--- -K---LDPFI L--LLP-S-- -YY-Y-GSLT -PPC-D-VEW IVF---V-IS --QL--FC-V

     CD45_MOUSE P1     .......... .......... .......... .......... .........M GLWLK.LLAF G.......FA LLDTEVFVTG QTPTP.....
       CD45_RAT P2     .......PQP DSQTPSARGA DTQTLSSQAD LTTLTAAPSG ETDPPGVPEE STVPE.TFPG G.......TP ILARNSTAPS PTHTSNVSTT
     CD45_HUMAN P3     .......THA DSQTPSA.GT DTQTFSGSAA NAKLNPTPGS NAISDVPGER ST.AS.TFPT DPVSPLTTTL SLAHHSSAAL PARTSNTTIT
          ChPTPλ P4     .......VAF LCQDAFAQAG NDNLTSASSL SSTLPTPTRS TSFSPP.... STTAG.VQPA STGASPTAST HLSTHS...G SGPTTGLGHL
           DPTP P5     FTEVYKKENG TYKCTVFDDT GLEITSKEIT LFVMEVPQVS IDFAKAVGAN KIYLNWTVND GNDPIQKFFI TLQEAGTPTF TYHKDFINGS
      Consensus        ---------- ---------- ---------- ---------- ---------- ---------- ---------- -L-------- ----------

   RPTP-μ_HUMAN Q1     QNVEVNAGQF ATFQCSAIGR TVAGDRLWLQ GIDVRDAPLK EIKVTSSRRF IASFNVVNTT KRDAGKYRCM IRTEGGVGIS NYAELVVKEP
   RPTP-μ_MOUSE Q2     QNVEVNAGQF ATFQCSAIGR TVAGDRLWLQ GIDVRDAPLK EIKVTSSRRF IASFNVVNTT KRDAGKYRCM ICTEGGVGIS NYAELVVKEP
   RPTP-κ_MOUSE Q3     GDVEVNAGQN ATFQCIATGR DAVHNKLWLQ RRNGEDIPVA QTKNINHRRF AASFRLQEVT KTDQDLYRCV TQSERGSGVS NFAQLIVREP
      Consensus        --VEVNAGQ- ATFQC-A-GR ------LWLQ -----D-P-- --K----RRF -ASF-----T K-D---YRC- ---E-G-G-S N-A-L-V-EP
```

```
           301                                                                351
NRTPVLAVLA SSSEIANQRF GMPDYFRQSG TYGSITVESK MTQQVGLGDG IMADMYTLTI REAGQKTISV PVVHVGNWPD QTAVSSEVTK ALASLVDQTA ETKRNMYESK
NRTPVLAVLA SSSEIANQRF GMPDYFRQSG TYGSITVESK MTQQVGLGDG IMADMYTLTI REAGQKTISV PVVHVGNWPD QTAVSSEVTK ALASLVDQTA ETKRNMYESK
ERTPIIVMIT .NIEEMNEKC TE...RWP.. .EEQVVHDGV EITVQKVIHT EDYRLRLISL RRGTEERGLK HYWFTSWPDQ KTPDRAPPLL HLVREVEEAA QQ........
EEVSLIVMLT .QLREGKEKC VH.,.YWP.. .TEEETYGPF QIRIQDMKEC PEYTVRQLTI QYQEERRSVK HILFSAWPDH QTPESAGPLL RLVAEVEESP .E........
YNVVIIVMAC REFEMGRKKC ER...YWPLY GEDPITFAPF KISCENEQAR TDYFIRTLLL EFQNESRRLY QFHYVNWPDH DVPSSFDSIL DMISLMRKYQ EH........
CGIQDIFVFC TRGELSK... .......... .......... .......... ..YRVPNLLD LYQQCGIITH HHPIADGGTP DIASCCEIME ELTTCLKNYR K.........
CGIQDIFVFC TRGELSK... .......... .......... .......... ..YRVPNLLD LYQQCGIITH HHPIADGGTP DIASCCEIME ELTTCLKNYR K.........
YGVTTIVRVC E......... .......... .......... .......... ....ATYDTT LVEKEGIHVL DWPFDDGAPP SNQIVDDWLS LVKIKFREEP G.........
ALAALGVKLD IDLRTADEEA QSPDLL.... ....ARDDRF DYQRISLMGT EKMDLQKMMT SFPD....SL GEAYVQWLGH SQPQFKQVFQ RIAAQQDGA. ..........
EDSPVIVMIT .KLKEKNEKC VL...YWP.. .EKRGTHGKV EVLVIGVNEC DNYTIRNLVL KRGSHTQHVK HYWYTSWPDH KTPDSAQPLL QLMLDVEEDR LA........
---------- ---------- ---------- ---------- ---------- ---------- ---------- ---------- ---------- ---------- ----------

.......... .......... .......... .......... .......... .......... .......... .......... .......... .MLYQLSKAT TRIRLKRQKA
.......... .......... .......... .......... .......... .......... .......... ...MDCATRK QQQLRAHHQQ QQIQIQTHGR KRQQLQKQRH
.......... .......... .......... .......... .......... .......... .......... .......... ...MKPAARE ARLPPRSPGL RWALPL....
.......... .......... .......... .......... .......... .......... .......... .......... .....MAGAG GGLGVWGN.. ..........
EVSNLKVTND GSLTSLKVKW QRPPGNVDSY NITLSHKGTI KESRVLAPWI TETHFKELVP GRLYQVTVSC VSGELSAQKM AVGRTFPDKV ANLEANNNGR MRSLVVSWSP
.......... .......... .......... .......... .......... .......... .......... .......... .......... .......... ..........
.......... .......... .......... .......... .......... .......... .......... .......... ...MKPAARE ARLPPRSPGL RWALPL....
.......... .......... .......... .......... .......... .......... .......... .......... .......... .......... ..........
.......... .......... .......... .......... .......... .......... .......... .......... .......... .......... ..........
---------- ---------- ---------- ---------- ---------- ---------- ---------- ---------- ---------- ---------- ----------

.......... .......... .......... .......... .......... .......... .......... .......... .......... ........ME GPVEGRDTAE
ARVPRLPEQG SSSRAEDSPE GYEKEGLGDR GEKPASPAVQ PDAALQRLAA VLAGYGVELR QLTPEQLSTL LTLLQLLPKG AGRNPGGVVN VGADIKKTME GPVEGRDTAE
---------- ---------- ---------- ---------- ---------- ---------- ---------- ---------- ---------- --------ME GPVEGRDTAE

.......... .......... ...DHSRVHL TPVEGVPDSD YINASFINGY QEKNKFIAAQ GPKEETVNDF WRMIWEQNTA TIVMVTNLKE RKECKCAQYW PDQGCWTYGN
NKTTPLLERR FIGKSNSRGC LSDDHSRVHL TPVEGVPDSD YINASFINGY QEKNKFIAAQ GPKEETVNDF WRMIWEQNTA TIVMVTNLKE RKECKCAQYW PDQGCWTYGN
.......... .......... ..NDHSRVIL SQLDGIPCSD YINASYIDGY KEKNKFIAAQ GPKQETVNDF WRMVWEQKSA TIVMLTNLKE RKEEKCHQYW PDQGCWTYGN
---------- ---------- ---DHSRV-L ----G-P-SD YINAS-I-GY -EKNKFIAAQ GPK-ETVNDF WRM-WEQ--A TIVM-TNLKE RKE-KC-QYW PDQGCWTYGN

RNVLELTDVK DSANYTCVAM SSLGVIEAVA QITVKSLPKA PGTPVVTENT ATSITVTWDS GNPDPVSYYV IEYKSKSQDG PYQIKEDITT TRYSIGGLSP NSEYEIWVSA
RNVLELTDVK DSANYTCVAM SSLGVIEAVA QITVKSLPKA PGTPVVTENT ATSITVTWDS GNPDPVSYYV IEYKSKSQDG PYQIKEDITT TRYSIGGLSP NSEYEIWVSA
RNVLELTDVK DSANYTCVAM SSLGVIEAVA QITVKSLPKA PGTPVVTETT ATSITITWDS GNPDPVSYYV IEYKSKSQDG PYQIKEDITT TRYSIGGLSP NSEYEIWVSA
RNVLELNDVR QSANYTCVAM STLGVIEAIA QITVKALPKP PGTPVVTEST ATSITLTWDS GNPGPVSYYI IQHKPKNSEE PYKEIDGIAT TRYSVAGLSP YSDYEFRVVA
RNVLELSNVV RSANYTCVAI SSLGMIEATA QVTVKALPKP PIDLVVTETT ATSVTLTWDS GNSEPVTYYG IQYRAAGTEG PFQEVDGVAT TRYSIGGLSP FSEYAFRVLA
RNVLQLINIQ ESANYTCIAA STLGQIDSVS VVKVQSLPTA PTDVQISEVT ATSVRLEWSY KGPEDLQYYV IQYKPKNANQ AFSEISGIIT MYYVVRALSP YTEYEFYVIA
RNVL-L---- -SANYTC-A- S-LG-I---- ---V--LP-- P------E-T ATS----W-- -------YY- I--------- ---------T --Y----LSP ---Y---V-A

FTTEQQDHVK SVEYLRNNFR PQQ.RLHDRV VSKSAVRDSW NHDMTDFLEN PLGTEASKVC SSPPIHMKVQ PLNQTALQVS WSQPETIYHP PIMNYMISYS WTKNEDEKEK
FTTEQQDHVK SVEYLRNNFR PQQ.ALNDRV VSKSAVRDAW NHDLADFLDN PLGTEASKVC SSPPIHMKVQ PLNQTALQVS WSQPETIYHP PIMNYMISYS WTKNEDEKEK
LTMQQSGYVM LMDYLQNNFR EQQYKFSRQV FSSYTGKEEI HEA....... ........VC SSEPENVQAD PENYTSLLIT WERPRVVYDT MIEKFAVLYQ PLEGNDQTKH
LTMQQSGYVM LMDYLQNNFR EQQYKFSRQV FSSYTGKEEI HEA....... ........VC SSEPENVQAD PENYTSLLIT WERPRVVYDT MIEKFAVLYQ PLEGNDQTKH
LTMQQSGYVM LMDYLQNNFR EQQYKFSRQV FSSYTGKEEI HEA....... ........VC SSEPENVQAD PENYTSLLVT WERPRVVYDT MIEKFAVLYQ QLDGEDQTKH
.......... .......... .......... .......... .......... .......... .......... .......... .......... .......... ..........
LT--Q---V- --DYL-NNFR -QQ--F---V -S-----E-- ---------- --------VC SS-P------ P-N-T-L-V- W--P--VY-- -I--F-V-Y- -----D----

..SDGASLTT LTPSTLG.LA STDPPSTTIA TT.TKQTCAA MFGNITVNYT Y..ESSNQTF KADLKDVQNA KCGNEDC.EN VLNNLEE... .......... CS.......Q
DISSGANLTT PAPSTLG.FA SNTTTSTEIA TPQTKPSCDE KFGNVTVRYI Y..DDSSKNF NANLEGDKKP KCEYTDC.EK ELKNLPE... .......... CS.......Q
ANTSDAYLNA SETTTLSPSG SAVISTTTIA TTPSKPTCDE KYANITVDYL Y..NKETKLF TAKLNVNENV ECGNNTCTNN EVHNLTE... .......... CK.......N
QHSSPAALTT RTLTAFHQTV SDYYSSTSLH NTTSPVITPA STETIPTSTI E..SATTTEE PCDNSIDYGN IEEKNNSAEV TLKNLKE... .......... NRIYDILLED
HTSYILDHFK PNTTYFLRIV GKNSIGNGQP TQYPQGITTL SYDPIFIPKV ETTGSTASTI TIGWNPPPPD LIDYIQYYEL IVSESGEVPK VIEEAIYQQN SRNLPYMFDK
---------- ---------- ---------- ---------- ---------- ---------- ---------- ---------- ------E--- ---------- ----------

PVPIAPPQLA SVGATYLWIQ LNANSINGDG PIVAREVEYC TASGSWNDRQ PVDSTSYKIG HLDPDTEYEI SVLLTRPGEG GTGSPGPALR TRTKCADPMR GPRKLEVVEV
PVPIAPPQLA SVGATYLWIQ LNANSINGDG PIVAREVEYC TASGSWNDRQ PVDSTSYKIG HLDPDTEYEI SVLLTRPGEG GTGSPGPALR TRTKCADPMR GPRKLEVVEV
PRPIAPPQLL GVGPTYLLIQ LNANSIIGDG PIILKEVEYR MTSGSWTETH AVNAPTYKLW HLDPDTEYEI RVLLTRPGEG GTGLPGPPLI TRTKCAEPMR TPKTLKIAEI
P-PIAPPQL- -VG-TYL-IQ LNANSI-GDG PI---EVEY- --SGSW---- -V----YK-- HLDPDTEYEI -VLLTRPGEG GTG-PGP-L- TRTKCA-PMR -P--L---E-
```

```
                       401                                                         451
YOP51      J1   GSSAVADDSK LRPVIHCRAG VGRTAQLIGA MCM.....ND SRNSQ..... ..LSVEDMVS QMRVQRNGIM VQKDE..QLD VLIKLAEGQG
YOP2B      J2   GSSAVGDDSK LRPVIHCRAG VGRTAQLIGA MCM.....ND SRNSQ..... ..LSVEDMVS QMRVQRNGIM VQKDE..QLD VLIKLAEGQG
STEP       J3   .....EGPHC SPIIVHCSAG IGRTGCFIAT SICCQQLRRE ...GV..... ..VDILKTTC QLRQDRG.GM IQTCE..QYQ FVHHAMS.LY
HePTP      J4   .....TAAHP GPIVVHCSAG IGRTGCFIAT RIGCQQLKAR ...GE..... ..VDILGIVC QLRLDRG.GM IQTDE..QYQ FLHHTLA.LY
RKPTP      J5   .....EDV.. .PICIHCSAG CGRTGAICAI DYTWNLLKAG KIPEE..... ..FNVFNLIQ EMRTQRH.SA VQTKE..QYE LVHRAIAQLF
Cdi-1      J6   .......... ..TLIHCYGG LGRSCLVAAC LLLY....., .LSDT..... ..ISPEQAID SLRDLRGSGA IQTIK..QYN YLHEFRDKL.
KAP        J7   .......... ..TLIHCYGG LGRSCLVAAC LLLY...... .LSDT..... ..ISPEQAID SLRDLRGSGA IQTIK..QYN YLHEFRDKL.
PRL-1      J8   .......... CCIAVHCVAG LGRAPVLVA. .LAL...... .IEGG..... ..MKYEDAVQ FIRQKR.RGA FNS.K..QLL YLEKYRPKM.
IphP       J9   .......... ..VLFHCTAG KDRTGIIAGL LLDLAGVPKA EIVHNYAISA HYLEGQPKDS DERADHGAGQ AEPGDRPQDG GHGRYRAGQH
PC12-PTP   J10  .....SEGR. GPVVVHCSAG IGRTGCFIAT SIGCQQLKEE ...GV..... ..VDALSIVC QLRVDRG.GM VQTSE..QYE FVHHALC.LF
Consensus       ---------- -----HC--G --R------- ---------- ---------- ---------- --R------- -------Q-- ----------

DPTP10D    K1   VPQHRW.... .......LWS LAFLAAFTLK D.......VR CADLAISIPN NPGLDDGASY RLDYSPPFGY PEPNTTIASR EI........
DPTP4E     K2   NHHHYYQNPQ QQQKHFVWLV VGILTIFLAR H.......AN AADLVINVPN ASS.NANAFY RIDYSAPFGF PEPNTTIPAS DI........
DEP-1      K3   .........L LLL....LRL GQILCAGGTP SPIPDPSVAT VATGENGITQ ISSTAESFHK QNGTGTPQ.. VETNTSEDGE SS........
SAP-1      K4   .......... LVL....LGL C.SWTGARAP APNPGRNLTV ETQTTSSISL SWEVPDGLDS QNSNYWVQCT GDGGTTETRN TT........
RPTP-β     K5   .......... .......... .......... .......... .......... .......... .......... .......... ..........
GLEPP-1    K6   PAGDWEQYRI LLFNDSVVLL NITVGKEETQ YVMDDTGLVP GRQYEVEVIV ESGNLKNSER CQGRTVPLAV LQLRVKHANE TSLSIMWQTP
RPTP-η     K7   .......... .......... .......... .......... .......... ISSTAESFHK QNGTGTPQ.. VETNTSEDGE SS........
RPTP-U2    K8   .......... .......... .......... .......... .......... .......... .......... .......... ..........
RPTP-BR7   K9   .........L LLL....LRL GQILCAGGTP SPIPDPSVAT VATGENGITQ .......... .......... .......... ..........
Consensus       ---------- ---------- ---------- ---------- ---------- ---------- ---------- ---------- ----------

ICA512     L1   LPARTSPMPG HPTASPTSSE VQQVPSPVSS EPPKAARPPV TPVLLEKKSP LGQSQPTVAG QPSARPAAEE YGYIVTDQKP LSLAAGVKLL
IA-2       L2   LPARTSPMPG HPTASPTSSE VQQVPSPVSS EPPKAARPPV TPVLLEKKSP LGQSQPTVAG QPSARPAAEE YGYIVTDQKP LSLAAGVKLL
Consensus       LPARTSPMPG HPTASPTSSE VQQVPSPVSS EPPKAARPPV TPVLLEKKSP LGQSQPTVAG QPSARPAAEE YGYIVTDQKP LSLAAGVKLL

RPTP-α_HUMAN  M1  IRVSVEDVTV LVDYTVRKFC IQ.QVGDMTN RKPQRLITQF HFTSWPDFGV PFTPIGMLKF LKKVKACNPQ YAGAIVVHCS AGVGRTGTFV
RPTP-α_MOUSE  M2  VRVSVEDVTV LVDYTVRKFC IQ.QVGDVTN RKPQRLITQF HFTSWPDFGV PFTPIGMLKF LKKVKACNPQ YAGAIVVHCS AGVGRTGTFV
RPTP-ε_HUMAN  M3  IRVCVEDCVV LVDYTIRKFC IQPQLPDGC. .KAPRLVSQL HFTSWPDFGV PFTPIGMLKF LKKVKTLNPV HAGPIVVHCS AGVGRTGTFI
Consensus         -RV-VED--V LVDYT-RKFC IQ-Q--D--- -K--RL--Q- HFTSWPDFGV PFTPIGMLKF LKKVK--NP- -AG-IVVHCS AGVGRTGTF-

LAR-PTP2_RAT  N1  VNSIGQGPPS ESVVTRTGEQ APASAPRNVQ ARMLSATTMI VQWEEPVEPN GLIRGYRVYY TMEPEHPVGN WQKHNVDDSL LTTVGSLLED
RPTP-σ_MOUSE  N2  VNSIGQGPPS ESVVTRTGEQ APASAPRNVQ ARMLSATTMI VQWEEPVEPN GLIRGYRVYY TMEPEHPVGN WQKHNVDDSL LTTVGSLLED
CRYPα         N3  VNSIGQGPPS ESVVTRTGEQ APASAPRNVQ GRMLSSTTMI IQWEEPVEPN GQIRGYRVYY TMEPDQPVSN WQKHNVDDSL LTTVGSLLED
RPTP-δ_MOUSE  N4  VNNIGRGPAS EPVLTQTSEQ TPSSAPRDVQ ARMLSSTTIL VQWKEPEEPN GQIQGYRVYY TMDPTQHVNN WMKHNVADSQ ITTIGNLVPQ
LAR_HUMAN     N5  VNSIGRGPPS EAVRARTGEQ APSSPPRRVQ ARMLSASTML VQWEPPEEPN GLVRGYRVYY TPDSRRPPNA WHKHNTDAGL LTTVGSLLPG
DLAR          N6  VNNIGRGPPS APATCTTGET KMESAPRNVQ VRTLSSSTMV ITWEPPETPN GQVTGYKVYY TTNSNQPEAS WNSQMVDNSE LTTVSDVTPH
Consensus         VN-IG-GP-S ------T-E- ---S-PR-VQ -R-LS--T-- --W--P--PN G---GY-VYY T--------- W--------- -TT-------

RPTP-γ_HUMAN          01  TFTKDSDKDL KATISHVSPD SLYLFRVQAV CRNDMRSDFS QTMLF..... .......... .......... .......... ..........
RPTP-γ_MOUSE          02  TFTKDSDKDL KATISHVSPD SLYLFRVQAV CRNDMRSDFS QTMLF..... .......... .......... .......... ..........
RPTP-ζ_RAT (full)     03  EFLTDGYQDL GAILNNLIPN MSYVLQIVAI CSNGLYGKYS DQLIVDMPTE DAELDLFPEL IGTEEIIKEE NYGKGNEEDT GLNPGRDSAT
RPTP-ζ_RAT (short)    04  EFLTDGYQDL GAILNNLIPN MSYVLQIVAI CSNGLYGKYS DQLIVDMPTE DAELDLFPEL IGTEEIIKEE NYGKGNEEDT GLNPGRDSAT
RPTP-ζ_HUMAN          05  EFLTDGYQDL GAILNNLLPN MSYVLQIVAI CTNGLYGKYS DQLIVDMPTD NPELDLFPEL IGTEEIIKEE EEGKDIEEGA IVNPGRDSAT
DPTP99A               06  .......... .......... .......... .......... .......... .......... .......... .......... ..........
Consensus                 -F--D---DL -A------P- --Y-L-I-AI C-N-L---YS --L------- ---------- ---------- ---------- ----------

CD45_MOUSE    P1  IKNISVSNDS C.A....... ......PATT IDLY.VPPGT DKFSLHDCTP KEKANTSICL EWKTKNLDFR KCNSDNISYV LH..CEPENN
CD45_RAT      P2  .KNVTLSNGS C.T....... ......PDKI INLD.VPPGT HNFNLTNCTP DIEANTSICL EWKIKNK..F TCDIQKISYN FR..CTPEMK
CD45_HUMAN    P3  .ASVSISHNS CTA....... ......PDKT LILD.VPPGV EKFQLHDCTQ VEKADTTICL KW..KNIETF TCDTQNITYR FQ..CG....
ChPTPλ        P4  GKSLSVNASN NIVMLNWCRR YTVQSRSCKV MYLT.IPPDE KRYTF.GAKS IGNDNATLRL NSLCIDCEDV CSNV.TVSCK TN..SINSGG
DPTP          P5  LKTATDYEFR VRACSDLTKT CGPWSENVNG TTMDGVATKP TNLSIQCHHD NVTRGNSIAI NWDVPK.... TPNGKVVSYL IHLLGNPMST
Consensus         ---------- ---------- ---------- ---------- ---------- ---------- ---------- ---------- ----------

RPTP-μ_HUMAN  Q1  KSRQITIRWE PFGYNVTRCH SYNLTVHYCY QVGGQEQVRE EVSWDTENSH PQHTITNLSP YTNVSVKLIL MNPEGRKESQ ELIVQTDEDL
RPTP-μ_MOUSE  Q2  KSRQITIRWE PFGYNVTRCH SYNLTVHYGY QVGGQEQVRE EVSWDTDNSH PQHTITNLSP YTNVSVKLIL MNPEGRKESQ ELTVQTDEDL
RPTP-κ_MOUSE  Q3  QARRIAVDWE SLGYNITRCH TFNVTICYHY .FRGHNESRA D.CLDMDPKA PQHVVNHLPP YTNVSLKMIL TNPEGRKESE ETIIQTDEDV
Consensus         --R-I---WE --GYN-TRCH --N-T--Y-Y ---G----R- ----D----- PQH----L-P YTNVS-K-IL -NPEGRKES- E---QTDED-
```

```
           501                                                                  551
RPLLNS.... .......... .......... .......... .......... .......... .......... .......... ........ J1  YOP51
RPLLNS.... .......... .......... .......... .......... .......... .......... .......... ........ J2  YOP2B
EKQLSLQSSE .......... .......... .......... .......... .......... .......... .......... ........ J3  STEP
AGQLPEEPSP .......... .......... .......... .......... .......... .......... .......... ........ J4  HePTP
EKQLQLYEIH GAQKITDGNE ISTGNMVSSI DSEKQDSPPP KPPRTRSCLV EGDAKEEILQ PPEPHPVPPI LTPSPPSAFP TLHCVARQ J5  RKPTP
AAHLSSRDSQ SRSVSR.... .......... .......... .......... .......... .......... .......... ........ J6  Cdi-1
AAHLSSRDSQ SRSVSR.... .......... .......... .......... .......... .......... .......... ........ J7  KAP
RLRFKDSNGH RNNCCIQ... .......... .......... .......... .......... .......... .......... ........ J8  PRL-1
GAVLAALHSQ YGGAEGYLKS IGVSEQEIQQ LKVRLGQAG. .......... .......... .......... .......... ........ J9  IphP
ESRLSPETVQ .......... .......... .......... .......... .......... .......... .......... ........ J10 PC12-PTP
---------- ---------- ---------- ---------- ---------- ---------- ---------- ---------- -------- Consensus

.......... .......... ......GDEI QFSRALPGTK YNFWLYYTNF THHDWLTWTV TITTAPDPPS NLSVQVRSGK NAIILWSPPT QGSYTAFKIK VLGLSEASSS
.......... .......... ......GKDI KNSRALPGTE YNFWLYYTNS THREQLTWTV NITTAPDPPA NLSVQLRSSK SAFITWRPPG SGRYSGFRIR VLGLTDL..P
.......... .......... ......GAND SLRTPEQGS. .........N GTDGASQ.KT PSSTGPSPVF DIKAVSISPT NVILTWKSND TAASEY.KYV VKHKMENEKT
.......... .......... ......ATNV TVDGLGPGSL YTCSVWVEKD GVNSSVG.TV TTATAPNPVR NLRVEAQTNS SIALTWEVPD GPDPQNSTYG VEYTGDGGRA
.......... LADRDLLLIH KSLSKDAKEF TFTDLVPGRK YMATVTSISG DLKNSS..SV KGRTVPAQVT DLHVANQGMT SSLFTNWTQA QGDVEFYQVL LIHENVVIKN
VAEWEKYIIS .......... ........MG HLPTRARGRR RLLPLLWLFV LLKTAAAFHV TVRDDNSIVV SLEASDVISP ASVYVVKITG ESKNYFFEFE ..........
           .......... ......GAND SLRTPEQGS. .........N GTDGASQ.KT PSSTGPSPVF DIKAVSISPT NVILTWKSND TAASEY.KYV VKHKMENEKT
.......... .......... ........MG HLPTGIHGAR RLLPLLWLFV LFKNATAFHV TVQDDNNIVV SLEASDVISP ASVYVVKITG ESKNYFFEFE ..........
.......... .......... .......... .......... .......... .......... .......... .......... .......... .......... ..........
---------- ---------- ---------- ---------- ---------- ---------- ---------- ---------- ---------- ---------- ----------

EILAEHVHMS SGSFINISVV GPALTFRIRH NEQNLSLADV TQQAGLVKSE LEAQTGLQIL QTGVGQREEA AAVLPQTAHS TSPMRSVLLT LVALAGVAGL LVALAVALCV
EILAEHVHMS SGSFINISVV GPALTFRIRH NEQNLSLADV TQQAGLVKSE LEAQTGLQIL QTGVGQREEA AAVLPQTAHS TSPMRSVLLT LVALAGVAGL LVALAVALCV
EILAEHVHMS SGSFINISVV GPALTFRIRH NEQNLSLADV TQQAGLVKSE LEAQTGLQIL QTGVGQREEA AAVLPQTAHS TSPMRSVLLT LVALAGVAGL LVALAVALCV

VIDAMLDMMH TERKVDVYGF VSRIRAQRCQ MVQTDMQYVF IYQALLEHYL YGDTELEVTS LETHLQKIYN KIPGTSNNGL EEEFKKLTSI KIQNDKMRTG NLPANMKKNR
VIDAMLDMMH SERKVDVYGF VSRIRAQRCQ MVQTDMQYVF IYQALLEHYL YGDTELEVTS LETHLQKIYN KIPGTSNNGL EEEFKKLTSI KIQNDKMRTG NLPANMKKNR
VIDAMMAMMH AEQKVDVFEF VSRIRNQRPQ MVQTDMQYTF IYQALLEYYL YGDTELDVSS LEKHLQTMHG TTTHFDKIGL EEEFRKLTNV RIMKENMRTG NLPANMKKAR
VIDAM--MMH -E-KVDV--F VSRIR-QR-Q MVQTDMQY-F IYQALLE-YL YGDTEL-V-S LE-HLQ---- --------GL EEEF-KLT-- -I----MRTG NLPANMKK-R

ETYTVRVLAF TSVGDGPLSD PIQVKTQQGV PGQPMNLRAE AKSETSIGLS WSAPRQ..ES VIKYELLFRE G.DRGREVGR TFDPTTAFVV EDLKPNTEYA FRLAARSPQG
ETYTVRVLAF TSVGDGPLSD PIQVKTQQGV PGQPMNLRAE AKSETSIGLS WSAPRQ..ES VIKYELLFRE G.DRGREVGR TFDPTTAFVV EDLKPNTEYA FRLAARSPQG
ETYTVRVLAF TSVGDGPLSD PIQVKTQQGV PGQPMNFRAE AKTETSIVLS WSPPRQ..EI IVKYELLFKE G.DHGREVPR NFEPTTSFTV EGLKPNTEYV FRLAARSALG
KTYSVKVLAF TSIGDGPLSS DIQVITQTGV PGQPLNFKAE PESETSILLS WTPPRS..DT IASYELVDRD G.DQGEEQRI TIEPGTSYRL QGLKPNSLYY FRLSATSPQG
ITYSLRVLAF TAVGDGPPSP TIQVKTQQGV PAQPADFQAE VESDTRIQLS WLLPPQ..ER IIMYELVYWA AEDEDQQHKV TFDPTSSYTL EDLKPDTLYR FQLAARSDMG
AIYTVRVQAY TSMGAGPMST PVQVKAQQGV PSQPSNFRAT DIGETAVTLQ WTKPTHSSEN IVHYELYWND TYANQAHHKR .ISNSEAYTL DGLYPDTLYY IWLAARSQRG
--Y---V-A- T--G-GP-S- --QV--Q-GV P-QP----A- ----T---L- W--P------ ---YEL---- ---------- ---------- --L-P---Y- --L-A-S--G

........QA NTTRIFQ... .......... .......... .......... .......... .......... .GTRIVKTGV PTASP..... .......... .....ASSAD
........QA NTTRIFQ... .......... .......... .......... .......... .......... .GTRIVKTGV PTASP..... .......... .....ASSAD
NQIRKKEPQV STTTHYNHMG TKYNEAKTNR SPTRGSEFSG KSDVLNTSLN PTSQQVAEFN PEREMSLPSQ IGTNLPPHSV EGTSASLNSG SKTLLVFPQM NLSGTAESLN
NQIRKKEPQV STTTHYNHMG TKYNEAKTNR SPTRGSEFSG KSDVLNTSLN PTSQQVAEFN PEREMSLPSQ IGTNLPPHSV EGTSASLNSG SKTLLVFPQM NLSGTAESLN
NQIRKKEPQI STTTHYNRIG TKYNEAKTNR SPTRGSEFSG KGDVPNTSLN STSQPVTKLA TEKDISLTSQ TVTELPPHTV EGTSASLNDG SKTVLRSPHM NLSGTAESLN
.......... .......... .......... .......... .......... .......... .......... .......... .......... .......... ..........
-------Q- -TT--Y---- ---------- ---------- ---------- ---------- ---------- --T-----V ---S------ ---------- -----A-S--

T......KCI RRNTFIPERC QLDNLRAQTN YTCVAEILYR GVKLVKNVIN ..VQTDLGIP ET.PKPSCGD PAARKTLVSW PEPVSKPESA S..KPHGYVL CYKNN.SEKC
T......FAL DKHGTL.... WLHNLTVRTN YTCAAEVLYN NVILLKQDRR ..VQTDFGTP EMLPHVQCKN STNSTTLVSW AEP......A S..KHHGYIL CYKKTPSEKC
N......MIF DNKEI..... KLENLEPEHE YKCDSEILYN NHKFTNASKI ..IKTDFGSP GEPQIIFCRS EAAHQGVITW NPP......Q R..SFHNFTL CYIKETEKDC
TGNLTGSYEL MKHDINADNI TILSLSSDSE YLCRVTVRFF EKNFTKEV.N ..ITTDYDAP KAPENLTV.H PTDRNVTVTW MKP......T G..TLEKHID GYTVECNNTS
VDREMWGPKI RRIDEPHHKT LYESVSPNTN YTVTVSAITR HKKNGEPATG SCLMPVSTPD AIGRTMWSKV NLDSKYVLKL YLPKISERNG PICCYRLYLV RINNDNKELP
---------- ---------- ---------- Y--------- ---------- ---------- ---------- ---------- --P------- ---------- ----------

PGAVPTESIQ GSAFEEKIFL QWREPTQTYG VITLYEITYK AVSSFDPEID LSNQSGRVSK LGNETHFLFF GLYPGTTYSF TIRASTAKGF GPPATNQFTT KISAPSMPAY
PGAVPTESIQ GSAFEEKIFL QWREPTQTYG VITLYEITYK AVSSFDPEID LSNQSGRVSK LGNETHFLFF GLYPGTTYSF TIRASTAKGF GPPATNQFTT KISAPSMPAY
PGPVPVKSLQ GTSFENKIFL NWKEPLEPNG IITQYEVSYS SIRSFDPAVP VAGPPQTVSN LWNSTHHVFM HLHPGTTYQF FIRASTVKGF GPATAINVTT NISAPSLPDY
PG-VP--S-Q G--FE-KIFL -W-EP----G -IT-YE--Y- ---SFDP--- -------VS- L-N-TH--F- -L-PGTTY-F -IRAST-KGF GP------TT -ISAPS-P-Y
```

```
                         601                                                                           651

         DPTP10D K1 YNRTFQVNDN .TFQHSVKEL TPGATYQVQA YTIYDGKESV AYTSRNFTTK PNTPGKFIVW FRNETTLLVL WQPPYPAGIY TH.YKVSIEP
          DPTP4E K2 FERSYSLEGN ETLQLSAKEL TPGGSYQVQA YSVYQGKESV AYTSRNFTTK PNTPGKFIVW FRNETTLLVL WQPPFPAGIY TH.YRVSITP
           DEP-1 K3 IT.V.....V HQPWCNITGL RPATSYVFSI TPGIGNETWG DPRVIKVITE PIPVSDLRVA HGCEEG.... CSLSWSNGNG TASCRVLLES
           SAP-1 K4 GTRS.....T AHTNITVDGL EPGCLYAFS. .......... .......... ......MWVG KNGINS.... SRETRNATTA HNPVRK.PES
          RPTP-β K5 ESIS.....S ETSRYSFHSL KSGSLYSVVV T.TVSGGISS RQVVVEGRTV PSSVSGVTVN NSGRNDYLSV SWLVAPGDVD NYEVTLSHDG
         GLEPP-1 K6 .....EFNST LPPPVIFKAN YHGLYYIITL VVVNGNVVTK PSRSITVLTK PLPVTSVSIY DYKPSPETGV LFEIHYPEKY NVFTRVNISY
          RPTP-η K7 IT.V.....V HQPWCNITGL RPATSYVFSI TPGIGNETWG DPRVIKVITE PIPVSDLRVA LTGVRK.... AALSWSNGNG TASCRVLLES
         RPTP-U2 K8 .....EFNST LPPPVIFKAS YHGLYYIITL VVVNGNVVTK PSRSITVLTK PLPVTSVSIY DYKPSPETGV LFEIHYPEKY NVFTRVNISY
        RPTP-BR7 K9 .......... .......... .......... .......... .......... .......... .......... .......... ..........
       Consensus    ---------- ---------- ---------- ---------- ---------- ---------- ---------- ---------- ----------

          ICA512 L1 RQHARQQDKE RLAALGPEGA HGDTTFEYQD LCRQHMATKS LFNRAEGPPE PSRVSSVSSQ FSDAAQASPS SHSSTPSWCE EPAQANMDIS
            IA-2 L2 RQHARQQDKE RLAALGPEGA HGDTTFEYQD LCRQHMATKS LFNRAEGPPE PSRVSSVSSQ FSDAAQASPS SHSSTPSWCE EPAQANMDIS
       Consensus    RQHARQQDKE RLAALGPEGA HGDTTFEYQD LCRQHMATKS LFNRAEGPPE PSRVSSVSSQ FSDAAQASPS SHSSTPSWCE EPAQANMDIS

    RPTP-α_HUMAN M1 VLQIIPYEFN RVIIPVKRGE ENTDYVNASF IDGYRQKDSY IASQGPLLHT IEDFWRMIWE WKSCSIVMLT ELEERGQEKC AQYWPSDGLV
    RPTP-α_MOUSE M2 VLQIIPYEFN RVIIPVKRGE ENTDYVNASF IDGYRQKDSY IASQGPLLHT IEDFWRMIWE WKSCSIVMLT ELEERGQEKC AQYWPSDGLV
    RPTP-ϵ_HUMAN M3 VIQIIPYDFN RVILSMKRGQ EYTDYINASF IDGYRQKDYF IATQGPLAHT VEDFWRMIWE WKSHTIVMLT EVQEREQDKC YQYWPTEGSV
       Consensus    V-QIIPY-FN RVI---KRG- E-TDY-NASF IDGYRQKD-- IA-QGPL-HT -EDFWRMIWE WKS--IVMLT E--ER-Q-KC -QYWP--G-V

    LAR-PTP2_RAT N1 LGAFTAVVCQ RTLQAKPSAP PQDVKCTSLR STAIL..... .......... .......... .......... .........I LLEALEKWTE
    RPTP-σ_MOUSE N2 LGAFTAVVRQ RTLQAKPSAP PQDVKCTSLR STAILVSWRP PPPETHNGAL VGYSVRYRPL GSEDPDPKEV NNIPPTTTQI LLEALEKWTE
           CRYPα N3 LGAFTPEVRE RTLQ...... .......... .......... .......... .......... .......... .......... ..........
    RPTP-δ_MOUSE N4 LGASTAEISA RTMQ...... .......... .......... .......... .......... .......... .......... ..........
       LAR_HUMAN N5 VGVFTPTIEA RTAQSTPSAP PQKVMCVSMG STTVRVSWVP PPADSRNGVI TQYSVAHEAV DGEDRGRHVV DGISREHSSW DLVGLEKWTE
            DLAR N6 EGATTPPIPV RTKQYVPGAP PRNITAIATS STTISLSWLP PPVERSNGRI IYYKVFFVEV GRED...DEA TTMTLNMTSI VLDELKRWTE
       Consensus    -G--T----- RT-O------ ---------- ---------- ---------- ---------- ---------- ---------- ----------

    RPTP-γ_HUMAN 01 MAPI...... .......... .......... ..SSGSSTWT SSGIPFSFVS MATGMGPSS. .......... .......... ....SGSQAT
    RPTP-γ_MOUSE 02 MAPI...... .......... .......... ..SSGSSTWT SSGIPFSFVS MATGMGPSS. .......... .......... ....SGSQAT
RPTP-ζ_RAT (full) 03 MVSITEYKEV SADLSEEENL LTDFKLDSGA DDSSGSSP.A SSTVPFSTDN LSHGYTSSSD TPEAVTYDVL RPESTRNALE DSAPSGSEES
RPTP-ζ_RAT (short) 04 MVSITEYKEV SADLSEEENL LTDFKLDSGA DDSSGSSP.A SSTVPFSTDN LSHGYTSSSD TPEAVTYDVL RPESTRNALE DSAPSGSEES
    RPTP-ζ_HUMAN 05 TVSITEY... .....EEESL LTSFKLDTGA EDSSGSSP.A TSAIPFISEN ISQGYIFSSE NPETITYDVL IPESARNASE DSTSSGSEES
        DPTP99A 06 .......... .......... .......... .......... .......... .......... .......... .......... ..........
       Consensus    ---I------ ---------- ---------- --SSGSS--- -S-IPF---- ---G---SS- ---------- ---------- ----SGS---

      CD45_MOUSE P1 KSLPNNVTSF E......VES LKPYKYYEVS LLAYVNGKIQ R...NGTAEK CNFHTKADRP DKVNGMKTSR PTDNSINVTC GPPYETNGPK
        CD45_RAT P2 ENLANDVNSF E......VKN LRPYTEYTVS LFAYVIGRVQ R...NGPAKD CNFRTKAARP GKVNGMKTSR ASDNSINVTC NSPYEINGPE
      CD45_HUMAN P3 LNLDKNLIKY D......LQN LKPYTKYVLS LHAYIIAKVQ R...NGSAAM CHFTTKSAPP SQVWNMTVSM TSDNSMHVKC RPPRDRNGPH
         ChPTPλ P4 QNVNRNETSF T......CGD LEPYSTGSVS VRAFKKSKYK NKNFEGEKVN GSFQTKPAKP ENVTDFKLTL LADNTVKVAC RSQ.KVYGNE
            DPTP P5 DPEKLNIATY QEVHSDNVTR SSAYIAEMIS SKYFRPEIFL GAEKRFSENN DIIRGNDEIC RKCLEGTPFL RKPEIIHIPP QGSLSNSDSE
       Consensus    ---------- ---------- ---Y-----S ---------- ---------- ---------- ---------- ---------- ----------

    RPTP-μ_HUMAN Q1 E.LETPLNQT DNTVTVMLKP AHSRGAPVSV YQIVVEEERP RRTKKTTEIL KCYPVPIHFQ NASLLNSQYY FAAEFPADSL QAAQPFTIGD
    RPTP-μ_MOUSE Q2 E.FETPLNQT DNTVTVMLKP AQSRGAPVSV YQIVVEEERP RRTKKTTEIL KCYPVPIHFQ NASILNSQYY FAAEFPADSL QAAQPFTIGD
    RPTP-κ_MOUSE Q3 EGVDASLNET ATTITVLLRP AQAKGAPISA YQIVVEQLHP HRTKREAGAM ECYQVPVTYQ NALSGGAPYY FAAELPPGNL PEPAPFTVGD
       Consensus    E-----LN-T --T-TV-L-P A---GAP-S- YQIVVE---P -RTK------ -CY-VP---Q NA------YY FAAE-P---L ----PFT-GD
```

```
701                                                751

PDANDSVLYV EKEGEPPGPA QAAFKGLVPG RAYNISVQTM SEDEI.SLPT TAQYRTVPLR PLNVTFDRDF ITSNSFRVLW EAPKGISEFD KYQVSVATTR RQSTVPR.SN
DDAIQSVLYV EREGEPPGPA QAAFKGLVPG REYNISVQTV SEDETSSVPT TARYLTYRSA WLNVTFDEAY TTSSSFRVRW EPPRTYSEFD AYQVMLSTSR RIFNVPRAAN
IGSHEELT.. .....QDSRL QVNISDLKPG VQYNINPYLL QSNKTKGDPL AQK..VAWMP AIQREAGQGA PPPLCMMSPF VGPVD..... .......... .PSSGQQSRD
GGSDH..... .......... QLHLPEL.GG PRWH.RPTEL DLLRTSALEM VAE..QRLET QQTPES.... .......... ..PVD..... .......... ..........
KVVQSLVI.. .....AKSVR ECSFSSLTPG RLYTV.TITT RSGKYENHSF SQE..RTVPD KVQGVSVSNS ARSDYLRVSW VHATGDFDHY EVTIKNKNNF IQTKSIPKSE
WEGKAFRTML YKDFFKG... KTVFNHWLPG ICYSNITFQL VSEATFNKST LVEYSGVSHE PKQHRTAPYP PRNISVRI.. .......... .....VNLNK NNWEEQSGSF
IGSHEELT.. .....QDSRL QVNISGLKPG VQYNINPYLL QSNKTKGDPL GTE..GGLDA SNTERSRAGS PTAPVHDESL VGPVD..... .......... .PSSGQQSRD
WGGKDFRTML YKDFFKG... KTVFNHWLPG MCYSNITFQL VCEATFNKST VVEYSGVSHE PKQHRTAPYP PQNISVRI.. .......... .....VNLNK NNWEEQSGNF
.......... .......... .......... .......... .......... .......... .......... .......... .......... .......... ..........
---------- ---------- ---------- ---------- ---------- ---------- ---------- ---------- ---------- ---------- ----------
```

```
TGHMILAYME DHLRNRDRLA KEWQALCAYQ AEPNTCATAQ GEGNIKKNRH PDFLPYDHAR IKLKVESSPS RSDYINASPI IEHDPRMPAY IATQGPLSHT IADFWQMVWE
TGHMILAYME DHLRNRDRLA KEWQALCAYQ AEPNTCATAQ GEGNIKKNRH PDFLPYDHAR IKLKVESSPS RSDYINASPI IEHDPRMPAY IATQGPLSHT IADFWQMVWE
TGHMILAYME DHLRNRDRLA KEWQALCAYQ AEPNTCATAQ GEGNIKKNRH PDFLPYDHAR IKLKVESSPS RSDYINASPI IEHDPRMPAY IATQGPLSHT IADFWQMVWE
```

```
SYGDITVELK KEEECESYTV RDLLVT.... .NTRENKSRQ IRQFHFHGWP EVGIPSDGKG MISIIAAVQK QQQQSGNHPI TVHCSAGAGR TGTFCALSTV LERVK.AEGI
SYGDITVELK KEEECESYTV RDLLVT.... .NTRENKSRQ IRQFHFHGWP EVGIPSDGKG MINIIAAVQK QQQQSGNHPI TVHCSAGAGR TGTFCALSTV LERVKWAEGI
THGEITIEIK NDTLSEAISI RDFLVTLNQP QARQEEQVRV VRQFHFHGWP EIGIPAEGKG MIDLIAAVQK QQQQTGNHPI TVHCSAGAGR TGTFIALSNI LERVK.AEGL
--G-IT-E-K -----E---- RD-LVT---- ----E---R- -RQFHFHGWP E-GIP--GKG MI--IAAVQK QQQQ-GNHPI TVHCSAGAGR TGTF-ALS-- LERVK-AEG-
```

```
YRVTAVAYTE VGPGPESSPV VVRTDEDVPS APPRKVEAEA LNATAIRVLW RSPTPGRQHG QIRGYQVHYV RMEGTEARGP PRIKDIMLAD AQ........ .EMVITNLQP
YRVTAVAYTE VGPGPESSPV VVRTDEDVPS APPRKVEAEA LNATAIRVLW RSPTPGRQHG QIRGYQVHYV RMEGAEARGP PRIKDIMLAD AQ........ .EMVITNLQP
.......... .......... .......... .......... .......... .......... .......... .......... .......... .......... ..........
.......... .......... .......... .......... .......... .......... .......... .......... .......... .......... ..........
YRVWVRAHTD VGPGPESSPV LVRTDEDVPS GPPRKVEVEP LNSTAVHVYW KLPVPSKQHG QIRGYQVTYV RLENGEPRGL PIIQDVMLAE AQWRPEESED YETTISGLTP
YKIWVLAGTS VGDGPRSHPI ILRTQEDVP. GDPQDVKATP LNSTSIHVSW KPPLEKDRNG IIRGYHIHAQ EL........ .RDEGKGFLN EPFKFDVVDT LEFNVTGLQP
---------- ---------- ---------- ---------- ---------- ---------- ---------- ---------- ---------- ---------- ----------
```

```
V..ASVVTST LLAG...... .......... .......... .......... .......... .......... ...LGFGGGG I......... .......... ..........
V..ASVVTST LLAG...... .......... .......... .......... .......... .......... ...LGFGGGG I......... .......... ..........
LKDPSLEGSV WFPGSTDLTT QSETGSGREG FLQVNSTDFQ VDESRETTET FSPDATASRG PSVTDMEMPH YSTFAYPPTE VTSHAFTPSS RPLDLAPTSN ILHSQTTQPV
LKDPSLEGSV WFPGSTDLTT QSETGSGREG FLQVNSTDFQ VDESRETTET FSPDATASRG PSVTDMEMPH YSTFAYPPTE VTSHAFTPSS RPLDLAPTSN ILHSQTTQPV
LKDPSMEGNV WFPSSTDITA QPDVGSGRES FLQTNYTEIR VDESEKTTKS FSAGPVMSQG PSVTDLEMPH YSTFAYFPTE VTPHAFTPSS RQQDLVSTVN VVYSQTTQPV
.......... .......... .......... .......... .......... .......... .......... .......... .......... .......... ..........
----S----- -F-------- ---------- ---------- ---------- ---------- ---------- ---F-Y---- V--------- ---------- ----------
```

```
TFYILV.... VRSGGSFVTN TTKTNCQFYG DNLYYSTDYE FLVSFHN... .......GVY EGDSVIRSES TNF....... NAKALIIFLV FLIIVTS... ..........
ARYILE.... VKSGGSLVKT FNQSTCKFVV DNLYYSTDYE FLVYFYN... .......GEY LGDPEIKPQS TSY....... NSKALIIFLV FLIIVTS... ..........
ERYHLE.... VEAGNTLVRN ESHKNCDFRV KDLQYSTDYT FKAYFHN... .......GDY PGEPFILHHS TSY....... NSKALIAFLA FLIIVTS... ..........
TKFKLS.... WNSSSNSGEN QRKNECNFTV RDLSYLTKYT FKISVFN... .......GVY TGDSVCEEIY TRY....... NSRALIIFLV FLIVVTS... ..........
LPILSEKDNL IKGANLTEHA LKILESKLRD KRNAVTSDEN PILSAVNPNV PLHDSSRDVF DGEIDINSNY TGFLEIIVRD RNNALMAYSK YFDIITPATE AEPIQSLNNM
---------- ---------- ---------- ---------- ------N--- ---------- -G-------- T--------- ---AL----- -----T---- ----------
```

```
NKTYNGYWNT PLLPYKSYRI YFQAASRANG ETKIDCVQVA TKGAAT..PK PVPEPEKQTD HTVKIAGVIA GILLFVIIFL GVVLVMKKRK LAKKRKETMS STRQEMTVMV
NKTYNGYWNT PLLPHKSYRI YYQAASRANG ETKIDCVRVA TKGAVT..PK PVPEPEKQTD HTVKIAGVIA GILLFVIIFL GVVLVMKKRK LAKKRKETMS STRQEMTVMV
NRTYKGFWNP PLAPRKGYNI YFQAMSSVEK ETKTQCVRIA TKAAATEEPE VIPDPAKQTD RVVKIAGISA GILVFILLLL VVIVIVKKSK LAKKRKDAMG NTRQEMTHMV
N-TY-G-WN- PL-P-K-Y-I Y-QA-S---- ETK--CV--A TK-A-T--P- --P-P-KQTD --VKIAG--A GIL-F----L -V----KK-K LAKKRK--M- -TRQEMT-MV
```

```
                                801                                                 851

         DPTP1OD K1  EPVAFSDFRD IAEPGKTFNV IVKTVSGKVT .SWPATGDVT LRPLPVRNLR SINDDKTNTM IITWEADPAS TQDEYRIVYH ELETFN....
          DPTP4E K2  GDSVYFDYSD ILEPGRTYEV VVKTIADNVN .SWPASGEVT LRPRPVRSLG GFLDDRSNAL HISWEPAETG RQDSYRISYH EQTNASEVPA
           DEP-1 K3  TEVLLVGL.. ..EPGTRYNA TVYSQAANGT EGQPQAIEFR TNAIQVFDV. TAVNISATSL TLIWKVSDNE SSSNYTYKIH VAGE......
           SAP-1 K4  ......GL.. ..GPGSLYTC SVWVEKDGVN SSSWRLVTST TAPNPVRNL. TVEAQTNSSI ALTWEVPDGP DPQNSTYGVE YTGD......
          RPTP-β K5  NECVFVQL.. ..VPGRLYSV TVTTKSGQYE ANEQG..NGR TIPEPVKDL. TLRNRSTEDL HVTWS.GANG DVDQYEIQLL FNDM......
         GLEPP-1 K6  PEESFMRSPE TIEKDRIFHF TEETPE.PSG NISSGWPDFN SSDY...... ...ETTSQPY WWDSASATPE SEDEFVSVLP MEYENNNTLS
          RPTP-η K7  TEVLLVGL.. ..EPGTRYNA TVYSQAANGT EGQPQAIEFR TNAIQVFDV. TAVNISATSL TLIWKVSDNE SSSNYTYKIH VAGE......
         RPTP-U2 K8  PEESFMRSQD TIGKEKLFHF TEETPEIPSG NISSGWPDFN SSDY...... ...ETTSQPY WWDSASAAPE SEDEFFSVLP MEYENNSTLS
        RPTP-BR7 K9  .......... .......... .......... .......... .......... .......... .......... .......... ..........
       Consensus     ---------- ---------- ---------- ---------- ---------- ---------- ---------- ---------- ----------

          ICA512 L1  SGCTVIVMLT PLVEDGVKQC DRYWPDEGAS LYHVYEVNLV SEHIWCEDFL VRSFYLKNVQ TQETRTLTQF HFLSWPAEGT PASTRPLLDF
            IA-2 L2  SGCTVIVMLT PLVEDGVKQC DRYWPDEGAS LYHVYEVNLV SEHIWCEDFL VRSFYLKNVQ TQETRTLTQF HFLSWPAEGT PASTRPLLDF
       Consensus     SGCTVIVMLT PLVEDGVKQC DRYWPDEGAS LYHVYEVNLV SEHIWCEDFL VRSFYLKNVQ TQETRTLTQF HFLSWPAEGT PASTRPLLDF

    RPTP-α_HUMAN M1  LDVFQTVKSL RLQRPHMVQT LEQYEFCYKV VQEYIDAFSD YANFK  M1  RPTP-α_HUMAN
    RPTP-α_MOUSE M2  LDVFQTVKSL RLQRPHMVQT LEQYEFCYKV VQEYIDAFSD YANFK  M2  RPTP-α_MOUSE
    RPTP-ε_HUMAN M3  LDVFQAVKSL RLQRPHMVQT LEQYEFCYKV VQDFIDIFSD YANFK  M3  RPTP-ε_HUMAN
       Consensus     LDVFQ-VKSL RLQRPHMVQT LEQYEFCYKV VQ--ID-FSD YANFK  Consensus

    LAR-PTP2_RAT N1  ETAYSITVAA YTMKGDGARS KPKVVVTKGA VLGRPTLSV. ..QQTPEGSL LARWEPPADA AEDPVLGYRL QFGREDAA.P ATLELAAWER
    RPTP-σ_MOUSE N2  ETAYSITVAA YTMKGDGARS KPKVVVTKGA VLGRPTLSV. ..QQTPEGSL LARWEPPADA AEDPVLGYRL QFGREDAA.P ATLELAAWER
           CRYPα N3  .......... .......... .......... .......... .......... .......... .......... .......... ..........
    RPTP-δ_MOUSE N4  .......... .......... .......... .......... .......... .......... .......... .......... ..........
       LAR_HUMAN N5  ETTYSVTVAA YTTKGDGARS KPKIVTTTGA VPGRPTMMI. ..STTAMNTA LLQWHPPKEL PGE.LLGYRL QYCRADEARP NTIDFGKDDQ
            DLAR N6  DTKYSIQVAA LTRKGDGDRS AAIVVKTPGG VPVRPTVSLK IMEREPIVSI ELEWERPAQT YGE.LRGYRL RWGVKDQALK EEMLSGPQMT
       Consensus     ---------- ---------- ---------- ---------- ---------- ---------- ---------- ---------- ----------

    RPTP-γ_HUMAN 01  ........SS FPSTVWPTRL P......... .TAASASKQA ARPVLATTEA LASPGPDGDS SPTKDGEG.. .......... ..........
    RPTP-γ_MOUSE 02  ........SS FPSTVWPTRL P......... .TASAASKQA GRTVLATTEA LASPGPDVHS APSKDSEG.. .......... ..........
RPTP-ζ_RAT (full) 03 YNGETPLQPS YSSEVFPLVT PLLLDNQTLN TTPAASSSDS A...LHATPV FPSVGVSFDS ILSSYDDAPL LPFSSASFSS DLFHHLHTVS
RPTP-ζ_RAT (short) 04 YNGETPLQPS YSSEVFPLVT PLLLDNQTLN TTPAASSSDS A...LHATPV FPSVGVSFDS ILSSYDDAPL LPFSSASFSS DLFHHLHTVS
    RPTP-ζ_HUMAN 05  YNGETPLQPS YSSEVFPLVT PLLLDNQILN TTPAASSSDS A...LHATPV FPSVDVSFES ILSSYDGAPL LPFSSASFSS ELFRHLHTVS
         DPTP99A 06  .......... .......... .......... .......... .......... .......... .......... .......... ..........
       Consensus     ---------S Y-S-VFP--- P--------- -T----S--- ----L--T-- F-S------S ---------- ---------- ----------

      CD45_MOUSE P1  .......... ....IALLVV LYKSYDLRKK RSSNLDEQQE LVE.RD.... .......... ......DEKQ LMDVEPIHSD ILLETYKRKI
        CD45_RAT P2  .......... ....IALLVV LYKIYDLRKK RSSNLDEQQE LVE.RD.... .......... ......EEKQ LINVDPIHSD LLLETYKRKI
      CD45_HUMAN P3  .......... ....IALLVV LYKIYDLHKK RSCNLDEQQE LVE.RD.... .......... ......DEKQ LMNVEPIHAD ILLETYKRKI
          ChPTPλ P4  .......... ....IALLLV LYKIYDLHQK KLSNSSEVIS LVAVKD.... .......... ......DERQ LLNIEPIPSE KLLETYKRKI
            DPTP P5  DYYLSIGVKA GAVLLGVILV FIVLWVFHHK KTKNELQGED TLTLRDSLSR ALFGRRNHNH SHFITSGNHK GFDAGPIHRL DLENAYKNRH
       Consensus     ---------- ---------V ---------K ---N------ -----D---- ---------- ---------- -----PI--- -L---YK---

    RPTP-μ_HUMAN Q1  NSMDKSYAEQ GTNCDE...A FSFMDTHNLN GRSVSSPSSF TMKTNTLSTS VPNSYYPDET HTMASDTSSL VQSHTYKKRE PADVPYQTGQ
    RPTP-μ_MOUSE Q2  NSMDKSYAEQ GTNCDE...A FSFMGTHNLN GRSVSSPSSF TMKTNTLSTS VPNSYYPDET HTMASDTSSL AQPHTYKKRE AADVPYQTGQ
    RPTP-κ_MOUSE Q3  NAMDRSYADQ STLHAEDPLS LTFMDQHNFS PRLPNDP... .....LVPTA V.....LDEN HSATAESSRL LDVPRY.LCE GTESPYQTGQ
       Consensus     N-MD-SYA-Q -T---E---- --FM--HN-- -R----P--- --------T- V------DE- H------S-L -----Y---E ----PYQTGQ
```

```
901                                               951

....GDTSTL TTDRTRFTLE SLLPGRNYSL SVQAVSKKME SNETSIFVVT RPSSP..IIE DLKSIRMGLN ISWKSDVNSK QEQYEVLYSR NGTSDLRTQK TKESRLVIKN
PFPVAAESQI TTNLTEYTLD SLLAGRRYLI AVQALSKGVA SNASDITRYT RPAAP..LIQ ELRSIDQGLM LSWRSDVNSR QDRYEVHYQR NGTREERTMA TNETSLTIHY
..TDSSNLN. VSEPRAVIPG .LRSSTFYNI TVCPVLGDIE GTPGFLQVHT PPVPVSDFRV TVVSTTE.IG LAWSSHDAE. .......... .......... ..........
..GGRAGTR. STAHTNITVD RLEPGCLWF  S\TWVGKNGIN SSRETRNATT APNPVRNLHM ETQTNSS.IA LCWEVPDGPY P......... .......... ..........
..KVFPPFHL VNTATEYRFT SLTPGRQYKI LVLTISGDVQ QS.AFIEGFT VPSAVKNIHI SPNGATDSLT VNWTPGGGDV DSYTVSAFRH SQKVDSQTIP KHVFEHTFHR
EAEKPTPAPF SFFPVQMILS WLPPKPPTAF DGFHIHIERE ENFTEYSTVD EEAHEFVAEL KEPGKYKLSV TTFSASGSCE TRESQSAKSL SF........ ..........
..TDSSNLN. VSEPRAVIPG .LRSSTFYNI TVCPVLGDIE GTPGFLQVHT PPVPVSDFRV TVVSTTE.IG LAWSSHDAE. .......... .......... ..........
ETEKSTSGSF SFFPVQMILT WLPPKPPTAF DGFHIHIERE ENFTEYLMVD EEAHEFVAEL KEPGKYKLSV TTFSSSGSCE TRKSQSAKSL SF........ ..........
.......... .......... .......... .......... .......... .......... .......... .......... MRRAVGFPAL CL........ ..........
---------- ---------- ---------- ---------- ---------- ---------- ---------- ---------- ---------- ---------- ----------

RRKVNKCYRG RSCPIIVHCS DGAGE..... ..DRHLHP.. .......... ...HRHGPEP HGKRSEGD.. .......... .........  L1  ICA512
RRKVNKCYRG RSCPIIVHCS DGAGRTGTYI LIDMVLNRMA KGVKEIDIAA TLEHVRDQRP GLVRSKDQFE FALTAVAEEV NAILKALPQ  L2  IA-2
RRKVNKCYRG RSCPIIVHCS DGAG------ --D--L---- ---------- ---H-----P ---RS----- ---------- ---------  Consensus


RFAAPA.HKG ATYVFRLAAR GRAGLGEEAS AALSIPEDAP RGFPQILGPA GNVSAGSVIL RWLPPVPAEG NGAIIKYTVS VREAGTPGPA TETELAAAAQ PGAETALTLQ
RFAAPA.HKG ATYVFRLAAR GRAGLGEEAA AALSIPEDAP RGFPQILGAA GNVSAGSVLL RWLPPVPAER NGAIIKYTVS VREAGAPGPA TETELAAAAQ PGAETALTLR
.......... .......... .......... .......... .......... .......... .......... .......... .......... .......... ..........
.......... .......... .......... .......... .......... .......... .......... .......... .......... .......... ..........
HFTVTGLHKG TTYIFRLAAK NRAGLGEEFE KEIRTPEDLP SGFPQNLHVT G.LTTSTTEL AWDPPVLAER NGRIISYTVV FRDINSQ... .....QELQN ITTDTRFTLT
KKRFDNLERG VEYEFRVAGS NHIGIGQETV KIFQTPEGTP GGPPSNITIR FQ.TPDVLCV TWDPPTREHR NGIITRYDVQ FHKKIDHGLG SERNM..... ..TLRKAVFT
---------- ---------- ---------- ---------- ---------- ---------- ---------- ---------- ---------- ---------- ----------

......TEEG EKDEKS.... .ESEDGEREH EEDGEKDSEK KEKSGVTHAA EERNQTEPSP T......... .......... ....PSSPNR T.A....EGG HQTI......
......TEEG EKEEKS.... .ESEDGEREH EEE.EKDSEK KEKSEATHTA AESDRTAPAP T......... .......... ....PSSPHR TAA....EGG HQTI......
QTLPQVTSAA ERDELSLHAS LLVAGGDLLL EPSLVQYSDV MSHQVTIHAA SDTLEFGSES AVLYKTSMVS QIESPSSDVV MHAYSSGPET SYAI...EGS HHVLTVSSSS
QTLPQVTSAA ERDELSLHAS LLVAGGDLLL EPSLVQYSDV MSHQVTIHAA SDTLEFGSES AVLYKTSMVS QIESPSSDVV MHAYSSGPET SYAI...EGS HHVLTVSSSS
QILPQVTSAT ESDKVPLHAS LPVAGGDLLL EPSLAQYSDV LS...TTHAA SETLEFGSES GVLYKTLMFS QVEPPSSDAM MHARSSGPEP SYALSDNEGS QHIFTVSYSS
.......... .......... .......... .......... .......... .......... .......... .......... .......... .......... ..........
------T--- E-D------- -----GD--- E------SD- -------H-A -E-------- ---------- ---------- -----S-P-- --A----EG- ----------

ADEGRLFLAE FQSIPRVFSK FPIKDARKPH NQNKNRYVDI LPYDYNRVEL SEINGDAGST YINASYIDGF KEPRKYIAAQ GPRDETVDDF WRMIWEQKAT VIVMVTRCEE
ADEGRLFLAE FQSIPRVFSK FPIKDARKSQ NQNKNRYVDI LPYDYNRVEL SEINGDAGST YINASYIDGF KEPRKYIAAQ GPRDETVDDF WKMIWEQKAT VIVMVTRCEE
ADEGRPFLAE FQSIPRVFSK FPIKEARKPF NQNKNRYVDI LPYDYNRVEL SEINGDAGSN YINASYIDGF KEPRKYIAAQ GPRDETVDDF WRMIWEQKAT VIVMVTRCEE
ADEGRLFLDE FQSIPRIFTK FPMKEAKRSH NQNKNRYIDI LPYDHNRVEL SEIPGDPGSD YINASYIDGF KEPRKYIAAQ GPKDETTDDF WRMIWEQKAT IIVMVTRCEE
KDTDYGFLRE YEMLPNRFSD RTTKNSDLKE NACKNRYPDI KAYDQTRVKL AVINGLQTTD YINANFVIGY KERKKFICAQ GPMESTIDDF WRMIWEQHLE IIVILTNLEE
-D----FL-E ----P--F-- ---K------ N--KNRY-DI --YD--RV-L --I-G----- YINA----G- KE--K-I-AQ GP---T-DDF W-MIWEQ--- -IV--T--EE

LHPAIRVADL LQHITQMKCA EGYGFKEEYE SFFEGQSAPW DSAKKDENRM KNRYGNIIAY DHSRVRLQTI EGDTNSDYIN GNYI...... DGYHRPNHYI ATQGPMQETI
LHPAIRVADL LQHITQMKCA EGYGFKEEYE SFFEGQSAPW DSAKKDENRM KNRYGNIIAY DHSRVRLQML EGDNNSDYIN GNYI...... DGYHRPNHYI ATQGPMQETI
LHPAIRVADL LQHINLMKTS DSYGFKEEYE SFFEGQSASW DVAKKDQNRA KNRYGNIIAY DHSRVILQPV EDDPSSDYIN ANYIDIWLYR DGYQRPSHYI ATQGPVHETV
LHPAIRVADL LQHI--MK-- --YGFKEEYE SFFEGQSA-W D-AKKD-NR- KNRYGNIIAY DHSRV-LQ-- E-D--SDYIN -NYI------ DGY-RP-HYI ATQGP--ET-
```

```
                         1001                                                1051

DPTP10D          K1 LQPGAAYELK VFAVSHDLRS EPHAYFQAVY PNPPRNMTIE TVRSNSVLVH WSPPESGEFT EYSIRY.... ...RTDSEQQ WVRLPSVRST
DPTP4E           K2 LHPGSGYEVK VHAISHGVRS EPHSYFQAVF PKPPQNLTLQ TVHTNLVVLH WQAPEGSDFS EYVVRY.... ...RTDAS.P WQRISGLHEN
DEP-1            K3 .......... .......... .......... .......... .....SFQMH ITQEGAGNSR VEI....... .......... ......TTNQ
SAP-1            K4 .......... .......... .......... .......... ..QDYTYWVG YTGDGGGTET RN........ .......... ......TTNT
RPTP-β           K5 LEAGEQYQIM IASVSGSLKN QINVVGRTVP ASVQGVIADN AYSSYSLIVS WQKAAGVAER YDILLL.... ...TENGILL RNTSEPATTK
GLEPP-1          K6 .......... .......YIS PTGEWIEELT EKPQH.VSVH VLSSTTALMS WTSSQENYNS TIVSVVSLTC QKQKESQRLE KQYCTQVNSS
RPTP-η           K7 .......... .......... .......... .......... .....SFQMH ITQEGAGNSR VEI....... .......... ......TTNQ
RPTP-U2          K8 .......... .......YIS PSGEWIEELT EKPQH.VSFH VLSSTTALMS WTSSQENYNS TIVSVVSLTC QKQKESQRLE KQYCTQVNSS
RPTP-BR7         K9 .......... .......LLN LHAA..GCFS RNNDHFLAIR QKKSWKPVFI YDHSQDIKKS LDIAQEAYKH NYHSPSE... ....VQISKH
Consensus           ---------- ---------- ---------- ---------- ---------- ---------- ---------- ---------- ----------

LAR-PTP2_RAT     N1 GLRPETAYEL RVRAHTRRGP GPFSPPL... RYR.LARDPV S......... .......... .......... .......... ..........
RPTP-σ_MOUSE     N2 GLRPETAYEL RVRAHTRRGP GPFSPPL... RYR.LARDPV S......... .......... .......... .......... ..........
CRYPα            N3 .......... .......... .......... ........SI L......... .......... .......... .......... ..........
RPTP-δ_MOUSE     N4 .......... .......... .......... ........SM F......... .......... .......... .......... ..........
LAR_HUMAN        N5 GLKPDTTYDI KVRAWTSKGS GPLSPSI... QSRTMPVEQV F......... .......... .......... .......... ..........
DLAR             N6 NLEENTEYIF RVRAYTKQGA GPFSDKLIVE TERDMGRAPM SLQAEATSEQ TAEIWWEPVT SRGKLLGYKI FYTMTAVEDL DDWQTKTVGL
Consensus           ---------- ---------- ---------- ---------- ---------- ---------- ---------- ---------- ----------

RPTP-γ_HUMAN     01 .......... .......... .......... ..PGHEQDHT AVPTDQTGGR RDAGPGLDPD MVTSTQ.... .......... ..........
RPTP-γ_MOUSE     02 .......... .......... .......... ..PGRRQDHS APATDQPG.. .HVAPDLDPL VDTATQ.... .......... ..........
RPTP-ζ_RAT (full)  03 AIPVHDSVGV ADQGSLLINP SHISLPESSF ITPTASLLQL PPALSGDGEW SGASSDSELL LPDTDGLRTL NMSSPVSVAD FTYTTSVSGD
RPTP-ζ_RAT (short) 04 AIPVHDSVGV ADQGSLLINP SHISLPESSF ITPTASLLQL PPALSGDGEW SGASSDSELL LPDTDGLRTL NMSSPVSVAD FTYTTSVSGD
RPTP-ζ_HUMAN     05 AIPVHDSVGV TYQGSLFSGP SHIPIPKSSL ITPTASLLQP THALSGDGEW SGASSDSEFL LPDTDGLTAL NISSPVSVAE FTYTTSVFGD
DPTP99A          06 .......... .......... .......... .......... .......... .......... .......... .......... ..........
Consensus           ---------- ---------- ---------- --P------- -------G-- -------E-- ---------- ---------- ----------

CD45_MOUSE       P1 GNRNKCAEYW PSMEEGTRAF KDIVVTINDH KRCPDYIIQK LNVAHKK... .EKATGREVT HIQFTSWPDH GVPEDPHLLL KLRRRVNAFS
CD45_RAT         P2 GNRNKCAEYW PCMEEGTRTF RDVVVTINDH KRCPDYIIQK LSIAHKK... .EKATGREVT HIQFTSWPDH GVPEDPHLLL KLRRRVNAFS
CD45_HUMAN       P3 GNRNKCAEYW PSMEEGTRAF GDVVVKINQH KRCPDYIIQK LNIVNKK... .EKATGREVT HIQFTSWPDH GVPEDPHLLL KLRRRVNAFS
ChPTPλ           P4 GNRNKCAQYG PSMENGSATY GDITVKINES KICPDYIIQK LHITNGR... .ERTSGRDVT HIQFTSWPDH GVPEDPHLLL KLRRRVNALS
DPTP             P5 YNKAKCAKYW PEKVFDTKQF GDILVKFAQE RKTGDYIERT LNVSKNKANV GEEEDRRQIT QYHYLTWKDF MAPEHPHGII KFIRQINSVY
Consensus           -N--KCA-Y- P--------- -D--V----- ----DYI--- L--------- -E----R--T ------W-D- --PE-PH--- K--R--N---

RPTP-μ_HUMAN     Q1 YDFWRMVWHE NTASIIMVTN LVEVGRVKCC KYWPDDTEIY KDIKVTLIET ELLAEYVIRT FAVEKRGVHE IREIRQFHFT GWPDHGVPYH
RPTP-μ_MOUSE     Q2 YDFWRMVWHE NTASIIMVTN LVEVGRVKCC KYWPDDTEIY KDIKVTLIDT ELLAEYVIRT FAVEKRGIHE IREIRQFHFT GWPDHGVPYH
RPTP-κ_MOUSE     Q3 YDFWRMVWQE QSACIVMVTN LVEVGRVKCY KYWPDDTEVY GDFKVTCVEM EPLAEYVVRT FTLERRGYNE IREVKQFHFT GWPDHGVPYH
Consensus           YDFWRMVW-E --A-I-MVTN LVEVGRVKC- KYWPDDTE-Y -D-KVT---- E-LAEYV-RT F--E-RG--E IRE--QFHFT GWPDHGVPYH
```

1101 1151

```
EADITDMTKG EKYTIQVNTV SFGVESPVPQ EVNTTVPPNP VSNI.IQLVD SRNITLEWPK PEGRVESYIL KWWPSDNPGR VQTKNVSENK SADDLS..TV RVLIGELMPG
EARIKDMHYG ERYLVQVNTV SFGVESPHPL ELNVTMPPQP VSNV.VPLVD SRNLTLEWPR PDGHVDFYTL KWWPTDEEDR VEFKNVTQ.. .LEDLSSPSV RIPIEDLSPG
SIIIGGLFPG TKYCFEIVPK GPNGTEGASR TVCNRTVPSA VFDIHVVNRR TTEMWLDWKS PDGASEYVY. ...HLVIESK HGSNHTSTYD KA........ .ITLQGLIPG
SVTAERLEPG TLYTFSVWA. EKNGARGSRQ NVSISTVPNA VTSLSKQDWT NSTIALRWTA PQGPGQSSYS YWVSWVREGM TDPRTQSTSG TD........ .ITLKELEAG
QHKFEDLTPG KKYKIQILTV SGGLFSKEAQ TEG.RTVPAA VTDLRITENS TRHLSFRWTA SEGELSW.YN IFLYNPDGNL QERAQVDPLV QS........ .FSFQNLLQG
KRIIENLVPG AQYQVVMYLR KGPLIGPPSD PVTFAIVPTG IKDLMLYPLG PTAVVLSWTR PYLGVFRKYV VEMFYFNPAT MTSEWTTYYE IAATVS.LTA SVRIANLLPA
SIIIGGLFPG TKYCFEIVPK GPNGTEGASR TVCNRTVPSA VFDIHVVYVT TTEMWLDWKS PDGASEYVY. ...HLVIESK HGSNHTSTYD KA........ .ITLQGLIPG
KPIIENLVPG AQYQVVIYLR KGPLIGPPSD PVTFAIVPTG IKDLMLYPLG PTAVVLSWTR PYLGVFRKYV VEMFYFNPAT MTSEWTTYYE IAATVS.LTA SVRIANLLPA
HQIINSAFPR PAYDPSL... ..NLLAESDQ DLEIENLPIP AANVIVVTLQ MDITKLNIT. .LLRIFRQGV AAALGLLPQQ VHINRLIEKK NQVELF.VSP GNRKPGETQA
---------- --Y------- ---------- -------P-- ---------- ---------- ---------- ---------- ---------- ---------- ----------

.......... .......... .......... .......... ...PKNFKVK MIMKTSVLLS WEFPDNYNSP TP...YKIQY NGLTLDVDGR T......... .........T
.......... .......... .......... .......... ...PKNFKVK MIMKTSVLLS WEFPDNYNSP TP...YKIQY NGLTLDVDGR T......... .........T
.......... .......... .......... .......... ...PKNFKVK MVTKTSVLLS WEFPENYNSP TP...YKIQY NGLNVDVDGR T......... .........T
.......... .......... .......... .......... ...AKNFHVK AVMKTSVLLS WEIPENYNPA ILSKFFMMMD GKMVEEVDGR A......... .........T
.......... .......... .......... .......... ...AKNFRVA AAMKTSVLLS WEVPDSYKSA VP...FKILY NGQSVEVDGH S......... .........M
TESADLVNLE KFAQYAVAIA ARFKNGLGRL SEKVTVRIKP EDVPLNLRAH DVSTHSMTLS WSPPIRLTPV N....YKISF DAMKVFVDSQ GFSQTQIVPK REIILKHYVK
---------- ---------- ---------- ---------- -----N---- -----S--LS W--P------ ---------- ------VD-- ---------- ----------

.......... .......... .......... .......... .......... .......... .......... .......... ........VP PTATEEQYAG SDPKRPEMPS
.......... .......... .......... .......... .......... .......... .......... .......... ........VP PTATEEHYSG SDPRRPEMPS
DIKPLSKGEM MYGNETELKM SSFSDMAYPS KSTVVPKMSD IVNKWSESLK ETSVSVSSIN SVFTESLVYP ITKVFDQEIS RVPEIIFPVK PTHTASQASG DTWLKPGLST
DIKPLSKGEM MYGNETELKM SSFSDMAYPS KSTVVPKMSD IVNKWSESLK ETSVSVSSIN SVFTESLVYP ITKVFDQEIS RVPEIIFPVK PTHTASQASG DTWLKPGLST
DNKALSKSEI IYGNETELQI PSFNEMVYPS ESTVMPNMYD NVNKLNASLQ ETSVSISSTK GMFPGSLAHT TTKVFDHEIS QVPENNFSVQ PTHTVSQASG DTSLKPVLSA
.......... .......... .......... .......... .......... .......... .......... .......... .......... .......... ..........
---------- ---------- ---------- ---------- ---------- ---------- ---------- ---------- --------V- PT-T-----G -----P-L--

NFFSGPIVVH CSAGVGRTGT YIGIDAMLEG LEAEGKVDVY GYVVKLRRQR CLMVQVEAQY ILIHQALVEY NQFGETQVNL SELHSCLHNM KKRDPPSDPS PREAEYQRL.
NFFSGPIVVH CSAGVGRTGT YIGIDAMLES LEAEGKVDVY GYVVNLRRQR CLMVQVEAQY ILIHQALVEY NQFGETEVNL SELHSCLQNL KKRDPPSDPS PLEAEYQRL.
NFFSGPIVVH CSAGVGRTGT YIGIDAMLEG LEAENKVDVY GYVVKLRRQR CLMVQVEAQY ILIHQALVEY NQFGETEVNL SELHPYLHNM KKRDPPSEPS PLEAEFQRL.
NFFSGPIVVH CSAGVGRTGT YIGIDAMLEG LDAEGRVDVY GYVVKLRRQR CLMVQVESQY ILIHQALVEY HQYGETEVSL SELHSYLNNL KRKDPPSEPS LLEAKFQRL.
SLQRGPILVH CSAGVGRTGT LVALDSLIQQ LEEEDSVSIY NTVCDLRHQR NFLVQSLKQY IFLYRALLDT GTFGNTDICI DTMASAIESL KRK.PNEGKC KLEMEFEKLL
----GPI-VH CSAGVGRTGT ----D----- L--E--V--Y --V--LR-QR ---VQ---QY I----AL--- ---G-T---- ---------- K---P----- --E-----L-

ATGLLGFVRQ VKSKSPPSAG PLVVHCSAGA GRTGCFIVID IMLDMAEREG VVDIYNCVRE LRSRRVNMVQ TEEQYVFIHD AILEACLCGD TSVPASQVRS LYYDMNKLDP
ATGLLGFVRQ VKSKSPPNAG PLVVHCSAGA GRTGCFIVID IMLDMAEREG VVDIYNCVRE LRSRRVNMVQ TEEQYVFIHD AILEACLCGD TSIPASQVRS LYYDMNKLDP
ATGLLSFIRR VKLSNPPSAG PIVVHCSAGA GRTGCYIVID IMLDMAEREG VVDIYNCVKA LRSRRINMVQ TEEQYIFIHD AILEACLCGE TAIPVCEFKA AYFDMIRIDS
ATGLL-F-R- VK---PP-AG P-VVHCSAGA GRTGC-IVID IMLDMAEREG VVDIYNCV-- LRSRR-NMVQ TEEQY-FIHD AILEACLCG- T--P------ -Y-DM---D-
```

```
                      1201                                              1251

DPTP10D  K1  V.....QYKF DIQTTSYGIL SGITSLYPRT MPLIQSDVVV ANGEKEDERD .TITLSYTPT PQSSSKFDIY RFSSG..DAE IRDKEKLAND
DPTP4E   K2  R.....QYRF EVQASSNGIR SGTTHLSTRT MPLIQSDVFI ANAGHEQGCD ETITLSYTPT PADSTRFDIY RFSMG..DPT IKDKEKLAND
DEP-1    K3  T.....LYNI TISPEVDHVW GDPNSTAQYT RPSNVSNI.. ....DVSTNT TAATLSWQNF DDASPTYSYC LLIEK..AGN S.........
SAP-1    K4  S.....LYHL TVWAERNEVR GYNSTLTAAT APNEVTDL.. ....QNETQT KNSVMLWWKA PGDPHSQLYV YWVQW..ASK GHPRRG....
RPTP-β   K5  R.....MYKM VIVTHSGELS NESFIFGR.T VPASVSHL.. ....RGSNRN TTDSLWF.NW SPASGDFDFY ELILY..NPN GTKKENWKDK
GLEPP-1  K6  W.....YYNF RVTMVTWGDP ELSCCDSSTI S..FITAPVA PEITSVE.YF NSLLYISWTY GDDTTDLSHS RMLHWMVVAE GKKKIKKSVT
RPTP-η   K7  T.....LYNI TISPEVDHVW GDPNSTAQYT RPSNVSNI.. ....DVSTNT TAATLSWQNF DDASPTYSYC LLIEK..AGN S.........
RPTP-U2  K8  W.....YYNF RVTMVTWGDP ELSCCDSSTI S..FITAPVA PEITSVE.YF NSLLYISWTY GDDTTDLSHS RMLHWMVVTE GKKKIKKSVT
RPTP-BR7 K9  LQAEEVLRSL NVDGLHQSLP QFGITDVAPE KNVLQGQHEA DKIWSKEGFY AVVIFLSIFI IIVTCLMIIY RLKERLQLSL RQDKEK....
Consensus    ---------- ---------- ---------- ---------- ---------- ---------- ---------- ---------- ----------

LAR-PTP2_RAT   N1  KKLITHLKPH TFYNFVLTNR GSSLG...GL QQTVTARTAF NMLSGKPSVA PKPDNDGSIV VYLPDGQSPV TVQNYFIVMV PLRKSRGGQF
RPTP-σ_MOUSE   N2  KKLITHLKPH TFYNFVLTNR GSSLG...GL QQTVTARTAF NMLSGKPSVA PKPDNDGFIV VYLPDGQSPV TVQNYFIVMV PLRKSRGGQF
CRYPα          N3  KKLITNLKPH TFYNFVLMNR GNSMG...GL QQNVAAWTAA NMLSRKPEVT HKPDADGNVV VILPDVKSSV AVQAYYIVVV PLRKSRGGQF
RPTP-δ_MOUSE   N4  QKLIVNLKPE KSYSFALTNR GNSAG...GL QHRVTAKTAP DVLRTKPAFI GKTNLDGMIT VQLPDVPANE NIKGYYIIIV PLKKSR.GKF
LAR_HUMAN      N5  RKLIADLQPN TEYSFVLMNR GSSAG...GL QHLVSIRTAP DLLPHKPLPA SAYIEDGRFD LSMPHVQDPS LVRWFYIVVV PIDRVGGSML
DLAR           N6  THTINELSPF TTYNVNVSAI PSDYSYRPPT KITVTTQMAA PQPMVKPDFY GVVNGEEILV ILPQASEEYG PISHYYLVVV PEDKSNLHKI
Consensus          ---I--L-P- --Y------- ---------- ---V----A- -----KP--- ---------- ---------- ---------V P---------

RPTP-γ_HUMAN        01  KKPMSRGDRF SE........ .......... .......... .....DSRFI TVNPAEKNTS GMISRPAPGR MEWIIPLIVV SA........
RPTP-γ_MOUSE        02  KKPMSRGDRF SE........ .......... .......... .....DSKFI TVNPAEKNTS GMLSRPSPGR MEWIIPLIVV SA........
RPTP-ζ_RAT (full)   03  NSEPALSDTA SSEVSHPSTQ PLLYEAASPF NTEALLQPSF PASDVDTLLK TALPSGPRDP VLTETPMVEQ SSSSVSLPLA SESASSKSTL
RPTP-ζ_RAT (short)  04  NSEPALSDTA SSEVSHPSTQ PLLYEAASPF NTEALLQPSF PASDVDTLLK TALPSGPRDP VLTETPMVEQ SSSSVSLPLA SESASSKSTL
RPTP-ζ_HUMAN        05  NSEPASSDPA SSEMLSPSTQ LLFYETSASF STEVLLQPSF QASDVDTLLK TVLPAVPSDP ILVETPKVDK ISSTMLHLIV SNSASSENML
DPTP99A             06  .......... .......... .......... .........M PRPQHHALLR AMLK...... .LLLFASIAE HCATALPTNS SNSPSSPSPF
Consensus               -------D-- S--------- ---------- ---------- -----D--L- T--P------ -L---P---- ---------- S---------

CD45_MOUSE   P1  .PSYRSWRTQ HIGNQGENKK KKRNSNVVPY DFNRVPLKHE LEMSKESEPE SDESSDDDSD SEETSKYINA SFVMSYWKPE MMIAAQGPLK
CD45_RAT     P2  .PSYRSWRTQ HIGNQEENKK KNRSSNVVPY DFNRVPLKHE LEMSKESEAE SDESSDEDSD SEETSKYINA SFVMSYWKPE MMIAAQGPLK
CD45_HUMAN   P3  .PSYRSWRTQ HIGNQEENKS KNRNSNVIPY DYNRVPLKHE LEMSKESEHD SDESSDDDSD SEEPSKYINA SFIMSYWKPE VMIAAQGPLK
ChPTPλ       P4  .PSYKGWRTQ NTGNREENKN KNRSANTIPY DFNRVPIRSE EEQSKEGEHD SEDSSDEDSD CEESSRYINA SFITGYWGPK AMIATQGPLQ
DPTP         P5  ATADEISKSC SVGENEENNM KNRSQEIIPY DRNRVILTPL PMR....... .......... ..ENSTYINA SFIEGYDNSE TFIIAQDPFE
Consensus        ---------- --G---EN-- K-R-----PY D-NRV----- ---------- ---------- --E-S-YINA SF---Y---- --I--Q-P--

RPTP-μ_HUMAN  Q1  QTNSSQIKEE FRTLNMVTPT LRVEDCSIAL LPRNHEKNRC MDILPPDRCL PFLITIDGES SNYINAALMD SYKQPSAFIV TQHPLPNTVK
RPTP-μ_MOUSE  Q2  QTNSSQIKEE FRTLNMVTPT LRVEDCSIAL LPRNHEKNRC MDILPPDRCL PFLITIDGES SNYINAALMD SYKQPSAFIV TQHPLPNTVK
RPTP-κ_MOUSE  Q3  QTNSSHLKDE FQTLNSVTPR LQAEDCSIAC LPRNHDKNRF MDMLPPDRCL PFLITIDGES SNYINAALMD SYRQPAAFIV TQYPLPNTVK
Consensus         QTNSS--K-E F-TLN-VTP- L--EDCSIA- LPRNH-KNR- MD-LPPDRCL PFLITIDGES SNYINAALMD SY-QP-AFIV TQ-PLPNTVK
```

```
1301                                                                          1351

TDRKVTFTGL VPGRLYNITV WTVSGGVASL PIQRQDRLYP EPITQLHATN ITDTEISLRW DLPKG..EYN DFDIAYLTAD N...LLAQNM TTRNEITISD LRPHRNYTFT
TERKLSFSGL TPGKLYNVTV WTVSGGVASL PVQRVYRLHP LPISDLKAIQ VAAREITLHW TAPAG.,EYT DFELQYLSAD EEAPQLLQNV TKNTEITLQG LRPYHNYTFT
.......... .......... ....SNATQV .VTDIGITDA T......... .......... .......... .......... .......... .......VTE LIPGSSYTVE
.......... .......... ...QDPQANW .VNQTSRTNE TWYK...... .......... .......... .......... .......... .......VEA LEPGTLYNFT
DLTEWRFQGL VPGRKNNWV VTHSGDLSNK .VTAESRTAP SPPSLMSFAD IANTSLAITW KGPPDWTDYN DFELQWLPRD ALTVFNPYNN RKSEGRIVYG LRPGRSYQFN
RNVMTAILSL PPGDIYNLSV TACTERGSNT SMLRLVKLEP APPKSLFAVN KTQTSVTLLW .......... .......... .......... .......... ..........
.......... .......... ....SNATQV .VTDIGITDA T......... .......... .......VTE LIPGSSYTVE
RNVMTAILSL PPGDTYNLSV TTCTERGSNT SMLRLVKLEP APPKSLFAVN KTQTSVTLLW .......... .......... .......... .......... ..........
.......... ..NQEIHLSP IARQQAQSEA KTTHSMVQPD QAPKVLNVVV DPQG...... .......... .......... .......... .......... ..........
---------- ---------- ---------- ---------- ---------- ---------- ---------- ---------- ---------- ---------- ----------

PILLGSPEDM DLEELIQDLS RLQRRSLRHS RQLEVPRPYI AARF..SILP AVFHPGNQKQ YGGFDNRGLE PGHRYVLFVL AVLQKN.EPT .FAASPFSDP FQLDNPDPQP
PVLLGSPEDM DLEELIQDIS RLQRRSLRHS RQLEVPRPYI AARF..SILP AVFHPGNQKQ YGGFDNRGLE PGHRYVLFVL AVLQKN.EPT .FAASPFSDP FQLDNPDPQP
LNPLGSPEEM DLEELIQDIA RLRRRSLRHS RQLDFPKPYI AARF..RSLP NHFVLGDMKH YDNFENRALE PGQRYVIFIL AVLQEP.EAT .FAASPFSDP IQLDNPDPQP
IKPWESPDEM ELDELLKEIS R.KRRSIRYG REVEL.KPYI AAHF..DVLP TEFTLGDDKH YGGFTNKQLQ SGQEYVFFVL AVMDHA.ESK MYATSPYSDP VVSMDLDPQP
TPRWSTPEEL ELDELLEAIE QGGEEQRRRR RQAERLKPYV AAQL..DVLP ETFTLGDKKN YRGFYNRPLS PDLSYQCFVL ASLKEPMDQK RYASSPYSDE IVVQVTPAQQ
......PDQF LTDDLLPGRN KPERPN.... ......APYI AAKFPQRSIP FTFHLGSGDD YHNFTNRKLE REKRYRIFVR AVVDTP.QKH LYTSSPFSEF LSLDMREAPP
------P--- ----L----- ---------- -------PY- AA-------P --F--G---- Y--F-N--L- ----Y--F-- A--------- ----SP-S-- ----------

.LTFVCLI.. .......... .......... .......... .......... .......... .......... .......... .........L LIAVLVYW.. ..........
.LTFVCLV.. .......... .......... .......... .......... .......... .......... .......... .........L LIAVLVYW.. ..........
HFTSVPVLNM SP.SDVHPTS LQRLTVPHSR EEYFEQGLLK SKSPQQVLPS LHSHDEFFQT AHLDISQAYP PKGRHAFATP ILSINEPQNT LINRLVYSED IFMHPEISIT
HFTSVPVLNM SP.SDVHPTS LQRLTVPHSR EEYFEQGLLK SKSPQQVLPS LHSHDEFFQT AHLDISQAYP PKGRHAFATP ILSINEPQNT LINRLVYSED IFMHPEISIT
HSTSVPVFDV SPTSHMHSAS LQGLTISYAS EKY.EPVLLK SESSHQVVPS LYSNDELFQT ANLEINQAHP PKGRHVFATP VLSIDEPLNT LINKLIHSDE ILTSTKSSVT
TVASLPPTTA SSSSSPAVIS TSSFDRNLA. ....DLVNPE AETSGSGWES LETEFNLATT VDSSTQKTAK EPVLGTAATS IEQQDQPPDV PATTLAFAN. ..........
--T-V----- ---------- ---------- ---------- ---------- ---------- ---------- ---------- ---------- LI--LVY--- ----------

ETIGDFWQMI FQRKVKVIVM LTELVNGDQE VCAQYWGEGK QTYGDMEVEM KDTNRASAYT LRTFELRHSK RKEPRTVYQY QCTTWK..GE ELPAEPKDLV SMIQDLKQKL
ETIGDFWQMI FQRKVKVIVM LTELMSGDQE VCAQYWGEGK QTYGDMEVML KDTNKSSAYI LRAFELRHSK RKEPRTVYQY QCTTWK..GE ELPAEPKDLV TLIQNIKQKL
ETIGDFWQMI FQRKVKVIVM LTELKHGDQE ICAQYWGEGK QTYGDIEVDL KDTDKSSTYT LRVFELRHSK RKDSRTVYQY QYTNWS..VE QLPAEPKELI SMIQVVKQKL
ETISDFWQMV FQRKVKVIVM LTELKEGDQE LCAQYWGEGR QTYDDIEVQV TDVNCCPSYT IRAFDVTHLK RKETQKVYQY QYHKWN..GL DVPEDPKDLV DMILSLKQKV
NTIGDFWRMI SEQSVTTLVM ISEIGDGPRK .CPRYWADDE VQYDHILVKY VHSESCPYYT RREFYVTNCK IDDTLKVTQF QYNGWPTVDG EVPEVCRGII ELVDQAYNHY
-TI-DFW-M- ----V---VM --E---G--- -C--YW---- --Y----V-- --------Y- -R-F-----K ------V-Q- Q---W----- --P------- ----------

DFWRLVLDYH CTSVVMLNDV DPAQLCPQYW PENGVHRHGP IQVEFVSADL EEDIISRIFR IYNAARPQDG YRMVQQFQFL GWPMYRDTPV SKRSFLKLIR QVDKWQEEYN
DFWRLVLDYH CTSVVMLNDV DPAQLCPQYW PENGVHRHGP IQVEFVSADL EEDIISRIFR IYNASRPQDG HRMVQQFQFL GWPMYRDTPV SKRSFLKLIR QVDKWQEEYN
DFWRLVYDYG CTSIVMLNEV DLSQGCPQYW PEEGMLRYGP IQVECMSCSM DCDVINRIFR ICNLTRPQEG YLMVQQFQYL GWASHREVPG SKRSFLKLIL QVEKWQEECE
DFWRLV-DY- CTS-VMLN-V D--Q-CPQYW PE-G--R-GP IQVE--S--- --D-I-RIFR I-N--RPQ-G --MVQQFQ-L GW---R--P- SKRSFLKLI- QV-KWQEE--
```

```
                   1401                                                  1451

 DPTP1OD K1 VVVRSG.... ......TESS VLRSSSPLSA SFTTNEAVPG RVERFHPTDV QPSEINFEWS LPSSEANGVI RQFSIAYTNI NNLTDAGMQD
  DPTP4E K2 VVVRSGSIQG TDFADVSVST LMRSSAPISA SYQTLTAPPG KVDYFQPSDV QPGEVTFEWS LEPAEQHGPI DYFRITCQNA DDAADVSSYE
   DEP-1 K3 IFAQVGDGIK SL.EPGRKSF CTDPASMASF DCEVVPKEPA LVLKWTCPPG ANAGFELEVS .......... ...SGAWNNA THLESCSSEN
   SAP-1 K4 VWAERNDVAS ST.QSLCAST YPDTVTITS. .CVSTSAGYG VNLIWSCPQG GYEAFELEVG .......... ...GQ..... .....RGSQD
  RPTP-β K5 VKTVSGDSWK TYSKPIFGSV RTKPDKIQNL HCRP.QNSTA IACSWIPPDS DFDGYSIECR .......... ...K...MDT QEVEFSRKLE
 GLEPP-1 K6 VEEGVADFFE VFCQQVGSGL ETKLQEPVAV SSHVVTISS. .......... .......... .......... .......... ..........
  RPTP-η K7 IFAQVGDGIK SL.EPGRKSF CTDPASMASF DCEVVPKEPA LVLKWTCPPG ANAGFELEVS .......... ...SGAWNNA THLESCSSEN
 RPTP-U2 K8 VEEGVADFFK VFFQHVGSSQ KTKLQEPVAV SPHVVTISS. .......... .......... .......... .......... ..........
RPTP-BR7 K9 .......... ....QCTPEI RNSTSTSVCP SPFRMKPIG. .......... .......... .......... .......... ..........
Consensus   ---------- ---------- ---------- ---------- ---------- ---------- ---------- ---------- ----------


      LAR-PTP2_RAT N1 IV........ .......... ..DGEEGLIW VIGPVLAVVF IICIVIAILL YKNK....PD SKRKDSEPRT KCLLNNADLA PHHPKDPVEM
      RPTP-σ_MOUSE N2 IV........ .......... ..DGEEGLIW VIGPVLAVVF IICIVIAILL YKNK....PD SKRKDSEPRT KCLLNNADLA PHHPKDPVEM
             CRYPα N3 II........ .......... ..DGEEGLIW VIGPVLAVVF IICIVIAILL YKNK....PD SKRKDSEPRT KCLLNNAEIT PHHPKDPVEM
      RPTP-δ_MOUSE N4 IT........ .......... ..DEEEGLIW VVGPVLAVVF IICIVIAILL YK........ RKRAESDSRK SSLPNSKEVP SHHPTDPVEL
         LAR_HUMAN N5 .......... .......... ..QEEPEMLW VTGPVLAVIL IILIVIAILL FKRKRTHSPS SKDEQSIGLK DSLLA..... ..HSSDPVEM
              DLAR N6 GERPHRPDPN WPAEPEVSVN RNKDEPEILW VVLPLMVSTF IVSTAL.IVL CVVKRRRQP. CKTPDQAAVT RPLMAADLGA GPTPSDPVDM
         Consensus    ---------- ---------- ----E----W V--P------ I------I-L ---------- -K-------- --L------- -----DPV--

      RPTP-γ_HUMAN 01 .......... .......... .......... .......... .......... .......... .......... .......... RGCNKIKSKG
      RPTP-γ_MOUSE 02 .......... .......... .......... .......... .......... .......... .......... .......... RGCNKIKSKG
 RPTP-ζ_RAT (full) 03 DKALTGLPTT VSDVLIATDH SVPLGSGPIS MTTVSPNRDD SVTTTKLLLP SKATSKPTHS ARSDADLVGG GEDGD.DYDD DDYDDIDSDR
RPTP-ζ_RAT (short) 04 DKALTGLPTT VSDVLIATDH SVPLGSGPIS MTTVSPNRDD SVTTTKLLLP SKATSKPTHS ARSDADLVGG GEDGD.DYDD DDYDDIDSDR
      RPTP-ζ_HUMAN 05 GKVFAGIPTV ASDTFVSTDH SVPIGNGHVA ITAVSPHRDG SVTSTKLLFP SKATSELSHS AKSDAGLVGG GEDGDTDDDG DDDDDRDSDG
          DPTP99A 06 .....AFPVP VAGEMGNGNG NYNDATPPYA AVDDNYVPSK PQNLTILDVS ANSITMSWHP PKNQNGAIAG YHVFHIHDNQ TGVEIVKNS.
         Consensus    ---------- ---------- ---------- ---------- ---------- ---------- ---------- ---------- -----I-S--

        CD45_MOUSE P1 P.KASPEGMK YHKHASILVH CRDGSQQTGL FCALFNLLES AETEDVVDVF QVVKSLRKAR PGVVCSYEQY QFLYDIIASI YPAQNGQVKK
          CD45_RAT P2 P.KSGSEGMK YHKHASILVH CRDGSQQTGL FCALFNLLES AETEDVVDVF QVVKSLRKAR PGMVGSFEQY QFLYDIMASI YPTQNGQVKK
        CD45_HUMAN P3 PQKNSSEGNK HHKSTPLLIH CRDGSQQTGI FCALLNLLES AETEEVVDIF QVVKALRKAR LGMVSTFEQY QFLYDVIAST YPAQNGQVKK
           ChPTPλ P4 PSRPASEDSR NSRSVPFVIH CCDGSQQTWC VLCLMTLLES AETEEVIDVF QVVKALRRSR LGVVSTFEQY QFLYDTIART YPAQNGQIKN
              DPTP P5 KNNKNSGC.. ...RSPLTVH CSLGTDRSSI FVAMCILVQH LRLEKCVDIC ATTRKLRSQR TGLINSYAQY EFLHRAIINY SDLHHIAEST
         Consensus    ---------- ---------H C--G------ ------L--- ---E---D-- -----LR--R -G------QY -FL------- ----------

      RPTP-μ_HUMAN Q1 GGEGPTVVHC LNGGGRSGTF CAISIVCEML RHQRTVDVFH AVKTLRNNKP NMVDLLDQYK FCYEVALEYL NSG Q1 RPTP-μ_HUMAN
      RPTP-μ_MOUSE Q2 GGEGPTVVHC LNGGGRSGTF CAISIVCEML RHQRTVDVFH AVKTLRNNKP NMVDLLDQYK FCYEVALEYL NSG Q2 RPTP-μ_MOUSE
      RPTP-κ_MOUSE Q3 EGEGRTIIHC LNGGGRSGMF CAIGIVVEMV KRQNVVDVFH AVKTLRNSKP NMVEAPEQYR FCYDVALEYL ESS Q3 RPTP-κ_MOUSE
         Consensus    -GEG-T--HC LNGGGRSG-F CAI-IV-EM- --Q--VDVFH AVKTLRN-KP NMV----QY- FCY-VALEYL -S- Consensus
```

```
1501                                                  1551

FESEEAFGVI KNLKPGETYV FKIQAKTAIG FGPEREYRQT MPILAPPRPA TQVVPTEVYR SSSTIQIRFR KNYFSDQNGQ VRMYTIIVAE .DDAKNASGL EMPSWLDVQS
FPVNATQGKI DGLVPGNHYI FRIQAKSALG YGAEREHIQT MPILAPPVPE PSVTPLEVSR TSSTIEISFR QGYFSNAHGM VRSYTIIIAE .DVGKIASGL EMPSWQDVQA
GTEYRTEVTY LNFSTSYNIS ITTVSCGKMA APTRNTCTTG ITDPPPP... ...DGSPNIT SVSHNSVKVK FSGFEASHGP IKAYAVILTT GEAG...... ..HPSADVLK
RSSCGEAVSV LGL....... .......... .......... .......... .......... .......... ........GP ARSYPATITT .......... ..........
KEKSLLNIMM LVPHKRYLVS IKVQSAGMTS EVVEDSTITM IDRPPPPPPH IRVNEKDVLI SKSSINFTVN CSWFSDTNGA VKYFTVVVRE ADGSDELKPE QQHPLPSYLE
.......... LLPATAYNCS VTSFSHDSPS VPTFIAVSTM VTEMNPNVVV ISVLAILSTL LIGLLLVTLI ILRKKHLQMA RECGAGTFVN FASLERDGKL PYN.......
GTEYRTEVTY LNFSTSYNIS ITTVSCGKMA APTRNTCTTG ITDPPPP... ...DGSPNIT SVSHNSVKVK FSGFEASHGP IKAYAVILTT GEAG...... ..HPSADVLK
.......... LLPATAYSCS VTSFSHDSPS VPTFIAVSTM VTEMNPNVVV ISVLAILSTL LIGLLLVTLI ILRKKHLQMA RECGAGTFAN CASLERDGKL PYNCRRSIFA
.......... LQERRGSNVS LTLDMSSLGS VEPFVAVST. .......... .......... .......... .........P REKVAMEYLQ SAS....... ..........
---------- ---------- ---------- ---------- ---------- ---------- ---------- ---------- ---------- ---------- ----------

RRINFQTPGM LSHPPIPITD MAEHMERLKA NDSLKLSQEY ESIDPGQQFT WEHSNLEANK PKNRYANVIA YDHSRVILQP LEGIMGSDYI NANYVDGYRR QNAYIATQGP
RRINFQTPGM LSHPPIPITD MAEHMERLKA NDSLKLSQEY ESIDPGQQFT WEHSNLEANK PKNRYANVIA YDHSRVILQP LEGIMGSDYI NANYVDGYRR QNAYIATQGP
RRINFQTPGM LSHPPIPVSE LAEHTEHLKA NDNLKLSQEY ESIDPGQQFT WEHSNLEVNK PKNRYANVIA YDHSRVILLP IEGIVGSDYI NANYIDGYRK QNAYIATQGP
RRLNFQTPGM ASHPPIPILE LADHIERLKA NDNLKFSQEY ESIDPGQQFT WEHSNLEVNK PKNRYANVIA YDHSRVLLSA IEGIPGSDYV NANYIDGYRK QNAYIATQGS
RRLNYQTPGM RDHPPIPITD LADNIERLKA NDGLKFSQEY ESIDPGQQFT WENSNLEVNK PKNRYANVIA YDHSRVILTS IDGVPGSDYI NANYIDGYRK QNAYIATQGP
RRLNFQTPGM ISHPPIPISE FANHIERLKS NDNQKFSQEY ESIEPGQQFT WDNSNLEHNK SKNRYANVTA YDHSRVQLPA VEGVVGSDYI NANYCDGYRK HNAYVATQGP
RR-N-QTPGM --HPPIP--- -A---E-LK- ND--K-SQEY ESI-PGQQFT W--SNLE-NK -KNRYANV-A YDHSRV-L-- --G--GSDY- NANY-DGYR- -NAY-ATQG-

FP........ .......... .......... .......... .......... .......... .......... .......... .......... .......... RRFREVPSSG
FP........ .......... .......... .......... .......... .......... .......... .......... .......... .......... RRSREVPSSG
FPVNKCMSCS PYRESQEKVM NDSDTQESSL VDQSDPISHL LSENTEEENG GTGV.....T RVDKSPDKSP PPSMLPQKHN DGREDRDIQM GSAVLPHTPG SKAWAVLTSD
FPVNKCMSCS PYRESQEKVM NDSDTQESSL VDQSDPISHL LSENTEEENG GTGV.....T RVDKSPDKSP PPSMLPQKHN DGREDRDIQM GSAVLPHTPG SKAWAVLTSD
LSIHKCMSCS SYRESQEKVM NDSDTHENSL MDQNNPISYS LSENSEEDNR VTSVSSDSQT GMDRSPGKSP SANGLSQKHN DGKEENDIQT GSALLPLSPE SKAWAVLTSD
.......... ..RNSVETLI HFELQNLRPY TDYRVIVKAF TTKNEGEPSD QIA....... ..QRTDVGGP SAPAIVNLTC HSQESITIRW RRP.YEFYNT IDFYIIKTRL
F--------- ---------- ---------- ---------- ---------- ---------- ---------- ---------- ---------- ---------- ---W-V--S-

TNSQ.DKIEF HNEVDGGKQD AN..CVRPDG PLNKAQEDSR GVGTPEPTNS AEEPEHAANG SASPAPTQSS P1  CD45_MOUSE
ANSQ.DKIEF HNEVDGAKQD AN..CVQPAD PLNKAQEDSK EVGASEPASG SEEPEHSANG PMSPALTPSS P2  CD45_RAT
NNHQEDKIEF DNEVDKVKQD AN..CVNPLG APEKLPEAKE QAEGSEPTSG TEGPEHSVNG PASPALNQGS P3  CD45_HUMAN
IH.QEDKVEF CNEVEKKDQE SDLITIDLTP STPEENDAPE CCDDFKAADT NKGTESSTNG PTTPVLT... P4  ChPTPλ
LD........ .......... .......... .......... .......... .......... .......... P5  DPTP
---------- ---------- ---------- ---------- ---------- ---------- ----------     Consensus
```

```
                           1601                                                                           1651

 DPTP1OD K1 YSVW.LPY.A IDPYYPF... .....ENRSV E...,.DFTI GTENCDNHKI GYCNGPLKSG TTIGVKVRRF TG........ ......ADKF
  DPTP4E K2 YTVW.LPYQA IEPYNPFLTS NGSRKSSLEA E.....HLTI GTANCDKHQA GYCNGPLRAG TTYRIKIRAF TD........ ......EDKF
   DEP-1 K3 YTYDDFKKGA SDTYVTYLIR TEEKGRSQSL SEVLKYEID. ..VGNESTTL GYYNGKLEPL GSYRACVAGF TNITFHPQNK GLIDGAESYV
   SAP-1 K4 .......... ........IW DGMKVVSHSV .......... .......... .......... .......... .......... ..........
  RPTP-β K5 YRHNASIRVY QTNYFASKCA ENPNSNSKSF NIKLGAEMES LGGKRDPTQQ KFCDGPLKPH TAYRISIRAF TQLF....DE DLKEFTKPLY
 GLEPP-1 K6 .......... .......... .WSKNGLK.. .......... .......... .......... .......... .......... ..........
  RPTP-η K7 YTYDDFKKGA SDTNRRYLIR TEEKGRSQSL SEVLKYEID. ..VGNESTTL GYLQWEAGTS GLLPACVAGF TNITFHPQNK GLIDGAESYV
 RPTP-U2 K8 FLTLLPSCLW TDYPLAFYIN PWSKNGLK..                       .......... .......... .......... ..........
RPTP-BR7 K9 .......... .......... .......... .......... .......... .......... .......... .......... ..........
Consensus   ---------- ---------- ---------- ---------- ---------- ---------- ---------- ---------- ----------


      LAR-PTP2_RAT N1 LPETFGDFWR MVWEQRSATV VMMTRLEEKS RVKCDQYWPN RGTETYGFIQ VTLLDTMELA TFCVRTFSLH KNGSSEKREV RHFQFTAWPD
      RPTP-σ_MOUSE N2 LPETFGDFWR MVWEQRSATV VMMTRLEEKS RIKCDQYWPN RGTETYGFIQ VTLLDTMELA TFCVRTFSLH KNGSSEKREV RHFQFTAWPD
             CRYPα N3 LPETFGDFWR MVWEQRSATI VMMTKLEEKS RIKCDQYWPG RGTDTYGMIQ VTLLDTIELA TFCVRTFSLH KNGSSEKREV RQFQFTAWPD
      RPTP-δ_MOUSE N4 LPETFGDFWR MIWEQ.EATV VMMTKLEERS RVKCDQYWPS RGTETHGLVQ VTLLDTVELH ILCPDICTLN .NGSSEKRKV RQFQFTAWPD
         LAR_HUMAN N5 LPETMGDFWR MVWEQRTATV VMMTRLEEKS RVKCDQYWPA RGTETCGLIQ VTLLDTVELA TYTVRTFALH KSGSSEKREL RQFQFMAWPD
              DLAR N6 LQETFVDFWR MCWELKTATI VMMTRLEERT RIKCDQYWPT RGTETYGQIF VTITETQELA TYSIRTFQLC RQGFNDRREI KQLQFTAWPD
         Consensus    L-ET--DFWR M-WE---AT- VMMT-LEE-- R-KCDQYWP- RGT-T-G--- VT---T-EL- --------L- --G----R-- ---QF-AWPD

      RPTP-γ_HUMAN 01 ERGEKG.... .......... .......... .......... .......... .......... .......... .......... ..........
      RPTP-γ_MOUSE 02 ERGEKG.... .......... .......... .......... .......... .......... .......... .......... ..........
 RPTP-ζ_RAT (full) 03 EESGSGQGTS DSLNDNETST DFSFPDVNEK DADGVLEADD TGIAPGSPRS STPSVTSGHS GVSNSSEAG. .......... ..........
RPTP-ζ_RAT (short) 04 EESGSGQGTS DSLNDNETST DFSFPDVNEK DADGVLEADD TGIAPGSPRS STPSVTSGHS GVSNSSEAG. .......... ..........
      RPTP-ζ_HUMAN 05 EESGSGQGTS DSLNENETST DFSFADTNEK DADGILAAGD SEITPGFPQS PTSSVTSENS EVFHVSEAEA SNSSHESRIG LAEGLESEKK
           DPTP99A 06 AGQDTHRDIR INASAKELET AMILQNLTTN SYYEVKVAAA T.......FS VINPKKIVLG KFSESRIIQL QPNCEKLQPL LRQSHNDYNL
         Consensus    E----G---- ---------- ---------- ---------- ---------- ---------- ---------- ---------- ---------
```

```
1701                                                          1751

TDTAYSFPIQ TDQDNTSLIV AITVPLTIIL VLLVT....L IFYKRRRNNC RKTT...... KDSRANDNM. SLPDSVIEQN RPILIKNFAE HYRLMSADSD FRFSEEFEEL
TDTVYSSPIT TERSDT.VIV AATVSAVLLV AMVLV....V VYCQHRCQLI RRAS...... KLARMQDELA ALPEAYITPN RPVHVKDFSE HYRIMSADSD FRFSEEFEEL
SFSRYSDAVS LPQDP..GVI CGAVFGCIFG ALVIVTVGGF IFWRKKRKDA KNNEVSFS.. .......... ...QIKPKKS KLIRVENFEA YFKKQQADSN CGFAEEYEDL
........VC HTESA..GVI AGAFVGIL.. .LFLILVGLL IFFLKRRNKK KQQKPELR.. .......... ...DLVFSSP GDIPAEDFAD HVRKNERDSN CGFADEYQQL
SDTFFSLPIT TESEPLFGAI EGVSAGLFLI GMLVAVVALL ICRQKVSHGR ERPSARLSIR RDRPLSVHLN LGQKGNRKTS CPIKINQFEG HFMKLQADSN YLLSKEYEEL
.......... .......... .......... .......... .......... .......... .......... .....KRKLT NPVQLDDFDA YIKDMAKDSD YKFSLQFEEL
SFSRYSDAVS LPQDP..GVI CGAVFGCIFG ALVIVTVGGF IFWRKKRKDA KNNEVSFS.. .......... ...QIKPKKS KLIRVENFEA YFKKQQADSN CGFAEEYEDL
.......... .......... .......... .......... .......... .......... .......... .....KRKLT NPVQLDDFDA YIKDMAKDSD YKFSLQFEEL
.......... .......... .......... .......... .......... .......... .......... .......RVL TRSQLRDVVA ........SS HLLQSEFMEI
---------- ---------- ---------- ---------- ---------- ---------- ---------- ---------- ---------- --------S- ----------


HGVPEYPTPF LAFLRRVKTC NPPDAGPVVV HCSAGVGRTG CFIVIDAMLE RIRTEKTVDV YGHVTLMRSQ RNYMVQTEDQ YSFIHEALLE AVGCGNTEVP ARSLYTYIQK
HG...YPTPF LAFLRRVKTC NPPDAGPIVV HCSAGVGRTG CFIVIDAMLE RIKTEKTVDV YGHVTLMRSQ RNYMVQTEDQ YGFIHEALLE AVGCGNTEVP ARSLYTYIQK
HGVPEYPTPF LAFLRRVKTC NPPDAGPIVV HCSAGVGRTG CFIVIDAMLE RIKHEKTVDI YGHVTLMRSQ RNYMVQTEDQ YSFIHDALLE AVACGNTEVP ARNLYTYIQK
HGVPEHPTPV PSFLTESQNL HPPDAGPMVV HCSAGVGRTG CFIVIDAMLE RIKHEKTVDI YGHVTLMRAQ RNYMVQTEDQ YIFIHDALLE AVTCGNTEVP ARNLYAYIQK
HGVPEYPTPI LAFLRRVKAC NPLDAGPMVV HCSAGVGRTG CFIVIDAMLE RMKHEKTVDI YGHVTCMRSQ RNYMVQTEDQ YVFIHEALLE AATCGHTEVP ARNLYAHIQK
HGVPDHPAPF LQFLRRCRAL TPPESGPVIV HCSAGVGRTG CYIVIDSMLE RMKHEKIIDI YGHVTCLRAQ RNYMVQTEDQ YIFIHDAILE AIICGVTEVP ARNLHTHLQK
HG----P-P- --FL------ -P---GP--V HCSAGVGRTG C-IVID-MLE R---EK--D- YGHVT--R-Q RNYMVQTEDQ Y-FIH-A-LE A--CG-TEVP AR-L----QK

.......... .......... .....SRKCF QTAHFYVEDS SSPRVVPNE. .SIPIIPIPD DMEAIPVKQF VKHIGELYSN NQ.....HGF SEDFEEVQRC TADMNITAEH
.......... .......... .....SRKCF QTAHFYVEDS SSPRVVPNE. .SVPIIPIPD DMEAIPVKQF GKHIGELYSN SQ.....HGF SEDFEEVQRC TADMNITAEH
.......... LTFICLVVLV GILIYWRKCF QTAHFYLEDN TSPRVISTP. .PTPIFPISD DIGAIPIKHF PKHVADLHAS NGFTEEFETL KEFYQEVQSC TVDLGITADS
.......... .......... .......... .......... .......... .......... .......... .......... .......... .......... ..........
AVIPLVIVSA LTFICLVVLV GILIYWRKCF QTAHFYLEDS TSPRVISTP. .PTPIFPISD DVGAIPIKHF PKHVADLHAS SGFTEEFETL KEFYQEVQSC TVDLGITADS
AVLVGIIFSC FGII..LIIM AFFLWSRKCF HAAYYYLDDP PHHPNAPQVD WEVPVKIGDE IRAAVPVNEF AKHVASLHAD GDI.....GF SREYEAIQNE CISDDLPCEH
---------- ---------- ------RKCF --A-FY-ED- ---------- ---PI----D ---AIPV--F -KHV--L--- ---------F ---Y--VQ-- --------E-
```

```
                  1801                                                    1851
```

```
 DPTP1OD K1 KHVGRDQPCT FADLPCNRPK NRFTNILPYD HSRFKLQPVD DDEG.SDYIN ANYVPGHNSP R.EFIVTQGP LHSTRDDFWR MCWESNSRAI
  DPTP4E K2 KHVGRDQACS FANLPCNRPK NRFTNILPYD HSRFKLQPVD DDDG.SDYIN ANYMPGHNSP R.EFIVTQGP FHSTREEFWR MCWESNSRAI
   DEP-1 K3 KLVGISQPKY AAELAENRGK NRYNNVLPYD ISRVKL.SVQ THST.DDYIN ANYMPGYHSK K.DFIATQGP LPNTLKDFWR MVWEKNVYAI
   SAP-1 K4 SLVGHSQSQM VASASENNAK NRYRNVLPYD WSRVPLKPIH EEPG.SDYIN ASFMPGLWSP Q.EFIATQGP LPQTVGDFWR LVWEQQSHTL
  RPTP-β K5 KDVGRNQSCD IALLPENRGK NRYNNILPYD ATRVKLSNVD DDPC.SDYIN ASYIPGNNFR R.EYIVTQGP LPGTKDDFWK MVWEQNVHNI
 GLEPP-1 K6 KLIGLDIPHF AADLPLNRCK NRYTNILPYD FSRVRLLSMN EEEG.ADYIN ANYIPGYNSP Q.EYIATQGP LPETRNDFWK MVLQQKSQMI
  RPTP-η K7 KLVGISQPKY AAELAENRGK NRYNNVLPYD ISRVKL.SVQ THST.DDYIN ANYMPGYHSK K.DFIATQGP LPNTLKDFWR MVWEKNVYAI
 RPTP-U2 K8 KLIGLDIPHF AADLPLNRCK NRYTNILPYD FSRVRLVSMN EEEG.ADYIN ANYIPGYNSP Q.EYIATQGP LPETRNDFWK MVLQQKSQII
RPTP-BR7 K9 PMNFVDPKEI ..DIPRHGTK NRYKTILPNP LSRVCLRPKN ITDSLSTYIN ANYIRGYSGK EKAFIATQGP MINTVNDFWQ MVWQEDSPVI
Consensus   ---------- ---------K NR----LP-- --R--L---- -------YIN A----G---- ----I-TQGP ---T---FW- ----------
```

```
LAR-PTP2_RAT N1 LAQVEPGEHV TGMELEFKRL ASSKAHTSRF ITASLPCNKF KNRLVNILPY ESSRVCLQPI RGVEGSDYIN ASFIDGYRQQ KAYIATQGPL
RPTP-σ_MOUSE N2 LAQVEPGEHV TGMELEFKRL ASSKAHTSRF ITASLPCNKF KNRLVNILPY ESSRVCLQPI RGVEGSDYIN ASFIDGYRQQ KAYIATQGPL
       CRYPα N3 LAQIEVGEHV TGMELEFKRL ANSKAHTSRF ISANLPCNKF KNRLVNIMPY ETTRVCLQPI RGVEGSDYIN ASFIDGYRQQ KAYIATQGPL
RPTP-δ_MOUSE N4 LTQIETGENV TGMELEFKRL ASSKAHTSRF ISANLPCNKF KNRLVNIMPY ESGRVCLQPI RGVEGSDYIN ASFLDGYRQQ KAYIATQGPL
   LAR_HUMAN N5 LGQVPPGESV TAMELEFKLL ASSKAHTSRF ISANLPCNKF KNRLVNIMPY ELTRVCLQPI RGVEGSDYIN ASFLDGYRQQ KAYIATQGPL
        DLAR N6 LLITEPGETI SGMEVEFKKL SNVKMDSSKF VTANLPCNKH KNRLVHILPY ESSRVYLTPI HGIEGSDYVN ASFIDGYRYR SAYIAAQGPV
   Consensus    L-----GE-- --ME-EFK-L ---K---S-F --A-LPCNK- KNRLV-I-PY E--RV-L-PI -G-EGSDY-N ASF-DGYR-- -AYIA-QGP-
```

```
      RPTP-γ_HUMAN 01 SNHPENKHKN RYINILAYDH SRVKLRPLPG KDSKHSDYIN ANYVDGYNKA KAYIATQGPL KSTFEDFWRM IWEQNTGIIV MITNLVEKGR
      RPTP-γ_MOUSE 02 SNHPDNKHKN RYINILAYDH SRVKLRPLPG KDSKHSDYIN ANYVDGYNKA KAYIATQGPL KSTFEDFWRM IWEQNTGIII MITNLVEKGR
 RPTP-ζ_RAT (full) 03 SNHPDNKHKN RYVNIVAYDH SRVKLTQLAE KDGKLTDYIN ANYVDGYNRP KAYIAAQGPL KSTAEDFWRM IWEHNVEVIV MITNLVEKGR
RPTP-ζ_RAT (short) 04 .......... .......... .......... .......... .......... .......... .......... .......... ..........
      RPTP-ζ_HUMAN 05 SNHPDNKHKN RYINIVAYDH SRVKLAQLAE KDGKLTDYIN ANYVDGYNRP KAYIAAQGPL KSTAEDFWRM IWEHNVEVIV MITNLVEKGR
           DPTP99A 06 SQHPENKRKN RYLNITAYDH SRVHLHPTPG .QKKNLDYIN ANFIDGYQKG HAFIGTQGPL PDTFDCFWRM IWEQRVAIIV MITNLVERGR
         Consensus    S-HPDNK-KN RY-NI-AYDH SRV-L----- ---K--DYIN ANYVDGY--- -AYI--QGPL --T-E-FWRM IWE----IIV MITNLVE-GR
```

1901 1951

```
VMLTRCFEKG REKCDQYWPN DTVPVFYGDI KVQILNDSHY ADWVMTEFML CRGSEQ...R ILRHFHFTTW PDFGVP..NP PQTLVRFVRA FRDRI..CAE QRPIVVHCSA
VMLTRCFEKG REKCDQYWPV DRVAMFYGDI KVQLIIDTHY HDWSISEFMV SRNCES...R IMRHFHFTTW PDFGVP..EP PLSLVRFVRA FRDVI..GTD MRPIIVHCSA
IMLTKCVEQG RTKCEEYWP. SKQAQDYGDI TVAMTSEIVL PEWTIRDFTV KNIQTSESHP L.RQFHFTSW PDHGVP..DT TDLLINFRYL VRDYMKQSPP ESPILVHCSA
VMLTNCMEAG RVKCEHYWPL DSQPCTHGHL RVTLVGEEVM ENWTVRELLL LQVEEQKTLS V.RQFHYQAW PDHGVP..SS PDTLLAFWRM LRQWLDQTME GGPPIVHCSA
VMVTQCVEKG RVKCDHYWPA DQDSLYYGDL ILQMLSESVL PEWTIREFKI CGEEQLDAHR LIRHFHYTVW PDHGVP..ET TQSLIQFVRT VRDYINRSPG AGPTVVYCSA
VMLTQCNEKR RVKCDHRWPF TEEPIAYGDI TVEMISEEEQ DDWAHRHFRI NYADEMQD.. .VMHFNYTAW PDHGVPTANA AESILQFVHM VRQQ..ATKS KGPMIIHCSA
IMLTKCVEQG RTKCEEYWP. SKQAQDYGDI TVAMTSEIVL PEWTIRDFTV KNIQTSESHP L.RQFHFTSW PDHGVP..DT TDLLINFRYL VRDYMKQSPP ESPILVHCSA
VMLTQCNEKR RVKCDHYWPF TEEPIAYGDI TVEMISEEEQ DDWACRHFRI NYADEMQD.. .VMHFNYTAW PDHGVPTANA AESILQFVHM VRQQ..ATKS KGPMIIHCSA
VMITKLKEKN E.KCVLYWP. .EKRGIYGKV EVLVTGVTEC DNYTIRNLVL KQGSHTQH.. .VKHYWYTSW PDH..KTPDS AQPLLQLMLD VEEDRLASEG RGPVVVHCSA
-M-T---E-- --KC--YWP- -------G-- ---------- ---------- ---------- ---------W PD-------- ---------- ---------- --P---HCSA
```

```
AETTEDFWRA LWENNSTIVV MLTKLREMGR EKCHQYWPAE RSARYQYFVV DPMAEYNMPQ YILREFKVTD ARDGQSRTVR QFQFTDWPEQ GAPKSGEGFI DFIGQVHKTK
AETTEDFWRA LWENNSTIVV MLTKLREMGR EKCHQYWPAE RSARYQYFVV DPMAEYNMPQ YILREFKVTD ARDGQSRTVR QFQFTDWPEQ GAPKSGEGFI DFIGQVHKTK
AETTEDFWRM LWENNSTIVV MLTKLREMGR EKCHQYWPAE RSARYQYFVV DPMAEYNMPQ YILREFKVTD ARDGQSRTVR QFQFTDWPEQ GVPKSGEGFI DFIGQVHKTK
AETTEDFWRM LWEHNSTIVV MLTKLREMGR EKCHQYWPAE RSARYQYFVV DPMAEYNMPQ YILRNSR.SR MPGIQSRTVR QFQFTDWPEQ GVPKSGEGFI DFIGQVHKTK
AESTEDFWRM LWEHNSTIIV MLTKLREMGR EKCHQYWPAE RSARYQYFVV DPMAEYNMPQ YILREFKVTD ARDGQSRTIR QFQFTDWPEQ GVPKTGEGFI DFIGQVHKTK
QDAAEDFWRM LWEHNSTIVV MLTKLKEMGR EKCFQYWPHE RSVRYQYYVV DPIAEYNMPQ YKLREFKVTD ARDGSSRTVR QFQFIDWPEQ GVPKSGEGFI DFIGQVHKTK
----EDFWR- LWE-NSTI-V MLTKL-EMGR EKC-QYWP-E RS-RYQY-VV DP-AEYNMPQ Y-LR------ -----SRT-R QFQF-DWPEQ G-PK-GEGFI DFIGQVHKTK
```

```
RKCDQYWPTE NSEEYGNIIV TLKSTKIHAC YTVRRFSIRN TKVKKGQKGN PKGRQNERVV IQYHYTQWPD MGVPEYALPV LTFVRRSSAA RMPETGPVLV HCSAGVGRTG
RKCDQYWPTE NTEEYGNIIV TLKSTKVHAC YTVRRLSVRN TKVKKGQKGN PKGRQNERTV IQYHYTQWPD MGVPEYALPV LTFVRRSSAA RMPDMGPVLV HCSAGVGRTG
RKCDQYWPTD GSEEYGSFLV NQKNVQVLAY YTVRNFTLRN TKIK...KGS QKGRSSGRLV TQYHYTQWPD MGVPEYSLPV LAFVRKTAQA KRHAVGPVVV HCSAGVGRTG
.......... .......... .......... .......... .......... .......... .......... .......... .......... .......... ..........
RKCDQYWPAD GSEEYGNFLV TQKSVQVLAY YTVRNFTLRN TKIK...KGS QKGRPSGRVV TQYHYTQWPD MGVPEYSLPV LTFVRKAAYA KRHAVGPVVV HCSAGVGRTG
PKCDMYWPKD GVETYGVIQV KLIEEEVMST YTVRTLQIKH LKLKK..... KKQCNTEKLV YQYHYTNWPD HGTPDHPLPV LNFVKKSSAA NPAEAGPIVV HCSAGVGRTG
-KCD-YWP-D --E-YG---V ------V--- YTVR-F---- -K-K------ -K-------V -QYHYT-WPD -G-PE--LPV L-FV-----A -----GPV-V HCSAGVGRTG
```

```
                      2001                                               2051

DPTP10D   K1 GVGRSGTFIT LDRILQQINT SDYVDIFGIV YAMRKERVWM VQTEQQYICI HQCLLAVLEG KENIVGPARE MHDNEGYEGQ QVQLDENGDV
DPTP4E    K2 GVGRGGTFIA LDRILQHIHK SDYVDIFGIV FAMRKERVFM VQTEQQYVCI HQCLLAVLEG KEHLLADSLE LHANDGYEVT KIYLERQPQT
DEP-1     K3 GVGRFGTFIA IDRLIYQIEN ENTVDVYGIV YDLRMHRPLM VQTEDQYVFL NQCVLDI... .......... VRSQKDSKVD LIYQNTTAMT
SAP-1     K4 GVGRTGTLIA LDVLLRQLQS EGLLGPFSFV RKMRESRPLM VQTEAQYVFL HQCICGS.SN S........Q PRPQPRRKSR MRMSKTSSTR
RPTP-β    K5 GVGRTGTFIA LDRILQQLDS KDSVDIYGAV HDLRLHRVHM VQTECQYVYL HQCVRDVLRA R........K LRSEQENPLF PIYENVNPEY
GLEPP-1   K6 GVGRTGTFIA LDRLLQHIRD HEFVDILGLV SEMRSYRMSM VQTEEQYIFI HQCVQLMWMK KKQQFCISDV IYENVSKS.. ..........
RPTP-η    K7 GVGRTGTFIA IDRLIYQIEN ENTVDVYGIV YDLRMHRPLM VQTEDQYVFL NQCVLDI... .......... VRSQKDSKVD LIYQNTTAMT
RPTP-U2   K8 GVGDTGTFIA LDRLLQHIRD HEFVDILGLV SEMRSYRMSM VQTEEQYIFI HQCVQLMWMK KKQQFCISDV IYENVSKS.. ..........
RPTP-BR7  K9 GIGRTGCFIA TSIGCQQLKE EGVVDALSIV CQLRVDRGGM VQTSEQYEFV HHALCLF... ......ESRL SPETVE.... ..........
Consensus    G-GR-G--I- ---------- ---------V ---R--R--M VQT--QY--- ---------- ---------- ---------- ----------


LAR-PTP2_RAT          N1 EQFGQDGPIS VHCSAGVGRT GVFITLSIVL ERMRYEGVVD IFQTVKVLRT QRPAMVQTED EYQFCFQAAL EYLGSFDHYA T
RPTP-σ_MOUSE          N2 EQFGQDGPIS VHCSAGVGRT GVFITLSIVL ERMRYEGVVD IFQTVKVLRT QRPAMVQTED EYQFCFQAAL EYLGSFDHYA T
CRYPα                 N3 EQFGQDGPIS VHCSAGVGRT GVFITLSIVL ERMRYEGVVD IFQTVKMLRT Q.PAMVQTED EYQFCYQAAL EYLGSFDHYA T
RPTP-δ_MOUSE          N4 EQFGQDGPIS VHCSAGVGRT GVFITLSIVL ERMRYEGVVD IFQTVKMLRT QRPAMVQTED QYQFCYRAAL EYLGSFDHYA T
LAR_HUMAN             N5 EQFGQDGPIT VHCSAGVGRT GVFITLSIVL ERMRYEGVVD MFQTVKTLRT QRPAMVQTED QYQLCYRAAL EYLGSFDHYA T
DLAR                  N6 EQFGQDGPIT VHCSAGVGRS GVFITLSIVL ERMQYEGVLD VFQTVRILRS QRPAMVQTED QYHFCYRAAL EYLGSFDNYT N
Consensus                EQFGQDGPI- VHCSAGVGR- GVFITLSIVL ERM-YEGV-D -FQTV--LR- Q-PAMVQTED -Y--C--AAL EYLGSFD-Y- -

RPTP-γ_HUMAN          01 TYIVIDSMLQ QIKDKSTVNV LGFLKHIRTQ RNYLVQTEEQ YIFIHDALLE AILGKETEVS SNQLHSYVNS ILIPGVGGKT RLEKQFKLVT
RPTP-γ_MOUSE          02 TYIVIDSMLQ QIKDKSTVNV LGFLKHIRTQ RNYLVQTEEQ YIFIHDALLE AILGKETAVS SSQLHSYVNS ILIPGVGGKT RLEKQFKLIT
RPTP-ζ_RAT (full)     03 TYIVLDSMLQ QIQHEGTVNI FGFLKHIRSQ RNYLVQTEEQ YVFIHDTLVE AILSKETEVP DSHIHSYVNT LLIPGPSGKT KLEKQFQLLS
RPTP-ζ_RAT (short)    04 .......... .......... .......... .......... .......... .......... .......... .......... ..........
RPTP-ζ_HUMAN          05 TYIVLDSMLQ QIQHEGTVNI FGFLKHIRSQ RNYLVQTEEQ YVFIHDTLVE AILSKETEVL DSHIHAYVNA LLIPGPAGKT KLEKQFQLLS
DPTP99A               06 TYIVLDAMLK QIQQKNIVNV FGFLRHIRAQ RNFLVQTEEQ YIFLHDALVE AIASGETNLM AEQVEELKNC ........TP YLEQQYKNII
Consensus                TYIV-D-ML- QI-----VNV FGFL-HIR-Q RNYLVQTEEQ YIF-HD-L-E AI---ET--- --------N- ---------- -LE-QF----
```

```
2101                                                             2151

.VATI..... .......... ........EG HLSHHDLQQA EAEAI..... .....DDENA AILHDDQQPL .......... .TSSFTGHHT HM.......P PTTSMSSFGG
KMGTLPIRAS LAMAEKLDAD LMTNKDEDED QEQQHEQQLQ LATEVKPKGS NDDEEDEEDD DDDDDDQQPL NNETTATLSS ASCSSSTHDV HVVLQEAIEQ PKQEQERICA
IYENLAPVTT FGKTNGYIA. .......... .......... .......... .......... .......... .......... .......... .......... ..........
TWPPSRPTSW RSK....... .......... .......... .......... .......... .......... .......... .......... .......... ..........
HRDPVYSRH. .......... .......... .......... .......... .......... .......... .......... .......... .......... ..........
.......... .......... .......... .......... .......... .......... .......... .......... .......... .......... ..........
IYENLAPVTT FGKTNGYIA.                                             .......... .......... .......... .......... ..........
.......... .......... .......... .......... .......... .......... .......... .......... .......... .......... ..........
.......... .......... .......... .......... .......... .......... .......... .......... .......... .......... ..........
---------- ---------- ---------- ---------- ---------- ---------- ---------- ---------- ---------- ---------- ----------
```

N1 **LAR-PTP2_RAT**
N2 **RPTP-σ_MOUSE**
N3 **CRYPα**
N4 **RPTP-δ_MOUSE**
N5 **LAR_HUMAN**
N6 **DLAR**
Consensus

```
QCNAKYVECF SAQKECNKEK NRNSSVVPAE RARVGLAPLP GMKGTDYINA SYIMGYYRSN EFIITQHPLP HTTKDFWRMI WDHNAQIIVM LPDNQSLAED EFVYWPSREE
QCNAKYVECF SAQKECNKEK NRNSSVVPAE RARVGLAPLP GMKGTDYINA SYIMGYYRSN EFIITQHPLP HTTKDFWRMI WDHNAQIIVM LPDNQSLAED EFVYWPSREE
QSNILQSDYS TALKQCNREK NRTSSIIPVE RSRVGISSLS G.EGTDYINA SYIMGYYQSN EFIITQHPLL HTIKDFWRMI WDHNAQLVVM IPDGQNMAED EFVYWPNKDE
.......... .......... .......... .......... .......... .......... .......... .......... .......... .......... ..........
QSNIQQSDYS AALKQCNREK NRTSSIIPVE RSRVGISSLS G.EGTDYINA SYIMGYYQSN EFIITQHPLL HTIKDFWRMI WDHNAQLVVM IPDGQNMAED EFVYWPNKDE
QFQPKDIHIA SAMKQVNSIK NR.GAIFPIE GSRVHLTPKP GEDGSDYINA SWLHGFRRLR DFIVTQHPMA HTIKDFWQMV WDHNAQTVVL LSSLDDINFA QF..WPDEAT
Q--------- -A-K--N--K NR---I-P-E --RV------ G--G-DYINA SY--GY---- EFIITQHPL- HT-KDFW-MI WDHNAQ-VVM ---------- -F--WP----
```

```
           2201                                              2251
```

```
   DPTP1OD K1  GGGGHTNVDA PDR....... .......... .......... K1  DPTP1OD
    DPTP4E K2  GTQSHADTES DNTDSDDDDE DGDGKVAKDG AVADEDGWWY K2  DPTP4E
     DEP-1 K3  .......... .......... .......... .......... K3  DEP-1
     SAP-1 K4  .......... .......... .......... .......... K4  SAP-1
    RPTP-β K5  .......... .......... .......... .......... K5  RPTP-β
   GLEPP-1 K6  .......... .......... .......... .......... K6  GLEPP-1
    RPTP-η K7  .......... .......... .......... .......... K7  RPTP-η
   RPTP-U2 K8  .......... .......... .......... .......... K8  RPTP-U2
  RPTP-BR7 K9  .......... .......... .......... .......... K9  RPTP-BR7
 Consensus     ---------- ---------- ---------- ----------     Consensus
```

```
      RPTP-γ_HUMAN  01  SMNCEAFTVT LISKDRLCLS NEEQIIIHDF ILEATQDDYV LEVRHFQCPK WPNPDAPISS TFELINVIKE EALTRDGPTI VHDEYGAVSA
      RPTP-γ_MOUSE  02  SMNCEAFTVT LISKDRLCLS NEEQIIIHDF ILEATQDDYV LEVRHFQCPK WPNPDAPISS TFELINVIKE EALTRDGPTI VHDEYGAVSA
 RPTP-ζ_RAT (full)  03  PINCESFKVT LMSEEHKCLS NEEKLIVQDF ILEATQDDYV LEVRHFQCPK WPNPDSPISK TFELISIIKE EAANRDGPMI VHDEHGGVTA
RPTP-ζ_RAT (short)  04  .......... .......... .......... .......... .......... .......... .......... .......... ..........
      RPTP-ζ_HUMAN  05  PINCESFKVT LMAEEHKCLS NEEKLIIQDF ILEATQDDYV LEVRHFQCPK WPNPDSPISK TFELISVIKE EAANRDGPMI VHDEHGGVTA
          DPTP99A   06  PIESDHYRVK FLNK.....T NKSDYVSRDF VIQSIQDDYE LTVKMLHCPS WPEMSNPNSI YDFIVDVHER CNDYRNGPIV IVDRYGGAQA
         Consensus      ----E-F-V- L--------- N----I--DF I----QDDY- L-V--F-CP- WP----P-S- ----I-V--- ----R-GP-I V-D--G---A
```

```
2301                                                        2351

GMLCALTTLS QQLENENAVD VFQVAKMINL MRPGVFTDIE QYQFIYKARL SLVSTKENGN GPMTVDKNGA VLIADESDPA ESMESLV... .......... ..........
GMLCALTTLS QQLENENAVD VFQVAKMINL MRPGVFTDIE QYQFVYKAML SLISTKENGN GPMTGDKNGA VLTAEESDPA ESMESLV... .......... ..........
GTFCALTTLM HQLEKENSMD VYQVAKMINL MRPGVFTDIE QYQFLYKVVL SLVSTRQEEN PSTSLDSNGA AL..PDGNIA ESLESLV... .......... ..........
.......... .......... .......... .......... .......... .......... .......... .......... .......... .......... ..........
GTFCALTTLM HQLEKENSVD VYQVAKMINL MRPGVFADIE QYQFLYKVIL SLVSTRQEEN PSTSLDSNGA AL..PDGNIA ESLESLV... .......... ..........
CTFCAISSLA IEMEYCSTAN VYQYAKLYHN KRPGVWTSSE DIRVIYN.IL SFLPGNLNLL KRTALRTEFE DVTTATPDLY SKICSNGNVP QHVILQQQQL HMLQLQQQHL
--FCA---L- --LE------ VYQ-AKM--- -RPGVF---E -----Y---L SL-------- ---------- ---------- ----S----- ---------- ----------
```

```
                        2401                                              2451
    RPTP-γ_HUMAN  01 .......... .......... .......... .......... .......... .......... .......... .......... ..........
    RPTP-γ_MOUSE  02 .......... .......... .......... .......... .......... .......... .......... .......... ..........
RPTP-ζ_RAT (full) 03 .......... .......... .......... .......... .......... .......... .......... .......... ..........
RPTP-ζ_RAT (short) 04 .......... .......... .......... .......... .......... .......... .......... .......... ..........
    RPTP-ζ_HUMAN  05 .......... .......... .......... .......... .......... .......... .......... .......... ..........
         DPTP99A  06 ETQQQQQQQQ QQQQQQQQTA LNETVSTPST DTNPSLLPIL SLLPPTVAPL SSSSSTTPPT PSTPTPQPPQ TIQLSSHSPS DLSHQISSTV
       Consensus     ---------- ---------- ---------- ---------- ---------- ---------- ---------- ---------- ----------
```

```
2501                                                2551

.......... .......... .......... .......... .......... .......... .......... .......... .......... .......... ..........
.......... .......... .......... .......... .......... .......... .......... .......... .......... .......... ..........
.......... .......... .......... .......... .......... .......... .......... .......... .......... .......... ..........
.......... .......... .......... .......... .......... .......... .......... .......... .......... .......... ..........
.......... .......... .......... .......... .......... .......... .......... .......... .......... .......... ..........
ANAASPVTPA TASASAGATP TTPMTPTVPP TIPTIPSLAS QNSLTLTNAN FHTVTNNAAD LMEHQQQQML ALMQQQTQLQ QQYNTHPQQH HNNVGDLLMN NADNSPTASP
---------- ---------- ---------- ---------- ---------- ---------- ---------- ---------- ---------- ---------- ----------
```

2601

```
     RPTP-γ_HUMAN  01 .......... .......... ........ 01 RPTP-γ_HUMAN
     RPTP-γ_MOUSE  02 .......... .......... ........ 02 RPTP-γ_MOUSE
 RPTP-ζ_RAT (full) 03 .......... .......... ........ 03 RPTP-ζ_RAT (full)
RPTP-ζ_RAT (short) 04 .......... .......... ........ 04 RPTP-ζ_RAT (short)
     RPTP-ζ_HUMAN  05 .......... .......... ........ 05 RPTP-ζ_HUMAN
          DPTP99A  06 TITNNNHITN NNVTSAAATD AQNLDIVG 06 DPTP99A
        Consensus     ---------- ---------- -------- Consensus
```